Studies in Logic
Logic and Argumentation
Volume 74

Dictionary of Argumentation
An Introduction to Argumentation Studies

Volume 64
Logic of Questions in the Wild. Inferential Erotetic Logic in Information Seeking Dialogue Modelling
Paweł Łupkowski

Volume 65
Elementary Logic with Applications. A Procedural Perspective for Computer Scientists
D. M. Gabbay and O. T. Rodrigues

Volume 66
Logical Consequences. Theory and Applications: An Introduction.
Luis M. Augusto

Volume 67
Many-Valued Logics: A Mathematical and Computational Introduction
Luis M. Augusto

Volume 68
Argument Technologies: Theory, Analysis, and Applications
Floris Bex, Floriana Grasso, Nancy Green, Fabio Paglieri and Chris Reed, eds

Volume 69
Logic and Conditional Probability. A Synthesis
Philip Calabrese

Volume 70
Proceedings of the International Conference. Philosophy, Mathematics, Linguistics: Aspects of Interaction, 2012 (PhML-2012)
Oleg Prosorov, ed.

Volume 71
Fathoming Formal Logic: Volume I. Theory and Decision Procedures for Propositional Logic
Odysseus Makridis

Volume 72
Fathoming Formal Logic: Volume II. Semantics and Proof Theory for Predicate Logic
Odysseus Makridis

Volume 73
Measuring Inconsistency in Information
John Grant and Maria Vanina Mrtinez, eds.

Volume 74
Dictionary of Argumentation. An Introduction to Argumentation Studies
Christian Plantin. With a Foreword by J. Anthony Blair

Studies in Logic Series Editor
Dov Gabbay   dov.gabbay@kcl.ac.uk

Dictionary of Argumentation
An Introduction to Argumentation Studies

Christian Plantin

with a Foreword by
J. Anthony Blair

© Individual author and College Publications, 2018
All rights reserved.

This *Dictionary of Argumentation* is a translation and adaptation of Christian Plantin, *Dictionnaire de l'argumentation*, Lyon: ENS Éditions, 2016.

ISBN 978-1-84890-271-8

College Publications
Scientific Director: Dov Gabbay
Managing Director: Jane Spurr

http://www.collegepublications.co.uk

Printed by Lightning Source, Milton Keynes, UK

All rights reserved. No part of this publication may be reproduced, stored in a retrieval system or transmitted in any form, or by any means, electronic, mechanical, photocopying, recording or otherwise without prior permission, in writing, from the publisher.

CONTENTS

Foreword, *by J. Anthony Blair*	iii
Acknowledgements for the English Translation	v
Preface to the French Edition	vii
Conventions	xii
Table of Entries	1
Entries	9
References	596

Foreword

By J. Anthony Blair

About ten years ago, obviously inspired by the *Stanford Encyclopedia of Philosophy*, and motivated by the evident need, I sat down at my computer and typed out "Windsor Encyclopedia of Argument and Argumentation; Terms, Concepts, Theories, Important historical and contemporary figures". Before too long, I compiled a list of close to 200 headings for entries. It struck me immediately that writing up those entries called for a team effort. Surely no one person, and certainly not I, had the necessary encyclopedic acquaintance with the field or the energy to acquire it. Over the years since then, I privately bemoaned the lack of such a reference work, however the time never seemed available to enlist a team of colleagues to undertake the task of writing it.

Then, in September of 2016, a copy of *Dictionnaire de l'argumentation, Une introduction aux études d'argumentation* arrived in the mail, the author's name in self-effacing tiny print under the title on the front cover—my old friend—Christian Plantin. I riffled through the pages. "Accident (fal.)" three-quarters of a page; "Ad hominem" four pages; "Définition" eleven and a half pages; "Éthos" ten pages; "Émotion" five and a half pages; "Dialectique" three and a half pages; and on and on. It has 248 main entries and 67 secondary entries and runs to 635 pages. Although it serves as a dictionary, and is restricted to listing the terms used in argumentation and argumentation theory, with no entries for the names of theorists or of their theories, it is in fact more like an encyclopedia. For in its main entries it refers to and discusses the various different theoretical treatments of these terms. Its list of the references alluded to in the text tops 600. And Plantin consulted some four dozen colleagues to check the accuracy

Foreword

of his accounts (they are listed). This is the reference book I had dreamed of, and Christian Plantin had accomplished it by himself.

There was just one problem: it is written in French. Like it or not, the lingua franca of argumentation studies these days is English, and even if many scholars are bilingual or multilingual, the sad fact remains that if the *Dictionnaire* were available only in French it would not get nearly the distribution or the usage it deserves. For it should be on the reference shelf of every argumentation scholar and every student of argumentation in the world.

So when I wrote to Christian to thank him for sending me a copy, I suggested that he should try to get the *Dictionnaire* translated into English. He replied that he agreed, but how to accomplish that enormous task was the problem. Only an expert could know how to translate the technical terms into their English equivalents. Moreover many French terms of art in the field of argumentation have no precise equivalent in English—*argument* itself is a prime example. There was really only one person eminently suited to the task, namely the author himself. Plantin's English is excellent and he has the requisite knowledge. So rather than relax and enjoy the much-deserved praise for having written the *Dictionnaire*, he turned to the gargantuan job of translating the book.

It remained to find a publisher. With the prices of books published by the commercial houses—the big scholarly presses even the prestigious university presses—in the stratosphere, if any of them published it, the book would not be affordable by its primary target audience, namely students. Plantin's subtitle is, after all, "An introduction to the study of argumentation". I contacted John Woods, a series editor at College Publications, to help us find out if they might be interested. A non-profit publisher dedicated to producing academic books of high quality and making them available at cost, it seemed an obvious choice. College Publications immediately welcomed the project. And here we have the wonderful result.

The *Dictionary of Argumentation* differs marginally from the *Dictionnaire de l'argumentation*. There are 303 entries, 225 main ones and 78 secondary entries. It is targeted at an Anglophone, not a Francophone audience. The author has taken advantage of the opportunity to make minor revisions and corrections.

I commend this book to students and established scholars of argumentation alike. All will discover new information in it. It bears the imprint of its author: astonishing erudition worn lightly; encyclopedic knowledge presented in an informal, accessible style; stuffed with eclectic examples; serious and amusing; with firm opinions and fair treatment of alternatives. It is a tour de force.

J. Anthony Blair
Center for Research in Reasoning, Argumentation and Rhetoric
University of Windsor, Canada
December 2017

Acknowledgements for the English Translation

Upon the 2016 publication of the *Dictionnaire de l'argumentation*, J. Anthony Blair pledged his support to the translation of the book. True to his word and with exceptional generosity, he took the time to read, correct and comment extensively on the various English versions of the book. I am convinced that without his initial impulse and continuous support, the English translation of the *Dictionnaire* would never have seen the light of the day.

Michael Baker offered his generous support from the very beginning of the *Dictionnaire* project. He backed the English publication, and gave me subtle and substantial corrections and comments on a large set of key entries. Many thanks Michael!

Thanks to Marianne Doury, Jean-Claude Guerrini and Claire Polo for their remarks and amendments.

John Woods accepted the *Dictionary* in the prestigious series "Logic and Argumentation", and offered judicious key improvements to the manuscript.

The first version of my translation was extensively corrected by Andrew Thatcher; the final version was expertly checked by Elsbeth Wright, and I'd like to extend my best thanks to both of them.

I am especially grateful to the ASLAN project (ANR-10-LABX-0081) of Université de Lyon, for its financial support within the program "Investissements d'Avenir" (ANR-11-IDEX-0007) of the French government operated by the National Research Agency (ANR).

About this Translation

This *Dictionary of Argumentation* is a translation and adaptation from the French *Dictionnaire de l'Argumentation*. Text and examples have been re-worked and extensively re-written, in order to meet the needs and expectations of a new broader audience.
Observations and criticisms made to the French edition have been integrated.

I am, of course, solely responsible for the remaining mistakes and shortcomings in form and substance. I have certainly not mastered the aura of English words and constructions, nor the multifarious inferences they project according to their uses.
A dictionary is an *ambiguous* object written under a double constraint, where one has to tread a fine line between over-refinement and over-simplification; I hope I have sometimes touched that winding road.

La Rochette, December 29, 2017

Preface to the French Edition
Translated by J. Anthony Blair

This *Dictionary* owes everything to Jean-Claude Anscombre, J. Anthony Blair, Oswald Ducrot, Frans van Eemeren, Jean-Blaise Grize, Rob Grootendorst, Charles L. Hamblin, Ralph Johnson, Lucie Olbrechts-Tyteca, Chaïm Perelman, Stephen E. Toulmin, Douglas Walton, John Woods — and many others. They introduced new ideas, reconceptualized the field, reconnected it to contemporary scholarship, and opened new fields of research and perspectives whose exploration is far from complete.

Aristotle, Cicero, Quintilian are the founding fathers of Western argumentation studies. The historical and cultural differences that separate us from them undoubtedly create an obstacle to reading them. No doubt influenced by the large body of contemporary American studies in rhetoric and argumentation, the definitions included in this *Dictionary* integrate their insights, at the same level as contemporary works.

*

The general vision employed in this work makes no claim to originality; it seems to me, largely a posteriori, to be the following. Argumentation is approached as *a linguistic activity*, and more fundamentally, as a semiotic activity, rooted in the ordinary exercise of language. Ordinary speech has first of all an *oral and dialogical* existence. Key concepts of *discourse and interaction studies* can be effectively implemented in the practical analysis of everyday argument. This *Dictionary* articulates the study of argumentation in the framework of *discourse studies*, under their two aspects, *monologal* and *interactional*. This position agrees, for example, with the framework of discourse analysis as it is elaborated in the *Dictionnaire d'Analyse du Discours* by Patrick Charaudeau and Dominique Maingueneau (Le Seuil, 2002), to which I contributed the entries concerning argumentation. I owe the idea for the present enterprise to their example.

Preface to the French Edition

Arguing is exercising *the critical function of language*. Full-blown argumentative situations have a characteristic *antiphonic* structure, where the participants express and balance the pros against the cons.

Argumentation is both monologue and dialogue, and both are *language and thought*. Argumentation as reasoning in ordinary language should not be seen as the inconclusive, vague, weak and easy counterpart of scientific reasoning. *Critical thinking* is at work in everyday private and public human affairs as well as in the most recondite scientific disciplines. The *acquisition of knowledge* begins with the tools of ordinary language and reasoning, and these are forgotten when they are no longer needed. It is an extraordinary characteristic of ordinary language to be thus capable of engendering other languages capable of going where it can never go itself.

*

This *Dictionary* is based on the experience acquired in teaching and research seminars on argumentation; certain propositions echo the discussions that took place there. The participants in those seminars were, as they no doubt will continue to be, a mix of experienced colleagues teaching and developing research programs in argumentation, junior researchers, and students beginning to develop their vision of the field. No doubt the odds are against appealing to these diverse groups at the same time. However, it is this tripartite audience that I constantly had in mind during the preparation of *this Dictionary*, with special emphasis on the last two.

I hope that consulting this *Dictionary* will prove useful not only to *argumentation theorists*, but also to the wide community of people wishing to better articulate their visions and practices of argumentation, and who, for that purpose need a *meta-language* of argumentation. To argue is, in effect, to express oneself – to speak or write, often both – in a space structured by a *question* defining an issue. This space is characterized by the presence of *opponents*, and the activity of arguing necessarily leads the speaker to refer to their discourses, that provide an alternative and distinctly different *answers* to the question. The arguer is inevitably led to *speak about* antagonistic discourses, whilst also developing "control loops" within his or her own argument.

Arguing thus implies meta-argumentative activities. The ordinary exercise of argumentation presupposes the systematic usage of a discourse about argumentation, or a sort of *ordinary meta-language* about argumentation, which theorists will develop into a full theory of argumentation. That's why we hope equally that the *practitioners* of argument no less than the *theoreticians* will take some interest in this *Dictionary*, and that the observations that it contains will be able to be *reinvested in argumentative practice*.

*

Beyond the requests for timely information, which find an answer on the internet, everyone working on argumentation, as in any other field of the human

Preface to the French Edition

sciences, finds himself or herself confronted by questions of *clarification*, of *definition*, and of *conceptual coherence*.

To answer these questions is not necessarily difficult in an isolated case. But the difficulties increase with the plurality of definitions of the same term, or the plurality of terms corresponding roughly to one and the same definition. Things are further complicated when these definitions overlap, and function in a shifting stylistic continuum, in which, moreover, one may take a certain pleasure. The case of the cluster constituted by the arguments *a pari*, from *similarity*, from *analogy*, from *categorization*, not to mention *per analogiam*, is an example of such a situation. If one wants not only to admire, but also to understand, one must sometimes resolve to give up this or that conceptual nuance and accept that certain labels are simple synonyms or translations of one another.

A second major difficulty is that of the global coherence of the definitions. To stick with the example of analogy, one encounters this issue when one adds to the preceding terms the *rule of justice* and the *precedent*. Without claiming to give the notional field of argumentation the kind of compact structure that one could dream of in the early days of structuralism, one must not only expose the *specificities* of the concepts but also their *commonalities*.

In trying to resolve the first difficulty one runs the risk of arbitrary simplification; to resolve the second, one risks imposing on these notions an arbitrary organization. If one fails in these two ways, one will simply have aggravated the malady for which one was claiming to bring the remedy.

*

This is not an encyclopedic dictionary that retraces the discussions about each concept, that presents each theory within its historical developments, its current structure and its research program, and that discusses the strengths and weaknesses of each author. The works cited do not claim to constitute a bibliography or a reading list of argumentation studies.

This *Dictionary* brings together a collection of relatively technical terms which form a vocabulary shared by argumentation studies and implemented in the analysis of argumentative texts and interactions. From *Argumentation* to *Topic* and *Waste*, their degree of technicality is very different.

Certain terms correspond to terms that are used outside the field of argumentation studies. Only the particular meaning that such terms have within the theory of argumentation feature in this *Dictionary*. In the entry "*Pragmatic*" one will not find general considerations on pragmatics as a philosophy or a branch of linguistics, but only a definition of *pragmatic argument*.

This *Dictionary* presents 303 entries, 225 basic entries, with the addition of 78 secondary entries.

A *main entry* defines, comments and illustrates a specific concept, and, when necessary a set of closely related concepts.

A *secondary entry* refers back to a main entry. The main entry may correspond:

Preface to the French Edition

(i) To a more usual label equivalent to, or a translation of the secondary entry, for example "*Ad Verecundiam* ▶ Modesty".

(ii) To an encompassing concept, for example "Amphiboly ▶ Ambiguity". The grouping of several secondary entries under the same main, uniting entry prevents dispersions and repetitions and favors the discussion of closely related concepts.

(iii) To a main entry grouping two correlative concepts, which are defined contrastively, for example the secondary entry "Conclusion ▶ Argument", "Argument" being an abbreviation referring unambiguously to the main entry, "Argument – Conclusion" (see *Conventions*, infra).

A system of *cross-references* connects the entries, to strengthen the conceptual coherence of the whole *Dictionary*.

The definitions are introductory. According to the fine catachresis used to refer to the items collected in a dictionary, the *entries* of this *Dictionary* should straightaway arrange an *entrée* to the idea. I have sometimes tried to add a bit of spice in the form of a commentary or a note that should open up the idea and prompts a questioning of it.

The examples are of various kinds: some are invented and only aim to give an idea of actual instances of the phenomenon under scrutiny. Others are borrowed from written texts; yet others come from oral exchanges, sometimes from recorded and referenced productions, sometimes simply caught on the fly and noted later; their oral indicators have been retained as much as possible.

The entries are listed according to alphabetical order. The numbering of some entries allows for certain thematic groupings, which should enable the reader to better follow the development of families of related key entries, for example regarding the large issues of argumentative *analogy* or *causality*.

One might find it strange that an entry is devoted to this or that minor form: that is because it is not so much minor as overlooked, and because it deserves its proper place in what can be considered the conceptual structure underlying argumentation studies.

The definitions, propositions and assertions presented in this *Dictionary* are certainly not intended to close down any discussion. They are rather trying to feed the debate, and sometimes to provoke it, pending criticism and improvement. I would be delighted if that were to happen.

Many dictionaries or logical and rhetorical lexicons define certain terms that are relevant to argumentation theory. To our knowledge, however — apart from *Sztuka argumentacji – Słownik terminologiczny* [*The Art of Arguing – Terminological Dictionary*] by Szymanek (2004) — there is hardly any other *Dictionary of Argumentation*.

Acknowledgements

I gratefully thank all those with whom I have been able to discuss and share seminars on argumentation.

Ruth Amossy, Tel Aviv
Vahram Atayan, Heidelberg
Marina Bondi, Modena
Dora Calderón, Bogotá
Sara Cigada, Milan
Emmanuelle Danblon, Brussels
Marc Dominicy, Brussels
Isabel Duarte, Porto
Ekkehard Eggs, Hannover
Anca Gata, Galati
Jean-Claude Guerrini, Saint-Étienne
Rob Grootendorst, Amsterdam
Thierry Herman, Neuchâtel
Catherine Kerbrat-Orecchioni, Lyon
Olga León Corredor, Bogotá
Kristine Lund, Lyon
Roberto Marafioti, Buenos Aires
María Cristina Martinez, Cali
Michel Meyer, Brussels
Nora Muñoz, Rio Gallegos
Constanza Padilla, Tucuman
Marie-Christine Pollet, Brussels
Matthieu Quignard, Lyon
Rui Ramos, Minho
Georges Roque, Paris
Hammadi Sammoud, Tunis
Véronique Traverso, Lyon

Elvira Arnoux, Buenos Aires
Michael Baker, Lyon
Christian Buty, Lyon
Claude Chabrol, Paris
Jacques Cosnier, Lyon
Joseph Dichy, Lyon
Marianne Doury, Paris
Frans van Eemeren, Amsterdam
Wander Emediato, Belo Horizonte
Silvia Gutiérrez, México
Rui Grácio, Minho
Silvia Gutiérrez Vidrio, México
Peter Houtlosser, Amsterdam
Roselyn Koren, Tel Aviv
Vincenzo Lo Cascio, Amsterdam
Anna Mankovska, Warsaw
Maria Aldina Marques, Minho
Davide Mazzi, Modena
Raphaël Micheli, Lausanne
Gerald P. Niccolai, Lyon
Chantal Plantin, La Rochette
Claire Polo, Lyon
Pierre-Yves Raccah, Limoges
Eddo Rigotti, Milan
Francisca Snoeck Henkemans, Amsterdam
Andrée Tiberghien, Lyon
Maria Zaleska, Warsaw

To my colleagues and the students of our Lyon Research Group, GRIC, then ICAR; to those who participated in the various argumentation seminars, courses and programs, a big thank you for your critiques and your sometimes unpredictable questions, for the challenges posed by your enthusiasm or irony, and for your exacting attention, always.

Conventions

ENTRIES
The entries are classified in alphabetical order.
The entries relating to the same topic have been regrouped. For example, the entries dealing with *Analogy* are grouped as follows.
> Analogy (I): Analogical Thinking
> Analogy (II): Intra-Categorical Analogy
> Analogy (III): Structural Analogy

— Main entry and secondary entry
A *main entry* is a standard entry.
A *secondary entry* has the following form:
> Accent ▶ Ambiguity
> (secondary entry) (see) (main entry)
> Read: "Accent is defined under the entry Ambiguity".

The arrow "▶" is found only in *secondary* entries, and reads as indicated.

— Latin labels
The Latin names of argument are listed as follows.
— When they have no specific English designation and are still used today, they are listed as main entries in the alphabetical order, ex: *Ad Hominem*.
— When there is an equivalent English designation, they are listed as secondary entries referring to the corresponding English entry.
Lists of Latin labels will be found under the entries:
> *Ab* — arguments
> *Ad* — arguments
> *Ex* — arguments

Conventions

CROSS-REFERENCES: "S." AND "@"

In an entry, cross-references can be made to another or other entries, where complementary information relevant for the current entry can be found. Such cross-references are signaled as follows:

"S. [for *See*] [followed by the name of the entry, in bold smaller characters]"
"S. Refutation" reads: "See under the main entry **Refutation**"

Due to space constraints, when the main entry referred to is a phrase, the reference is made to the first word (or sub phrase) designating this entry unambiguously:

"S. Argumentation (I)" is equivalent to "S. Argumentation (I): Definitions"

"@"

To save place and avoid repetitions, when the reference is made to an entry unambiguously mentioned in the text, the reference is made by a superscript @ following the word corresponding to that entry. For example, the reference in the following quotation is made according to the preceding convention:

"This causal argument can be supplemented by a pragmatic@ argument".

The word used for the cross reference may correspond to several entries:

"S. Analogy" or "analogy@" refer to the three entries Analogy (I), Analogy (II), Analogy (III).

Cross-references are made only when deemed *useful*. Not all occurrences of the words "refutation" or "rebuttal" are referred back to the entry "Refutation" or "Layout of Argument".

A cross-reference can be made to several entries, for example "**S. Persuasion; Argumentation; Demonstration**". The order of the cross-referenced entries is arbitrary.

REFERENCES

— References to dictionaries

The references to dictionaries are made as follows:

(Author(s) or Title of the dictionary), (entry)
(Gaffiot, *Ludicrum*)
(MW, *Authority*)

Greek words are transliterated between square brackets:

Bailly, [*Antilepsis*])

Full references of the quoted dictionaries are found in the Reference section.

— References to theoretical texts

References to theoretical texts are made as follows:

(Author) (Year), (Page number)
Walton 2006, p. 125

Conventions

When the quotation comes from a re-publication of the work, the reference includes the original work and is made as follows:
>(Perelman & Olbrechts-Tyteca [1958], p. 5)
>(Author(s) [date of the original publication], [page number])

The quoted edition is to be found in the reference list as:
>Perelman, Olbrechts-Tyteca [1958] = (1969). *The New rhetoric*, etc.

For ancient texts, the reference is made as follows:
>(Aristotle, *Rhet.*, II, 23, 1397b15)
>([Author] [Abridged title of the text] [Classical reference system, when available from the quoted edition]

When different translations of the same text have been used, the reference includes the name of the translator:
>(Aristotle, *Rhet.*, II, 23, 1397b15; Rhys Roberts)

The quoted edition will be found in the reference list as:
>Aristotle, *Rhet* = (1926) *Rhetoric*. Trans. by Rhys Roberts, etc.

— References to examples
References for the example follow the given example. Additional information, including the page number, is given in a note.

— Web references
References to websites are given in notes, followed by the date of the access in the format (mm-dd-yyyy).

— *Ibid.* and *id.*
Ibid. is used to refer to exactly the same source and page as the last one previously mentioned: same author; same year, that is, same work; same page.
Id. is used to refer to the same source as previously mentioned (same author; same year, that is, same work) and is followed by a new page number.
Their capitalization depends on the preceding punctuation mark.

QUOTATIONS
Examples of argument schemes or argumentative strategies may be taken from theoreticians arguing within their field or they may *not*. For example, Chomsky is quoted as a theoretician, under the entry *Ambiguity* and as an arguing theoretician under the entry *Counter-Argumentation*.
Theoretical considerations relevant for argumentation studies relate first from authors recognized for their work in the field of argumentation, from Aristotle to van Eemeren.
They may also originate from outstanding arguers, commenting on their current practice. For example, Erasmus is quoted under the entry *Conditions of discussion*, for a theoretical remark he developed during his dispute with Luther.

Conventions

This case is typical of what happens in argumentative situations, where theory of argumentation develops along with practice.

Quotations respect, as far as possible, the spelling, punctuation and typography of the quoted texts.

TRANSLATIONS
Theoretical texts and examples from languages other than English are quoted in translation. The translator's name is mentioned in reference (theoretical texts) or in a note (examples). Translations not accompanied by the translator's name come from texts for which I was unable to find any translation, and the translation is mine, CP.

ARROWS: ▸, →

The arrow "▸" is used in secondary entries. The word or expression at its left is defined under the word or expression at its right.

The arrow "→" is used to refer to an illative reasoning move, such as:
— The logical implication: "A → B"
— The argument-conclusion relation.

ASTERISK: *
The asterisk is used before a word or a sentence, to signal that this word or sentence is problematic from a linguistic point of view: agrammatical, alexical…
Before the conclusion of a syllogism, the asterisk signals that this conclusion does not follow from the preceding premises.

EXAMPLES NUMBERING
When necessary, examples are numbered (1), (2)…
These numbers are used to refer to the example in the surrounding text
This numbering is strictly local, and begins again with (1) in a new context.

GRAMMATICAL ABBREVIATIONS
Abbreviation such as "Sg, Pl, … Adj, V…" refer to the current grammatical abbreviations.

TRANSCRIPTIONS OF ORAL TALK
A simplified set of symbols is used for the transcription of oral talk.
\ marks falling (closing) intonation, corresponding to a period in written language, and is characteristically used in categorical affirmation.

/ marks rising intonation, used, among other cases, in questions.

- hyphen is used to mark an incomplete word.

— Em dash notes an unfinished sentence (a syntactic rupture in the text).

Conventions

:: notes a lengthened syllable.

(.), (..), ... note the length of a silence.

"this is REALLY important": capitalization notes a segment said with a strong voice.

A meaningless sequence of letters notes approximately an uninterpretable sequence of sounds.

S1, S2, ... refer to speakers in reconstructed dialogues. **S1** is the first speaker, **S2** the second speaker, etc.

The partners of authentic interactions are given a name, their name in the case of public interactions, or a pseudo in the case of private interactions.

S1_1 , S1_2,... note respectively the first, second, ... speech turn of speaker **S1**. Idem for **S2_1, S2_2,** ... and speaker **S2**.

HISTORICAL PERIODIZATION

Antiquity, ancient Greece and Rom: from the origins to the sixth century AC
Middle Ages: from the 6th Century AC to the Renaissance
Modern times: From the Renaissance to the nineteenth century
Contemporary times: From the nineteenth century to present times

SYLDAVIA is a fictional country. The name comes from the comic series *The Adventures of Tintin*, by Hergé.

Table of Entries

A Comparatione
A Conjugata ▶ Related Words
A Contrario ▶ Opposite
A Fortiori
A Pari
A Priori, A Posteriori
A Repugnantibus
A Simili
Ab — Arguments (*A Contrario*…)
Ab —, *Ad* —, *Ex* —: Latin Labels
Ab Exemplo
Abduction
Absurd
Accent ▶ Ambiguity
Accident
Ad — Arguments (*Ad Ignorantiam*…)
Ad Baculum ▶ Threat
Ad Hominem
Ad Incommodum
Ad Judicium ▶ Matter
Ad Personam ▶ Personal Attack
Ad Populum
Ad Quietem ▶ Calm
Ad Rem ▶ Matter
Ad Verecundiam ▶ Modesty
Affirming the Consequent ▶ Deduction

Table of Entries

"After as Before" ▶ Consistency
Agreement
Alignment ▶ Orientation
Ambiguity
Amphiboly ▶ Ambiguity
Analogy (I): Analogical Thinking
Analogy (II): Intra-Categorical Analogy
Analogy (III): Structural Analogy
Antanaclasis, Antimetabole, Antiparastasis ▶ Orientation Reversal
Antithesis
Apagogic
Aporia ▶ Assent; Stasis
(To) Argue, Argument, Argumentation, Argumentative: The Words
Argument — Conclusion
Argumentation (I): Definitions
Argumentation (II): Key Features and Issues
Argumentation Studies: Contemporary Developments
Argumentativity ▶ Argumentation (II)
Assent
Association ▶ Dissociation
Audience ▶ Rhetorical Argumentation; Persuasion
Authority
Autophagy and Retaliation
Backing ▶ Layout; Scheme
Bandwagon ▶ Consensus
Begging the Question ▶ Vicious Circle
Beliefs of the Audience
Bias ▶ Orientation
Burden of Proof
Calm
Case-by-Case Argument
Categorization and Nomination
Causality
Cause – Effect: The Causal Link
Cause To Effect Argumentation
Circumstances
Collections (I) and Typologies of Arguments Schemes
Collections (II): From Aristotle to Boethius
Collections (III): Modernity and Tradition
Collections (IV) : Contemporary Innovations and Structurations
Common Place
Common Sense ▶ Doxa; Authority; Common place
Comparison
Completeness
Composition and Division
Concession
Conclusion ▶ Argument — Conclusion

Table of Entries

Conditions of Discussion
Conductive Argument
Connective
Consensus
Consequence and Effect
Consistency
Contradiction
Contrary and Contradictory
Convergent
Convergent — Linked — Serial
Conversion
Cooperative Principle
Coordinate Argumentation ▶ Linked
Correlative Terms
Counter-Argumentation
Counter-Discourse ▶ Counter-Argumentation
Counter-Proposition ▶ Counter-Argumentation
Criticism — Rationalities — Rationalizations
Debate
Deduction
Default Reasoning
Definition (I): Definition and Argument
Definition (II): Argumentation Justifying a Definition
Definition (III): Argumentations Based on a Definition
Definition (IV): Persuasive Definition
Demonstration and Argumentation
Denying
Denying the Antecedent ▶ Deduction
Derived Words
Destruction of Speech
Dialectic
Diallel ▶ Vicious Circle
Dilemma
Direction ▶ Gradualism; Slippery Slope
Disagreement
Dismissal
Dissensus
Dissociation
Distinguo
Division ▶ Case-by-Case; Composition
Doubt
Doxa
Ecthesis ▶ Example
Effect to Cause Argument ▶ Consequence and Effect
Emotion
Enantiosis ▶ Disagreement
Enthymeme

3

Table of Entries

Epicheirema
Epitrope
Ethos
Etymology ▶ True Meaning of the Word
Evaluation and Evaluators
Evidentiality
Ex — Arguments (*Ex Concessis*…)
Ex Concessis
Ex Datis
Exaggeration and Euphemization
Example
Exemplum
Explanation
Expression
Faith
Fallacies (I): Contemporary Approaches
Fallacies (II): Aristotle's Foundational List
Fallacies (III): From Logic and Dialectic to Science
Fallacies (IV): A Moral and Anthropological Perspective
Fallacies as Sins of the Tongue
False Cause ▶ Cause-Effect
Figure
Follow-the-Leader ▶ *Ad Populum*; Consensus
Force
Forum
Generality of the Law
Genetic Argument ▶ Intention of the Legislator; Fallacy (I)
Genus
Gradualism and Direction
Hasty Generalization ▶ Induction
Historic Argument (Law) ▶ Legislator Intent
Ignorance
Ignorance of Refutation, *Ignoratio Elenchi* ▶ Relevance
Imitation - Paragon – Model
Implication ▶ Inference; Deduction; Connective
Index ▶ Natural Sign
Indicator
Induction
Inference
Intention of the Legislator
Interaction, Dialogue, Polyphony
Interpretation
Interpretation, Exegesis, Hermeneutics
Irony
Juridical Arguments: Three Collections
Justice: Rule of Justice
Justification and Deliberation

Table of Entries

Kettle Argumentation
Laughter and Seriousness
Laws of Discourse ▶ Scale
Layout of Argument (Toulmin)
Legal Syllogism ▶ Layout; Categorization; Definition
Likely ▶ Probable
Linked Argumentation
Logic: A Branch of Mathematics, an Art of Thinking
Logics for Dialogues
Logos - Ethos – Pathos
Manipulation
Many Questions
Map ▶ Script
Matter
Metaphor, Analogy, Model
Metonymy, Synecdoche
Moderation and Radicalism
Modesty
Motives and Reasons
Multiple Argumentation ▶ Convergent
Natural Signs
Nature; "Naturalistic Fallacy" ▶ Weight of Circumstances
Negation ▶ Denying
Non-Contradiction Principle
Novelty ▶ Progress
Number ▶ Consensus; Authority
Object of Discourse
Objection
Ontological Argument ▶ *A Priori, A Posteriori*

Opponent ▶ Roles
Opposite
Opposite: Refutation by the—
Opposites and *A Contrario*
Orientation
Orientation Reversal
Orienting Words
Ornamental fallacy?
Paradiastole ▶ Orientation Reversal

Paradoxes of Argumentation
Paralogism
Pathetic Argument
Pathos
Personal Attack
Persuading, Convincing
Persuasion
Petitio Principii ▶ Vicious Circle

Table of Entries

Plausible ▶ Probable
Politeness
Political Arguments: Two Collections
Polysyllogism ▶ Sorite; Serial
Pragmatic Argument
Precedent
Presupposition
Priming and Stages ▶ Gradualism
Probable, Plausible, True
Progress
Prolepsis
Proof and the Arts of Proof
Presence ▶ Object of discourse
Proper Name
Proponent ▶ Roles
Proportion and Proportionality
Proposition
Pseudo-Simplicity ▶ Fallacies (I)
Psychological Argument (in Law) ▶ Intention of the Legislator
Quasi-Logical Arguments
Question
Question: Argumentative Question
Rationality ▶ Criticism
Reciprocity
"Red Herring"
Reflexivity ▶ Relations
Reformulation ▶ Vicious Circle
Refutation
Related Words
Relations
Relevance
Repetition
Respect
Resumption ▶ Straw Man
Retaliation ▶ Autophagy
Rhetorical Argumentation
Rich and Poor
Right Balance Argument ▶ Moderation and Radicalism
Roles: Proponent Opponent, Third Party
Rules
Scale: Argument Scale and Laws of Discourse
Schematization
Scheme: Argument scheme
Scheme, Schema, Schematization
Script
Self-Argued Claim
Self-Evidence

Table of Entries

Serial Argumentation
Signs ▶ Natural Signs
Silence
Slippery Slope
Sophism, Sophist
Sorite
Stasis
Strategy
Straw Man
Strict Meaning
Structures of Argumentation
Subject Matter of the Law
Subjectivity
Subordinate Argumentation ▶ Serial Argumentation
Superfluity of the Law
Syllogism
Symmetry ▶ Relations
Synecdoche ▶ Metonymy
Systemic
Syzygy
Tagging
Taxonomies and Categories
"Technical" and "Non-Technical" Evidence
Testimony
Third Party ▶ Roles
Threat
Title
Topic, Topos, Commonplace, Argument Scheme, Argument Line
Topos in Semantic
True Meaning of the Word
Truth ▶ Probable, Plausible, True
"Tu Quoque!" ▶ *"You too!"*
Two-Term Reasoning
Value
Verbiage
Vertigo
Vicious Circle
Warrant ▶ Layout of Argument; Topic
Waste
Weight of Circumstances
Whole and Parts ▶ Composition and Division
Words as Arguments
"You too!"

A

A Comparatione

Lat. *comparatio*, "comparison".

The label "argument *a comparatione*" refers to two argument schemes:
1. Most often, to the argument by comparison: **S. Comparison;** ***A fortiori***.
2. Sometimes to the argument *a pari*: **S. *A pari*.**

A Conjugata ▶ **Related Words**

A Contrario ▶ **Opposites**

A Fortiori

Lat. *a fortiori ratione*, "for a stronger reason". *Ratio*, "reason"; *fortis*, "strong", *fortior* "stronger".

The argument *a fortiori* applies in two directions:
(i) "From bigger to smaller" (Lat. *a maiori ad minus*). This formula allows inferences from *more* to *less*:

The hook can hold a load of up to 20kg, so it can support 10kg.
If he is capable of killing someone, he is capable of striking someone.

Other expressions to the same effect: "*for stronger reason*"; "*all the more reason to/for*"; "*those who can do hard things can readily do easy ones*"...

(ii) "From the smaller to the greater" (Lat. *a minori ad maius*). This formula rejects inferences from less to more:
> The hook cannot hold a load of more than 20kg, so it certainly cannot support a 30 kg burden.
> If one has no right to strike, one has no right to kill.

Other expression to the same effect: *"still / much / even less"*…

This scheme can be specified in a discursive domain, for example as a consolation discourse:
> The idea that "death should spare young people" is more acceptable (more normal) than "death should spare the elderly". And you know that around you many younger people have died. Therefore accept death.

This form underlies the statement *"others died much younger"*, said to comfort the living for the death of an elderly relative.

1. *A fortiori*, a transcultural rule
The *a fortiori* argument scheme is a clear example of a cross-cultural interpretative-argumentative rule, S. **Interpretation**.

1.1 Greco-Latin tradition
In the Greco-Latin tradition all collections of argument schemes throughout the history of Western argumentation mention the *a fortiori* rule. Aristotle illustrates this rule via the following examples:
> If even the gods are not omniscient, human beings are certainly not. (*Rhet*, II, 23, 1397b15, RR, p. 359)
> A man who strikes his father also strikes his neighbors […] for a man is less likely to strike his father than to strike his neighbors (*ibid.*).

The second argument can be used in the following situation. Somebody was assaulted. Who is guilty? We know that someone in the victim's neighborhood committed violence against his own father. The *a fortiori* line casts suspicion upon he who has already committed more strongly prohibited forms of violence. The conclusion is that the police should question him.

1.2 Muslim legal argumentation
In Muslim legal argumentation, the *bi-l-awla* argument corresponds exactly to the *a fortiori* argument. The problem is discussed in the Koran (Sura 17, verse 24), dealing with the respect that a child owes to his parents:
> Do not say *"pfff!"* to them!

The prohibition refers to a minimal impolite retort of a child shrugging off the words of his parents, or obeying them reluctantly, puffing out a sigh of exasperation. The *a fortiori* principle extends the prohibition to all disrespectful behavior: *"since it is forbidden even to say "pfff!" to one's parents, it is all the more forbidden to say harsh words to them, to bully or to hit them"*.

The prohibition takes its support on the lowest point on the scale, the epsilon of disrespect. Commentators have noticed that *a fortiori* argument can be a case of semantic deduction (Khallâf [1942], p. 216).

1.3 Talmudic exegesis
The rules of Talmudic exegesis have been established by various authors since Hillel (1st century CE). The entry "Hermeneutics" of the *Encyclopædia Judaïca*, enumerates the thirteen interpretation rules of Rabbi Ishmael. The first is the rule *qal va-homer*, "how much more", going *a fortiori* from the "minor" (*qal*) to the "major" (*homer*) (Jacobs & Derovan 2007, p. 25). The rule helps to determine what is lawful and what is not, for example the conditions under which the Easter sacrifice, *Pesach*, should be offered. The Bible asks that Pesach be offered at Easter. Some actions are forbidden on the Sabbath, so what is one to do when Pesach coincides with the Sabbath? The calculation *a fortiori* gives the answer: the sacrifice *Olat Tamid* ("daily burnt-offering"[1]) is offered every day, including Shabbat. *Pesach* is more important than *Tamid* (proof: if one does not respect *Tamid*, one does not incur penalties; if one does not respect *Pesach*, the sanctions are severe). Since not to celebrate *Pesach* is more serious than not to celebrate *Tamid*, and *Tamid* is lawful when Easter falls on the day of Shabbat, it is therefore *a fortiori* lawful to proceed to sacrifice *Pesach* when Easter falls on the day of Sabbath.

2. Nature of gradation
The application of the *a fortiori* rule presupposes both that the facts put in relation fall within a certain category and that they are hierarchically positioned within this category. This gradation may follow very different principles:
— Objective gradation: "*he can hardly go from his bed to the window, and you would like to take him shopping downtown?*"
— Socio-semantic gradation: "*even grandparents sometimes make big mistakes, so their grandchildren...*"
— Gradation based on the authority of the sacred book: "*the Pesach sacrifice is more important than the Tamid sacrifice*".
When there is a consensus on the gradation, ratified by the dictionary, the argumentative or interpretive deduction is purely semantic, **S. Definition.**

In the Argumentation within Language theory (Ducrot 1973) the concept of a *graduated* category is represented as an argumentative scale@, the *a fortiori* rule being an operator of reasoning on such scales.

3. *A fortiori* in paragon scales
Some of these scales are topped by an ultimate individual, the most excellent specimen of the category, the *paragon*. The absolute degree in the category is

---

[1] After https://www.jewishvirtuallibrary.org/tamid (11-08-2017)

established in terms of comparability with the paragon: "*sly as a fox*".
These paragon scales are effective in rejecting a complaint: "*You say that what happens to you is unjust. That's true. But consider that Christ is the Innocent par excellence. Now, you are not Christ, and Christ accepted an unjust death. You must therefore accept this injustice.*"

> *An episode of the Spanish Civil War (1936-1939). Paco, a somewhat turbulent villager, turns himself in after the war, at the request of Mosén Millán, a priest. Mosén Millán assures him that he will be convicted, but that his life will be saved. Paco surrenders, and now he is to be shot along with his companions.*
>
> — Why do you want to kill me? What did I do? We didn't kill anyone. Tell them I've done nothing wrong. You know very well that I'm innocent; that we're all innocent.
> —Yes, my son. You are all innocent. But what can I do?
> — They want to kill me because I fought back at Pardinas. Fair enough, but the other two did nothing wrong."
> Pedro clung to the cassock of Mosén Millán, and repeated: "They did nothing, and they are going to kill them. They did nothing." Moved to tears, Mosén Millán said to him:
> — Sometimes, my son, God allows the death of an innocent. He allowed it for his own son, who was more innocent than you three.
> On hearing these words, Paco remained paralyzed and mute. The priest said nothing either.
>
> Ramón J. Sender, [*Requiem for a Spanish Peasant*], [1953][1]

A Pari

    Lat. *a pari*, or *a pari ratione*, "for the same reason": *par*, "equal, same" *ratio*, "reason".

A distinction must be drawn between two kinds of *a pari* arguments, depending on whether they deal with *individuals* or *classes of individuals*.
1. When the argument concerns *individuals*, the *a pari* argument includes an individual **x** in a *category* **C**. The individual becomes (is identified as) a member of the category, in logical symbols < x ∈ C >, **S. Categorization.**
2. When the argument concerns *classes* of individuals, the *a pari* argument reorganizes the category system (or *taxonomy*). It reduces two formerly distinct categories (*class, species*) to one, on the basis that they belong to the same super category (*genus*). This entry deals with this second definition.
The vocabulary of analogy and the label "argument *a comparatione*" are sometimes used to refer to the argument *a pari*, in both of its forms. **S. Taxonomies.**
The *a pari* argument "[applies] to another species of the same genus what can be asserted about some particular species." (Perelman, Olbrechts-Tyteca [1958],

---

[1] Quoted after Ramón J. Sender, *Requiem por un Campesino Español*. Barcelona: Destinolibro, 7th ed. 1981. P. 100-101.

p. 241);
> *A pari* reasons by equality of the cases if a parricide deserves death, the same applies to matricide. (Chenique 1975, p. 358)

The *a pari* argument transfers a property (a quality, a right, a duty…) (here "*— deserves death*") originally attached to a *species* **A** (here: "*— is a parricide*") to another *species* **B** (here: "*— is a matricide*"), arguing that they belong to the same *genus* (here: "*— is a murderer of a parent*"). The reasoning is as follows:
> The trend is towards severity
> The penalty for matricide is life imprisonment.
> Let's strengthen the punishment of matricide!
> The penalty for parricide is death.
> Parricide and matricide are crimes of the same genus (type, genre, kind…).
> The penalty for matricide should be death!

Two different situations should be distinguished for the discussion of *a pari*:
— Situations of complete knowledge, where the truth is fully known and can be fully contemplated; then, syllogistic reasoning applies.
— Situations where the truth is debatable and a concrete decision has to be made, i.e., argumentative situations.

1. Syllogistic *a pari*

From the point of view of absolute knowledge, the *a pari* argument is either a truism or a paralogism, depending on whether or not the property considered is generic, S. **Taxonomies and categories.**

(i) If the property *is* generic, then it is true of all species attached to the genus, and particularly true of the two species involved in the *a pari* argument. The syllogism runs as follows:
> Having a constant body temperature is a *generic* property of mammals.
> Whales, humans… are mammals
> *SO* Whales, humans… have a constant body temperature.

The corresponding *a pari* argument is:
> Both men and whales are mammals *("belong to the same genus", here mammals)*
> Men have a constant body temperature *("what is true of a species", here humans)*
> So whales (must) have a constant body temperature *("is applied to another species", here whales)*.

(ii) If the property is *not* generic, then, the inference is a paralogism:
> Labradors, poodles… are dogs
> Labradors are gun dogs
> So, poodles are / must be gun dogs.

The corresponding *a pari* argument is:
> Both labradors and poodles are dogs *("belong to the same genus" here dogs)*
> Labradors are gun dogs *("what is true of a species", here labradors)*
> *SO* poodles are gun dogs *("is applied to another species", here gun dogs)*.

But poodles are not gun dogs. The property "— *is a gun dog*" is not a generic property, it is attached to labrador as a species, not to the genus "dogs". It follows that this property cannot be safely transferred to poodles.

In short, a property can be transferred from a species to another species belonging to a same genus only if the said property is generic. The validity of the argument depends on the quality of the taxonomy it exploits, and the argument will be considered convincing only if people agree on the taxonomy@.

2. The seeming deadlocks *a pari* vs. *a contrario* and *a pari* vs. *a pari*

Two paradoxes are attributed to *a pari* argument. In the same situation:
    (i)   *a contrario* and *a pari* cancel each other out;
    (ii) *a pari* can destroy *a pari*.

2.1 *A contrario* against *a pari*

(i) *A pari* extends to the **A**s the treatment given to the **B**s, arguing that both are attached to a common super-category:

    (1) The **A**s are like the **B**s! They should be treated as **B**s!

(ii) *A contrario*, the argument from the opposite, justifies the difference in treatment of the **A**s and **B**s, arguing that they are *indeed* opposites:

    (2) The **A**s and **B**s are different, so they are rightly treated as such!

In both cases, the question is whether a difference between **A** and **B** should be preserved: *a contrario* answers "yes!", *a pari* answers "no!".

2.2 *A pari* against *a pari*

A pari argument extends to **B** a characteristic of **A**, or to **A** a characteristic of **B**. It can be objected to (i):

    (3) If the **A**s are like the **B**s, then the **B**s are like the **A**s; the **B**s are the ones which should be treated like **A**s!

Here, proponent and opponent refer to the same data and use the same rule to support contrary claims. They agree on the necessity to re-categorize **A**s and **B**s into just one category, but disagree about which should prevail.

Hence the conclusion may be reached that all this maneuvering is useless (in the following quotation "analogy" means *a pari*):

> That the *argumentum a contrario* and analogy as means of interpretation are entirely worthless can be seen from the fact that both lead to opposite results, and that no criterion exists to decide when the one and when the other should be applied. (Kelsen 1967, p. 352)

This is the case for an abstract, syllogistic situation, where:
— *A contrario* is actually logically invalid, S. *a contrario*.
— *A contrario* can be systematically opposed to *a pari*.
— As a "bidirectional" argument scheme, *a pari* can always be opposed to *a pari*.

3. Argumentative *a pari* and the situated condition of argument
Let us schematize a situation in which **G**s and **B**s are treated differently. *A pari* can be used to support the claims "All **G**s!" or "All **B**s!" and *a contrario*, to rebut both.

Present situation: **G ≠ B** and **G** and **B** are treated differently

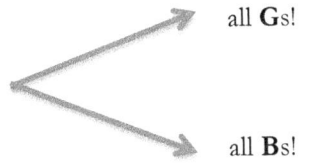

all **G**s!

all **B**s!

— *A contrario* is the *status quo* argument, which may be set up against both *a pari* alignments. Those arguing for a *status quo* do not support the burden of proof, they can simply reformulate and amplify the current "doxical" discourse, to maximize the opposition between **A**s and **B**s, and so to justify *a contrario* the difference in treatment:
> It is not by chance that **A**s and **B**s are called **A** and **B**, precisely because they *are* **A** and **B**, and not something else!

— The proponent of either *a pari* alignments of categories must undermine this discourse, showing that the difference formerly considered as essential should now be considered a mere accident. These minimizing strategies accompanying *a pari* depend on the characteristics of the specific situations.
In a slightly more complicated move, the individual forwarding an *a pari* argument may attempt to show that it is possible to construct a super category, including both **A** and **B**. This solution implies that the former definitions of both categories need to be amended.

The problem with the syllogistic approach of *a priori* is that it does not take the argumentative situation into account, whereas there are preferences and impossibilities enshrined in any such situation. These contextual conditions systematically eliminate one or the other application of *a pari*. *A pari* is logically bi-directional and contextually mono-directional, as can be seen in the following cases.

3.1 Military service: girls / boys
Context: a country where boys, but not girls, complete compulsory military service. Applying *a pari* to the boys, that is, claiming that they should *not* do their military service, amounts to requesting the dissolution of the army, its professionalization, or similar. This would be the real issue, and not that of equal treatment of boys and girls.
So, the *pari* argument can only be advanced by the girls, or by the military administration willing to incorporate girls. The relevant argumentative question can only be *"should the girls do military service too?"*, and *a pari* backs a positive answer very well.

3.2 Murders: patricide / matricide

Context: a social situation in which a "civilizing process" is taking place; there is a clear effort being made to eliminate all forms of violence. In such a situation, an *a pari* generalization of the death penalty is "out of the question". The only relevant issue can be *"should we reduce the penalty for patricide?"*, with *a pari* being used to back a positive answer.

If the social climate is oriented towards the strengthening of penalties, the situation is the same, with *a pari* serving as justification for the positive answer to the question *"should we increase the penalty for matricide?"*.

3.3 Employment contracts: fixed-term / permanent Contract

Context: Some employees receive a Permanent Contract (PC), while others have a Fixed Term Contract (FTC), the former kind of contracts being considered better than the latter from the point of view of the workers. In a period of prosperity and full employment, PCs are the golden standard, the *a pari* alignment of FTCs with PCs is on the agenda. The issue of a possible alignment of PCs with FTCs is irrelevant. The difference will be denied through argumentations such as:

> People with FTCs are exploited, we are all workers, everyone should be able to get a PC!

In less favorable economic conditions, FTCs become the norm, and their alignment with PCs is not on the agenda. The difference will be denied by argumentations such as:

> People with PCs are privileged, privileges should be ended, everyone should be put on an FTC!

3.4 Going out at night: girls / boys

Let us consider a family of consisting of both teenage boys and teenage girls, two species of the same genus. The boys are authorized to go out in the evening, the girls however are not. Let us suppose that this prohibition annoys the girls. They can argue their point in many ways. They might, for example draw on the positive outcomes that going out at night will have on their social awareness, **S. Pragmatic;** they might also point out that their brothers go out at night, in an elliptic *a pari*:

> G — *But the boys do go out at night!*

Unsurprisingly, the parents argue *a contrario*:

> P — *Yes but you are a girl...*

To strengthen their case and eliminate the difference, girls might stress the common features characterizing the new catchall category *"as the boys"*:

> Boys and girls receive the same education; they have access to the same media; they practice judo; they follow the same courses with the same requirements; they share the same tasks at home...

and minimize the gender gap:

> We are mature, we know how to avoid trouble...

A Priori, A Posteriori

>Lat. *prior*, "superior, anterior, older, better, first"; *posterior*, "coming after, behind, later; second".

In ordinary language, the modifier *a priori* is equivalent to "at first sight, before any thorough examination"; the expression is sometimes used to refer to *biased* thought. *A posteriori* currently refers to "on second thoughts; after the event"

1. *A priori / a posteriori*

The *a priori / a posteriori* distinction expresses an epistemological issue. *A posteriori* knowledge is concrete knowledge, built from sense data extracted from the world through observation and practice. In contrast, *a priori* intellectual knowledge is based only on knowledge of language (natural or formal), perhaps coupled with an intuition of essences.

In philosophy, the distinction *a priori / a posteriori* is linked to the *necessary / contingent*, and the *analytic / synthetic* oppositions.

1.1 *A posteriori*

A posteriori argument starts from an element of experience and reconstructs its material causes or origin. Alternatively, it functions via an abduction@ process, attaching this experience to a general explanation or a law accounting for the existence of the fact. Arguments from consequences@ to causes or principles; induction@; arguments based on a natural@ sign or a concrete example@, are cases of *a posteriori* argumentation.

When investigating the "origin and foundation of inequality among men", Rousseau highlights the difference between what would be a historical, *a posteriori*, approach to this topic and his own philosophical, *a priori* inquiry:

>Let us begin therefore by laying aside Facts, for they do not affect the Question. The Researches, in which we may engage on this occasion, are not to be taken for Historical Truths, but merely as hypothetical and conditional Reasonings, fitter to illustrate the Nature of Things, than to show their true Origin, like those systems, which our Naturalists daily make of the Formation of the World.
>Jean-Jacques Rousseau, *A Discourse upon the Origin and Foundation of Inequality Among Mankind.* [1755][1].

1.2 *A priori*

Unlike *a posteriori* argumentation, *a priori* argumentation is carried out without any explicit consideration of what exists. It starts from what is considered to be *deep, first, essential, superior* in an intellectual, religious or metaphysical order, and

---

[1] Quoted after John James Rousseau, *A Discourse upon the Origin and Foundation of Inequality among Mankind.* London: R. and J. Dodsley, 1761. P. 10.

develops its consequences in order to account for apparent, second order, derived, subordinated phenomena.

A priori argumentation may be based on foundations of various kinds.

— Causal *a priori* argumentation. Causes are considered as primary, as *conditioning*, with respect to the effect, which are secondary, that is to say, *conditioned*. *A priori* argumentation then corresponds to the cause@ to effect argumentation (or argumentation *propter quid*).

— Essentialist a *priori* argumentation is the fruit of pure contemplation and intellectual activity. It assumes that the human mind has the capacity to come into contact with (to apprehend) the essence; that is to say, the hidden and true reality of things, and to adequately express their concept in substantial definitions. Basic concepts are considered as *primary* in relation to their mundane incarnations. Practically, such argumentation starts with the definition of a concept corresponding to an object of investigation. The deduction then progresses analytically from one intellectual evidence to the other, all the while remaining in the domain of the *a priori*.

A priori argumentation corresponds to various kinds of deduction@ which start from principles, from language definitions@ or from axioms, in order to identify their consequences.

In a Platonic ontology, the ordered contemplation of essences defines supreme knowledge, and an *a priori* argument, which bears on the being of things, is the most valued form of argumentation.

2. *Propter quid* and *quia* argumentations

Lat. *propter quid*, "on account of which"; *quia*, "that"'.

The distinction *propter quid* / *quia* / proposed by Thomas Aquinas (*ST* 1st Part, Q. 2, 2; *Com*. NE, 4, § 51) is close to the *a priori* / *a posteriori* relation, and covers the same kind of argumentation respectively.

The proof *quia* is primary in relation to us, starts from what is better known to us, whereas the proof *propter quid* is primary in the absolute.

This distinction expresses the difference between a *cause to effect*, that is a "*propter quid*" *because*:

>The lawn is wet because it is raining
>Why is the lawn wet? — Because it is raining

and an *effect to cause*, that is, a "*quia*" *because*:

>It is raining, because the lawn is wet
>*Why is it raining? — Because the lawn is wet
>Why do you say it's raining? — Because the lawn is wet

In theology, the *a priori* - *propter quid* proof corresponds to the *ontological* argument for the existence of God, whose existence is deduced from the *a priori* perfection attributed to him. The ontological proof of the existence of God consists in defining God as an infinitely perfect being, in order to deduce that he necessarily exists, this conclusion being reached, as St. Anselm says "by arguing silently with [one]self" (*Pros.*, Preface).

The proof *quia* of the existence of God corresponds to the argument from the world itself (effect) to a creator (cause), as in the Voltairian metaphor:

> The universe embarrasses me, and I cannot imagine
> That such a clock should exist without a clockmaker.
>
> Voltaire, [*The Cabals*], 1772[1].

A Repugnantibus

> Lat. *a repugnantibus*, lat. *repugnans* "contradictory; resistant, contrary, incompatible". *Repugnant* meaning "disgusting" is also derived from this source; but the argument *a repugnantibus* is not the "argument of disgust", and refutation by the unpleasant consequences corresponds more closely to the refutation *ad incommodum*. *A repugnantibus* is closer to "revolting, unacceptable", the second meaning of *repugnant*.

1. In Cicero's *Topica*, the *a repugnantibus* argument is based on logically "contradictory" things (Cicero, *Top*, XII, 53; p. 420). In her translation of Boethius, Stump translates *a repugnantibus* as "from incompatible" (Boethius *Top*. p. 64), S. **Contradiction; Opposites.**

2. Bossuet defines the argumentation *a repugnantibus* as a contradiction between act and speech: "your conduct does not suit your speech" ([1677], p. 140), which corresponds to the third type of *ad hominem* argument, S. ***Ad hominem.***

A Simili

> Argument *a simili*, Lat. *similis*, "similar, looking like, identical".
> Argument *per analogiam*, Lat. *analogia*, "resemblance, analogy".

Perelman defines the argument *a simili* or "by analogy" as follows:

> A legal proposition being given, which affirms a legal obligation relative to a subject or a class of subjects, this same obligation exists with regard to any other subject or class of subjects having with the first subject (or class of subjects) sufficient analogy so that the reason which determined the rule with respect to the first subject (or class of subjects) is valid with respect to the second subject (or class of subjects). Thus, the fact of having forbidden a traveler to climb on the steps accompanied by a dog leads us to the rule that it is also necessary to forbid it to a traveler accompanied by an equally inconvenient animal. (1979, p. 56)

As the extension clause "an equally inconvenient animal" shows, *a simili* argu-

---

[1] Quoted in Pierre Hadot, *The Veil of Isis*. Cambridge, MAS & London, England: Harvard UP, 2008. P. 127.

ment is based on categorization@ mechanisms. It covers the same kind of reasoning as *a pari@*, and the rule of justice@. The terminology seems somehow redundant. S. **Analogy (I); Genus.**

By application of the *a fortiori@* rule, travelers may be accompanied by a *less* inconvenient animal than a dog (maybe a cat?), but not by a *more* inconvenient animal (a goat?)

Ab — Arguments (*A Contrario*...)

Some argument schemes are designated by Latin labels, S. *Ab* —; *Ad* —; *Ex* —. This entry lists the labels using the Latin preposition *a / ab*.
The same Latin preposition has two forms, *a* or *ab*: in general, *a* is used before a noun beginning with a consonant, and *ab* is used before a noun beginning with a vowel, for example "*a contrario* argument"; "*ab auctoritate* argument".

1. The construction
In classical Latin, the *a / ab* preposition means "separation; away from" and governs the ablative case. Grammatically, *ab / a* introduces only a circumstantial clause of a verb, indicating the origin. This means that the Latin construction "*argumentum ab* + N" is to be interpreted as elliptical for "*argumentum [ducetur*, "drawn"] *ab* ["from"] N". Latin texts regularly use expressions of this type, Cicero for example, wrote in the *Topics*:

cum autem a genere *ducetur argumentum* (my emphasis) (IX, 39; p. 411);

that is "when, however, an argument is drawn from *genus*". *Genere* is the ablative case of the noun *genus*; the construction is "*argumentum [ducetur] a genere*", "argument [provided by, taken] [from] the *genus*". Similarly, the rhetoric *Ad Herennius* suggests that, in order to amplify the charge, the orator has to look first for an argument drawn from authority:

primus locus ab auctoritate *sumitur* (my emphasis) (*Ad Her.*, II, 48; p. 147);

that is "the first commonplace [*primus locus*] is taken [*sumitur*] from [*ab*] authority [*auctoritate*]". *Auctoritate* is the ablative case of *auctoritas*, "authority". *Locus* means literally "place", and is taken here metaphorically as "inferential commonplace" or "argument scheme", S. **Topic, Topos, Commonplace.**

2. List of the "*ab* (*a*) + N" arguments
The set of "*ab / a* + N" arguments belongs to the original stock of Latin argument labels; its core is drawn from the Ciceronian typology, passed on to the Middle Ages by Boethius, up to modern times S. **Collections (II).**
In stark contrast to the list of "*ad* + N" arguments (S. *Ad* — **arguments**), the following list contains no label referring to feelings or subjective beliefs.

Ab — Arguments (A Contrario...)

Table: First column: Latin name of the argument
Second column:
- Meaning of the word(s) (based on Gaffiot).
- (When necessary a word-for-word translation)
- Reference for the corresponding entry

Latin name of the argument	• **Latin term(s) and their English equivalent(s)** — • **(Global translation)** — • **Corresponding entry/ies.**
ab auctoritate	Lat. *auctoritas*, "authority" — S. **Authority; Humility.**
a carcere	Lat. *carcer*, "jail" S. **Punishments and Rewards; Threat; Emotion.**
a cohaerentia	Lat. *cohærentia*, "coherence, consistency" - S. **Consistency.**
a comparatione	Lat. *comparatio*, "comparison; confrontation" S. **Comparison;** *A fortiori;* **Analogy**
a completudine	Lat. *completus*, "complete" — S. **Completeness**
a conjugata	Lat. *conjugatus* "belonging to the same family" S. **Related Words**
a contrario (sensu) (or: *ex contrario*)	Lat. *contrarius* "opposite, contrary" — S. **Opposite.**
a consequentibus	Lat. *consequens* "close; what logically follows" S. **Circumstances; Consequences.**
a fortiori *a fortiori ratione*	Lat. *a fortiori ratione*, "for a stronger reason"; *ratio*, "reason"; *fortior = fortis* + higher degree comparative "stronger" — S. *A fortiori.*
a generali sensu	Lat. *generalis*, "general"; *sensus* "meaning, point of view" — S. **Generality of the law.**
a genere	Lat. *genus*, "genus" — *Argument from genus* S. **Taxonomies and categories; Definition;** *A pari.*
a pari	Lat. *par*, "equal, same" — S. *A pari.*
a posteriori	Lat. *posterus*, "which comes after" S. *A priori; A posteriori*
a priori	Lat. *prior*, "the first of two, superior" S. *A priori; A posteriori*

Ab —, *Ad* —, *Ex* —: Latin Labels

a repugnantibus	Lat. *repugnans*, from *repugnare* "contradictory; contrary; incompatible" — S. ***A repugnantibus***; **Opposite**.
a rubrica	Lat. *rubrica*, "title of the section (law)" — S. **Title**
a silentio	Lat. *silentium*, "silence" — S. **Silence**.
a simili	Lat. *similis*, "resembling, similar" — S. **Analogy**; *A pari*.
ab absurdo [or: *ad absurdum*]	Lat. *absurdus*, "absurd" — S. **Absurd**.
ab adjunctis	Lat. *adjuncta*, "attached to" — *argumenta ex adjunctis ducta*, arg. from circumstances — S. **Circumstances**.
ab antecedentibus	Lat. *antecedens*, "preceding" — S. **Circumstances**.
ab consequentibus	Lat. *consequens*, "following" S. **Circumstances; Consequences**.
ab auctoritate (or: *ad auctoritatem*)	Lat. *auctoritas*, "authority" — S. **Authority**.
ab enumeratione partium	Lat. *enumeratio* "enumeration"; *pars*, "part" *arg. from enumeration of parts* S. **Whole and parts; Case-by-case; Definition**.
ab exemplo	Lat. *exemplum* "example" S. **Example**; *Exemplum*; **Precedent**.
ab inutilitate	Lat. *inutilitas*, "useless, dangerous" — S. **Superfluity**.
ab utili	Lat. *utilitas*, "useful, beneficial" — S. **Pragmatic**.

Ab —, *Ad* —, *Ex* —: **Latin Labels**

Latin labels are used to name arguments or fallacies. This practice, although not systematic, is common in modern texts, not exceptional in law, and some traces remain in contemporary usage. A few of these labels belong to the usual vocabulary of argumentation theory:
>argument *ad hominem, a fortiori, a contrario, a pari...*

The English counterpart of the Latin word is often transparent:
>argument *e silentio*, argument from silence.

Nonetheless, some labels remain opaque when one is not familiar with Latin:
>argument *ad crumenam*, argument to the purse.

The English translation of these Latin labels may be questionable. The label argument *ad verecundiam* is often translated as "argument from authority", while the Latin word *verecundia* means "modesty, humility". For Locke, who introduced this label, the *ad verecundiam* argument is not precisely a sophism of *authority* but of *submission* to authority, **S. Modesty**.

This terminology is no longer spontaneously understood. In many cases, this piecemeal Latin appears gibberish and even ridiculous, particularly when well established, or more readily understood English terms can be used to refer to the same argument scheme. This continued use of Latin labels, however, is due to the power of Latin as the language of law, theology, philosophy and traditional logic. This designation system for argumentation parallels the one which is well established and currently used for the designation of rhetorical figures. Latin has provided a common technical language for everyday reasoning, whilst giving the theoretical discourse some fragrance of Ciceronian authority. This use of Latin is altogether comparable to the contemporary use of English in the non-English speaking world.

Three main types of Latin phrases can be distinguished.

1. Prepositional labels using the prepositions *ab* (*a*); *ad*; or *ex* (*e*)

Some arguments or fallacies are designated, in contemporary texts, by prepositional phrases having the following structure:

>   Latin Preposition + Latin Noun + *argument*

Sometimes, the Latin word *argumentum* replaces *argument*.

Latin is an inflected language; in prepositional phrases, the preposition imposes a specific grammatical case on the following noun, marked by a morphological variation at its end.

The three most used prepositions are *ab, ex,* and *ad.*

— The preposition *ab* (*a* before consonant) means "from, pulled of, drawn from":

>   *a contrario* argument, argument from the contrary.

— The preposition *ad*, means "to, towards, for":

>   *ad personam* argument, argument to the person.

— The preposition *ex* means "from, out of", indicating the origin:

>   Argument *ex datis*: argument drawn from what is admitted by the audience.

Ex labels are less common. Occasionally, other prepositions can be found:

>   *per*: *per analogiam* argument, argument by analogy;
>   *in*: argument *in contrarium*, argument from the opposites;
>   *pro*: argument *pro subjecta materia*, argument relative to the subject matter. **S. Subject matter.**

From a semantic point of view, there is a directional contrast, *origin vs. purpose*, between the prepositions *ab* and *ex* on the one hand, and *ad* on the other hand:

>   *ab, ex* + Latin noun + argument = argument based on —, using —
>   *ad* + Latin noun + argument = argument targeting —.

Ab —, Ad —, Ex —: Latin Labels

Ab, ad and *ex* compete in the designation of some arguments, with the same meaning:
> *ab auctoritate* or *ad auctoritatem* argument;
> *ab absurdo* or *ad absurdum* or *ex absurdo* argument.

The argument schemes designated by each of these labels have no common semantic basis. Many *ad* tags have been introduced in the modern period. Sometimes, these *ad* tags refer to very specific contents, in particular, to appeals to emotion or to a subjective position, whilst the labels *ab* and *ex* are never used in this sense.

The following entries list the Latin labels according to the preposition head of the noun phrase, give some equivalent of the Latin terms, and refer to the corresponding entry or entries:
> S. *Ab* — Arguments (*A Contrario*, etc.)
> S. *Ad* — Arguments (*Ad Ignorantiam*, etc.)
> S. *Ex* — Arguments (*Ex Concesso*, etc.)

These lists are taken from Bossuet ([1677]), Locke ([1690]), Bentham ([1824]), Hamblin (1970); Perelman & Olbrechts-Tyteca ([1958]), and from the Internet. They do not claim to be exhaustive.

Modern Latin labels are presented along with ancient ones, as they were used by Cicero, Quintilian and Boethius, and sometimes incorporated unchanged by modern authors. Examples of this original terminology may be found under the entry **Typologies (I): Ancient**.

2. Other Latin phrases

Less frequently, various Latin phrases are used to refer to classical Aristotelian fallacies:

— Fallacy of omission of relevant qualification or circumstances; undue generalization of a limited claim:
> Fallacy *a dicto secundum quid ad dictum simpliciter*: a reasoning concluding from a qualified statement (limited in scope) to a generalizing statement (absolute).
> Lat. *dictum* "word; maxim; sentence" here: "assertion"; Lat. *secundum quid* "according to something"; Lat. *simpliciter*, from *simplex*, "simple".

This formula is abbreviated as "*secundum quid* fallacy", **S. Circumstances**.

— Fallacies of false cause, that is to say, of poor construction of the causal relation, **S. Causation; Cause**:
> *Non causa pro causa*: "a non-cause is taken for a cause". **E1** is said to be the cause of **E2**, although this is not the case.
> *Cum hoc, ergo propter hoc*: "At the same time as, thus because of". From the fact that **E1** and **E2** are concomitant, one wrongly infers that they are causally linked.
> *Post hoc, propter hoc ergo*: "later, thus because of": from the fact that **E1** always occurs before **E2**, one wrongly infers that **E2** is due to **E1**.

— **Fallacy of vicious@ circle**, or *petitio principii*:
> Lat. *petitio*, "demand"; *principium* "principle": "request to grant (something equivalent to) the claim which is actually disputed"

The language of law uses Latin phrases and expressions to refer to argumentative principles, for example:
> *eiusdem generi*: lat. *idem*, "the same"; *genus*, "genus". Argument from the identity of genus; **S. Genus; Juridical arguments.**

3. A mocked pattern

In *Tristram Shandy*, Sterne mentions the arguments *ad verecundiam*, *ex absurdo*, *ex fortiori*, *ad crumenam* and the *argumentum baculinum* (*ad baculum*) and asks to add to this list the *argumentum fistulatorium*, which he claims to have invented.

> — There lies your mistake, my father would reply; — for in *Foro Scientiae* there is no such thing as MURDER, —'tis only DEATH, brother.
>
> My uncle *Toby* would never offer to answer this by any other kind of argument, than that of whistling half a dozen bars of *Lillabullero*.——You must know it was the usual channel thro' which his passions got vent, when any thing shocked or surprised him; — but especially when any thing, which he deem'd very absurd, was offerd.
>
> As not one of our logical writers, nor any of the commentators upon them, that I remember, have thought proper to give a name to this particular species of argument, — I here take the liberty to do it myself for two reasons. First, That in order to prevent all confusion in disputes, it may stand as much distinguished for ever from every other species of argument — as the *Argumentum ad Verecundiam*, *ex Absurdo*, *ex Fortiori*, or any other argument whatsoever: — And, secondly, That it may be said by my children's children, when my head is laid to rest, — that their learn'd grandfather's head has been busied to as much purpose once, as other people's; — That he had invented a name, — and generously thrown it into the TREASURY of the *Ars Logica*, for one of the most unanswerable arguments in the whole science. And, if the end of disputation is more to silence than convince, — they may add, if they please, to one of the best arguments too.
>
> I do therefore, by these presents, strictly order and command, That it be known and distinguished by the name and title of the *Argumentum Fistulatorium*, and no other; — that it rank hereafter with the *Argumentum Baculinum* and the *Argumentum ad Crumenam*, and for ever hereafter be treated of in the same chapter.
>
> As for the *Argumentum tripodium* [...]
>
> Laurence Sterne *The Life and Opinions of Tristram Shandy, Gentleman* [1760][1]

Lillibullero is a famous Irish march; the *fistula* is a panpipe (Gaffiot, *Fistula*). Uncle Toby's maneuver is an excellent, although rude, strategy to annihilate a discourse, **S. Destruction of discourse; Dismissal.**

---

[1] In *The Complete Work of Laurence Sterne*. Delphi Classics, 2013. P. LV

Ab Exemplo

Lat. *exemplum*, "example".

In law, the label *ab exemplo* refers to an argument that interprets the law according to:
1. A previous case, S. **Precedent**;
2. A traditional interpretation, "the doctrine generally accepted" (Tarello, quoted in Perelman 1979, p. 59).

The argument *ab exemplo* is therefore distinct from the argument *from example*, S. **Example**.

Abduction

Lat. *Abductio*, "action of taking", by an outwardly directed movement (see infra, meaning 2).

1. Abduction as inference from facts to hypothesis
The concept of abduction was introduced in modern philosophy by the philosopher Charles Sanders Peirce. According to Peirce, there are two kinds of inferences: *deductive* inference and *abductive* inference or *abduction*. Abduction starts from the observation of a fact "contrary to what we should expect" Peirce ([1958], § 202), that is to say, a fact that does not fit into an available explanatory system. Abduction is a kind of inference by which one proposes a hypothesis accounting for this fact.

This hypothesis is not the product of the application of a "discovery algorithm", but the fruit of a creative process, "abduction is, after all, nothing but guessing" (Peirce [1958], § 219).

Abduction is not an issue in logic, but rather a scientific method (*id.*, Chap. 6). Scientific work consists in proposing, on the basis of facts, plausible hypotheses "suggested" by these facts. Abduction is the first step in this process. The practice of abduction is not guided by logical rules but by general principles, such as the principle of exclusion of so-called *metaphysical hypotheses*, that is to say, hypotheses which would have no experimental consequences, or the principle according to which *every fact has an explanation*: an abducted hypothesis is interesting "if it seems to make the world reasonable" (*id.*, §202).

Unlike abduction, which starts from facts in search of theory, the Peircian *deduction* starts from a theory in search of facts; that is, it seeks to identify the crucial experimental consequences of a hypothesis.

Much more than a form of deduction or induction, argumentation should be seen as a form of abduction: because the light is on, "I abduct", I make the hypothesis, that there is someone in the room; but this hypothesis still needs to be checked, S. **Probable**.

Woods redefines abductions as "responses to *ignorance-problems*. An agent has an ignorance-problem in relation to an epistemic target that cannot be hit by the cognitive resources presently at his command, or within easy and timely reach of it" (Woods, 2009; Gabbay & Woods, 2005). The study of argument as an abductive process has proved especially fruitful in the fields of medicine, science and law (Walton 2004).

2. Abduction as reduction of uncertainty
In its Peircian sense, abduction is a kind of inference by which one arrives at a hypothesis accounting for this fact. Aristotle defines abduction as a kind of dialectical syllogism (Aristotle, *PA*, II, 25), whose major premise is true, the minor just probable, and, consequently, the conclusion also probable. The conclusion alone, without the minor, is more improbable than the minor. The minor therefore strengthens the relative acceptability of the conclusion. This situation recalls the Ciceronian definition of argumentation, S. **Argumentation** (I).

The question is: "can virtue be taught?" By combining:
  1. A true premise: it is clear that *science can be taught*;
  2. A doubtful premise: *virtue is a science*;
  3. Conclusion: *virtue can be taught*.

Though uncertain, the veracity of the second premise is still less in doubt than the conclusion "virtue can be taught". This second premise may therefore serve as an argument for the conclusion. We find this montage in speeches such as:
  Citizenship can be taught.
  Citizenship is essentially a set of social knowledge and practices.
  Knowledge is being taught and all practical skills can be improved by teaching
  So, citizenship can be taught.

Argument functions "for want of better". Reduction of uncertainty serves to modify relevantly the epistemic status of a belief. This is a logic not of *elimination* but of *reduction* of doubt and uncertainty, S. **Default reasoning**.

Absurd

Lat. *absurdus*, "absurd". Argument *ad absurdum, ab absurdo, ex absurdo*; or *reductio ad absurdum*, "reduction to absurdity", under different forms: *reductio ad impossibile*, "reduction to the impossible"; *r. ad falsum*, "r. to the false"; *r. ad ridiculum*, "r. to the ridicule"; *r. ad incommodum*, "r. to the undesirable".

1. The scheme
Argumentation *from the absurd* is a form of indirect evidence based on contradiction. This label includes a family of arguments concluding that a proposal should be rejected on the basis of the indefensible consequences which would

result from its adoption. The general operation of reduction to the absurd corresponds to the following mechanism:
>   1. A claim is put forward, as a hypothesis, a possibility…
>   2. Consequences are drawn from this proposition, whatever they may be, causal, logical…
>   3. One of these consequences is deemed to be "absurd" in relation to some criteria, cf. infra
>   4. The initial proposal or hypothesis is rejected.

2. Varieties of absurdities

There are as many kinds of reduction to absurdity as modes of deduction and reasons to evaluate a consequence as inadmissible. The qualification as absurd may thus apply to:

— Mathematical consequences. One clearly sees the variety and the diversity of what is called the "absurd" in argumentation by contrasting these forms with the demonstration from the absurd, where *absurd* means "contradictory", cf. infra.

— Logical or semantical consequences. The consequences analytically derived, from the very meaning of the expression lead to a semantic difficulty, **S. Dialectic; Opposites; Consequence**.

— Causal consequences. In the physical domain and natural experience, the effects predicted by the hypothesis are not attested, **S. Causality**. The refutation by an attested fact, different from the theoretically expected fact, is a kind of refutation from the absurd.

As soon as one turns from the scientifically established causal link to the "causal story" as constructed in a pragmatic@ argument, however, the speaker intervenes through his or her positive or negative valuation of the consequences. The consequence is then:

— Contrary to the intended goals, the effects of the proposed action are *perverse*; the measure is counterproductive, contrary to various interests, **S. Pragmatic**:

— Inadmissible from the point of view of law, morality or common sense, **S. Apagogic;** *Ad Incommodum*.

Pragmatic refutation by negative consequences is opposed to a measure by showing that it will have negative consequences unforeseen by the individual who proposes the measure, and that these drawbacks will prevail over any possible advantage. We approach demonstration by the absurd if we can show that the measure will have effects *diametrically opposed* to those which it proposes, and that it will in fact increase the evil it is intended to combat, **S. Pragmatic**.

Argumentation *to the absurd* is not an argument *from ignorance*. An argument from ignorance concludes that **P** is true because we have failed to prove **not-P**, whilst an argument to the absurd concludes that **P** is true because it has been shown that the proposition **not-P** is false, and that between **P** and **not-P**, only one can hold true. This corresponds to a case-by-case argument in a situation where the number of cases is reduced to two: **P** is true or **not-P** is true; but **not-P** is false, so **P** is true. S. **Apagogical; Contradiction; Ignorance.**

3. Demonstration by reduction to the absurd

Proof by the absurd, or by contradiction, is based on the principle of the excluded middle, according to which "**A** or **not-A**" is necessarily true. The reasoning is based not on the proposition **A** that we want to prove, but on its negation, **not-A**.

The negation, **not-A**, is provisionally admitted and its consequences are deduced; these consequences lead to statement **A**. But the conjunction "**A** and **not-A**" contravenes the principle of contradiction; thus, **not-A** is false, and **A** is necessarily true.

In the language of implication, we are in a situation where "**A** $\rightarrow$ **non-A**'. According to the principle of "one can deduce anything from the false", this implication is true only if **A** is false.

Let's show by reduction to the absurd that "the square root of 2 (the number whose square is 2, denoted by $\sqrt{2}$) is not a rational number" (proposition **A**).

(1) Suppose that "the number corresponding to $\sqrt{2}$ is a rational number" (proposition **not-A**).
(2) By definition, a rational number can be written in the form of a fraction "**p** / **q**", where **p** and **q** are prime (admit only **1** as a common denominator).
(3) "2 = p / q" therefore "$p^2 = 2q^2$"; therefore p^2 is even; and we know that if the square is even, the root is even. Therefore **p** is even.
(4) **p** being even, it can be written: "$p = 2k$", and its square "$p^2 = 4k^2$".
(5) We know from (3) that "$p^2 = 2q^2$".
(6) Therefore $2q^2 = 4k^2$; $q^2 = 2k^2$. So the square of **q** is even, so **q** is even.
(7) **p** and **q** are even; Therefore they admit **2** as a common divisor, which is contradictory to the initial hypothesis.
(8 Conclusion: hypothesis (1) is false, and $\sqrt{2}$ is not a rational number.

Demonstration by the absurd is an indirect way of demonstrating a proposition. It has not been proved that **A** is true, but only that **not-A** is false. This reasoning is by no means permitted by all specialists, "if the classical mathematicians consider the proof by the absurd as valid, the intuitionists reject it: in order to prove **a**, they say, it is not enough to establish that **not-(not-a)**" (Vax 1982, *Absurd*). We see that the demonstrative character of a demonstration can be discussed.

Accent ▶ Ambiguity

Accident

The fallacy of accident is the first on Aristotle's list of fallacies independent of discourse, S. *Fallacious* (III). The idea is that a valid syllogistic inference develops in the same category domain, for example, the class of animals:

> Socrates is a man, man is a mammal, so Socrates is a mammal,

whereas the following fallacious inference develops from an accident:

> Socrates is white, white is a color, so Socrates is a color.

The word *accident* is understood in its philosophical meaning, which contrasts *accident* with *essence*. A being is characterized by a set of essential features that determine its place in a scientific taxonomy@: its *generic* features express its genus and its specific *difference* indicates its species. Unlike "— *is a mammal*", which is *constantly* true of all dogs, the truth of the accidental predicate "— *is tired*" is circumstantial, it may be true of a dog at a given time but become false as soon as the dog's condition changes.

The fallacy of accident occurs when an accidental characteristic of a being is mistaken for an essential one. In a definition, the corresponding defect consists in defining a being by a feature which belongs to it only accidentally. So for example, "— *wanders off in the middle of the road*" is not a trait likely to define a dog, "— *is a good time for having a nap*" is not a defining feature of "afternoon", S. **Two-term Reasoning**.

The charge of committing the fallacy of accident is possible only if the accuser can refer to a solid and stabilized categorization, corresponding to a set of essentialist definition, S. **Definition (II)**.

The ethical value of a profession is evaluated on the basis of an examination of the moral worth of its values and practices. In a classical democratic regime, a politician can be honest or dishonest without ever ceasing to be a politician. Dishonesty is not a necessary condition for becoming a politician; it is an accidental feature; "*he is an honest politician*" is not an oxymoron, "*he is a dishonest politician*" is not tautologically true. For those sharing this vision of things and people, characterizing political activity as an intrinsically dishonest activity, is committing the fallacy of accident. The person blamed for committing the fallacy might retort that the argument is not based on any transcendental organization of things, but on an inductive generalization, from "*a number of politicians we all know very well*"; or on the actual structural condition of our political system.

A contrario@ argument plays with the essential *vs.* accidental character of the differences between two categories of beings, "*boys can go out at night, so girls should not go out, well, you know, girls are different from boys*". It is refuted by demoting the difference from essential to accidental. The same strategy applies to the distinctions between the defining features of a fact, and its circumstantial, contextual characteristics.

Dissociated from the strict Aristotelian ontology, the "essence *vs.* accident" opposition corresponds to the distinction between *central* traits and *peripheral* traits, and, in everyday life, to the distinction between the *important* and the *incidental*.

Ultimately, in the absence of backing by an accepted ontology, the so-called fallacy of accident functions as a refutation arguing *from the incidental nature* of an element, and finally corresponds to a strategy of *minimization* of the disputed character.

Ad — Arguments (*Ad Ignorantiam*...)

Some argument schemes are designated by Latin labels, S. *A/Ab* —; *Ad* —; *Ex* — This entry lists the labels using the Latin preposition *ad*. In classical Latin, the preposition *ad* is constructed with the accusative and introduces a goal complement; the phrase "argument *ad hominem*" reads "argument addressing the person".

According to Hamblin, the oldest scheme in this grouping is *ad hominem*, which appears in the Latin translations of Aristotle; this naming method was popularized by Locke ([1690]) and by Bentham ([1824]), and most of these terms seem to be nineteenth or twentieth century creations (Hamblin 1970, p. 41; p. 161-162).

1. List of the "*ad* + N" arguments

Latin name of the Argument	• Meaning of the Latin word(s)Latin • (When necessary a word-for-word translation) • (English equivalent(s)) • Reference to the corresponding entry/ies
(*reductio*) *ad absurdum* (also: *ab absurdo*)	Lat. *absurdus*, "false, unpleasant, absurd" — reduction to the absurd — S. Absurd
ad amicitiam	Lat. *amicitia*, "*friendship*" — appeal to friendship — S. Emotion.
ad antiquitatem	Lat. *antiquitas*, "antiquity, tradition" — appeal to antiquity, to tradition — S. Authority.
ad auditorem (pl. *ad auditores*)	Lat. *auditor*, "hearer, audience" — S. Beliefs of the audience.
ad baculum	Lat. *baculus*, "stick" — S. Punishment and Reward.
ad captandum vulgus	Lat. *captare*, "try to seize ... by insinuation, by guile"; *vulgus* "crowd, ordinary people" — playing to the gallery ; playing to the crowd — S. Orator; Emotion; *Ad populum* ; Laughter and Seriousness.

ad consequentiam	Lat. *consequentia*, "following, consequence" — **S. Consequence.**
ad crumenam	Lat. *crumena*, "purse" — argument to the purse — **S. Emotion ; Punishments and Rewards**
(*reductio*) *ad falsum*	Lat. *falsum*, "false" — reduction to a falsehood — **S. Absurd**
ad fidem	Lat. *fides*, "faith" — **S. Faith**
ad fulmen	Lat. *fulmen*, "thunderbolt" — argument from thunderbolt — **S. Punishment and Rewards ; Threat**
ad hominem	Lat. *homo*, "man, human being" — **S. *Ad hominem***
ad ignorantiam	Lat. *ignorantia*, "ignorance" — **S. Ignorance**
ad imaginationem	Lat. *imaginatio*, "picture, vision" — appeal to imagination — **S. Subjectivity**
(*reductio*) *ad impossibile*	Lat. *impossibile* "impossible" — reduction to the impossible — **S. Absurd**
(*deducendo, reductio*) *ad incommodum*	Lat. *incommodum* "unfortunate, disadvantageous" — *reduction to the uncomfortable* — **S. *Ad incommodum***
ad invidiam	Lat. *invidia*, "hate, envy" — *appeal to envy* — **S. Emotion**
ad iudicium	Lat. *iudicium*, "sentence, judgment, opinion" — *arg. appealing to the judgment*; *to common sense* — **S. Matter**
ad lapidem	Lat. *lapis*, "stone; (symbol of stupidity, insensibility)" — arg. by dismissal — **S. Dismissal**
ad Lazarum	Lat. *Lazarus*, character of the Bible, paragon of the destitute — *arg.* ad Lazarum — **S. Rich and Poor**
ad litteram	Lat. *littera*, "letter" — **S. Strict Sense**
ad ludicrum	Lat. *ludicrum*, "public game (theater, circus...)" — appeal to the gallery — **S. Emotion; Orator; *Ad populum* ; Laughter and Seriousness**
ad metum	Lat. *metus*, "fear, apprehension" — appeal to fear — **S. Threat; Punishment and Reward**
ad misericordiam	Lat. *misericordia*, "compassion, pity" — appeal to pity — **S. Emotion**
ad modum	Lat. *modus* "measure, just measure, moderation" — arg. of gradualism — **S. Just proportion**

ad naturam	Lat. *natura*, "nature" — appeal to nature ; naturalistic fallacy — **S. Necessity**
ad nauseam	Lat. *nausea*, "nausea, seasickness" — proof by assertion — **S. Repetition**
ad novitatem	Lat. *novitas*, "novelty, innovation; unexpected thing" — appeal to novelty — **S. Progress**
ad numerum	Lat. *numerus*, "number, great number" — arg. from number — **S. Authority**
ad odium	Lat. *odium*, "hate" — appeal to hatred, to spite — **S. Emotion**
ad orationem	Lat. *oratio*, "language, comments, speech, discourse" — **S. Matter**
ad passionem (pl. *ad passiones*)	Lat. *passio*, "passivity; passion, emotion" ; *appeal to passion, to emotion* — **S. Pathos** ; **Emotion**
ad personam	Lat. *persona*, "mask; role; person" — abusive *ad hominem* — **S. Personal Attack;** *Ad hominem*
ad populum	Lat. *populus* "people" — appeal to people, arg. from popularity — **S.** *Ad populum*
ad quietem	Lat. *quies* "rest; political neutrality; calm; peace", tranquility" — *appeal for calm, conservatism*, **S. Peace**
ad rem	Lat. *res*, "thing, being, reality ; judicial matter, issue" — *arg. addressed to the thing, to the point, dealing with the matter at hand* — **S. Matter**
ad reverentiam	Lat. *reverentia*, "respectful fear; deference" — **S. Respect**
ad ridiculum	Lat. *ridiculus*, "funny; ridicule" — appeal to ridicule, appeal to mockery — **S. Absurd** ; **Laughter and seriousness**
ad socordiam	Lat. *socordia*, "stupidity; indolence" — appeal to weak-mindedness — **S. Subjectivity**
ad superbiam	Lat. *superbia*, "pride" — *appeal to pride; arg. of popular corruption* — **S. Emotion;** *Ad populum*
ad superstitionem	Lat. *superstitio*, "superstition" — **S. Subjectivity**
ad temperantiam	Lat. *temperantia*, "moderation, restraint" — **S. Proportion**
ad verecundiam	Lat. *verecundia*, "respect, modesty, discretion ; fear of shame" — *arg. from modesty* ; *arg. from authority* — **S. Subjectivity ; Humility ; Authority**
ad vertiginem	Lat. *vertigo*, "rotation, dizziness" — **S. Vertigo**

2. Characteristics of the "*ad* + N" family

2.1 A productive pattern
There are many more "*ad* +N" arguments than there are "*a / ab* + N" arguments. Only the "*ad* +N" construction is still productive; the pattern is popular and mocked (*ad bananum* argument).

2.2 Origin of the labels
Some of these names have been defined and used by Locke and Bentham, S. Collections (III).
Locke has defined the arguments:

    ad hominem   *ad judicium*
    ad ignorantiam   *ad verecundiam*

Bentham has defined the arguments:

ad amicitiam	*ad judicium*	*ad socordiam*
ad ignorantiam	*ad metum*	*ad superbiam*
ad imaginationem	*ad odium*	*ad superstitionem*
ad invidiam	*ad quietem*	*ad verecundiam*

2.3 Semantic subsets of "*ad* + N" arguments
These arguments refer to very different strategies. Nonetheless, some groupings can be proposed according to their semantic content.

(i) Arguments bound to affects, emotions, often via positive interest (rewards) or negative results (threats):

ad amicitiam	*ad misericordiam*	*ad personam*
ad captandum vulgus	*ad novitatem*	*ad populum*
ad invidiam	*ad numerum*	*ad superbiam*
ad ludicrum	*ad passionem*	*ad verecundiam*
ad metum (ad carcerem, ad baculum, ad fulmen, ad crumenam)	*ad odium*	
	ad quietem	

(ii) Arguments involving a limited, subjective system of beliefs, not universal, questionable:

ad consequentiam	*ad ignorantiam*	*ad socordiam*
ad fidem	*ad imaginationem*	*ad superstitionem*
ad hominem	*ad incommodum*	*ad vertiginem*

Categories (i) and (ii) list arguments often considered as misleading, insofar as they express the subjectivity of the speaker. In other words, they are related to the ethotic and pathemic components, S. Subjectivity; Ethos; Pathos; Emotion.

(iii) Arguments opposed to the subjective series (i) and (ii) and **dealing with the substance of the issue:**

    ad iudicium   *ad rem*

Ad Baculum ▶ Threat

Ad Hominem

Lat. *homo*, "human being".

1. *Ad hominem* as personal attack, *ad personam*

Today, *ad hominem* is commonly used to mean *ad personam*, but classical *ad hominem* argument is quite distinct from personal attack (or *ad personam* attack), which seeks to disqualify the person in order to get rid of the arguments.

2. *Ad hominem* as self-contradiction or inconsistency

The concept of the *ad hominem* strategy is to be found in Aristotle's *Rhetoric*, topic n° 22:
> Another line of argument is to refute your opponent's case by noting any contrast or contradiction of dates, acts or words that it anywhere displays. (1400a15; RR p. 373).

Under that name, the *ad hominem* argument is defined by Locke as a discussion technique by which the speaker "[presses] a man with consequences drawn from his own principles or concessions. This is already known under the name of *argumentum ad hominem*". ([1690], p. 411)
The term "principle" can be taken in the moral or intellectual sense of "first principles". In both cases, the speaker rearticulates the system of beliefs@ and values of the opponent, in order to identify a contradiction. Locke rejects this form of argument as fallacious, insofar as it is based on the specific belief structure of a person, without relevance for the discussion of the truth *per se* of the thesis under debate, "[it does not] follow that another man is in the right way, because he has shown me that I am in the wrong" (*ibid.*). The *ad hominem* argument is of no force and plays no role as an alethic instrument, in the process of establishing truth, **S. Collections (III)**.

In regard to this definition, Leibniz notes that:
> The argument *ad hominem* has this effect, that it shows that one or the other assertion is false and that the opponent is deceived whatever way he takes it. ([1765], pp. 576-577)

He thus recognizes the merits of this form of argument in the context of a discussion, as an epistemic instrument, urging a reorganization of a system of knowledge.

Under Locke's presentation, *ad hominem* argument bears on explicit propositions as put forward in a knowledge acquisition dialogue and is clearly deductive and propositional.
In general terms, *ad hominem* argumentation occurs in a dialogue when the speaker builds a discourse, referring not only to propositional beliefs but also to the behavior and actions of his or her opponent, in order to point out some contradiction. This has the effect of embarrassing the opponent and causing

him or her to reconsider his or her speech, positions or actions.

Ad hominem argumentation typically results in the feeling of "embarrassment", considered as a basic emotion by Ekman (1999, p. 55). The production of such an emotion is not an accidental by-product of *ad hominem*, but is built into it, as revealed by the verb "to press", that is "to assail, harass; afflict, oppress". "Embarrassment" is typically a cognitive-emotional feeling, as is the basic argumentative emotion, "doubt". Nonetheless, *ad hominem* is not emotional in the same vein as personal abuse can be, **S. Personal Attack**.

3. Setting up the words against the words

We have a reply *ad hominem* in the following case:
>Proponent: — **P**. *I propose* **P**
>Opponent: — *Before, you proposed entirely different things.*

>Issue: — *Should the term of the presidential mandate, currently five years, be reduced to four years?*
>Proponent (former President): — *I am for a reduction to four years.*
>Opponent: — *But in an earlier statement, while you were president yourself, you yourself argued that five years were necessary for the proper functioning of our institutions. Please, clarify.*

The quoted statement which opposes the present one may be drawn not only from what has been said by the opponent in the past, but also from what has been said by "his or her people", that is to say, by members of the discursive community sharing the same argumentative orientations: people of the same party, religion, scientific trend, etc., that cannot be easily disavowed.

The *ad hominem* reply allows the speaker to intervene in a discourse in the third party's mode, that is, without committing himself to the substance of the debate. He does not explicitly take on the role of an opponent, but speaks simply as a participant in good faith, seeking clarification.

In an accusatory context, the charge of narrative incoherence allows the accused to reject the accusatory narrative, **S. Consistency**.

Reactions to *ad hominem* refutation on what has been said before — The target of the *ad hominem* argument can choose to sacrifice the former position, to reject the contradiction, or to accept it.

(i) Sacrifice the former position:
>— *Circumstances have changed, we must follow our times.*
>— *I have developed my system*
>— *I have changed, only madmen never change their mind; do you prefer psychorigid people?*

(ii) Use a direct rebuttal. The opponent elicits the contradiction: "you say both **A** and **Z**, which is inconsistent"; the force of this argument is derived from the quotation mechanism. The proponent did not necessarily say **A** or **Z** but something else, **A'** or **Z'**, that the opponent paraphrases, rephrases or reinterprets as **A** or **Z**. The contradiction may therefore proceed from a reworking of the

speech, **S. Straw Man**. It follows that the proponent can reply *to the letter*, and reject the key *ad hominem* phrase "you yourself admitted" in his or her second turn:

> *— You make me out to say what I have never said, you distort my words*

In other cases, the precise relation between **A** and **Z**, that is, the nature and degree of the inconsistency, might be disputable, **S. Denying; Opposites**.
The *ad hominem* imputation can be directly dismissed on these two counts.

(iii) Accept the contradiction. The *ad hominem* reply seeks an individual free from contradiction. By a classic maneuver in stasis theory, the recipient may choose to assume what he or she has been criticized for, thus making contradiction a system of thought, **S. Stasis; Contradiction**:

> *— I fully accept my inconsistencies. I love rain and good weather.*

4. Setting up the beliefs against the words

In the preceding case, there was direct opposition between a present claim and an earlier assertion. Consider the issue of the withdrawal of troops sent to intervene in Syldavia:

> Q: *—Should we withdraw our forces from Syldavia?*
> S1: *— Yes!*

Let us suppose however that **S1** has been led to admit **A**, **B**, and **C**; or, at least that **S2** speaks as if he sincerely believed that **S1** supports these propositions:

> S2: *— But you said yourself that (A) the Syldavian troops are badly trained, and (B) that the political unrest in Syldavia is likely to extend to the whole region, there is a real contagion risk. You will agree that such an extension would threaten our own security (C); and no one denies that we must intervene if our security is threatened. So you have to admit that we have to stay in Syldavia.*

S1 therefore claims that **P**; **S2** argues *ex datis@*, that is, on the basis of beliefs held by **S1** (or attributed to him), and concludes **not-P**. This is the case considered by Locke. Must **S1** admit that he or she has made an error, and that we should not withdraw the troops? Obviously not; **S2** simply showed by his objection that one cannot support both {**A, B, C**} and **not-P**.

Reactions to the *ad hominem* refutation on reconstructed beliefs — S1 can re-adjust and rearticulate all the key components of **S2**'s discourse. He can argue that **A, B, C** are abusive reformulations of his beliefs, or that the full analysis of the Syldavian situation is much more complex than these three assertions. If **S1** accepts such a reconstruction of his speech and beliefs, then he or she must reform one or more of these propositions, rejecting for example the idea that the troubles in Syldavia can extend to the whole region. **S1** is expected only to correct, clarify or explain more thoroughly why this system of beliefs {**A, B, C**} cannot be expanded into **non-P**. This is precisely the point the argument *ad hominem* is getting at. In this function, *ad hominem* replies are a powerful educational tool.

5. Setting up the prescriptions and practices against the words

A contradiction can also be raised between, on the one hand, what I require from others, what I prescribe or forbid them, and, on the other hand, what I'm doing myself, the kind of example I set. There is some paradox in asking others not to smoke, while I smoke myself. In our culture, acts are considered "to speak louder than words", and injunctions are systematically flouted if the speaker does not comply with them himself:

> *Doctor, heal thyself!*
> *He's not a good marriage counselor, he's always arguing with his wife!*
> *You claim to teach argumentation and you are unable to argue yourself!*
> *You advocate for the rights of women and at home you never do the dishes.*

Note that, in the last two arguments, the conjunction *and* coordinates two *anti-oriented* statements, and not, as is more commonly the case, two *co-oriented* statements, **S. Orientation**.

The *ad hominem* game can be played in several moves:

> Question: *Should hunting be prohibited?*
> S1:   — *yes, hunters kill animals for pleasure!*
> S2:   — *but you eat meat, don't you?*

L2's argumentation can be reconstructed as "*We must prohibit, suppress hunting. Hunters kill for pleasure. That's awful!*". The opponent constructs an *ad hominem* argument:

> *You say killing animals for pleasure is wrong. But you eat meat, which presupposes that animals are killed for you. You condemn the hunters and you support the butchers. There is a contradiction here.*

In his follow up, **S1** can retort that there is a decisive difference. The hunter kills for pleasure, the butcher by necessity; and **S2** can refute this refutation by arguing that there is no need to eat meat, whereas it is quite necessary to have fun.

This last form of *ad hominem* corresponds to what Bossuet calls an *a repugnantibus@* argument: "*Your conduct does not suit your speech*" ([1677], p.140), **S. *A repugnantibus***

The expression "circumstantial *ad hominem*" refers to cases in which the speaker the notices a contradiction between his or her opponent's speech and his or her personal circumstances, material welfare, lifestyle or personal position. **S. Circumstances**.

Defense against such an accusation — The preacher of virtue, to whom one points out that his or her practices do not support his or her counsels, finds support in the Lockian analysis of *ad hominem*, declared inherently fallacious:

> *My personal circumstances have no bearing on the truth or moral validity of my preaching.*

Such a person may add that he or she has a *divided personality*:

> *It is true, I am a sinner, but it is from the depths of darkness that one feels best the necessity of light.*
> *This is natural, the cobbler's children go barefoot.*

Nonetheless, this form of argumentation is feared by preachers, who are expected to preach by example@, S. *Exemplum*.

The real impact of *ad hominem* argument is not on the truth of what is said, but on the *right to say* what is said. The next reply may be *"What you say is probably true and right, but I do not want to hear it from you"*, or *"That's true, but it's not for you to say"*.

6. Setting up facts against words, S. Irony

7. Argumentation upon the beliefs of the partner

Whereas *ad hominem* argument goes after possible inconsistencies in the discourse of the opponent, arguments built upon the beliefs@ of the opponent or of the audience are as a positive form of exploitation of the partner's belief system, considered as a coherent whole, S. *Ex datis*; *Ex concessis*

Ad Incommodum

Lat. *incommodum*, "inconvenience".

Bossuet defines the argument *ad incommodum* as "the argument that brings about an inconvenience" ([1677], p. 131). This is a variant of the refutative use of the pragmatic@ argument, and can be considered as a kind of argumentation from the absurd@.

Bossuet illustrates this scheme via an example designed to prove the necessity of absolute political power and absolute religious power. He argues that the negation of these authoritarian postulates would have *"pernicious"* consequences, respectively *"men would devour one another"*, which is certainly not a desirable state, and *"there would be as many religions as heads"*, which is deemed undesirable by Bossuet:

> If there were no political authority which one obeys without resistance, men would devour one another. And if there were no ecclesiastical authority to which individuals were obliged to submit their judgment, there would be as many religions as heads. Now, it is false that men should devour one another, and that there be as many religions as heads. Therefore, we must necessarily admit a political authority to which we obey without resistance, and an ecclesiastical authority to which individuals submit their judgment. ([1677], p. 131)

Ad Judicium ▶ Matter

Ad Personam ▶ Personal Attack

Ad Populum

Lat. *populus*, "people".

The label "populist speech" is both descriptive and evaluative. Such speech is stigmatized and is widely considered to be used to promote negative values, xenophobia and other irrational and brutal phobia; to call for action on the basis of non-controlled emotions and poor analysis as opposed to argued rational conclusions; and to make indiscriminate promises, suggesting that the proposed solutions are the only ones possible, easy to implement, that they will work miracles, and will have no negative consequences.

Populist discourse appeals to immediate satisfaction, and is opposed to the hardship discourse of perseverance and slow improvements: "*If you vote for me, you will have to accept sacrifices. But, later, may be…*"

"Populist" is the new label for ancient and modern "demagogues", developing, for the sake of pure short-term electoral benefits, a discourse which they know is untenable.

1. The call to the beliefs of a group

The *ad populum* argument is sometimes defined as an argument derived from premises admitted by the audience, rather than from universal premises. Such an argument would therefore aim to achieve adherence rather than truth (Hamblin 1970, p. 41, Woods and Walton 1992, p. 211).

According to the Socratic criticism of assembly discourse as focusing on social persuasion when addressing the audience about their everyday affairs and worries, to the detriment of transcendental truth, all political speech would be inherently populist, **S. Probable**. In this sense, all rhetorical or dialectical arguments would be *ad populum*. The argument *ad populum* is then no different from the argumentation on the beliefs@ of the audience's interests, beliefs and passions, abundantly referred to as *ex concessis*, *ex datis*, or *ad auditores* argument.

2. An appeal to emotion

"We can define the paralogism known as *argumentum ad populum* as an attempt to win the popular assent to a conclusion by arousing the emotion and enthusiasm of the masses" (Copi 1972, p. 29; quoted in Woods and Walton 1992, p. 213). The *ad populum* argument is negatively related to hatred and fanaticism, and not always positively to enthusiasm: it is caught in the general condemnation of passions, without taking into account the fact that on the one side, emotions may or may not be justified, and that, on the other side, good and bad arguments may be based on strong emotions, **S. Emotion**.

This definition corresponds to the designation *ad captandum vulgus* "playing to the gallery", in other words, to theatrical oratory, not an exclusive characteristic of politicians. The orator becomes an actor. The criticism of *ad populum* joins the moral criticism of flattering discourse, and the critique of enthusiasm, conformism and group effects in general, as "bandwagon fallacies" and alignment with the majority crowd (*ad numerum*), S. **Pathos; Emotions; Laughing; Consensus.**

As in all cases of appeal to the passions, we might suspect substitution of the passions for the logos, hence a lack of *relevance* (Woods, Walton 1992, p. 215), S. **Begging the question**.

3. Argumentative orientation of the word *people*

The word *people* can take two opposite argumentative orientations. The individualist, who believes that all virtue resides in the individual, may conclude, by application of the scheme of the opposite, that the crowd is inherently corrupt, and that all argumentation appealing *to popular sentiment* is therefore fallacious. The people are always the populace.

On the other hand, the adage *vox populi vox dei*, "the voice of the people, is the voice of God" gives the people a degree of infallibility. The *popular corruption* argument mirrors the *ad superbiam* fallacy, that is the accusation of pride (*ad superbiam*), a sin committed by people who consider themselves to be superior to an inherently corrupt people, S. **Contempt; Typologies, II.**

Boldly relying on an effect of composition@ backed by two analogies, Aristotle supports the superiority of the Many over the One:

> According to our present practice assemblies meet, sit in judgment, deliberate, and decide, and their judgments all relate to individual cases. Now any member of the assembly, taken separately, is certainly inferior to the wise man. But the state is made up of many individuals. And as a feast to which all the guests contribute is better than a banquet furnished by a single man, so a multitude is a better judge of many things than any individual.
> Again, the many are more incorruptible than the few; they are like the greater quantity of water, which is less easily corrupted than a little. The individual is liable to be overcome by anger or by some other passion, and then judgment is necessarily perverted; but it is hardly to be supposed that a great number of people would all get into a passion and go wrong at the same moment.
>
> Aristotle, *Politics*, III, 15. Jowett, p. 99

— Maybe "hardly to be supposed", nonetheless historically well documented.

4. *Populum* and *plebs*: The people and the crowd

In republican Rome, the appeal to the people, *provocatio ad populum*, was a right of appeal (*jus provocationis*) in criminal trials, a basic human right of the defendant. As a last resort, an accused Roman citizen would be able to bring his case before the *populus*. The *populus* is the assembled people, constituted as a political-judicial body, in the *comitia centuriata*, the solemn assembly of the people, in which full citizens vote and make decisions. In these assemblies, the gods themselves speak via the *voice of the people*. The *populus* is therefore very distinct from the *vulgus* or the *plebs* as haphazard, unorganized wholes.

This right is linked to Republican institutions: "tradition claims that the *provocatio ad populum* was created by a law of the consul Publicola the same year the Republic was created" (Ellul [1961], 278). With the Empire, "the *provocatio ad Cæsarem* evicted *the provocatio ad populum*" (Foviaux 1986, p. 61), that is to say, that Caesar replaced the People.

Ad Quietem ▶ Calm

Ad Rem ▶ Matter

Ad Verecundiam ▶ Modesty

Affirming the Consequent ▶ Deduction

"After as Before" ▶ Consistency

Agreement

Agreements can be considered under four perspectives.

(1) In general, fully developed argumentative interactions are characterized by a *preference for disagreement*, which distinguishes them from consensual interactions, governed by a *preference for agreement* (Bilmes 1991), **S. Disagreement; Politeness.**

(2) *The existence of "preliminary agreements"* (Perelman, Olbrechts-Tyteca) in regard to both the organization of the discussion and the issues to be discussed, can be considered as a necessary condition for the fruitful conclusion of argumentation. In a dialectical exchange, previous specific agreements are *imposed* on the participants, as the rules of a game are imposed on the players. In a rhetorical address, the orator seeks *a priori* areas of agreement with the audience. In civil life, argumentative encounters (courts, conciliation offices, parliaments, decisional meetings…) follow pre-established standard procedures upon which *volens nolens*, the participants must agree and comply with, whether they find them fair or not, **S. Rules; Conditions of discussion.**

(3) *The production of an agreement* can be regarded as the ideal purpose of argumentative interactions. In combination with (2), this makes argumentation a

technique for transforming preliminary agreements into a final consensus. S. **To persuade, to convince; Persuasion.**

(4) *The existence of a consensus can be exploited as an argument.* In argumentations that justify a proposal by claiming that it is the subject of general consensus agreed on by everyone. The actual opponent to the claim appears therefore as an isolated eccentric individual, excluded from "our community". His or her opinion is disqualified, and can be dismissed without taking the trouble to refute or even consider his or her arguments, S. **Dismissal.**

Alignment ▶ Orientation

Ambiguity

Ambiguity (N), *ambiguous* (Adj) come from the Latin verb *ambigere*, "to discuss, to be in controversy": *qui ambigunt* 'those engaged in a discussion' (Cic. *Fin.* 2,4)" (Gaffiot, *Ambigo*). To refer to the issue, the point upon which the partners disagree, Cicero uses the expression "*illud ipsum de quo ambiguebatur*", "precisely that - about which - [they] dissent" (*ibid.*). *Ambiguitas* means "doubt"; the answers given by the Oracles were *ambiguous* in this sense.
The word *amphiboly* is sometimes used in the discussion of the Aristotelian fallacies of ambiguity. It adapts a Greek word [*amphibology*] composed of *amphi* "on both sides"; *bolos* "throwing on all sides"; *logos*, "word", and means "having a double meaning, equivocal". Literally, an *amphiboly* is an "explosion of meaning".

The word *ambiguity* may be used to refer to three fallacies "dependent on language", *homonymy, amphiboly, accent.* These fallacies are defined as violations of the rule of syllogism@ or of dialectical reasoning, which require that language be univocal, S. **Dialectic; Fallacy (II).**
Issues of ambiguity arise at the *word level* (homonymy, accent), at the *sentence level* (syntactic ambiguity), or at the level of discourse. Such issues combine with the fact that *non*-ambiguous sentences may have several layers of signification, S. **Presupposition; Words as Arguments.**

1. Syntactic ambiguity
Sentence ambiguity, discussed by Aristotle from the perspective of a grammar of argumentation, is now seen as a syntactical issue. The famous Chomskyan ambiguous statement "*flying planes can be dangerous*" can be paraphrased as:
    In some circumstances, *flying planes* is a dangerous activity
    Planes are dangerous when they are *flying*.

These paraphrases are non-equivalent. The no less famous statement "*the teacher says the principal is an ass*" is syntactically ambiguous, it admits of two syntactic structures whose difference is marked by intonation or punctuation:

> "*The teacher*", says the principal, "*is an ass*"
> The teacher says: "*The principal is an ass*".

Ambiguity is sometimes a de-contextualization artifact, produced for the sake of grammatical or logical theory. In practice, the addition of a sufficient amount of left and right context suffices to clarify the intended meaning, as shown by the re-contextualization of the sentence "*we saw her duck*" (Wikipedia, *Ambiguity*), which is four times ambiguous when decontextualized:

> we saw her duck *swimming in the pool*
> we saw her duck *to pick up something on the floor*
> we have no knife, *so we saw her duck*
> she is a smart bridge player, *we saw her duck*

Serious ambiguity occurs when context does not disambiguate the sentence.

The reduction of ambiguity to univocity is no less important for the interpretation of texts, sacred and others, than it is for logic, **S. Interpretation**. In *De Doctrina Christiana*, St Augustine specifies a rule to be applied when trying to interpret religious texts:

> But when proper words make Scripture ambiguous, we must see in the first place that there is nothing wrong in our punctuation or pronunciation. Accordingly, if, when attention is given to the passage, it shall appear to be uncertain in what way it ought to be punctuated or pronounced, *let the reader consult the rule of faith which he has gathered from the plainer passages of Scripture, and from the authority of the Church.*
>
> Augustine, [397] *On Christian Doctrine, in Four Books*, (our emphasis)[1]

The interpretive rule in the emphasized passage appeals to the consistency@ of the field of theological argument. It applies to the interpretation of the first verse of the first chapter of the St John Gospel. The issue is nothing less than the very concept of God. It must be shown that the correct "punctuation", that is the correct reading of this verse, coincides with the orthodox conception of the Trinity, which affirms the divine identity and equality of the Father, the Son and the Holy Spirit. The reading which attributes a syntax of coordination to the utterance results in denying the identity of the Word, that is the Holy Spirit, with God; so, is must be considered heretical and rejected as such.

> 3. Now look at some examples. The heretical [punctuation], "*In principio erat verbum, et verbum erat apud Deum, et Deus erat*" ("In the beginning was the Word, and the Word was with God, and God was"); so as to make the next sentence run, "*Verbum hoc erat in principio apud Deum*" ("This word was in the beginning with God"), arises out of unwillingness to confess that the Word was God. But this must be rejected by the rule of faith, which, in reference to the equality of the Trinity, directs us to say: "*et Deus erat verbum*" ("and the Word was God"); and then to add: "*hoc erat in principio apud Deum*" ("the same was in the beginning with God"). (*Id.*, Chap. II, 3)

---

[1] Bk III, Chap. 2, 2. No pag. Quoted after https://www.ccel.org/ccel/augustine/doctrine.txt . (11-08-2017)

It thus follows that, for Augustin, the orthodox punctuation and construction of the verse is:
> In principio erat verbum, et Verbum erat apud Deum, et Deus erat Verbum. (*Biblia Sacra*...Parisiis, Letouzey et Ané, 1887).

This is a case of *argumentative interpretation*. The starting point is a sentence taken from the sacred text:
> *et verbum erat apud Deum et Deus erat*
> the Word was with God$_1$ and God$_2$ was

First reading, *God$_2$* resumes (is co-referential with) *God$_1$*. This is a mere case of repetition, a kind of stylistic anaphora.
> the Word was with God$_1$ and [God$_1$] was.

The following argued interpretation might be developed from this reading:
> (i) *Data*: (1) **B** does exist
> (2) **A** is with **B**.
> (ii) *Semantic rule*: if **A** is with **B**, then **A** is not **B**; that is,
> **A** and **B** are two different entities.
> (iii) *So, conclusion, by instantiation of the rule*
> The Word is not God.

To sum up, God exists, and He is unique (not Trinitarian). According to Augustine, this first interpretation is heretic.

Second reading, *God$_2$* is co-referential with *the Word*:
> the Word {[was with God] and [was God]}

Now, the Logos is God. This is the basis of the orthodox concept of the Trinity. The first reading is deemed *fallacious*, that is to say *heretical*. The alleged semantic rule (iii) is disposed of in the name of the mysterious nature of the Trinitarian link.

An interpretation is based upon a reading of the text; when necessary, this reading must itself be based upon a grammatical argument, the conclusion of which may or may not be decisive. Disambiguation is the founding operation for the vast and important domain of interpretive argumentation.

2. Homonymy, polysemy

Two words are homonymous when they have the same signifier (same spelling (homographs), same pronunciation (homophones) or both of these, yet have entirely different meanings. Homonymous words are listed as different entries in the dictionary:
> *Mine*: "that which belongs to me." (MW, *Mine*)
> *Mine*: "a pit or excavation in the earth from which mineral substances are taken" (*ibid.*).

Polysemous words are semantic particularizations or acceptations of the same signifier within the same grammatical category. In the dictionary, they are listed under the same entry, and correspond to the first subdivision of meaning:

> Mine: **1** a: a pit or excavation in the earth from which mineral substances are taken. b: an ore deposit.
> **2**: a subterranean passage under an enemy position.
> **3**: an encased explosive that is placed in the ground or in water and set to explode when disturbed.
> **4**: a rich source of supply (*id.*)

When two different lines of derived words stem from the same root word, this word is in a process of splitting into two homonyms; this is the case of the three series derived from the word *argument*, S. ***To Argue, Argument***.

2.1 Paralogism and sophisms of homonymy

A syllogism is fallacious by homonymy when it articulates not three but four terms, one of the terms being taken in two different senses, S. **Paralogism**.

In the *Euthydemus*, Plato provides an example of sophisticated practice using a very special kind of homonymy. Euthydemus the sophist, the eponymous character of this dialogue, asks Clinias "who are the men who learn, the wise or the ignorant?" (*Euth.*, 275d; p. 712). Poor Clinias blushes and answers that "the wise [are] the learners"; and six turns of speech later, he must agree that "it is the ignorant who learns" (*Euth.*, 276a - b; p. 713). The young Clinias is quite stunned, and Euthydemus' followers "broke into applause and laughter" (*ibid.*). Such sophisms are not intended to deceive their victims, but to destabilize their naive certainties about the language. By this salutary shock, the public becomes aware of the opacity and the proper form of language, S. **Persuasion; Sophism**. As Socrates later explains, "the same word is applied to opposite sorts of men, to both the man who knows and the man who does not" (*id.*, 278a, p. 715).

Generally, the subject and object of a verb cannot be permuted; the situation where "**A** loves **B**" is different from the situation where "**B** loves **A**". As *to learn, to be the host of, to rent* present this property:

> **to rent** 1. pay someone for the use of (something, typically property, land, or a car). 2. (of an owner) allow someone to use (something) in return for payment.) (MW, *Rent*)

2.2 Homonymous and polysemous shifts

The plurivocity of words is blamed as a major source of confusion. Scientific language prohibits polysemy as well as homonymy, and calls for the use of univocal, well-defined terms stabilized in their meaning and syntax, in a given scientific field. Homonymy between a scientific term and a current word is harmless. In physics, the use of the word *charm* to refer to a particle, the *charm quark* creates no ambiguity.

In a reasoning using natural language, the meaning of terms is constructed and recomposed in the course of discourse, **S. Object of discourse**. The meaning of a word used by the same speaker may change from one stage of the argument to the following one. This results from a variety of mechanisms, such as the use of homonymous or closely similar words, or the use of a word in its literal and figurative senses in the same discourse. The discussion about the *credit* to be given to a person may, for example, subtly shift between *setting the amount of a loan* and *trusting* that person. In German, it seems that the economic discussion of *financial debt* remains linked to the discussion of *moral fault*, the same signifier, *Schuld*, having these two meanings. (Reverso, *Schuld*).

Homonymy and Polysemy may be re-adjusted by the operation of *distinguo@*.

3. "Accent" and paronomasia

In a language where word stress is linguistically relevant, shifting the stress from one syllable to another may change the meaning of the word, for example in Spanish (my underlining):

Hacía: stress on the second syllable, means "did".
Hacia: stress on the first syllable, means "to", preposition.

The words seem homonymous save for the *accent* (verbal and written), but are in reality two different words. Much like the fallacy of homonymy which shifts the meaning of a single signifier, the fallacy of accent also shifts the meaning of the word via a minimal but crucial supra-segmental change. This process occurs as though the difference between the signifiers is not considered salient enough to discriminate between the variations of meaning.

This is a special case of *paronomasia* (or *annominatio*), defined as a:

> (pseudo-) etymological play on the slightness of the phonetic change on the one hand and the interesting range of meaning which is created by means of the change on the other. The range of meaning can in such cases be raised to the level of paradox. (Lausberg [1960], §637)

Generally speaking, paronomasia creates a meaning generating cell, by contrasting or assimilating a word (signifier) W_0 with a minimally different word (signifier) W_1.

In dialogue, the paronomastic resumption of a term used by the opponent operates as a rectification, breaking the orientation of this discourse, **S. Orientation Reversal**, "*this is not a crisis of conscience, this is a crisis of confidence*".

Amphiboly ▶ Ambiguity

Analogy (I): Analogical Thinking

From an anthropological perspective, analogy is a form of thought that posits that things, people and events are reflected in each other. For *analogical thinking*, knowing is deciphering similarities; analogy unveils a world of secret links underlying reality, and generates a "cosmic feeling where triumph order, symmetry, perfection", a closed world (Gadoffre *& al.* 1980, p. 50); thus conceived, analogy is the foundation of gnosis. From the perspective of the history of ideas, this form of thinking culminated in the Renaissance, when our "sublunary" world was, by analogy, mapped with the heavenly spheres, and with the divine world more generally.

In one of its manifestations, the doctrine of analogical correspondences validates the following type of argument:

>Data: *This plant looks like such or such part of the human body.*
>
>Conclusion: *This plant has a hidden virtue, effective to cure the ills that affect the corresponding part of the body.*
>
>Warrant: *If the shape of a plant is like a body part, then it cures ailments affecting that body part.*
>
>Backing: *This is a divine provision.*

This form of analogical thinking postulates that plants have hidden medicinal properties. The plant bears a divine *signature*, that is, a representation of the human body part that it can heal. This signature or "analogical sympathy" is a motivated signifier, a similarity or "resemblance" of the given body part. God, in his benevolence, has imposed this signature on particular plants in order to make them of use to us. A plant resembling the eyes, therefore might cure eye irritation.

Since the skin of the quince is covered with small hairs, it bears the "signature" of the hair, and eating the quince can make your hair grow. In the wording of Oswald Crollius [1609]:

>Data: '*This downy hair growing around quinces [...] represents hair in some way.*" (*id.*, p. 41)
>
>Conclusion: "*So, their decoction makes hair grow, which fell because of the pox or another similar illness.*" (*ibid.*)
>
>Warrant: the healing power of plants "*can be recognized more easily by the signature or analogical and mutual sympathy with the members of the human body with these plants than by anything else.*" (*id.*, p. 8)
>
>Backing: "*God gave an interpreter to each plant so that its natural virtue (but hidden in its silence) can be recognized and discovered. This interpreter can be nothing else than an external signature, that is to say a resemblance of form and figure, true indications of the goodness, essence and perfection thereof.*" (*id.*, p. 23)

Oswald Crollius, [*Treatise on Signatures, or the True and Living Anatomy of the Big and the Small World*]; [1609][1]

From this doctrine derives a research program for "those who want to acquire the true and perfect science of medicine", "they should devote all their efforts to the knowledge of signatures, hieroglyphs and characters" (*id.*, p. 20). Training will enable them to recognize "at first glance, on the surface of the plants, what faculties they are endowed with" (*id.*, p. 9).

The knowledge of the medicinal properties of plants is acquired by learning how to read and understand the "discourse of nature", that is to say, by mastering the signs scattered around the world. Such an analogical reading of the world is opposed to empirical causal investigation, which consists of observation and experience, practicing dissection or prescribing a concoction to the patient and then finding out if he or she is better, dead, or neither better nor worse. Analogical knowledge is a specific mode of thought, constitutive of magical thinking that substitutes for causal knowledge mysterious correspondences conveying influences, and bypasses the hierarchical system of categories organized according to genus and species, for which it substitutes a similarity network.

Analogy (II): Intra-Categorical Analogy

Intra-categorical analogy draws on the relationship between individuals belonging to the same category. For a definition of the concept of category, the categorization process of individuals; the organization of categories in taxonomies and the corresponding forms of syllogistic reasoning, S. **Categorization and Nomination**.

1. From identity to intra-categorical analogy and circumstantial analogy
1.1 Individual identity
An individual is identical to itself (not similar nor resembling); it is not "more or less" identical to itself. This self-evidence establishes the *principle of identity* "**A = A**".

1.2 Identity of indiscernibles
Two different individuals perfectly identical, for example products taken out of the same industrial production chain, are materially identical, that is perceptually indistinguishable. All that can be said of one can be said of the other; their descriptions coincide, they share all their properties, *essential (categorical)* or *accidental*.

---

[1] Quoted after Oswald Crollius, *Traicté des Signatures ou Vraye et Vive Anatomie du Grand et Petit Monde*. Milan: Archè, 1976.

Discernibility depends on the observer, the layman does not see any difference, and believes that *"it's all the same"*, whereas the specialist will make crucial distinctions.

1.3 Intra-categorical analogy

Intra-categorical analogy is the relationship between the members of a category **C**. All members share, by definition, the characteristics defining the category. The phrase "another **C**" refers to another member of the same class **C**. Two beings belonging to the same category are *identical for this category*; a whale and a rat are identical from the point of view of the category *"— be a mammal"*. This categorical identity is a partial identity, compatible with major differences; two beings of the same category are said to be *analogous* or *similar*. They are *comparable* in respect of their other non-categorical properties. Chicken eggs are all similar as eggs; an egg is identical to another egg; it is comparable to all other eggs in terms of freshness, size, color, etc. **S. Comparison.**

1.4 Circumstantial analogy

An individual **a** possessing the features (**x, y, z, t**), is similar to all individuals who have any of those features, whether it be an essential or accidental feature. The descriptors of two objects define the *point of view* under which they are equivalent; two beings are similar if their descriptions overlap, contain a common part, which may or may not include all or some of their essential features. In other words, this common part generates a category, which may or may not make sense. One might speak of *circumstantial analogy*. Alice and a snake are identical from the standpoint of the category *"— is a long-necked pigeon egg eater"*, **S. Definition.**

2. Intra-categorical analogy as induction or deduction

Intra-categorical analogy can be reconstructed as an induction or a deduction:

2.1 As an induction

>   **O** is similar to **P**
>   **P** has the properties **w, x, y, m**
>   **O** has the properties **w, x, y**
>   So **O** probably has also property **m**.

From an overall judgment of analogy between two beings, based on the shared features **w, x, y** ... we conclude that if one has the property **m** then the other most probably also possesses **m**. In other words, analogy is pushed towards identity.

2.2 As a deduction

>   **O** is similar to **P**
>   **P** has the property **m**
>   Conclusion: **O** probably has the property **m**.

O is similar to **P**. This means that they share a common set of features, and therefore belong to the category **C** defined by those features. In conclusion, as members of the same category **C**, **O** and **P** probably share other properties, among them **m**. This means that the predicate *"— is like"* is to be interpreted as a weaker form of *"— is the same as"*; analogy is seen as a weakened identity.

Deduction and induction are considered valid forms of reasoning. The purpose of the discussion about the possibility of reducing analogy to deduction or induction is to determine whether or not analogy is also valid as a form of reasoning. Reasoning by analogy is sometimes used to prove the existence of God, the ideological stakes of this issue are therefore high.

These formulations of the argument by analogy in the form of a dialectical syllogism are rather sterile because they do not emphasize *the warranting operations*, that contain all the interesting problems. The formulation of the conclusion not as a secure finding but as the product of a heuristic rule of thumb, however, is of great value. The conclusion should be written not as something "probable", that is a kind of belief, but as a suggestion *to do* something:

> it might be interesting to test **P** for property **m**.
> it might be interesting to see whether **O** and **P** share other properties.

3. Arguments based on intra-categorical analogy

— Categories as a whole are structured according to their respective definition; two individuals belong to the same category if they have the same definition, S. **Definition**.

— Categories may be gradual, S. **Rule of Justice**

— Categorical analogies may be restructured S. *A pari*; **Definition (III)**.

4. Refutation of categorical analogy

In one or other aspect, everything is like everything else, and analogies can be more or less "far-fetched". Any rejected categorical analogy will be dubbed fallacious and denounced as a confusion, an *amalgam* (Doury 2003, 2006).

Intra-categorical analogy can be refuted by showing that the category created from those two beings is not based on essential features, but on some accidental property; in general, the generated class is deemed irrelevant. The nonsensical analogy "Chinese ~ Butterfly", ironically discussed by Musil, illustrates the perils of circumstantial analogy, based on the arbitrary choice of a nonessential feature, here the "lemon yellow" color.

> There are lemon yellow butterflies; there are also lemon yellow Chinese people. So in a sense, butterflies can be defined as miniature winged Chinese people. Butterflies and Chinese people are symbolic of sensual pleasure. Here we can see for the first time a glimmer of a possible match, never considered before, between the great period of the moth fauna and Chinese civilization. The fact that butterflies have wings and not the Chinese people is only a superficial phenomenon. [...] Butterflies did not invent powder: precisely because the Chinese have done it before them. The suicidal predilection for the lights of some nocturnal species is still an artifact of the past, which is difficult to explain in view of the daylight understanding of this morphological relationship be-

tween butterflies and China.

Robert Musil, [*Spirit and Experience*], [1921][1]

The analogy relationship has difficulties with transitivity, **S. Relation.** Intra-categorical analogy is transitive: if **A** and **B** on the one hand, **B** and **C** on the other hand, are said to be similar because they possess the same essential features, **A** is thus similar to **C**. Circumstantial analogy is not transitive: nothing proves that if, on the one hand, the descriptions of **A** and **B** have common parts, and, on the other hand, the description of **B** and **C** have common parts, then the description of **A** and **C** will also have also common parts. Khallaf invokes a traditional analogy to criticize the concatenation of analogies:

> A man is walking on the beach trying to find similar shells; once he finds a shell similar to the original, he throws away the original shell and goes on to find a seashell which resembles the second, and so on. When she has found the tenth shell, she should not be surprised to see that it is totally different from the first in the series. (Khallâf [1942], p. 89)

Analogy (III): Structural Analogy

1. Terminology

Structural analogy connects two complex domains, each articulating an indefinite and unlimited number of objects and relationships between these objects. It combines *intra-categorical analogy* (a property of objects) with *proportional analogy* (a property of relations). One could also speak of *formal analogy* (the areas have the same shape) or borrow the mathematical term *"isomorphism"*, **S. Intra-categorical analogy; Proportion.**

The expression *"physical analogy"* refers to the relationship between two objects when one is a *replica* of the other. The concept covers different phenomena, such as the relationship between a model and its *original*, or the relationship between a *prototype* and the object to be manufactured. The reasoning based on the model or prototype is then applied to the original.

Structural analogy is involved in the two following situations.

(i) A, B, C ... are similar — To establish if the complex objects or domains **A, B, C** are *similar*, one has to *compare* their components and the relations between them. The conclusion of this investigation will be a claim such as "**A, B, C...** are similar"; "**A, B**, are indeed similar, but **C** is something different", etc. One may ask if the 1929 Great depression, the Lost Decade of Japan during the 90s, and the Argentinian Crisis in 2001 share some significant characteristics. The whole purpose of the investigation may be to establish a typology of economic crisis, without — as far as possible — drawing on preconceived ideas of how politicians will use the conclusions of this investigation.

---

[1] Quoted in Jacques Bouveresse, *Prodigies and Dizziness of Analogy*. Paris: Raisons d'Agir, 1999. P. 21-22.

The areas are symmetrical from the viewpoint of the investigation, which does not favor one of the areas over the others, but only focuses on their relationships.

(ii) A is similar to B — *A contrario*, the importance of the previous situation appears when the series involves the 2008 crisis. Given the actuality of this last crisis, it will certainly be tempting to see if we can "learn lessons" from the previous crises and to apply them to the 2008 case, with the intention of making provisions for the current situation. If the proponent uses the analogy 1929 ~ 2008 to predict a third world war, her opponent can rebut the inference by showing that the domains are not similar, and that it is therefore impossible to rely on the first instance, in 1929, to make inferences about something about what will happen in 20** and after (see farther).

The difference in status between the two areas is expressed in different ways. In his analysis of the metaphor, Richards opposes *Tenor* and *Vehicle* (1936); Perelman & Olbrechts-Tyteca speak of *Theme* and *Phore* ([1958], p. 501). A simple way to name these domains may be *comparing domain / compared domain*; or, in view of the analysis of argument, *Resource domain / Target domain*.

The argument by analogy works on the asymmetry of the compared areas; that is why these two areas will be designated, when necessary by the letters of alphabets, **R**, as Resource field and **Π** (capital Greek letter "pi"), the *Problematic field*, targeted by the investigation. The field **R** is the source or the *Resource* on which the arguer relies to make changes in the Targeted area **Π**, or to derive from **R** certain consequences about **Π**. In other words, the Resource field **R** is the *argument* domain and the Targeted field **Π** is the *conclusion* domain. The two fields are differentiated from epistemic, psychological, linguistic and argumentative perspectives.

— In *epistemic* terms, the Resource field is the best-known area; the Target field is the area under exploration.

— In *psychological* terms, intuition and values operating in the Resource field are put to work in the Target field.

— In *linguistic* terms, the Resource field is well covered by a stabilized, well-known and easily spoken language; the Target domain is not.

— In *practical* terms, we know what to do within the Resource field whereas in the Target domain, we do not.

2. Explicative analogy

In the well-known analogy proposed by Ernest Rutherford between the atom and the solar system, the Resource field is the solar system, the Target field is the atom:

> the atom is like the solar system.

This is a didactic analogy, intended to provide a first intuitive understanding of the atomic structure, taking advantage of a (supposed) better understanding of the solar system. The asymmetry of the areas is obvious: the Resource field, the solar system, has been known and understood for a long time. The Targeted field, the atom, is new, poorly understood, inaccessible to direct perception, enigmatic.

The explanatory analogy retains some educational merits even when partial. A comparison is not identification, and two systems can be compared simply in order to identify the limits of the comparison, that is, the irreducible specificities of each field, cf. infra, §6.

The analogy has explanatory value in the following situation:

> In the world **Π**, the proposition π is poorly understood. In a world **R**, there is no debate over **r**. **Π** is isomorphic to **R** (structural, systemic analogy). The position of π in **Π** is the same as that of **r** in **R**. So, the knowledge, images, obligations... attached to **r** are now transferred to π; π is now slightly better understood; we know how to do with π.

The analogy relationship integrates the unknown on the basis of the known. As causal explanations, explanations by analogy break the insularity of the facts.

The analogy is an invitation to see and handle the Problem through the Resource. The Resource domain is considered to be a model of the Target domain. The relation of the domain under investigation to the Resource domain is treated like that of the domain of investigation to an abstract representation of this domain. Otto Neurath uses a maritime metaphorical analogy to explain his vision of epistemology:

> There is no *tabula rasa*. We are like sailors at sea, who must rebuild their ship without ever bringing her to a dock to be disassembled and rebuilt it with better items. (Otto Neurath, [*Protocol Statement*],1932/3.[1])

The analogy can be translated word for word: "*There is no ultimate foundation of knowledge from which we could, without any presuppositions, re-build the whole of our present knowledge.*" This resource is extremely powerful; the image could also be applied to social life: "*There is no 'good explanation' (meaning "good discussion of our disagreements") that permits reconstructing a damaged relationship and re-start from scratch.*"

3. Arguments based on structural analogy

In ordinary situations, analogy is used argumentatively, as in the following case:

> In the world **Π**, we are in a difficult situation; what should we do? Should we accept or reject perspective π?
> But we know for sure what happened in a world **R**.
> Fortunately, **Π** is isomorphic to **R** (structural, systemic analogy); if necessary we can argue for that.
> The position of π in **Π** is the same as that of **r** in **R**.

---

[1] Otto Neurath, "Protokollsätze". *Erkenntnis* 3 (1932/3), p. 206. Quoted in A. Beckermann "Zur Inkohärenz und Irrelevanz of Wissensbegriffs". *Zeitschrift für Philosophische Forschung* 55, 2001. P. 585.

So we can act, in world **Π**, on the basis of the knowledge, images, obligations… attached to **r** — that is to say, we can now decide about **π**.

This argumentative operation argues that "if the domains are analogous, so are their corresponding elements and the relations between them", which may prove true or false under further investigation. The analogy gives us something to think about, but proves nothing; the conclusion projected upon **Π** may be false or ineffective.

4. From analogy to metaphor and back

A *language* is attached to the Resource domain. For example, the *human body* is referred to in a language that may be incomplete and fairly incoherent, but commonly understood, the language of the flow of organic matter, of popular physiology, of good health and sickness, life and death. This language synthetizes and builds a common intuition of the body. Other unfamiliar areas are not equipped with such a dense, effective and functional language. The analogy projects the language of the Resource area, the human body, onto the Problematic field, the society. As a result, the target can be problematized in a familiar, non-controversial language; so that social *convulsions* can be discussed and a *cure* found. The analogy is an invitation to see the problem through the lens of the resource; full metaphorization enables us to forget the glasses.

The following apologue is based on the analogy *"society is like a body"*, as expressed in the metaphorical expression *"social body"*. Note the explicitness of the vocabulary of analogy in the final commenting section.

> The senate decided, therefore, to send as their spokesman Menenius Agrippa, an eloquent man, who was also accepted by the plebs as being himself of plebeian origin. He was admitted into the camp, and it is reported that he simply told them the following fable in primitive and uncouth fashion. *'In the days when all the parts of the human body were not as now agreeing together, but each member took its own course and spoke its own speech, the other members, indignant at seeing that everything acquired by their care and labour and ministry went to the belly, whilst it, undisturbed in the middle of them all, did nothing but enjoy the pleasures provided for it, entered into a conspiracy; the hands were not to bring food to the mouth, the mouth was not to accept it when offered, the teeth were not to masticate it. Whilst, in their resentment, they were anxious to coerce the belly by starving it, the members themselves wasted away, and the whole body was reduced to the last stage of exhaustion. Then it became evident that the belly rendered no idle service, and the nourishment it received was no greater than that which it bestowed by returning to all parts of the body this blood by which we live and are strong, equally distributed into the veins, after being matured by the digestion of the food.'* By using this comparison, and showing how the internal disaffection amongst the parts of the body resembled the animosity of the plebeians against the patricians, he succeeded in winning over his audience.
>
> Titus Livius, *The History of Rome*, Vol. 1, Bk 2; between 27 and 9 BC.[1]

---

[1] Trans. by Rev. Canon Roberts; Ed. by Ernest Rhys. J. M. Dent & Sons, Ltd., London, 1905. Quoted from; http://mcadams.posc.mu.edu/txt/ah/Livy/Livy02.html. No pag. (11-08-2017)

The resource does not necessarily preexist its use in an analogy. An analogy can create *ex nihilo* a self-evident resource, as in the following analogy, proposed by Heisenberg in 1955. The danger mentioned in the first line refers to the cold war era, and the resource term is *"a ship built with such a large quantity of steel and iron that its compass, instead of pointing to the North is oriented towards the iron mass of the ship."* Note that, once again, there is no clear-cut frontier between structural analogy and metaphor@. Heisenberg refers to the situation he imagines as a metaphor; and in the next line, he uses a construction expressing an analogy: "humanity is in the position of a captain…".

> Another metaphor might make such a danger even clearer. By the seemingly unlimited growth of its material power, humanity is might be compared to a captain whose ship has been built out of such a large quantity of steel and iron that its compass, rather than pointing to the North, orients towards the huge iron mass of the ship. Such a ship would get nowhere. It would be blown off course and led in circles.
>
> But back to the situation of modern physics: we must admit that the danger exists only if the captain does not know that his compass no longer responds to the magnetic force of the earth. By the time he understands this, the danger is already halved. Because the captain who, not wishing to turn around, wants to achieve a known or unknown purpose, will find a way to steer the boat, either by using new modern compass that does not react to the iron mass of the boat, or by steering in relation to the stars as sailors once did. It is true that the visibility of stars does not depend on us, and perhaps today do we see them only rarely. Despite this, our awareness of the limits of our hope in progress supposes the desire not to go in circles, but to achieve a goal. Once recognized, this limit becomes the first fixed point which allows a new orientation.
>
> Werner Heisenberg, [*Nature in Contemporary Physics*], [1955][1]

5. Structural analogy as an epistemological barrier

Analogy is fertile to stimulate discovery and invention, useful for teaching and popularizing knowledge. Yet it becomes an epistemological obstacle when the proposed explanation by analogy seems so clear and satisfying that it hinders further research:

> For example, blood flow like water. Canalized water irrigates the ground, so blood should also irrigate the body. Aristotle was the first to assimilate the distribution of blood from the heart to the body with the irrigation of a garden by canals (*De Partes Animalium*, III, v, 668 a 13 et 34). Galen did not think otherwise. But to irrigate the soil, it is ultimately to get lost in the soil. And here is exactly the main obstacle to a proper understanding of blood circulation.
>
> Georges Canguilhem, [*The Knowledge of Life*], 1951.[2]

---

[1] Quoted after Werner Heisenberg (1962) *La Nature dans la Physique Contemporaine*. Paris: Gallimard, 1962. P. 35-36.

[2] Quoted after Georges Canguilhem, *La Connaissance de la Vie*. Paris: Vrin, 1965. P. 26-27.

6. Refutation of structural analogies

6.1 Vain analogy
In an explanation, the explanation (*explanans*) must be clearer than the thing to explain (*explanandum*). Analogical explanation must also satisfy this condition, and if the resource area is even less well known than the area under investigation the analogy does not help in the understanding of things.

The analogy is also vain when used to impress the audience and display the grandstanding of the speaker as familiar with the Resource domain. Gödel's theorem is used extensively for this purpose (Bouveresse [1999]).

6.2 False analogy
An argument by analogy can be rejected by showing that there are *critical differences* between the Resource domain and the Target domain, prohibiting the projection of the former upon the latter so that no lesson can be learned from the supposed Resource domain. In the following passage for example, it is argued that the comparison of the 2008 and 1929 crisis is marred by the facts that the present situation in Germany has nothing to do with its situation after 1918 and the coming years. Furthermore, it is argued that there is nothing similar to Hitler and Nazism in the European landscape in 2009:

> *Jean-François Mondot — Does the economic crisis weaken our civilization? We sometimes hear intellectuals and columnists making analogies with the 1929 crisis that led to World War II.*
>
> *Pascal Boniface* — We often make the mistake of thinking that history repeats itself, and so make very risky comparisons. Russia bangs his fist on the table, everybody immediately talks about the Cold War. An economic and financial crisis erupts on Wall Street, and immediately an analogy is drawn with 1929, the suggestion being that Hitler could come to power as a result of these difficulties. Yet the political circumstances are obviously very different, insofar as no great country is now humiliated as Germany was after 1918, leaving it wishing to take revenge. This comparison is easy to make, but it has no basis, neither strategic nor intellectual.
>
> Pascal Boniface, [*The clash of civilizations is not inevitable*], 2009.[1]

6.3 Partial analogy
Partial analogy ("misanalogy" Shelley, 2002, 2004) is an analogy that has been criticized and recognized as limited. The two domains cannot be equated. Nonetheless, partial analogy still has a pedagogical use, as seen in the case of the analogy between the solar system and the atom (cf. supra §2):

> A central body: the sun, the nucleus of the atom.

---

[1] Pascal Boniface, "Le clash des civilisations n'est pas inévitable". Interview by J.-F. Mondot, *Les Cahiers de Science et Vie*, 2009. www.iris-france.org / Op-2009-03-04.php3] (09-20-2013)

Peripheral elements: the planets, the electrons.
A central mass much larger than peripheral masses: the mass of the sun is larger than the planets; the mass of the core is larger than that of electrons. — etc.

Differences (analogy breaks):
The nature of the attraction: electrical for the atom, gravitational for the solar system.
There are identical atoms, each solar system is unique.
There may be several electrons in the same orbit, whereas there is only one planet in the same orbit. — etc.

The fact that the limits of analogy are precisely known prohibits any automatic transposition of the knowledge gained in one field into the other field.

6.4 Reversed analogy

A conclusion **C1** has been established for a Target resource on the basis of an analogy drawn from the Resource domain **R**. The opponent argues that the same analogy drawn from the same domain **R** leads to another conclusion **C2** about the same Target domain, that is incompatible with **C1** ("disanalogy" Shelley, *ibid.*). These two contradictory conclusions prohibit the use of the Resource domain to argue in the Target domain.

This is particularly effective because the opponent concedes to playing on her adversary's home ground. The opponent accepts and examines more closely the analogy advanced by the proponent, in order to neutralize his or her conclusions. This strategy is exploited in the refutation of argumentative metaphors.

Argument: — *This area lies at the heart of our discipline.*
Refutation: — *That's true. But disciplines also need eyes to see clearly, legs to move in, hands to act, and even a brain to think.*
Other refutation — *That's true, but the heart can very well keep beating preserved in a jar.*

A supporter of hereditary monarchy speaks against universal suffrage:
Argument:— *An elected president, that's absurd, we do not elect the driver.*
Rebuttal: — *Nor are there natural born drivers.*

Both sides enact the same metaphorical field. This form of rebuttal has the strength of an *ad hominem@* refutation, based on the own beliefs of the speaker: "*You are your own refuter*".

Counter-analogy — As with any argument, one can oppose an argumentation by analogy by putting forward a counter-argumentation (an argumentation whose conclusion is incompatible with the original conclusion). This counter-argumentation can be of any kind, including another argument by analogy, taken from another Resource domain; an analogy equilibrates another analogy:

Antanaclasis, Antimetabole, Antiparastasis
▶ *Orientation Reversal*

Argument: — *The university is (like) a company, so ...*
Rebuttal: — *No, it is (like) a daycare, an abbey ...*

Antanaclasis, Antimetabole, Antiparastasis
▶ **Orientation Reversal**

Antithesis

The rhetoric of figures defines the antithesis as an opposition between two terms (words or phrases) of opposite meanings, entering into parallel syntactic constructions. The argument scheme of the opposites@ materializes discursively as an antithesis.

1. Antithesis as argumentative diptych

An argumentative situation emerges with the appearance of a point of confrontation ratified as such, a stasis@. It develops into a diptych, characterized by the confrontation of two schematizations@, that is to say two sets of descriptions, narrations and argumentations supporting two opposing conclusions. At this stage, the two discourses develop at cross-purposes, without explicitly taking this opposition into account, S. **Stasis**. This elementary argumentative situation corresponds to a discursive *antithesis*.

Such a confrontation might be taken up in a structured monologue juxtaposing the two sides of the issue. Such a monologic diptych features an "antiphony", that is two voices putting forward incompatible arguments with respect to the same issue. This is typically seen when an individual having a vested interest in an issue engages in inner deliberation, and oscillates between two points of view, acting actually as a third party. This situation is elaborated as a *dilemma*@ whose anti-oriented horns are articulated by an *and*:

> I admire your courage and pity your youth.
> Corneille, *Le Cid* 2, 2, verse 43. Quoted by Lausberg [1960], §796

When the speaker clearly identifies with one of the two voices, the balance of the two voices is broken in favor of one of the positions. The *and* dilemma transforms into a *but* opposition, overcoming the antithesis:

> ... but I pity your youth; so I won't accept your challenge to duel.

2. Antithesis, figure and argument

The following argumentation is structured by the scheme of the opposite:

> (D1) He is submissive to the privileged; I would not like to confront him in a weak position.

exactly as the self-argued description:

> (D2) He is submissive to the privileged and powerful, and hard with the weak.

Whereas in (D1), the second member of the scheme "*he must be hard with the weak*", remains implicit, (D2) corresponds to a complete expression of the topos. But the two discourses are based on the same mechanisms, the argumentation is "valid" or acceptable insofar as the portrait sounds "true"; both are "convincing". Description and argument are rooted in the same figure or scheme.

Apagogic

An *apagogic argument* is a form of argument by the absurd, which argues that unreasonable interpretations of the law must be rejected:

> The apagogic argument assumes that the legislator is reasonable and could not have admitted an interpretation of the law that would lead to illogical or unfair consequences. (Perelman 1979, p. 58)

It parallels the *psychological argument*, presupposing that the legislator is rational and benevolent, **V. Absurd; Juridical arguments.**

According to Alexy, the apagogic argument is one of the four types of arguments prevailing in law, the others being the arguments by analogy, *a contrario* and *a fortiori*, (1989, quoted in Kloosterhuis 1995, p. 140).

Aporia ▶ Assent; Stasis

(To) Argue, Argument, Argumentation, Argumentative: The Words

1. The English words

1.1 *To Argue*

The verb *to argue* has two different accepted meanings which will be referred to, respectively, as *to argue₁* and *to argue₂*:

— *To argue₁*: "to put forth reasons for or against; debate"
— *To argue₂*: "to engage in a quarrel; dispute: *We need to stop arguing and engage in constructive dialogue*" (tfd, *Argue*).

The morphological, syntactic, and semantic differences between these meanings are crucial and clear.

— **Morphology:** The word *argumentation* derives from *to argue₁* via *argument₁*; it refers only to a speech in which a conclusion is supported by good reasons.

— **Syntax**
• *To argue_1* is followed by a *that* clause: "*A* argues that **P**"; **P** is the *claim*.
• *To argue_2* is followed by a double indirect complementation, "*A* argues with **B** about *Q*". **Q** is neither **A**'s nor **B**'s claim, but refers to the issue of the dispute.

(To) Argue, Argument, Argumentation, Argumentative: The Words

— Semantics:

- *To argue_1* means "to give reasons" (MW, *Argue*), and refers to a semiotic activity (verbal and co-verbal).
- *To argue_2* means "to have a disagreement a quarrel, a dispute" (*ibid.*), and refers to the broad field of interactions ranging from a lively discussion to outright pugilism, as shown in the following passage, in which the detective Ned Beaumont questions an informant, Sloss:

> Ned Beaumont nodded. 'Just what did you see?'
> 'We saw Paul and the kid standing there under the trees arguing'
> 'You could see that as you rode past?'
> Sloss nodded vigorously again.
> 'It was a dark spot,' Ned Beaumont reminded him. 'I don't see how you could've made out their faces riding past like that, unless you slowed up or stopped.'
> 'No, we didn't, but I'd know Paul anywhere,' Sloss insisted.
> 'Maybe, but how'd you know it was the kid with him?'
> 'It was. Sure it was. We could see enough of him to know that'
> 'And you could see they were arguing? What do you mean by that? Fighting?'
> 'No, but standing like they were having an argument. You know how you can tell when people are arguing sometimes by the way they stand'
> Ned Beaumont smiled mirthlessly. 'Yes, if one of them's standing on the other's face.' His smile vanished.
>
> Dashiell Hammett, *The Glass Key*, [1931][1].

1.2 *Argument*

The noun *an argument* inherits the two meanings of *to argue*; an *argument_1* is a "good reason", an *argument_2* is a "dispute", possibly containing *argument_1*. Grimshaw's book, *Conflict talk. Sociolinguistic investigations of arguments in Conversation* (1990), exclusively deals with *arguments_2* "dispute", not at all with *arguments_1*, "good reasons".

A third, specific, meaning adds to these two inherited meaning, *argument_3*, as "the abstract, the theme, the subject matter" (of a literary work, etc.).

"Argument is War" — Lakoff and Johnson have proposed the famous equivalence "argument is war":

> Let us start with the concept ARGUMENT and the conceptual metaphor ARGUMENT IS WAR. This metaphor is reflected in our everyday language by a wide variety of expressions:
>
> Your claims are *indefensible*.
> He *attacked* every weak point in my argument.
> His criticisms were right on *target*.
> I *demolished* his argument. [...]"

---

[1] Quoted after Dashiell Hammett, *The Four Great Novels*. Picador, 1982. P. 725-726.

(To) Argue, Argument, Argumentation, Argumentative: The Words

"We can actually *win* or *lose* arguments" (1980, p. 4)

Lakoff and Johnson refer to the "concept argument". If the preceding conclusion is correct, there is not just one but two concepts of argument. To $argue_2$ and $argument_2$ may be associated with a kind of war; but what about $argument_1$ and to $argue_1$?

If interlinguistic comparisons can tell something about words used as concepts, note that, in French, the first series of metaphors easily translates word-for-word; but the expression *"we can actually win or lose arguments"*, does not. The words *to argue, argument, and argumentation* have clearly recognizable counterparts in French or Spanish, or in the Romance languages at large:

French *argumenter, argument, argumentation*
Spanish *argumentar, argumento, argumentación*

This graphic illustration of the proximity of these words certainly favors the internationalization of the concept. Yet there are deep differences between their respective meanings, which can be roughly represented as follows:

English	dispute	good reason	topic
French		good reason	topic
Spanish		good reason	topic

The French word *argument* and the Spanish word *argumento* never refer to a dispute. The field of argumentation studies develops from the shared meaning of $argument_1$, "good reason".

This shows that the meaning of *to $argue_2$, $argument_2$* in a language is *independent* of the concept referred to by the family *to $argue_1$, $argument_1$, argumentation*.

1.3 *Argumentative*

According to the Merriam-Webster Dictionary, the adjective *argumentative* shares the two meanings of its morphological base, *argument*: "controversial" and "disputatious" (MW, *Argumentative*). The *Merriam-Webster Learner's Dictionary*, however, is more categorical (MWLD, *Argumentative*):

Argumentative: tending to argue; having or showing a tendency to disagree or argue with other people in an angry way: QUARRELSOME.
an *argumentative* person
he became more *argumentative* during the debate.
an *argumentative* essay.

In this dictionary, *argumentative* will be attached by default to the family "*argumentation*", thus a semantically derived of $argument_1$ "good reason", unless otherwise specified. An *argumentative essay* will be taken as "an essay developing an argumentation"; if referring to "a polemical essay", its quarrelsome character will be explicitly mentioned.

2. Differential orientations: the French words *arguer, argutie*.

From a morphologic point of view, the French verb *arguer* is the basic verb from which all the *argu-* words derive:

arguer → *un argument* → *argumenter* → *une argumentation, etc.*
"an argument" "to argue" "an argumentation", etc.

But *arguer*$_F$ must be set apart; *to argue* matches *argumenter*$_F$, nor *arguer*$_F$. There is a semantic discontinuity between *arguer*$_F$ and *argumenter*$_F$. When **S1** says:

S: — *Pierre argumente en faveur de P*, "Peter argues that P"

S recognizes that Peter *does* give arguments. When he or she says:

S: — *Pierre argue que...* "Peter argue$_F$ that..."

S just quotes the argumentative discourse of Peter without taking a position on the validity of the arguments he offers, and even suggesting that they might be *fallacious*. In a democratic or republican newspaper the construction:

the extreme right argue$_F$ that...

introduces an argumentation considered as weak or invalid. That is, the verbs *arguer*$_F$ and *argumenter*$_F$ have opposing *orientations*. The former values discourse as *argumentative*, whereas the latter suggests that it posits only pseudo-arguments.

A *quibble* may translate in French as an *argutie*$_F$, a word derived from *arguer*$_F$:

These people are the manipulated agents of subversion, performing instructions and *rehashing quibbles* ["répétant des arguties"].

Arguer$_F$ and *argutie*$_F$ are only used occasionally. *Arguer*$_F$ might be replaced by *argument*$_F$ between quotation marks. So a pro-wind farm group quotes the arguments of its opponents, the anti-wind farm group, as follows:

Let's look at some of the 'arguments' put forward by anti-wind farms (Complete example, **S. Convergence**)

The *concept* of arguments, and argumentation studies, benefit from the strong positive orientation that the *words argument* and *argumentation* have in ordinary language. The case is the same for the word and the concept of *dialogue*, **S. Interaction.**

Argument — Conclusion

1. Argument

The word *argument* is used in different domains, in grammar, logic, literature, and argumentation, with quite different meanings.

— In logic and mathematics, the *arguments of a function f* are the empty places x, y, z… characterizing the function; the independent entities (variables) organized by the function.

— By analogy, in grammar, the verb plus its subject and object(s) can be considered the counterpart of a function. *To give* for example, corresponds to a predicate governing three arguments "**x** gives **y** to **z**"; *to love* to a two-argument predicate, "**x** loves **y**". By substituting adequate phrases (i.e., respecting the semantic relationship characterizing the verb) for each of these variables, we form a proposition@: "*Adam gives Eve an apple*", **S. Classical logic (II)**.

— In literature, the central *argument* of a play or a novel corresponds to the plan, the summary, or the guiding principle of the plot. With this meaning, the word *argument* is morphologically and semantically isolated; *argument* as "summary" bears no relation to *conclusion*, nor to *to argue, argumentation*.

2. Argument and argumentation

The words *argument* and *proof* are used to translate the Greek word *pistis* and the Latin word *argumentum*.

2.1 Argument ~ argumentation

By synecdoche, *argument* often means *argumentation*: "*let the best argument prevail!*"

2.2 Premise, data, argument

— In logic, the *premises* of the syllogism lead to a *conclusion*. The premises are propositions expressing true or false judgments. The conclusion is a proposition which is different from the premises and which is derived exclusively from their combination, without the surreptitious introduction of implicit background information into the reasoning, **S. Syllogism**. A premise is not an argument but a constituent of an argument; the argument is constructed by combining the two premises.

— In argumentation, the conclusion is derived from an item of information combined with an inferential topic. The situation is the same in Toulmin's layout@ of argument, where the data becomes an *argument* when combined with an often implicit system warrant / backing. The word *argument* is routinely used to refer to the *data* element as the head of such combinations. **S. Topic**.

— In *analytical and immediate inferences*, the conclusion is derived directly from a single statement, which is an argument in itself. The conclusion is derived from the form or the semantic contents of the statement argument, **S. Logic (II)**.

Argument and *conclusion* are correlative terms. The "argument — conclusion" relationship is expressed, more or less accurately by expressions such as those listed below. If necessary, "is" may be replaced by "is presented as such by the speaker" (as in line 1, etc.).

Argument — Conclusion

The argument	The conclusion
— is a consensual statement, *or presented as such by the arguer)*	— is a dissensual, challenged, disputed statement
— is more likely than the conclusion	— is less likely than the argument
— is the cognitive starting point in deliberative argumentation — is the end point in justificatory argumentation	— is the end point of deliberative argumentation — is the starting point in justificatory argumentation
— expresses a reason	— is in search of a reason
— does not carry the burden of proof	— carries the burden of proof
— is oriented towards the conclusion	— is a projection of the argument
— (*in a functional perspective*) determines legitimizes the conclusion	— (—) determined, legitimized by the argument
— (*in a dialogical perspective*) accompanies the answer given to the argumentative question	— (—) is the proper answer to the argumentative question

2.3 Argument: true, probable, plausible, accepted, conceded…

A statement is considered (or presented) as a certain truth and may function as an argument on very different bases.

— The argument conveys a well-known fact, an intellectual self-evidence, S. **Self-Evidence**

> the heat of the wax dilates the pores and pulling up is thus less painful (Linguee)

— The partners have explicitly agreed on the statement, for example as part of a (quasi-) dialectical agreement:

> We agree that now Syldavia cannot leave the Eurozone, so we can place further requirements upon them.

— The speaker has chosen his argument from those considered to be true by the audience, even if he or she has personal doubts about its validity, S. *Ex datis*:

> You think that Syldavia will never leave the Eurozone, so…

— A simple fact: the statement is challenged neither by the opponent nor by the audience.

The audience's acceptance of stable statements that may serve to support the conclusion, is always precarious. The opponent's belief in the truth of a given statement is even less stable. The choice of what will be considered a valid argument is thus a strategic choice which will change in view of the circumstances, **S. Strategy.**

Challenging the argument — If the *argument* is disputed, it must itself be legitimized. As part of this operation, the argument takes the status of a *claim* put forward by the proponent and supported by a series of arguments. These new arguments serve as sub-arguments supporting the overarching claim, S. **Linked; Epicheirema**. If no agreement can be reached on any statement, things can, theoretically, go back indefinitely and the debate may continue without end. The risks associated with such "deep disagreement" should not be considered to invalidate argumentation as a useful social tool to deal with social incompatibilities, as far as *third parties* play their role in well-regulated settings.

3. Claim, thesis, conclusion, point of view, standpoint

In argumentation, the *conclusion* is also called the *claim*, or *point of view*. A philosophical conclusion is often called a *thesis*, S. **Dialectic**. The set of conclusions drawn from complex data at the end of an abduction process can be a full-blown theory, S. **Abduction**.

3.1 Point of view, viewpoint, standpoint

In the socio-political domain, a *point of view* is an "opinion", possibly justified by arguments. The pragma-dialectical program is aimed at reducing, resolving, or eliminating differences of *opinions*. The corresponding expressions *"resolving... differences of conclusions, claims, thesis..."* are not in use.

An argument as a point of view, an opinion, a perspective... conveyed in just one sentence is a very special case. Points of views and opinions are generally expressed in complex discourses, supported by equally complex argumentative sub-discourses. The expression *point of view* can be used to refer to a whole speech, including the point of view and the good reasons supporting it.

In ordinary language, the concept of point of view organizes the perceptual reference system of the speaker:

> On the other side of the hedge, was a gardener.
> On the other side of the hedge, we saw a road.

In one case, the speaker is *outside* the garden, in the other *in* the garden. The concept of *point of view* used in argumentation is strongly metaphorical. It frames the argumentative situation according to the visual metaphor of a spectator within a landscape, which would be the reality, inaccessible as such, if not represented on a map.

The spectator's vision provides a slice of reality restructured according to the laws of perspective. The reality referred to by the point of view is only so with regard to a, by definition, unstable focus. In this sense, a point of view is either questionable as it functions as blinkers; or valuable, because it protects one from the objectivist illusion produced by consensus, and from the paranoia of absolute knowledge.

An affirmation corresponds to a point of view if it is brought back to one subjective source, while absolute truth, or vision, is independent of any source, or has a universal, absolute source.

The point of view is an inescapable starting point. Points of view are comparable and assessable. We cannot be *without* perspective-point of view, yet we are able to define a *better* point of view; *change* our point of view, and *multiply* our points of view. In order to eliminate differences in points of view, one would have to eliminate subjectivity, or the plurality of voices, and *de*-contextualize the discourse. Scientific discourses do that routinely, but, as far as argumentative discourse seeks to deal with human affairs, involving (legitimate) interests, values, and their affective correlates, argumentation analysis cannot align itself with scientific language without changing the nature of its objects and objectives. The radical elimination of points of view would require the resurrection of the absolute rational Hegelian subject, or of the objective and omniscient narrator of nineteenth century novels.

3.2 *Conclusion*

The opening section of a discourse is its *introduction*, its closing section its *conclusion*. The *argumentative conclusion* is distinct from the *material conclusion* ending an intervention. The argumentative conclusion can be stated, or repeated, in any part of speech, at the beginning or at the end, or both.

The *argumentative conclusion* is defined in correlation with the argument (see Table above). In an argumentative monological text, the conclusion is the assertion according to which the discourse is organized; towards which it converges; in which its *orientation* materializes; the *intention* which gives the discourse its meaning, and the ultimate core of the text obtained by condensing it.

The conclusion is more or less detachable from the arguments supporting it. Once we have reached the conclusion that "*probably, Harry is a British citizen*", we can, by default, act on the basis of this belief. But, as far as the modal *probably* expresses clear reservations on the whole inferential process, the claim will remain *revisable* if conditions change. The "fire and forget" principle[1] does not work well in argumentation. The conclusion is never fully detachable from the speech used in its construction.

A statement **S** becomes a claim in the following dialogical configuration
    (1) — **S** is put forward by a speaker (as something essential for him, or merely anecdotal)
    (2) — **S** is not ratified by the addressee: *not preferred second turn*
    (3) — **S** is re-asserted, possibly reformulated by the speaker
    (4) — **S** is explicitly rejected by the dialogue partner.
        Re-statement not ratified: disagreement ratified
    (5) — Emergence of pro- and contra-arguments.

At stage (3), the disagreement emerges. At stage (4) the disagreement is ratified as such, a stasis is formed, and **S** is now a *Claim* put forward by the first speaker. At stage (5), the stasis begins to develop

---

[1] "(Of a missile) able to guide itself to its target once fired." (EOD, *fire-and-forget*) (11-08-2017)

Stage (1) is not a dialectical "opening stage". The speaker does not necessarily intend to open a dispute. Non-ratification can occur at any time in an interaction, and may concern any foreground or background statement, S. **Negation; Disagreement**. In other words, *being a claim* is not a property of a statement, but is attached to the treatment of a statement in an interactive configuration.

Argumentation (I): Definitions

Argumentation analysis has been intensely and specifically investigated since the post-second world war period. The bi-millennial framework of logic as an "art of thinking" in natural language have been taken up and reworked in the new intellectual framework of the post-Fregean mathematical logic as a *Substantial Logic*, an *Informal Logic*, or a *Natural Logic* (references infra).
A new vision of argumentation as discourse orientation has been developed in the semantic theory of *Argumentation within Language*.
Ancient rhetoric has been reshaped into a *New Rhetoric*. Dialectics has been revisited in relation to pragmatics and speech acts theories, and expanded into a powerful critical instrument within the *Pragma-dialectic* framework.
The prospects of rhetoric and dialectic are now ubiquitous in contemporary studies and teaching programs on argumentation. The links between rhetoric, text linguistics and discourse analysis have been recognized and rearticulated.
The spectacular results obtained in interaction analysis have opened the immense field of everyday conversational interactions as a specific investigation domain, where *argument* as "dispute" intertwines with *argument* as "good reason". The various theories of argumentation developed in the late twentieth century are based on different visions and definitions of their objects, their methods and their goals. Given this diversity, and the apparent and real discrepancies between definitions, there is a real temptation of synthesis, that is, to look for a definition which, while not trivial, will restore order, unity, simplicity and consensus.
Experience shows, however, that many of the new definitions meant to supplant older ones, are merely added to the existing list, thereby further aggravating the problem that they are intended to solve. Another solution could be to start with things as they are, that is, to admit that the field of argumentation studies develops not in the hypothetico-deductive style, starting from an overwhelming "master definition" and deriving its consequences, but in a more empirical, data driven, manner. In practice, this means starting with a corpus of working definitions of the concept of argumentation, and stressing the various insights in the field that have proved to be of interest and use.

Argumentation (I): Definitions

1. Rhetorical argumentation as an instrument of persuasion

Socrates considers and rejects rhetoric as an enterprise in social persuasion through speech. He shares this definition with his opponents, in particular Gorgias:

> Gorgias — I'm referring to the ability to persuade by speeches judges in a law court, councilors in a council meeting, and assemblymen in an assembly or in any political gathering that might take place. (Plato, *Gorgias*, 452e; p. 798)
>
> Socrates — Well, then isn't the rhetorical art, taken as a whole, a way of directing the souls by means of speech, not only in the law courts and on other public occasions, but also in private? (Plato, *Phaedrus*, 261a ; *CW*, p. 537)

This defines the common *use* of the word *rhetoric* in ancient Greece, what people *call* rhetoric. Now what rhetoric *is*, in its *substance* — or lack of substance — is another story:

> By my reasoning, oratory is an image of a part of politics. (Plato, *Gorgias*, 463d; *CW*, p. 807)

Politics is defined as the craft of addressing "the soul" *(ibid*, 464b, p. 808), and rhetoric is disposed of as an unsubstantial "image", an *eidolon*, a counterfeit of politics. Socrates unreservedly condemns rhetorical discourse aimed at persuasion, as a lie, an illusion, a manipulating enterprise, antagonistic to truth-seeking philosophical discourse. This unqualified and irrevocable condemnation of rhetoric as a fake is at the root of the current negative acceptance of the word, and obviously includes argumentative rhetoric. The criticism of rhetoric is part of the field of rhetoric, and the same applies to the field of argument.

Aristotle positions rhetoric not as a counterfeit but as "the counterpart of dialectic" (*Rhet*, I, 1, 1354a1; RR p. 95) and defines it as an empirical *techne*, a craft, oriented towards the study of specific cases:

> Rhetoric may be defined as the faculty of observing in any given case the available means of persuasion (*Rhet*, I, 2, 1355b25;. RR, p. 105).

Cicero follows this functional definition:

> Cicero Junior: — *What is an argument?*
> Cicero Father — *A plausible device [probabile] to obtain belief.*
> Cicero, *Part.*, II, 5; p. 315

> Crassus — *As becomes a man well born and liberally educated, I learned those trite and common precepts of teachers in general; first, that it is the business of an orator to speak in a manner adapted to persuade.* (Cicero, *De Or.*, I, XXXI ; p. 40)

Likewise, **Perelman & Olbrechts-Tyteca** "New Rhetoric" focuses on persuasion:

> The object of the study of argumentation is the study of the discursive techniques allowing us *to induce or to increase the mind's adherence to the theses presented for its assent.* ([1958], p. 4; italics in the original)

By focusing on "discursive techniques" and on "the mind's adherence", this definition re-builds argumentation studies on the same basis as those of the Aristotelian argumentative rhetoric, persuasive speech. It re-connects contemporary understanding of argumentation with the experience gained throughout two millennia.

Thesis, mind, presented, assent, discursive techniques: this definition articulates the core concepts of what could be called "the argumentation movement" as a vision of man and discourse in modern democratic societies.

— The claims are *theses*. This is a philosophical term; the issues covered by argumentative interventions are complex and high level, "the most rational" (*id.*, p. 7). The *Treatise* keeps its distances from everyday argument and minds: it does not address the ignoramus, and more: "there are beings with whom any contact may seem superfluous or undesirable…" (*id.*, p. 15).

— These theses are *presented* and not *imposed* on the audience.

— Moreover, they are presented to the audience's *mind*, that is to say to men and women endowed with a choice and decision-making capacity; and living under social conditions that allow them to fully exercise this capacity. This action upon *minds* can be opposed to the manipulation of *souls* and *bodies*: souls with their capacities of emotion and sensibility / sensitivity to romantic or mystical appeals; bodies which can be forced to march or vibrate in unison under a musical mantra or image.

— The *assent* results from an explicit judgment of a free and conscious mind. Assent can be given or withdrawn. Expressing one's assent is opposed to producing a response under the causal pressure of a stimulus, S. **Assent**.

— Finally, argumentation is a *discursive technique*, that is, a form of speech in which speakers can practice and improve.

— The *Treatise* does not deal with *fallacies*, but the evaluation of argument is a key issue of the book. The sound criticism and evaluation of arguments is not a matter for the orator, but for the partner audiences, particular and universal, S. **Persuasion**

2. Argumentation as a way to deal with stasic situation.

The *Rhetoric to Herennius* by an unknown author of the first century BC (formerly attributed to Cicero) articulates argumentative rhetoric with the key concept of stasis. In court, the contradiction brought by one party to another party determines the "point to adjudicate" and produces a *stasis*, which defines an argumentative situation:

> The Point to adjudicate is established from the accusation and the denial, as follows: Accusation: '*You killed Ajax.*' Denial: '*I did not.*' The point to adjudicate: Did he kill him? (*To Her.*, I, 17; p 53)

Argumentation is thus defined as an instrument developed institutionally to deal with stasic situations and leading to their legal settlement. S. **Question; Stasis**.

3. Argumentation as "substantial logic" and default reasoning

According to Toulmin's "layout of argument", the argumentative passage is defined by its structure.

— A speaker puts forwards a *Claim*, based on *Data* oriented by general rules or principles, the *Backing*, and the *Warrant*, defining the *monologic assertive component* of argumentation.

— The *Claim* is defeasible under certain *Rebuttal* conditions, expressed by a *Modal* affecting the Claim. This *reservation component* refers to a *dialogic* and *critical* approach of argumentation.

The combination of these two components into an "argumentative cell", both linguistic and cognitive, defines reasonable-rational discourse. S. **"Layout; Categorization; Definition**. This Toulminian complex layout is often reduced to the main parts of its assertive component "Data, Claim":

> Slavery was abolished, why not prostitution? I do believe in the progress of civilization.
> When snakes come out, it's going to rain. We know that from experience.

Toulmin makes no reference to rhetoric. But as Bird has pointed out (1961), with his warrant and backing, Toulmin has "re-discovered" the more than two-thousand-year-old concept of *topic*, fundamental to the rhetorical theory of argument. This approach is entirely compatible with a class of classical definitions of rhetorical argument, such as the following:

> Cicero Senior — I take it that what you desire to hear about is ratiocination, which is the process of developing the arguments. [...]
> Cicero Junior — Clearly that is exactly what I require.
> C. Senior — Well then, ratiocination as I said just now is the process of developing the argument; but this process is achieved when you have assumed indubitable or probable premises from which to draw a conclusion that appears in itself either doubtful or less probable.
>
> Cicero, *Part.*, XIII, 46; p. 345-347; my italics

How to make the *doubtful* a little *less doubtful*? Like Toulmin, Cicero sees argumentation ("ratiocination") as a technique to *reduce uncertainty*.

4. Argumentation as saying and schematizing

According to Jean-Blaise Grize,

> As I understand it, argumentation considers the interlocutor not as an object to manipulate but as an alter ego with whom a vision has to be shared. To work on him means to try to change the various representations attributed to him, by highlighting certain aspects of things, hiding others, proposing him new perspectives, and all this with the help of an appropriate schematization.
>
> Grize 1990, p. 40

Arguing consists in schematizing the world for the interlocutor; such a generalization extends the concept of *argumentation* over the whole act of *saying* something:

> Arguing amounts to putting forward some assertions that we choose to compose in a discourse. Conversely, asserting (saying) amounts to arguing, simply

because we choose to say and put forward some meanings rather than others. (Vignaux 1981, p. 91)

This vision of saying as essentially a rhetorical argumentative activity has deep roots in the rhetorical tradition. It may be compared with what Quintilian presents as the essence of rhetorical argumentation:
> The art of speaking well. (*IO*, II, 15, 37)

This famous formula is often quoted in Latin, rhetoric is the "ars bene dicendi"; the definition is complemented by the definition of the orator as "a good man speaking well". Argumentative rhetoric becomes the legislative technique of speech, guaranteed by the quality of the person using it, S. **Ethos; Persuasion.** These definitions make rhetoric the backbone of classical humanities.

Compared with Grize — who, to my knowledge, never quoted Quintilian, no more than Toulmin referred to the classical science of topoi — the only difference is that Quintilian stresses the educative dimension of rhetoric, whereas Grize simply analyzes argumentation as found in natural discourse. This line of thought generalizes rhetoric to all forms of *controlled expression*, thus founding a *Rhetorik der Sprache* (Kallmeyer 1996), a "rhetoric of speech".

5. Argumentation as orientation

Anscombre and Ducrot's theory of Argumentation within Language is based on the fact that, in natural language, the argument as a statement is linguistically linked to the conclusion, defined as the following statement:
> A speaker argues when he presents a statement **S1** (or a set of statements) as intended to make acceptable a new one (or a set of new ones), **S2**. Our thesis is that there are linguistic constraints governing this presentation. For a statement **S1** to be given as an argument in favor of a statement **S2**, it is not sufficient that **S1** gives reason to admit **S2**. The linguistic structure of **S1** must also meet certain conditions to be able to constitute, in a speech, an argument for **S2**. (Anscombre & Ducrot 1983, p. 8)

This approach results in a redefinition of the concept of topos, as a semantic link between two predicates, S. **Topos in Semantics.**
By re-defining the argumentative constraint as an inter-statements linguistic constraint, Anscombre and Ducrot generalize the concept of argumentation as a property of the linguistic system (*langue* and not *parole* "speech", as defined by de Saussure).
S. **Orientation; Argumentative scale; Argumentative Marker.**

6. Argumentation between monologue and dialogue

> Argument seems to be a mode of discourse which is neither purely monologic nor dialogic. (Schiffrin 1987, p. 17)
> [I have defined argument as] a discourse through which speakers support disputable positions. (*Id.*, p. 18)

Argumentation (I): Definitions

Schiffrin's work is not primarily devoted to argument. This succinct definition does, however, perfectly express the mixed character of the argumentative activity.

7. Argumentation, a discourse submitted to a rational judge

> Argumentation is a verbal and social activity, aiming to strengthen or weaken the acceptability of a controversial point of view from a listener or reader, advancing a constellation of proposals to justify (or disprove) that view before a rational judge. (van Eemeren *& al.* 1996, p. 5)

This definition summarizes the rhetorical and dialectical positions. It re-defines the position of the third party, the judge, not as an empirical, institutional figure, arguing on the basis of the legal corpus of law and jurisprudence shaped by history and sociology, but instead as a normative rational figure, arguing on the basis of a set of independently defined rational principles, S. **Norms; Evaluation and Evaluators**.

8. Guidelines adopted in this dictionary

(i) An *argumentative situation* is defined in the *Ad Herennium* style: a complex dialogic situation opened by an argumentative question.

(ii) An *argumentative question* is a question to which the arguers (the debaters) give argued answers, possibly both sensible and reasonable, but incompatible, organized in pro- and a contra-discourse.

(iii) These *answers* express the *conclusions* (points of view) of the arguers about the issue. The elements of pro- and counter-discourse which support these conclusions have the status of *argument* for their respective conclusions.

(iv) Argumentative situations come in a variety of *degrees and types of argumentativity*, according to the kinds of relationship established between the pro- and counter- discourses and to the interactional and institutional parameters framing the exchanges.

Points (i) to (iv) define the *external argumentative relevance*, as the relevance of a conclusion for a question.

(v) An *argumentation*, in the monologic sense is defined as the "argumentative cell", as represented in Toulmin's layout. In the broad sense, the word argumentation covers all the verbal and semiotic activities produced in an argumentative situation.

(vi) An *argument* is an implicit or explicit combination of statements supporting a conclusion.

(vii) The *internal argumentative relevance*, as the relevance of an argument for a claim is defined in relation to *an argument scheme*.

Argumentation (II): Key Features and Issues

The explosion in theoretical questioning of the notion of argumentation at the end of the twentieth and beginning of the twenty-first centuries (van Eemeren & al. 1996; 2014), and the multiplicity of disciplines interested in the topic encourage the characterization of the domain according to an underlying system of key features, issues and orientations.

The following table proposes a possible organization of the field according to the role of language and the kind of speech situation which is given theoretical prominence. This hypothesis makes it possible to represent the various concepts of argument as a tree structure, where the nodal points correspond to research questions, or crossroad questions, which articulate the field. Such a representation illustrates that what could at first sight seem to be an arbitrary dispersion of options, in fact reflects the necessity of taking the complex range of argumentative situations into account. A vision of argumentation might be characterized as a structured choice between the various options opened by the following questions (other possible points of departure are suggested in §2).

1. Key issues about the role of language

> Table (p. 75):
> *Key features and issues about the role of language in argumentation*

(2) *vs.* (3) *vs.* (4): The cognitive, linguistic and multimodal dimensions of argument

Various general questions might be taken as points of departure, and each question would produce a different mapping of the field. This map is born of the general question: *is argumentation basically a language activity or a cognitive activity — or both?*

If argumentation were defined as a *pure activity of thought*, expressed in a perfectly transparent language, argumentation studies would correspond to a psychology of reasoning without language.

But, in the same way as everyday argumentation, mathematical thinking and scientific reasoning require a language. Language-based approaches to argumentation deal with the cognitive component within the linguistic component. Such approaches are compatible with various positions on the question of thinking and reasoning. Classical logic, Natural Logic, Informal Logic and cognitive approaches stress the articulation of thought and language in the argumentative activity.

Argumentation (II): Key Features and Issues

	as a thought activity (2)	study of reasoning as a pure psycho-cognitive process (2a)		
argumentation (1)	as a linguistic cognitive activity (3)	extended (5)		form of language (7)
				general form of discourse (8)
		situated (6)	monologue (9)	non polyphonic (11) — logic, as an art of thinking (9a)
				polyphonic (12) — "bene dicendi" rhetoric (10a)
			dialogue (10)	without turn-taking (13) — persuasion rhetoric (11a)
				with turn-taking (14) — dialogue logic (15)
				interaction (16)
	as a multimodal activity (4)			

75

Argumentation is unanimously considered to be a discursive practice. The consideration of still and moving images raises questions about how argumentative meanings are able to invest nonverbal semiotic supports. Research on argumentation in working situations also demands that we take the signifying intention steering both the action and the argument into account. In both cases, it is necessary to reconsider what exactly constitutes a well-built corpus within the field of argumentation.

(5) *vs.* (6) — Argumentation as a linguistic-cognitive activity: Extended or situated?
Should argumentation, as a linguistic-based cognitive process, be considered a *local* or a *generalized phenomenon*?

(7) *vs.* (8) — Extended argumentation: Saussurian *langue* or discourse?
Two different theories have extended the concept of argumentation to all linguistic activities, the theory of *Argumentation within Language* (Anscombre, Ducrot 1983) and the theory of argumentation as a *Natural Logic* (Grize 1982).

The former generalizes the concept of argumentation at the level of language (of Saussurian *langue*), whereas the latter enacts the same generalization at the level of speech (*parole*).

> **(7) Argumentation, as a condition on well-formed linguistic chain {E1, E2}:** *S. Orientation*
>
> **(8) Argumentation as a schematization of the situation,** *S. Schematization*

(9) *vs.* (10) — Situated argumentation: Monologue or dialogue?
If argumentation is limited to some characteristic forms of discourse, then in which kind of discourse is it best exemplified, in monological discourse, or in dialogue?

(11) *vs.* (12) — Monologue: Logic or rhetoric?

> **(11) Logic,** *S. Logic*
>
> **(12) *Bene dicendi* rhetoric,** *S. Rhetoric*

(13) *vs.* (14) — Dialogue: With or without turn-taking?
According to the externalization principle (van Eemeren and Grootendorst 1992, p. 10), dialogic theories consider either that dialogue is the basic form of argumentative activity, or that it is in the form of a dialogue that argumentative mechanisms of argument, can be most clearly seen.
Within this set of dialogic approaches, there are distinctions. Has the dialogue an exchange structure or not? Does the dialogue admit turns of speech? Do all the participants have equal possibility of taking the floor in the same conditions?

(13) Argumentation, a dialog without exchange structure: The rhetorical address

The rhetorical address is a special kind of dialogue, having a polyphonic structure; the voices of the others, especially the voice of the opponent, are re-built into the discourse of the speaker who holds the floor. The audience will give its answer only later and indirectly, as a judgment on the case or a decision on the policy.

(15) vs. (16) — A turn-taking dialogue: Dialogue logic or natural interaction?

In the case of a dialog in which there is a possibility of exchange, one of the two following poles will provide the appropriate baseline, 1) a logical approach to formal dialogues, or 2) an empirical approach to natural interactions.

(15) Argumentation, a formalized critical dialogue

Since the 1970s the Informal Logic and the Pragma-Dialectic theories have re-orientated argumentation studies by giving the priority to the study of argumentation as a kind of dialogue.

Dialectical critical theories of argumentation strengthen the constraints on the dialogue either by means of a system of rules designed to embody a rational standard, as in Pragma-Dialectic, or by means of a system of critical questions, as in Informal Logic. **S. Norms**.

(16) Argumentation, a kind of ordinary interaction

Proto-argumentative activity is triggered by a lack of ratification by the addressee. Depending on the reaction of the interaction partners, conversational disorder might pass quickly, being absorbed into the flow of the on-going task they are engaged in. Otherwise, the interaction might develop into a fully-fledged argumentative situation. In all cases, the argumentative situation is basically ruled by interactional principles.

This vision is compatible with the ancient theory of "argumentative questions" (or *stasis*, or *point to adjudicate*). **S. Stasis; Question; Dialectic**.

For each of these points, the question is not which to adopt and which to exorcise, but to clearly articulate the contrast between the approaches they define.

2. Other points of departure

The above table develops from the question of language. Other questions might give rise to alternative maps of the field.

2.1 Kind of rationality?

Truth and rationality can be considered:
(i) As an attribute of a *well-thought monological discourse*, best exemplified in *logic*,

as an art of thinking;
(ii) As the *consensus* of the properly defined universal audience, within the prospect of a rhetoric of persuasion;
(iii) As a social production, the result of a well organized *critical dialog* to reach the best possible true and rational answer in the course of a dialectical process;
(iv) A a progressive construct, through a closer contact with *scientific* results, thought and method.

In complete opposition to these guidelines, generalized theories of argumentation maintain an agnostic perspective on rationality, and question the very possibility of reaching it through ordinary discourse.

2.2 Form or function?

Is argumentation (first, better) defined by its *function* or by its *form*? This question opposes two theoretical families, one focusing on *persuasion*, and the other focusing on the structural *description and formal representation* of argumentative episodes. These two starting points themselves give rise to symmetrical questioning: how to deal with functional aspects in the latter case? What are the structural criteria that ensure the descriptive adequacy of the in the former case?

2.3 Argumentativity, a binary or gradual concept?

For *extended* theories of argumentation, language (Ducrot) or discourse (Grize) are basically argumentative, S. **Orientation; Schematization.**

In the case of *restricted* theories of argumentation, however, some discursive genres (deliberative, epideictic, judicial) or, more broadly, certain kinds of discursive sequences are argumentative and opposed to other non-argumentative genres or other types of sequences. These definitions tend to consider that argumentativity is a binary concept: a sequence is or is not argumentative.

In reference to the language exchanged between partners defending contrasting positions, the argumentativity of a situation is not an all or nothing concept; various forms and degrees of argumentativity can be distinguished.

— A given linguistic situation begins to become argumentative when opposition emerges between two lines of speech, quite possibly without reference to each other, as in an argumentative diptych. This is most probably the basic argumentative structure, each partner repeats and restates his position. S. **Disagreement.** We can thus go beyond the opposition between narrative, descriptive or argumentative sequences. When a description or a narration is developed in support of an answer to an argumentative question, this narration or description should be considered as fully argumentative and evaluated as such.

— Communication is fully argumentative when the difference is problematized as an argumentative question, with the participants taking roles as proponent, opponent, or third party, S. **Question; Roles.**

2.4 Central objects?

The various approaches to argumentation are characterized by the nature of their *internal assumptions* and *external assumptions*. The former correspond to the organization of the *concepts* postulated in the system, and the latter, to the kinds of *objects* taken into consideration. Both types of hypotheses are bound.

The extremities of the branches in any of the preceding "decision trees" represent a pole articulating theoretical views with specific "preferred" objects. To satisfy the requirement of *descriptive adequacy* each theory must combine its *central* objects with what it posits as *peripheral* objects. Decisions as to what is to be considered as central and as peripheral (derived or secondary) data, fall within the domain of external assumptions. Such choices are never self-evident and require justification. So, for example, the decision to give priority to dialogue or to take as reference monologal syllogistic discourse, correspond to two distinct external assumptions regarding the structure of the argumentation field, and clearly put to the fore quite different kinds of data.

This does not imply that second level (often annoying) facts and data are excluded, rather that all phenomena cannot be put on the same level; data must be ordered, and prioritized. In practice, the problem is to determine how the results established on the basis of central facts can be expanded to peripheral data.

Some major types of coupling of internal and external assumptions:
— Rhetorical argumentation, and planned monological speech.
— Dialectical argumentation, and conventionalized dialogues.
— Argumentation as orientation, and pairs of statements.
— Argumentation as schematization, and texts, etc.

Argumentation Studies: Contemporary Developments

The long history of argumentation studies cuts across the history of rhetoric, dialectic and logic. Argumentation studies appeared as autonomous field only after the Second World War; it is nevertheless possible to note inflections during this short history.

1. The long history: dialectics, logic, rhetoric
Greek and Latin Antiquity — From the perspective of classical disciplines, argumentation studies are related to *logic*, "art of thinking correctly"; to *rhetoric*, "art of speaking well and addressing a group"; and to *dialectics*, "art of interacting well, articulating one's intervention and thought with those of others". This triad is the basis of the system in which argumentation was conceptualized, from the time of Aristotle until the late nineteenth century. Argumentation is seen as a theory of convincing reasoning in ordinary language. The central issues are argument scheme theory, and validity and soundness theory, depend-

ing on the quality of the premises and the reliability of the principles used to derive conclusions from these premises. S. **Dialectic; Logic; Rhetoric**.

Modern Times — Walter Ong has commented upon the decline of dialectical practices (1958) since the Renaissance, the reduction of rhetoric to figures of speech and considerations of literary style, and the critique and rejection of the Aristotelian logic as an exclusive or essential instrument of scientific thought. New scientific methods based on observation and experimentation, making increasing use mathematics, are looked for.

Late nineteenth, early twentieth century — At the end of the nineteenth century rhetorical argument is delegitimized as a source of knowledge. Logic is formalized and becomes a branch of mathematics. The tradition of argumentation studies remains active in law and theology.

2. A symptom: the titles

In French, until the publication of Perelman & Olbrechts-Tyteca's *Treatise on Argumentation*, the books entitled *Argumentation* were pamphlets containing arguments about specific topics, not theoretical books about argumentation in general, as shown by their complete titles:

> 1857 - *Discussion About Etherization Considered from the Standpoint of Medical responsibility — Argumentation*. By Marie Guillaume Alphonse Devergie.
> 1860 - *Arguments on Administrative Law of the Municipal Administration*. By Adolphe Chauveau.
> 1882 - *The Issue of Water Before the Medical Society of Lyon. Argumentation in Response to Mr. Ferrand*. By Mr Chassagny. P.-M. Perrellon.
> 1922 - *Argumentation of the Polish Proposal About the Border in the Industrial Section of High-Silesia.*

The substance and field of the argument is specified by an additional subtitle: *argumentation on, about* ... The title *Argumentation* corresponds to modern titles such as "*An Essay on —*" or "*Thesis*"; it refers to a textual genre. Thus, it seems that the emergence of the genre "*[Theoretical work on] Argumentation*" came with the disappearance of the genre "*Argumentation [on —]*".

In English – Toulmin's book "*The Uses of Argument*" (1958) comes apparently in a traditional line of books titled "Argument". Some of these books offer "an argumentation" in support of a position, such as the following:

> Yale C., *Some Rules for the Investigation of Religious Truth; and Some Specimens of Argumentation in its Support*, 1826.

Others are textbooks for composition and debate teaching:

> Brewer E. C., *A Guide to English Composition: And the Writings of Celebrated Ancient and Modern Authors, to Teach the Art of Argumentation and the Development of Thought*, 1852
> Foster, W. T., *Argumentation and Debating*, 1917.
> Baird A. C., *Argumentation, Discussion and Debate*, 1950.

Lever R., *The Arte of Reason, Rightly Termed Witcraft; Teaching a Perfect Way to Argue and Dispute*, 1573.

The best known may be:
Whately R., *Elements of Rhetoric Comprising an Analysis of the Laws of Moral Evidence and of Persuasion, with Rules for Argumentative Composition and Elocution*, 1828.

In the first half of the twentieth century, many such books are published, where didactic purposes mingle with more theoretical considerations. But the work of Toulmin does not fit at all in this tradition, linked to the practices of the Speech Communication Departments or of the English Departments in the United States. No book of that kind is listed in his bibliography, and he quotes no work coming from the field of rhetoric.

Actually, Toulmin and Perelman both break with a modern tradition and establish a new foundation in the treatment of the concept of argument.

3. 1958 and after: Constitution of the field of argumentation studies

3.1 A key date, 1958

Chaïm Perelman, Lucie Olbrechts-Tyteca, 1958, *Traité de l'Argumentation. La Nouvelle Rhétorique* = 1969, *The New Rhetoric — A Treatise on Argumentation*.

Stephen E. Toulmin, 1958, *The Uses of Argument*.

These two titles are the best known in an impressive constellation of works that all help define, positively or negatively, the new field of argumentation studies.

— On "Public Relations": a non rhetorical and non argumentative perspective on persuasion:
Vance Packard, 1957, *The Hidden Persuaders*.

— On the language of propaganda:
Sergei Chakhotine, 1939, *Le Viol des foules par la Propagande Politique*.
= 1940, *The Rape of the Masses - The Psychology of Totalitarian Political Propaganda*.

Jean-Marie Domenach 1950. *La Propagande Politique* [Political Propaganda]

— In law:
Theodor Viehweg, 1953, *Topik und Jurisprudenz. Ein Beitrag zur rechtswissenschaftlichen Grundlagenforschung* = 1993, *Topics and Law. A Contribution to Basic Research in Law*.

— On the rhetorical foundations of literature and Western culture:
Ernst Robert Curtius, 1948, *Europäische Litteratur und Lateinisches. Mittelalter*.
= 1953, *European Literature and the Latin Middle Ages*.

— An historical and systematic reconstruction of the field of rhetoric
Heinrich Lausberg, 1960, *Handbuch der literarischen Rhetorik*.
= 1998, *Handbook of Literary Rhetorik. Foundation for Literary Study*.

— A history of the adventures of dialectic and rhetoric at the time of the Renaissance
>   Walter J. Ong, 1958, *Ramus. Method and the Decay of Dialogue*.

3.2 Extended theories of argumentation
These theories have been developed since the 1970s, mainly in French:
— In a linguistic perspective:
>   Oswald Ducrot, 1972, *Dire et ne pas Dire* [To Say and Not To Say]
>   — 1973, *La Preuve et le Dire* [Proving and Saying]
>   — *& al.* 1980, *Les Mots du Discours* [The Words of Discourse]
>   Jean-Claude Anscombre et Oswald Ducrot, 1983, *L'Argumentation dans la Langue* [Argumentation within Language]

— In a discursive and cognitive perspective:
>   Jean-Blaise Grize, 1982, *De la Logique à l'Argumentation* [From Logic to Argumentation]

3.3 The dialectical and critical approaches
Perelman & Olbrechts-Tyteca work is considered to be a revival of rhetorical argumentation, originating in Aristotle's *Rhetoric*. Along the same line, Hamblin's foundational work revived argumentation as a dialectical and critical thinking, based on concept of fallacies, and originating in Aristotle's *On Sophistical Refutations*.
>   Charles L. Hamblin, 1970, *Fallacies*

3.4. The Pragma-Dialectical trend
From the 1980s on, Frans van Eemeren and Rob Grootendorst have developed the "Pragma-dialectical" approach. They recast the study of argumentation in terms of speech acts, linguistic pragmatics and a new conception of dialectic. They elaborated a powerful system of guidelines for the evaluation of arguments as a system of rules for the rational resolution of differences of opinion, **S. Norms; Rules; Evaluation.**
>   Frans H. van Eemeren & Rob Grootendorst, 1984, *Speech Acts in Argumentative Discussions A Theoretical Model for the Analysis of Discussions Directed Towards Solving Conflicts of Opinion*.
>   Frans H. van Eemeren & Rob Grootendorst, 1992, *Argumentation, Communication, and Fallacies*.
>   Frans H. van Eemeren & Rob Grootendorst, 2004, *A Systematic Theory of Argumentation - The Pragma-Dialectical Approach*.

Since 1986, every four years, a reference conference on argumentation is organized in Amsterdam. The series of *Proceedings* propose an up to date vision of the discipline (van Eemeren & *al.* (1987, 1991, 1995, 1999, 2003, 2006, 2010).

3.5 The Informal Logic trend
The "Informal Logic" of Anthony Blair, Ralph Johnson, Douglas Walton and John Woods connects argumentation studies to a logic and to a philosophy

which take into account the ordinary dimensions of speech and reasoning. The focus is on the evaluation of the arguments and their educational applications in the development of critical thinking. The concept of argument scheme has been defined so as to integrate their corresponding counter-arguments, and developed on this basis a new approach to argument criticism.

> Howard Kahane, 1971, *Logic and Contemporary Rhetoric The Use of Reason in Everyday Life*.
> Ralph H. Johnson & J. Anthony Blair, 1977, *Logical Self Defense*.
> Ralph H. Johnson, 1996, *The Rise of Informal Logic*.
> J. Anthony Blair & Ralph H. Johnson, 1980, *Informal Logic - The First International Symposium*.
> John Woods & Douglas Walton, 1989, *Fallacies. Selected Papers 1972-1982*.
> R. Douglas Walton, Chris Reed & Fabrizio Macagno, 2008, *Argumentation Schemes*.
> J. Anthony Blair, 2012, *Groundwork in the Theory of Argumentation*.

3.6 Argumentation and ordinary interactions

The Pragma-Dialectic and the Informal Logic schools of argumentation give special importance to dialog. The first papers integrating the perspective of conversation and interaction analysis are found in:

> J. Robert Cox & Charles A. Willard (eds), 1982, *Advances in Argumentation Theory and Research*.
> Moeschler J. (1985). *Argumentation et Conversation*. [Argumentation and Conversation]
> Frans H. van Eemeren & al. (eds), 1987, *Proceedings of the [ISSA] Conference on Argumentation 1986*.

4. Relations with other disciplines

The leading research programs maintain different relationships with the rhetorical, dialectical and logical heritage, as well as with language studies philosophy and education. The table below tries to give an idea of these links.

> **0**: no significant link
> **+**: the number of stars indicates the importance of the link

	New Rhetoric	Arg. within Language	Natural Logic	Fallacies (Hamblin)	Pragma-dialectics	Informal Logic
Rhetoric	+++	+	+	0	++	+
Dialectic	+	0	0	+++	+++	+++
Classical Logic	0	0	+++	+++	++	+++
Grammar, Linguistics	0	+++	++	0	++	+
Philosophy	+++	+	+	++	+	+++
Teaching, Education	++	0	0	0	+	+++

5. Dialogues between main trend theories

The arrows represent commonalities, solidarities or affiliations between different schools

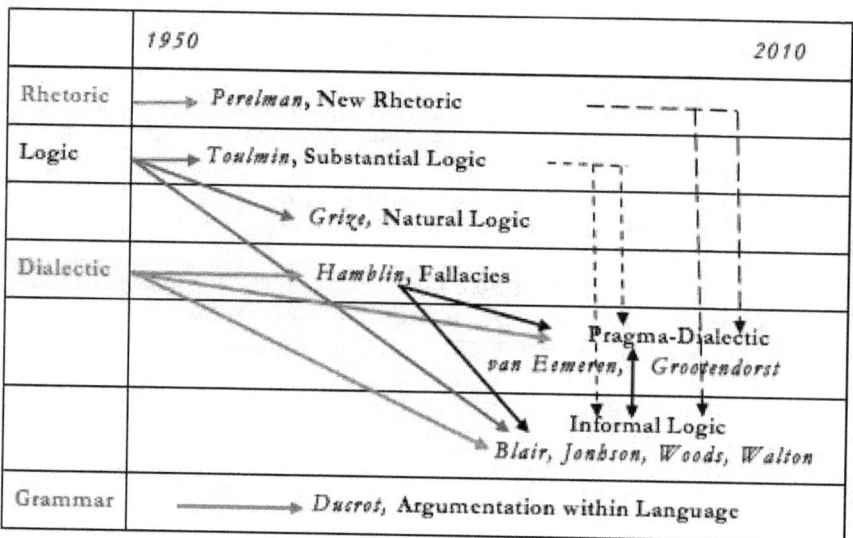

6. Argumentation studies, argumentation scholars: How to name the field and its specialists?

The talk about of the "revival of the field of argumentation" in the fifties should be taken with precaution. Firstly, the expression is ambiguous: the talk is not about the field of argumentative *practices*; but about the *theory* of argumentation, the meta-language used to study this practice. Secondly, it is also slightly simplistic: although discontinuous, reflections on argumentation have been underway for more than two millennia, not half a century. The point is that, since the fifties, a learning community has formed around a vast and differentiated corpus of studies taking for object a set of practices directly characterized as argumentative.

How to designate a field of study, its object and its specialists? The situation is clear when each of these distinct realities is designated by a specific term. This is the case for example with the *economists*, specialists of *economics*, whose object is the study of *economy* (production and consumption of goods and services). But the term *argumentation* refers to both the *object* of study, as in "everyday argumentation", and to the *study* itself, when, especially in the titles of books where "argumentation" shortens "theory of argumentation".

The spectacular appearance of papers and books entitled "... *Argumentation* ..." hides a deeper reality, the change in the disciplinary status of logic. All ancient books entitled *Logic*, dealing with the logic of terms, quantifiers, connectors, analyzed and non-analyzed propositions, etc., are actually theories, logic-based treatises on argumentation, as, for example the Port-Royal *Logic, or The Art of Thinking* ([1662]). Basically, we now use the word *argumentation* to refer to a field of study or to a theoretical book because, since the mathematization of logic in the late nineteenth century, the title *Logic* can only be used in the domain of formal logic, and is no longer available as referring to natural language argument. Exceptions are rare. In French, one can think of works such as the *Elements of classical logic* (François Chenique 1975, vol. I: *The art of thinking and judging*; t II. *The art of reasoning*), or especially Jacques Maritain's *Introduction to Logic* ([1923]), which is perhaps one of the last books providing under the heading *Logic* a traditional "art of thinking", inspired by neo-Thomist philosophy. This logic is, in this respect, the first in the series of "non formal", "substantial", "natural" logics... that flourished at the end of the last century; it is a treatise of argumentation as a theory of logical reasoning within natural language.

So we are left with the problem of naming the field by a single unambiguous term. Following the example of *polemology*, that is war studies, it might be *argumentology*. Along the same line, the corresponding professionals would be called *argumentologists*, a figure clearly distinct from that of the *arguers*. But these words sound jargon-ridden and slightly ridiculous. Anyway, usage will have the last say, and presently nobody seems to feel an urgent need for such words. *Argumentology* does not appear in the monumental and fundamental *Proceedings on the Fourth International Conference of the International Society for the Study of Argumentation* of 1999; one case in 2003, one in 2007; and no occurrence of *argumentologist* or any derivative name of that kind (van Eemeren *& al.* (eds.), 1999, 2003, 2007).

Argumentativity ▶ Argumentation (II)

Assent

Perelman & Olbrechts-Tyteca discuss the effects of argumentation on the basis of an opposition between *to persuade* and *to convince*, the former being a *local* achievement involving a *particular* audience, while the latter is a *global* achievement involving the *universal* audience. The functional definition of argumentation provided at the opening of the *Treatise*, however, does not use these concepts but speaks of "adherence of minds" and "assent". In this passage, argumentation is seen as an activity aiming "to induce or to increase the mind's adherence" to "theses" that are "presented for its assent" (Perelman & Olbrechts-Tyteca, 1958, p. 4). The concept of assent refers to Newman's *Grammar of Assent* (1870).

The Stoic theory of knowledge defines *assent* as a voluntary act of the soul which occurs when the soul receives a true impression; this process implies a pre-established harmony between the will and the mind. "The soul wants truth", and truth is *index sui*, its own mark. The mark of the true impression is the *assent* granted to it. The skeptics reject this harmony between true representation and assent; truth is not capable of self-certification, that is, one can give its assent to *false* representations.

The suspension or abstention of assent, is the basis of the skeptical method to achieve tranquility (*ataraxia*):

> The Skeptic Way is called [...] aporetic either, as some say, from its being puzzled and questioning about everything or from its being at a loss as to whether to assent or dissent. (Sextus Empiricus, *Outlines*, I, iii)

Assent may be granted or refused by an act of will:

> I think it a very great exploit to resist one's perceptions, to withstand one's vague opinions, to check one's propensity to give assent to propositions; [...] Carneades achieved a Herculean labor when, as it had been a savage and formidable monster, he extracted assent, that is to say, vague opinion and rashness from our minds. (Cicero, *Ac.* II, 34; Trans. Yonge, p. 74)

Skepticism characterizes the argumentative situation as a standoff between two equal (isosthenic) and opposed discursive forces, which imposes a suspension of assent, **S. Force; Stasis.**

Common language considers assent to be an action. Assent can be given or suspended, in the same way that one can give or suspend an agreement or an authorization. From a rhetorical point of view, the problematic of assent makes the concept of persuasion more complex, by granting some activity to the recipient. Whilst people are *passively* persuaded, they *actively* grant their assent. This maintains a balance between the speaker and the audience, in that the speaker's effort to persuade his or her audience corresponds the audience's capacity to grant or to refuse his or her assent. Withheld assent plays a role in all varieties of rational exchanges, as it brings about a state of *doubt*@ which characterizes the third party position, **S. Roles.**

The assent granted in regard to a proposition is characterized by varying degrees, as one moves from opinion to belief and knowledge:
— The lowest degree corresponds to *opinion*, defined as a belief accompanied by the awareness that there are other equally valid opinions.
— The intermediate degree is that of *belief*. There are other beliefs, considered not false, but less valid than one's own belief.
— The strongest degree is that of *conviction*. The convinced party considers that the proposition to which he or she adheres is true and that opposing arguments are fallacious, perverse or insane.
According to Perelman & Olbrechts-Tyteca, *persuading* produces *belief*, while *convincing* produces a generalized belief, defining social, legitimized knowledge.

Association ▶ Dissociation

Audience ▶ Rhetorical Argumentation; Persuasion

Authority

1. *Auctoritas*, authority, authoritarian, authoritative

The word *authority*, and hence elements of the problematic of authority, originate in Latin and Roman law. According to Benveniste, the words *auctor*, "author", *auctoritas*, "authority" are linked to the primary meaning of *augere*, "to bring out, to promote" ([1969], no pag.); *augeo* is the first person singular of the present indicative of *augere*:

> In its oldest uses, *augeo* denotes not the increase in something which already exists, but the act of producing from within itself; a creative act which causes something to arise from a nutrient medium and which is the privilege of the gods or the great natural forces, but not of men (*ibid.*).

The speech delivered with *auctoritas* is creative:

> The primary sense of augeo is discovered in *auctoritas* with the help of the basic term *auctor*. Every word pronounced with authority determines a change in the world; it creates something. This mysterious quality is what *augeo* expresses, the power which causes plants to grow and brings a law into existence. That one is the *auctor* who promotes, who alone is endowed with the quality [...]. Obscure and potent values reside in this *auctoritas*, this gift which is reserved to a handful of people who can cause something to come into being and can literally 'bring into existence (*ibid.*).

Ellul describes the institutional exercise of the *auctoritas* as follows:

> The *auctoritas* is the quality of the *auctor*. It gives its support, its approval to the act done by another person. At first it was probably an act of sacred law: an

individual makes the legal act, and another validates this act by an intervention which manifests the agreement of the gods. (Ellul [1961], p. 248-249)

The *auctoritas* is held by the father, the priest, the judge; its use is foundational for family life, as well as for religious and legal life:

> The *auctoritas* appears as the authority of a person who serves as a basis for a legal act. This act has value and efficiency only by the *auctoritas*. [...] The *pater* ["father"] gives his auctoritas to the marriage of his son. In religious life, the *auctoritas* of the priest delimits the domain of the sacred, and draws the boundaries of the profane. In juridical life, the *auctoritas* delimits the domain of what is legitimate, separating it from the illegitimate (*ibid*).

The *auctor* does not back up a statement by authority, but *creates* a reality by his *auctoritas*; this situation is quite different from what we call argument from authority.

The *author-authority* relation is now distorted, an author may have not so much authority, and the person with authority is not necessarily an author.
Authoritarian and *authoritarianism* develop along a lexical line which stigmatizes authority.
In contrast, *authoritative* as "possessing recognized or evident authority" (MW, *Authority*) develops from the positively oriented lexical line associated with *authority*.

2. Authority as a social issue

The concept of authority is redefined and discussed in all the fields of the human sciences, in relation to *submission* and in opposition to *freedom* or *freedoms*. Major studies on authority, power and totalitarianism marked the last century: in psychology, particularly since the resounding experiences of Stanley Milgram on "*Obedience to Authority*" (1974); in philosophy, with the study of the "*The Authoritarian Personality*" of Theodor Adorno (1950); in history with Hannah Arendt's "*The Origin of Totalitarianism*" (1951); in sociology with Max Weber (1922), whose distinctions between *traditional*, *charismatic*, and *rational-legal* authority are now a part of common knowledge.

In our society, basic authority is expressed through various standards regulations and norms, enforced by law, backed by the police and the legal security services, in relation with the current political authorities. In addition, every organization enforces its rules, and, at the more local level, places have rules, however informal they may be, **S. Rules**.

These authorities and their associated coercive powers, the violence of which is not only symbolic, rules everyday modern life to a great extent. The criticism of authority is an everlasting enterprise, quite inseparable from resistance to authority.

3. Appeals to Authority

Along with the issue of authority, the study of discourse engages in multidisciplinary reflection on the *epistemic level* (non-truth conditional conditions of acceptability of statement); on *social influence* (management of the powers of discourse); on *interpersonal relationships* (manifestations and effects on the interaction of the relative positions of authority of the participants).

In the specific field of argumentative rhetoric, the notion of authority is considered in relation to speech: What is an appeal to authority? In which identifiable ways, from implicit evocation to explicit invocation, can authority invest a statement? What are the types of critical responses to authoritarian or authoritative speeches? To the extent that it refers to reason and free inquiry, argumentation is antithetical to authority and violence, even if they avail themselves of legal and even moral legitimacy.

Argumentative speech, however, operates on a knife-edge: as critical speech, it *denounces the discourse of authority*; as strong affirmative discourse, it impacts upon the others' minds and representations in the name of reason. Argumentation has to find a way *to be authoritative*, without being *authoritarian*.

Claiming to be the instrument of reason, argumentation studies develop towards reflection on how this argumentative reason interacts with legitimate social authority, a fundamental element of social life, **S. Agreements; Roles; Persuasion; Evaluation**. The ideal of rational persuasion and consensus served by argumentation is invoked, but one must bear in mind that the decision rests with the legal power that be, and that the best argument may or may not be reflected in the voter's decision.

Appeals to authority structure different forms of argumentation:
— Authority, or lack of authority, may be *self-attributed*, incarnated and manifested in the speaker's speech and attitudes, **S. Ethos; Modesty**.
— The authority of the *testimony*@ is supported by the character and reputation of the witness, and is thus connected with ethos.
— The authority of the *precedent*@ rests upon an earlier judgment (in all the senses of the word *judgment*). The cause may also have been decided in the fable or parable; **S. Example;** *Exemplum*.
— Dialectic@ problematizes discourses supported by various kinds of social authority, **S. Doxa**.

The following paragraphs develops the following aspects of authority:
— The speaker's inherent authority
— The argument from legal authority
— The classical appeal to authority and its criticism.

3.1 Taking his or her word for it: the speaker's inherent authority

The speaker holds uncontested authority over various classes of statements. According to Austin [1962], the performative utterance produces the reality that it states: by saying, "*I promise*", I promise; the speaker is the *auctor* of the reality created.

Ordinarily, no argument is needed in order to make somebody believe something, it must simply be stated. If a speaker says, "*hello!*", even if his friendliness is actually feigned, the default belief is that this is true friendly behavior.

As a general rule, the speaker will be taken at face value; what he or she says will be believed and acted upon without hesitation. When somebody is asked "*What time is it?*", the answer is accepted, no need to check the respondent's watch. Assertions about inner states, "*I feel in good shape today*", are generally accepted without question, as are assertions made by individuals with special access to the facts under discussion. If having authority means having power to successfully transmit one's representations to the listeners, this is the most common form of linguistic authority, based on the preference for agreement@, **S. Testimony.**

This basic linguistic authority combines with other, social authorities, which are attributed to the speaker according to the various social identities and roles he or she plays. These identities and roles converges in the *shown authority* of the authoritative speaker, precisely as defined by the theory of ethos, **S. Ethos.**

Nonetheless, the preference for agreement is not automatic; recipients routinely disagree, and if not, they may be to blame, **S. Modesty.**

3.2 Legal argument form authority

Authority, in the most common sense of the term, is defined by its claim to compliance and obedience; orders are obeyed by virtue of their source, without being systematically backed by a lengthy justification.

> Context: **L** holds the power and means of coercion, reward and punishment in domain **D**
> **L** tells **O** to do **F** (**F** is in the area of **D**)
> **O** does **F**.

The ideal of authoritarian authority is to exert a direct, causal influence on the behavior of others. If the tyrant's subjects are not submissive to his good reasons or charisma, he can still opt for a hard punishments@ or a sweet reward. Radical authority demands that the person who receives the order obey "like a corpse" (*perinde ac cadaver*, according to the metaphor Ignatius of Loyola uses to illustrate the perfection of the virtue of obedience), as a pure instrument, without the intervention of free will.

Conversely, the order is invoked as a sufficient justification for the action: "*I obeyed the orders*". Such an appeal to authority is diametrically opposed to the philosophy of argumentation, which universalizes the imperative of justification and individual responsibilities. It can be challenged by appealing to the international conventions on Human Rights and the Geneva Convention.

Everyday democratic authority is the authority of legal and regulatory norms, backed by the monopoly of legal violence, enforced by the powers that be, and implemented by the person legally in charge. In such a context, the basic expression of a valid legal and democratic argument from authority can be schematized as follows:

> *Context:* There is a system of norms **N**. One of these norms empowers a judge to enforce this system and gives him or her the means of coercion necessary for its application.
> Person **P** has done action **A**.
> The judge shall assess, in a procedure conforming to the requirements of **N**, whether or not **A** constitutes a breach of a norm.
> If it does, the judge sentences **P** to **F**, considering that **R** (justification of the decision)
> Willingly or not, **P** complies with **F**.

Sentences are about "making do", not "making believe" or convincing the convict. The recipients of the judge's good reasons are much more the judge's colleagues, or **P**'s counsel, than **P** himself or herself. **P** may be convinced of the legitimacy of the punishment by the good reasons given by the judge, but this psychological condition is not necessary. **P** must only comply with the judge's decision, willingly or not. One cannot ask everyone to share the theory of redeeming punishment, and to gladly submit to a condemnation, even a democratic one.

The relevant regulation as expressed in the grounds of the judgment backs the argument. This is the basic and current form of argumentation by authority in our societies.

Authority cannot force someone to believe something. But, as belief is manifest in words and behaviors, "make do" may be indistinguishable from "make believe": "*kneel down, pray, and you will believe.*"

4. Classical appeal to authorities

Critical studies of argumentation draw a distinction within ethotic authority, rejecting as fallacious its seductive *charismatic* component, to discuss only its *expert* component, S. **Ethos**.

In the case of the classical argument from authority, the speaker legitimizes the argument by referring to a preexisting source, external from the speaker: authority is *hetero-founded*. The technical study of the concrete part of such external authority in argumentative discourse thus lies within the more general framework of discourse repetition, reformulation, reinterpretation, S. **Straw Man**.

4.1 The authority store

Authority is at the foundation of topos No. 11 of Aristotle's *Rhetoric*:

> Another line of argument is founded upon some decision already pronounced, whether on the same subject, or on one like it, or contrary to it. Such a proof is most effective if everyone has always decided thus; but if not everyone, then at any rate, most people; or if all, or most, wise or good men have thus decided, or the actual judges of the present question, or those whose authority they

accept, or people whose decision they cannot gainsay because they have complete control over them, or those whom it is not seemly to gainsay, as the gods, or one's father, or one's teachers. (*Rhet.*, II, 23, 1398b15-30, RR, p. 365)

We note that the "decision" to be made is of both an intellectual and judicial nature.

On this basis, later rhetoricians list the authorities likely to be called upon to strengthen the position of a party. In the judicial field, the *Rhetoric to Herennius* proposes ten "formulae" [*loci comunes*, "common places", **S. Topic**) "to amplify an accusation":

> The first commonplace is taken from authority, when we call to mind of what great concern the matter under discussion has been to the immortal gods, to our ancestors, or kings, states, barbarous nations, sages, the Senate; and again, especially how sanction has been provided in these matters by laws.
>
> *Ad Her.*, II, 48.

These authorities are distinct from the judicial precedent@ (**S. *Ab exemplo***), and can support any form of speech. Quintilian, for the same judicial situation, considers as authoritative "whatever can be adduced as expressing the opinions of nations or people, or of wise men, eminent political characters, or illustrious poets. 37. Nor will common sayings, established by popular belief, be without their use in this way" (*IO*, V, 11, 36-37).

This authority store will be extensively used, with some adjustments; *Gods* should read *God*:

— Authority of Books, tradition, ancestors (*ad antiquitatem*); the argument of Progress is opposed to this form of authority.
— The famous verses, proverbs, fables, parables...
— The Chinese, the Americans...
— Authority of the media, professionals, scientists, professors...
— Truths from the mouths of children, the rich, the poor... **S. Rich and Poor**.
— Authority of large numbers, prestige of the majority consensus, of a particular group... **S. Consensus; Doxa**.

These forms of authority are cumulative: the scientific authority of the Master is sometimes mitigated by the charismatic authority of the Guru.

All these varieties of authority can be *quoted*; some can be *incarnated* by the speaker as a Chinese, an expert, a poor, a member of a distinguished community.

4.2 Invoked authority

The classical argument of authority exploits an authority taken from the authority store. It is based on a quotation, and can be schematized as follows (see Hamblin 1970: 224 et seq.):

S: — *A is an authority*, *A says that P; therefore P is true and indisputable.*

Or, put simply, "**A** *says that* **P**", when the context clearly establishes that **A** is an authority, and that **S** itself defends **P**, or a position cooriented with **P**. The prototypical example in this category is that of Pythagoras quoted by his disciples, "*he said it himself*" ("*ipse dixit*"). Pythagoras has of course nothing to do with the matter; it is the speaker who quotes him as an authority. Authority can justify ways of doing, beliefs, or combine both:

S: — *That's how they hold their fork and knife in New York.*
S: — *The Master said that pity is fallacious*
S: — *I never give money to homeless people, I read in a book that's just encouraging laziness.*

4.3 Evoked authority

When analyzing discourse backed by an external authority one must take into account the fact that the quotation is not always direct and open. The speaker can also proceed by *allusion* referring indirectly to a discourse, considered as authoritative because dominant, prestigious or associated with an expert. By subtly using the expressions "*discursive formation*", "*ideological state apparatuses*"; "*great other*"... I suggest my knowledge and complicity respectively with the thinking of Michel Foucault, Althusser, Lacan – or *Games of Thrones*.

Quoting an authority in support of a proposition has ethotical repercussions. When Orestes says to Pyrrhus, "*All the Greeks speak to you by my voice*"[1], the speaker does more than quote, he incarnates the authority he quotes. Self-quotation does not grant much authority to what is said, quoting a prestigious authority however does improve the personal authority of the speaker. The Master's voice being heard from the speaker's mouth, the speaker identifies with Him, reframes the exchange accordingly, and hopes that the audience will follow.

The philosophy of argumentation invokes a Popperian ideal of exposure to refutation, according to which it is perfectly legitimate to argue by authority, if the argument is explicit, if one knows exactly who said what and when. This rational requirement of making explicit is opposed to the burying of authority into the depths of discourse in order to shield it against a possible refutation.

5. Evaluation and criticism of expert authority

From a logical-scientific point of view, a discourse is admissible if it collects and articulates true propositions, in order to deduce a new true proposition, according to procedures accepted in the relevant community. In argumentation, the acceptance of a statement or a global vision is based on authority if it is not based upon a review of the good reasons supporting it, or upon a direct examination of the statement's conformity with things themselves, but relies on the source and channel through which the information was communicated. The

---

[1] Racine, *Andromache*, 1667. I, 2. Quoted after
http://www.poetryintranslation.com/PITBR/French/AndromacheActI.htm#anchor_Toc169494154 (11-08-2017)

argument from authority substitutes peripheral, indirect evidence for direct evidence or examination, which is considered inaccessible, too costly, or too tiring. Such daily practice is justified by a principle of economy, division of labor, or simply because someone else was more qualified, or in a better position to tell how events have gone. It works quite well and rationally, as a *default argument*, which can be edited when more information becomes available. Seen from this perspective, authority subtracts nothing and nobody from dispute, it simply shifts the burden of proof to the person who challenges it, **S. Dialectic**.

The argument of authority is therefore a form of argumentation when it exposes the authority which it claims. One could oppose the *authoritarian support* of a statement, as backed by the socio-discursive position of the speaker, to the *argument of authority*, hetero-founded, whose source is clearly exposed. In other words, when invoked to open the debate, the argument of authority is neither authoritarian nor fallacious, but it is if it claims to close the discussion, **S. Modesty**.

The counter-discourse method provides a principle of evaluation and criticism of arguments from authority. Referring to the structure of the argument of authority, discourses against authorities are directed as follows:

(i) Against the quotation itself, preserving the status of the person quoted as an authority:

> Authority **A** is not interpreted correctly; **A** did not say that, or mean that; **P** is not quoted correctly, has been diverted from its context, has been reformulated, reoriented tendentiously…

(ii) Against the authority quoted:

— **A** has no direct evidence.

— By application of the *ad hominem* argument to the source: **P** is incompatible, contradictory, with other assertions (or prescriptions) of **A**.

— **A** has evolved on this point, as testified by his or her more recent statements.

— **A** has spoken outside of his area of expertise; he or she is not an expert in the precise field covered by the **P**-type claim.

— There is no consensus among experts.

— **A** is not an expert, his or her views are outdated; he or she is mistaken, and has often been mistaken in the past. **A** is biased, manipulated, paid to say what he says. Launching a personal attack (*ad personam*), the opponent can utterly dismiss@ **A**: "*A is not an expert but a jester*".

One can distinguish between two distinct strategies dealing with authority: arguments *establishing* an authority as such, and arguments *exploiting* an established authority. This opposition has a general value, **S. Causality, Definition, Analogy**. The first discourse against authority attacks the use made of authority, whereas the second discourse attacks the authority itself. It follows that the discourse (ii) against authority mirrors a discourse defining a legitimate expert: "**A** speaks in his sphere of competence, and is aware of the state of the matter; **A**'s system is coherent; **A** has direct evidence, serious experts agree with what **A** says; the previous anticipations made by **A** are proven correct."

(iii) Against the person who submits to authority
The interaction framework shifts the focus from the statement of authority itself to the *relationship of authority*. Criticism is now aimed at the pusillanimity of the interlocutor, S. **Modesty**.

(iv) Counter-argumentation
Finally, better arguments can be directly opposed to **P**, direct argument dealing with the matter at hand, drawn not from authority but from scientific reason, or from historical knowledge, considered as superior to a lazy appeal to authority.

6. Refutative uses of authority

6.1 Refutative uses of positive authority
The preceding paragraphs address authority inasmuch as it serves as support for an affirmation. An appeal to authority is used for rebuttal when the authoritative assertion can be opposed to the opinion to rebut:
S1:   — *P!*
S2:   — *X says the opposite, and he knows what he is talking about!*

If **X** and **S1** are of the same affiliation, the refutation combines authority and *ad hominem*, S. ***Ad hominem***. Positive authority can also be used to destroy not the content of what is said, but the *claim to authority* and therefore the competence of the person holding the discourse:
S1:   — *P!*
S2:   — *That's exactly what Perelman says!*
        — *We've known that since Aristotle!*

6.2 Negative authority
Negative authority is used to rebut the saying in the following case:
S1:   — *P!*
S2:   — *H says exactly the same thing!*

H is a person, a party rejected by the speech community to which **L2** belongs, or by the third parties arbitrating the discussion, or possibly **L1**; **H** is an anti-authority, an anti-model, S. **Imitation**
In the case of a positive authority, the proponent connects the statement with an authority; here, the connection of the statement he or she disputes with the negative authority is made by the opponent. Hitler is the paragon of the negative authorities, whose words cannot be repeated. The *reductio ad Hitlerum* puts an end to all argument.

> Last year, you may recall, a number of financial-industry barons went wild over very mild criticism from President Obama. They denounced Mr. Obama

as being almost a socialist for endorsing the so-called Volker rule, which would simply prohibit banks backed by federal guarantees from engaging in risky speculation. And as for their reaction to proposals to close a loophole that lets some of them pay remarkably low taxes — well, Stephen Schwarzman, chairman of the Blackstone Group, compared it to Hitler's invasion of Poland.

Paul Krugman, "Panic of the Plutocrats", 2011.[1]

Autophagy and Retaliation

A statement can be *self-justified*: **S. Self-Argued Claim**. This self-defense is made possible by the multi-layered semantic structure of language, and in particular by the fact that words have an orientation@, which may be well grounded on implicit arguments, **S. Words as Arguments**. Just as it can be self-justified, a statement can be *self-defeated*. A statement is self-defeated when it expresses a logical or material impossibility, or when it involves a pragmatic contradiction between what is said and the act of saying it.

This phenomenon is also called *autophagy*. Perelman defines autophagy as a contradiction arising from the fact that "the assertion of a rule or a principle is incompatible with the conditions or with the consequences of its assertion or application. Such arguments can be called *autophagy*. *Retaliation* is the argument that attacks the rule by highlighting the autophagy" (Perelman 1977, p. 72-73).

The assertion is incompatible with the fact asserted, "the very act implies what the words denies" (*id.* p. 73). Perhaps the best-known case of autophagy is that of the Cretan Epimenides affirming that *"all the Cretans are liars"*:

> There are no more cannibals, we have eaten the last one.

> S1 — *All statements can be questioned.*
> S2 — *I question this statement.*

Retaliation is a kind of refutation reconstructing a claim as pragmatically self-defeating on the basis of its very content, and in virtue of its own principles. In philosophy, this strategy, known as the *epitrope*, is applied by Socrates to refute Protagoras' thesis according to which:

> Man is the measure of all things: of the things which are, that they are, and of the things which are not, that they are not. (Plato, *Theaethetus*, 152a; *CW*, p. 169)

This doctrine exhibits that "most exquisite feature" that if true, it is false:

> *Socrates*: — [...] Protagoras admits, I presume, that the contrary opinion about his own opinion (namely, that it is false) must be true, seeing he agrees that all men judge what is.
> *Theodorus*: — Undoubtedly.

---

[1] www.nytimes.com/2011/10/10/opinion/panic-of-the-plutocrats.html?_r = 1&ref=global-home (11-08-2017)

Socrates: — And in conceding the truth of the opinion of those who think him wrong, he is really admitting the falsity of his own opinion?
Theodorus: — Yes, inevitably. (*Id.*, 171a-b; *OC*, p. 190)

This refutation is based on the principle of non-contradiction@; to maintain consistency, a Skeptic will have to doubt this principle. S. *Ad hominem*; *Ex datis*.

B

Backing ▶ Layout; Scheme

Bandwagon ▶ Consensus

Begging the Question ▶ Vicious Circle

Beliefs of the Audience

Arguments based *on the beliefs and character of the audience* are opposed to those based on the *substance of the issue*.

In classical rhetorical argumentation, the orator must not only know the case and the law, but also the judge, that is the people he or she intends to address and convince and the opponent he or she is going to face and refute. Before engaging in the quest for arguments, he or she must gather information about the *ethos*@ of the audience and of the opponent; that is about their beliefs, habits and general character, including their previous discourses and positions. The orator exploits this information either *positively* to confirm his or her position, or *negatively* to reject the opponent's position:

— The *ex datis* argument positively exploits the natural ethos of the audience to infer a positive conclusion, S. **Ex datis**.

— The *ad hominem* argument exploits the information about the opponent's discourses and beliefs in a negative way, by exhibiting their inconsistencies, S. *Ad hominem*.

Bias ▶ Orientation

Burden of Proof

> Lat. *onus probandi*; Lat. *onus* "charge, burden"; *probandi*, from *probare* "to make believable, to make accept, to prove".

The burden of proof plays a fundamental role in argument. It is a conservative principle, like the principle of inertia in physics: "*I keep doing business as usual unless I have a good reason to change*". Mill reports an anecdote that vividly illustrates the heaviness of the burden of proof imposed by a conservative society upon social innovators, S. **Calm**:

> The propounder of a new truth, according to this doctrine should stand, as stood, in the legislation of the Locrians, the proposer of a new law, with a halter round his neck, to be instantly tightened if the public assembly did not, on hearing his reasons, then and there adopt his proposition. People who defend this mode of treating benefactors, cannot be supposed to set much value on the benefit; and I believe this view of the subject is mostly confined to the sort of persons who think that new truths may have been desirable once, but that we have had enough of them now.([1859]. p. 88)

In court, the burden of proof is expressed by the presumption of innocence "*a person is presumed innocent until proved guilty*"; that is, the accusation must provide positive evidence of the accused's guilt. The stabilization of the burden of proof is an institutional decision, organizing the situation; the last word is left to the defendant.

In informal social debates, there is no clear preliminary agreement about who supports the burden of proof, and the proponent can try to shift it onto the adversary. It becomes a stake of the debate.

The *doxa*@ can be defined according to the same principle: an *endoxon*, that is a fragment of the doxa, is best defined not as a "probable" belief, but as a belief which is not subject to the burden of proof, and is, accordingly, considered to be "normal" by the given group. The individual challenging an *accepted* proposition bear the burden of proof, and has to provide good reasons. This is why Descartes, willing to reject all his pre-established beliefs, must back this radical doubt by the hypothesis of the Evil Genius (Descartes [1641], *First Meditation*).
S. **Rules**.

When it comes to current trends and fashions, the burden of proof is reversed: "*it is new, it has just come out!*" is a direct argument for buying the product in question. Good reasons are instead needed for *not* following fashion, *not* adopting new theories, and *not* voting for a new candidate.

Burden of proof and Initiative — Hamblin has redefined the burden of proof in a language game as attributed to the player taking the initiative, that is, making the first move. This definition can be transposed to highly argumentative multi-speaker interactions, where the first turn is generally allocated to the person supporting the proposal to be discussed. In a debate on the legalization of drugs, the facilitator addresses the first question to a supporter, not to an opponent of legalization.

The burden of proof relates to a question and a proposal. If the opponent makes a counter-proposal, he will bear the corresponding burden of proof.

The burden of proof may vary with the group involved, and where the debate takes place. If the doxa of the group is that no prohibition should apply to drug consumption, then, in this group, the *supporter* of the prohibition will have to justify his stance.

C

Calm

Lat. *ad quietem* arg.; *quies*, "rest; in politics, peaceful period; neutrality".

Calm is the emotional and cognitive state of a person having no reason for concern, in particular, having no urgent issue to address.

1. Calm and emotionality
The Aristotelian list of socio-rhetorical emotions opposes *calm* to *anger*, **S. Emotion**. In fact, calm may be opposed to any strong positive or negative emotion. Strong emotions are characterized by a marked variation of *arousal*. Specific actions, speech and arguments might be used to reduce such excitation and re-instill a quieter mood, that is to *calm down* overexcited people, be they a group of enthusiasts enraged by the prospect of a war, or children throwing a tantrum.

2. Appeal to tranquility
In the political sphere, the *"leave us alone!"* maneuver has been identified and named *ad quietem* by Bentham (1824; **S. Political Arguments**). It is defined as an attempt to postpone the discussion of a problem in the hope that the issue will never be addressed. This maneuver is revealed by discourses amplifying the following topics:

this issue is not so important, already settled, we have other priorities, we'll discuss that later, you are the only one to see that as a problem...

A meta-discussion about the relevance and timing of the discussion is substituted for the discussion itself. Bentham regards this maneuver as fallacious, and classifies it in the category of "fallacies of delay", directed against freedom of proposition and political innovation.

The appeal to calm and tranquility ("*leave us in peace!*") values calm as a peaceful conservative political state, which may side with apathy, inertia and laziness. Such a state is threatened by dissatisfied proponents, willing to demand changes and commence an argument, which will in turn provoke a disturbing surge of adrenaline, excitation, anger or anxiety within the group. The burden@ of proof is the price paid by the proponent for disturbing the *tranquility* of the group. Tranquility may be invoked as an argument for not participating in political and social life:

> Voting concerns only men, since women — fortunately for their tranquility — do not have political rights.
> Clarisse Juranville, [*Handbook of Moral Education and Civic Instruction*], [1911].[1]

The following intervention is taken from a debate on immigration and French nationality. It is made by a female student at the very beginning of the discussion. First, she gives a carefully worded and slightly oriented description of the two parties and of their positions, **S. Orientation**. She then takes an implicit but clear stand in favor of the party holding that "*the government currently has other priorities that are more important and that it [is] not necessary to go back to this point*", on the basis of a perfect "*leave us in peace*" argument:

Prof:	*then you say nothing stay mute/ you learned nothing from all that, nothing struck you/ — you what are the points/ — so let's start listing them\ you can give them/ yes/*
Student:	already two points of view actually, finally
Prof:	*there are two points of view you have seen that there was yes/*
Student:	two parties that oppose well those who want to— as the petition of all the screen actors and filmmakers etcetera who want that: im- well the nationality code be unlimited\ and that all the— undocumented people be regularized\ therefore hmm without any limit
Prof:	*hm hm hm hm*
Student:	and the second point of view is those who say that for there to be a right of the people there must be:: a right of state\ therefore precisely there must be limits and that:: and also these people are those who say that the government currently has other priorities that are

---

[1] Quoted after Clarisse Juranville, *Manuel d'éducation morale et d'instruction civique*, Paris: Vve P. Larousse. Quoted after the 5ᵉ ed., 1ʳᵉ part *Éducation morale* ["Moral Education"]; chap. *Le vote* ["The Vote"]; § *Les femmes et la politique* ["Women and Politics"]. No Date. No pag.

> more important and that it was not necessary to go back to that point\
Prof: OK
>> Corpus *On Immigration and French Nationality*, Student Workshop.[1]

Case-by-Case Argument

1. Definition

Case-by-case argumentation is a technique of inquiry developing in several stages, from questions like "*What happened, what can happen?*":
— First, make an inventory of possible cases.
— Second, consider each of these cases.
— Third, sum up the cases considered and see if examination leads to the elimination of all possible cases but one.
— Fourth, conclude that the last remaining case is real and true.

> S1 — *All this money, either comes from a legacy or is your labor, or has been stolen. If it comes from your labor income or from a legacy, it'll be easy for you to prove it by showing us the relevant documents. No documents of that kind available? So you stole it.*

This argument illustrates the classical law of negation of a disjunction, S. **Connectives**:

> "**P** or **Q** or **R**" is true; but **P** is false and **Q** is false; so necessarily, R is true.

Definitions can be given on a case-by-case basis. A crime such as impiety might be defined as a lack of respect for either the gods, their priests or their shrines. To accuse someone of impiety (or to exonerate oneself from that crime) one must show that one of the three parties listed above has been disrespected (or not) (after Aristotle, *Rhet.*, II, 23, 1399a5; RR p. 367).

2. Argument by division

Argument by division is illustrated by the following example:

> The tyre exploded because it was worn out, because there were nails on the road, or because of a manufacturing defect. Now, the tyre had just been bought and no nails were found in it. So there was bad workmanship. (Perelman, 1977, p. 65)

This shows that the label "argumentation by division" is homonymic: it can refer either to the argumentation by *composition@ and division*, or to the case-by-case argument.

---

[1] Corpus "Débat sur l'immigration — Débat étudiants". Clapi Base, http://clapi.univ-lyon2.fr/V3_Feuilleter.php? Num_corpus = 35] (07-30-2013).

3. Refutation of the case-by-case argument

A case-by-case argument is perfectly conclusive if all cases have been considered; it can be rejected on the same case-by-case basis by showing that the enumeration of cases is incomplete:

    S2 (as a reply to **S1**, supra): — *No Sir, I just won the lottery, here is the winning ticket!*

    S3 (as a reply to Perelman, supra) — *Well, Sir, here are some other possibilities. The tire might have exploded because it was badly inflated, because there was a pothole on the road, because it hit the curb, because it has been overheated (if the driver happens to have just used a torch to unscrew a wheel bolt), because the brake was glued, because it had been brought into contact with an electrical source, because the car was too loaded or was running too fast... My conclusion is that the investigation must go on.*

Categorization and Nomination

The term *categorization* refers to the various cognitive and practical operations through which an individual is integrated into a *category* and designated by the *name* attached to that category:
- What is this? *Identification process*
- This is a X *Name of the object*

The name can be taken from the current lexicon or from a scientifically controlled taxonomy or theory. Categorization as a *cognitive and empirical* operation cannot be dissociated from *nomination*, a *linguistic* operation.

The classical example illustrating Toulmin's layout@ of argument is an example of an administrative categorization: the individual Harry is categorized as a *British citizen* on the basis of the criterion, *"— to be born in Bermuda"*.

Categorization is the first step to implement an argumentation by definition, *"he is a British citizen, so ..."* **S. Definition**. In law, categorization corresponds to the legal qualification of an act (*is it a crime or an accident?*); it determines the law applicable to the case, **S. Stasis**.

1. Categorization tests: distinctive features and global analogy

An individual is given a name and integrated in a category mainly on the basis of a set of *distinctive features* or out of a global *analogy* with an outstanding member of the category.

— The categorization by *distinctive features* is based upon a definition. A definition of a noun is a set of heterogeneous features that can be used to test an individual for the corresponding category, **S. Definition**. If a significant number of these distinctive features fit with the description of the individual, then this individual belongs to this category, and can be given the corresponding name.

If the categorization-nomination is based on unsystematic, anecdotal features the category is inconsistent: *"the bird is gray, the sky is gray, the bird is a cloud, the cloud is a bird"* **S. Intra-categorical analogy**.

— The categorization *by analogy* is based on a common global form (Gestalt) shared by the individual under consideration and a prototypical member of the category: this mushroom looks like a Scotch bonnet, it is a Scotch Bonnet. The prototypical species is the species with which the community is best acquainted with.

The concrete task of nomination–categorization combines the two sets of tools, distinctive features and analogy. The distinctive features can be drawn from the stereotype rather than from any kind of definition; all the features found on the stereotype tend to be considered as essential for the definition of the category, **S. Imitation.**

Binary and gradual categorization — The categorization made on the basis of essential, distinctive features entails that category predicates are binary: an individual is a member of a category or is not.
If membership within a category is determined simply by stacking any sufficient number of features, category predicates are gradual; the richer the combination of features, the stronger the link with the category. Similarly, a bird which looks more like the prototypical bird than another is "more" a bird than the other one. Category membership becomes gradual, and its top members cannot be transcended; this can be the meaning of the juvenile expression "more **X** than him, you die", "*cooler than him, you die*" in other words, one comes out of the category upwards.

Categorization mistake? — In *Alice in Wonderland*, the pigeon wrongly categorizes Alice as a serpent:

> 'Serpent!' screamed the pigeon.
> 'I'm not a serpent', said Alice indignantly. 'Let me alone!' […]
> 'A likely story indeed!' said the Pigeon in a tone of the deepest contempt. 'I've seen a good many little girls in my time, but never *one* with such a neck as that! No, no! You're a serpent; and there is no use denying it. I suppose you'll be telling me next that you never tasted an egg!'
>
> Lewis Carroll, *Alice in Wonderland*. [1865] [1].

The pigeon wrongly categorizes Alice as a serpent on the basis of the long neck she is developing in this episode. For the pigeon, this characteristic evokes a snake, so that the pigeon fears for its eggs; and in addition, Alice eats eggs, a feature perhaps inessential for the categorization of beings, but which reinforces the pigeon's conclusion.
From an *essentialist* view, the pigeon miscategorizes Alice; "*having a long neck*" is not a specific difference nor a characteristic proper of snakes; giraffes, herons, swans... are also animals with long necks. Actually, the pigeon classifies Alice from a *functional* point of view. From the pigeon's perspective, a long neck is a

---

[1] Quoted after Lewis Carroll, *Alice in Wonderland*, BookVirtual digital edition. P. 71; 72-73. https://www.adobe.com/be_en/active-use/pdf/Alice_in_Wonderland.pdf (11-08-2017).

natural sign of danger and it is wise to apply a precautionary principle, that is to shout "*snake!*" as people shout "*wolf!*" when perceiving a strange creature lurking behind the house.

2. Technical categorization

The categorization-nomination can be expressed via a simple judgment about an individual "X *is a bastard, it shows immediately*"; most designations are not the result of a careful examination of the relevant criteria, but if in doubt, the availability of such criteria proves essential. The mushroom picker who has doubts about the nature of the mushroom he has just picked must engage in a careful process of categorization; the same goes for the municipal employee seeking to determine the rights of an individual applying for social security benefits. First of all, they must refer to the criteria enumerated in the relevant reference books: the encyclopedia of mushrooms in the first case; the decrees and dispositions defining the terms and conditions of attribution of social security benefits in the other. A well-conducted process of categorization will lead to reasoned conclusions, such as:

Y is / is not a *marasmius oreades*, i.e., a Scotch bonnet.
X is / is not a single parent in the administrative sense of the expression.

The investigating parties will then take the relevant action: keeping the mushroom for eating or throwing it away; accepting or denying the application for social security benefits.

Social Categorization — A parent is defined as "*a parent or a person who bears the financial burden of one or more children*". "*To be single*" is defined as: "*to be widowed, divorced, separated or unmarried not cohabiting*". The meaning of parent is finally extended to include "pregnant" and "people having the legal responsibility of a child".

Natural Categorization — Wikipedia describes the Scotch Bonnet as follows:
Marasmius oreades, the **Scotch bonnet**, is also known as the **fairy ring mushroom** or **fairy ring champignon**. The latter name tends to cause some confusion, as many other mushrooms grow in fairy rings (such as the edible *Agaricus campestris*, the poisonous *Chlorophyllum molybdyte*, and many others).
Distribution and habitat — *Marasmius oreades* grows extensively throughout North America and Europe in the summer and autumn (fall) (June - November in the UK), or year-round in warmer climates. It loves grassy areas such as lawns, meadows, and even dunes in coastal areas.
Description — It grows gregariously in troops, arcs, or rings (type II, which causes the grass to grow and become greener). The cap is 1-5 cm across; bell-shaped with a somewhat inrolled margin at first, becoming broadly convex with an even or uplifted margin, but usually retaining a slight central bump — an "umbo"; dry; smooth; pale tan or buff, occasionally white, or reddish tan; usually changing color markedly as it dries out; the margin sometimes faintly lined.
The bare, pallid stem grows up to about 7cm by 5mm in diameter.

> The gills are attached to the stem or free from it, fairly distant (rather a distinctive character), and white or pale tan, dropping a white spore-print. The spores, themselves, are 7-10 x 4-6 μ; smooth; elliptical; inamyloid. Cystidia absent. Pileipellis without broom cells.
> This mushroom can be mistaken for the toxic *Clitocybe rivulosa* which lacks an umbo, is white to grey in color, and has closely spaced decurrent gills.
>
> Wikipedia, *Marasmius oreades*

If the harvested object thing complies with this description, then it is a Scotch Bonnet. Categorization is achieved on the basis of a set of quite different procedures: observing whether the key elements of a definition by description apply to the individual; looking carefully at the picture showing a prototypical Scotch Bonnet; testing the object for its "elasticity under finger pressure". Some features of the definition can be checked immediately, for example, by looking at the surroundings:

> grassy area —grows gregariously in troops, arcs, or rings (*ibid.*);

or at the mushroom itself:

> a slight central bump: an 'umbo' (*ibid.*);

or practicing a small experimentation:

> usually changing color markedly as it dries out (*ibid.*)

These are positive criteria, that, if met, justify the claim "this is a *M. oreades*".

Of special importance for the task of categorizing and giving names, are the distinctive criteria; the umbo criteria proves essential, and, for some other species, vital:

> This mushroom can be mistaken for the toxic Clitocybe rivulosa which lacks an umbo, is white to grey in color, and has closely spaced decurrent gills (*id.*)

In contrast the name-derived criteria "*fairy ring mushroom*" seems to be a necessary, not sufficient criteria, very risky since it is shared by both edible and toxic species. These are key criteria in the case of categorization issues (cf. infra, §3).

Notably, other parts of the definition may remain puzzling for many: "*inamyloid. Cystidia absent. Pileipellis without broom cells*". Categorization is commonly achieved on the basis of a selection of criteria. Once categorization has been performed in view of a reasonable set of elements, it is possible to allocate to the object under examination all of the features mentioned in the definition. It is in this way that categorization connected with definition becomes a powerful argumentative machine, *argumentation by definition*:

> it is a Scotch Bonnet, SO "*inamyloid*, etc.

or, more realistically perhaps:

> "Many mushroom connoisseurs are fond of *M. oreades*" SO, let's cook it at once!

Over time and with growing experience, this knowledge, manipulations and, most importantly, *reasoning* will be incorporated in *perception*, and the mushroom picker will immediately *see and recognize Marasmius oreades* as such: "*look, Scotch Bonnets!*".

3. Categorization Issues

The fact that categorization is an argumentation-based process is clearly illustrated by borderline cases, in which the individual or situation under consideration meets some, but not all of the criteria defining the given category.

Let us consider the above-mentioned case of social security benefits, provided by the state to help a single parent to raise a child. The municipal employee receives the following application:

> I am currently separated from my husband, who has moved out of the conjugal home, leaving with another woman. We will be taking steps to divorce, but in the meantime, I am living alone with my daughter.

This woman is not divorced, but is apparently engaged in court proceedings, or at least plans to file for divorce. Does she therefore qualify for *immediate* financial support?

A stasis or conflict of categorization occurs when discourse and counter-discourse are based on conflicting categorizations of the same event, action, or person:

S1_1 — *he is a poor guy*
S2 — *no, he's a real bastard*
S1_2 — *no, he is a poor guy, we should pity him*

S1_1 — *Syldavia is now a great democracy!*
S2_1 — *how can you talk about democracy in a country that does not respect the rights of minorities?*
S1_2 — *there are tons of democracies that do not respect the rights of minorities.*

Such antagonistic categorizations occur frequently in conversations.
— In dialogue (1), the antagonistic categorizations of the same individual as *a poor guy* vs. *a bastard*, are just stated and repeated.
— In dialogue (2), **S2_1** rejects the categorization of Syldavia as a democracy, arguing that *protecting the right of the minorities* is a necessary feature to qualify for *being a democracy*. **S1_2** maintains and backs up his or her appreciation, arguing that democratic regimes, as they are, often fail to respect minority rights. In a very common opposition, **S1** categorizes Syldavia on an *essentialist criterion*, **S2** on an *empirical criterion*, which opens a perfect argumentative situation.

Causality

1. The causal relationship and its expression

The notion of cause is central in daily argument as well as in scientific argument. It is considered a primitive, intuitively clear notion; this means that ordinary language defines cause only through notions which are equally complex. Let us consider some possible ways to refer to and think about *causal* links and processes:

— The cause *explains, accounts for* its effect; it gives *the why*, the *reason* of things. The effect is *understood* when its cause is *known*.

— The cause of something is its *principle; origin, basis, foundation, grounds*; its *occasion*. The cause is a *motor*, which *triggers, starts* a series of effects.

— Humans act as *cause*; they are *agent, maker, author, creator, inspirer, instigator, promoter, producer...*; their *aims, purposes, intentions, motives* and *motivations...* are considered as causes. Their *incitements, inducements instigations*, are second-level causes.

— Metaphorically, the cause is thought of as a *spark, a ferment, a germ; a root, a seed; a source, a spring*. Their cause is the *mother* of things as they are.

Beyond the specific verbs corresponding to the preceding nouns, different kinds of causal relations are associated with very general verbs such as *bring (about), to give (rise to), to make, procure, lift...*

Like the logical relation of implication, the causal relation can be associated with passages articulated by conjunctions or adverbs:

Since, because...; as soon as...; so ; when; if ... then...

All these terms and constructions might point to some kind of causal relation, and can therefore be considered as causal indicators@ of a sort, being kept in mind that they can also express other functional relations. A spontaneous "causal impulse" always suggests a causal relation behind a purely temporal succession, or concomitance (see infra).

It would be difficult, and is not necessary, to identify and reconstruct all of the multi-level, potential causal relations in a text. Relevant and indisputable causal argumentative causal relations are substantial, in the foreground of the discussion, articulated and thematized in the argumentative lines developed by the participants in the discussion.

2. Time, causal, logical series

Let us consider the causal, logical and temporal series. In the physical world, the *cause* precedes its *consequence* (this is not, however, always straightforward). In the logical world the *antecedent* is to the left of the logical connective '→' and the *consequent* is to its right; in the world at large, events simply follow one another.

Causal series	*cause*	*effect, consequence*
Logical series	*antecedent*	*consequent, consequence*
Time series	*prior, previous*	*posterior, later*

The time series includes three terms:

before… / during… / after…
prior, anterior, previous… / simultaneous… / posterior, later, subsequent…

The word *consequence* is thus used to designate the *effect*, linked to its *cause*, or the *consequent*, linked to its logical *antecedent*. In general, logical relations develop the *consequences* of hypotheses or postulates. If the length of the side of the square is doubled, its surface is multiplied by four: this result is a *consequence*, linked to a *cause* which is *a mathematical reason*.

> Mind your words, you speak of the *birth* of the gods, so you suggest that at one time, the gods did not exist?

This is not a causal, but a *semantic* consequence, based on the linguistic meaning of the word "birth".

3. Argumentations about causes, mobiles, reasons… and effects

The terminology of argumentation involving a causal relation might be confusing. We will distinguish between, on the one hand, argumentation *establishing* a causal relationship, and, on the other, argumentation *exploiting* a previously established causal relationship.

(i) The *cause-effect@ argumentation* establishes a causal relationship between two facts and eliminates "false causes".

(ii) Several kinds of arguments exploit a pre-established causal relationship. In this second case, we will distinguish between:

— *Cause@ to effect* argumentation, going *forward* from the cause to the effect. A fact-argument considered to be a *cause*, is claimed to have such effect.

— *Effect@ to cause* argumentation, goes in the opposite direction, from the effect to cause. A fact-argument to which a status of *effect* is attributed, is claimed to have such cause.

— *Pragmatic@ argumentation* develops first from cause to effect, before returning to the cause. In order to make a decision about a practical measure (assimilated to a cause), one develops its possible positive or negative effects, before arguing back to the cause.

— Argumentations based on *motives@* align the cause-effect relation with the relation from a motive to do something to the corresponding action.

— *A priori@* and *a posteriori* arguments, *propter quid* and *quia*, exploits causal and logical links.

Cause – Effect: The Causal Link

1. Causal argumentation

Causal argumentation establishes a causal link between two different kinds of facts. For example, we notice that, on the one hand, (1) that the use of pesticides is intensifying, and (2) that bees are disappearing. Is there a causal relationship between these two facts, are the following statements true?

The use of pesticides causes the disappearance of bees.
Pesticides are used and bees disappear (with a causal implication).

There may be disagreement about this kind of conclusion, even if there is agreement on the facts under consideration:

We use pesticides and the bees disappear, that's true. But…

The causal investigation starts with a salient fact, as "bees disappear", "the climate seems to be changing", and the cause of this is problematic. Generally, several facts can be evoked as possible causes, and possible explanations of the phenomenon. This creates a stasis of causality, expressed via the confrontation of these two hypotheses, for example in the case of climate change, taken as a fact:

S1: — *the increase in solar activity causes the change of climate.*
S2: — *the increasing emission of greenhouse gases causes climate change.*

These explanatory causes integrate themselves within broader theories on the climatic equilibrium of the terrestrial globe. Broad conceptions of the physical and social world are in confrontation through such local causal affirmations.

Affirmation of causal relationships are therefore based on elaboration of crucial experiments and the retrieval of key observations. Causes are determined according to the methodology relevant to the given domain. Ordinary causal experimentation also involves observation and experience. So for example, if I suffer a mild allergy reaction, I must consider what the possible allergens might be which have caused it. I might observe that yesterday I went to the swimming pool and ate strawberries. There are two possible allergenic, strawberries or chemical products used in the pool. I might conduct the following checks, eating strawberries without going swimming, and going swimming without eating strawberries. If I am unlucky, I'll have to investigate further and perhaps see a specialist, who will proceed in much the same way. If I am lucky, however, I'll suffer a (controlled) mild allergic reaction in one case and not in the other, and will be able to identify the allergen. As the allergic reaction is undesirable, I pragmatically reason, in view of the negative consequence, I change my behavior, and so eliminate the cause.

2. Refutation of causal assertions

The correct establishment of causal relationships is a fundamental requirement, both in science and in ordinary life. The priority given to the correct determination of causal relations is the basis of Aristotelian thought. The "false cause" fallacy is committed when a causal relation is asserted between two phenomena that in fact have no causal relation between them. This fallacy is sometimes designated by its Latin name *non-causa pro causa*, "'non-cause' taken for a cause", **S. Fallacious (III).**

"*Smoking causes cancer*": strictly speaking, the positive existence of such a relationship is difficult to establish. It can only be considered as a remainder, persisting when all other possibilities have been discarded. Causal imputation might be revised. If we are to confirm that a link of the causal type does exist between two facts, it is necessary to answer a set of standard objections which oppose the existence of a causal relation.

2.1 The alleged effect does not exist

The causal assertion *"the use of pesticides is the cause of the disappearance of the bees"* is refuted by showing that although the bees have disappeared from a certain area, there are still as many bees as before if a larger, more general area is considered. The bees have not disappeared, they have simply migrated.

The facts must be confirmed, before looking for and discussing their causes. This methodological rule is well illustrated by the famous case of the golden tooth, described by Fontenelle.

> Let us be well assured of the matter of fact, before we trouble our selves with inquiring into the cause. It is true, that this method is too slow for the greatest part of mankind, who run naturally to the cause, and pass over the truth of the matter of fact; but for my part, I will not be so ridiculous as to find out a cause for what is not.
>
> This kind of misfortune happened so pleasantly, at the end of the last age, to some learned Germans, that I cannot forbear speaking of it. *"In the year 1593, there was a report that the teeth of a child of Silesia of seven years old dropped out, and that one of gold came in the place of one of his great teeth. Horstius, a professor of physic in the university of Helmstad, wrote in the year 1595, the history of this tooth, and pretended that it was partly natural and partly miraculous, and that it was sent from God to this child, to comfort the Christians who were then afflicted by the Turks."* Now fancy to your self what a consolation this was, and what this tooth could signify, either to the Christians or the Turks. In the same year, (that this tooth of gold might not want for historians) one *Rullandus* wrote the history of it: two years after, *Ingolsteterus*, another learned man, wrote against the opinion of *Rullandus* concerning this golden tooth; and *Rullandus* presently makes a fine learned reply. *Libavius*, another great man, collected all that had been said of this tooth, to which he added his own opinion. After all, there wanted nothing to so many famous works, but the truth of its being a tooth of gold. When a Goldsmith had examined it, he found that it was only a leaf of gold laid on the tooth with a great deal of art. Thus they first compiled books, and then they consulted the goldsmith.
>
> Nothing is more natural than to do the same thing in all other cases. And I am not so much convinced of our ignorance, by things that are, and of which the reasons are unknown, as by those which are not, and for which we yet find out reasons. That is to say, as we want those principles that lead us to the truth, so we have others means that not only do we not have the principles that lead to truth, but we have others which are exceeding well with that which is false.
>
> Bernard Le Bouyer of Fontenelle, *The History of the Oracles* [1686][1],

---

[1] Bernard Le Bouyer of Fontenelle, *The History of the Oracles*. Glasgow: R. Urie, 1753. P. 14-15.

2.2 The effect exists independently of the alleged cause
The determining cause has a consistent impact. If **C** is the cause of **E**, we cannot have **C** without **E**. If a metal is heated, it necessarily expands. It follows that a causal statement can be rejected by showing that the effect persists when the cause is absent. To refer again to the example above, if it can be shown that bees also disappear from areas where pesticides are not used, pesticides cannot be considered to blame for the fall in bees' number.

2.3 There is no causality but concomitance
In that case, **A** both regularly accompanies and precedes **B** without being the cause of **B**. The cock sings regularly before the break of day, but it is not the cause of the sunrise. Taking an antibiotic might be accompanied by a feeling of exhaustion, but the cause of this exhaustion is not the antibiotic but the infection that it fights. The general principle to check whether a causal relation exists is to suppress the agent which is the suspected cause; if the so-called effect is still there, there is no causal link between the two facts. If the cock is eliminated, for example, the sun still rises; if we do not take antibiotics, we will still be exhausted and perhaps even more.
The use of pesticides is concomitant with the disappearance of bees; but in areas where pesticides cease to be used, bees' numbers continue to fall at the same rate. The cause is to be sought elsewhere: perhaps climate change is to blame?
Such erroneous causal imputation are well identified in the ancient theory of fallacies, which denotes them by two Latin expressions:
> — Fallacy of the antibiotic: *cum hoc, ergo propter hoc*:
> "with, therefore because of": **A** accompanies **B**, so **A** is cause of **B**.

> — Fallacy of the cock: *post hoc, ergo propter hoc*:
> "after, therefore because": **B** appears after **A**, so **A** is cause of **B**.

2.4 Another cause may have the same effect
One can be tired because one has been physically exhausted, because one has an infection, or because one is depressed.

2.5 Not one, but several causes: complex causality.
It may be necessary for several causes to exist in conjunction in order that they produce some effect. This is the case of economic crises, or lung cancer.
The determination of causes establishes the responsibility of the human agents who have set the causal machinery in motion. If the causality is complex, it is possible for the defendants to argue that they are responsible only for a causal factor, which would not alone have given rise to the relevant problem. Upon being arrested, a person dies. The autopsy shows that this person was suffering from a weak heart:
> Lawyer: — *If the police had treated him gently, he would not have died. The police are responsible.*

> Police: — *If he had not been sick before, he would not have died. The police are not responsible.*

In cases of heavy pollution, the authorities apologize to people suffering from respiratory diseases: "*people without such respiratory issues have no problem*".

2.6 The effect feeds the cause
Feedback is a sort of causal circle: atomic fusion raises the temperature and the rise of temperature accelerates fusion. In the social field, this kind of mechanism is invoked to reject a particular measure, arguing that it will not alleviate the issue in question, but rather aggravate it:

> L1: — *To fight recession, public services must be strengthened / reduced.*
> L2: — *But the strengthening / reduction of public services will reinforce the recession.*

One can always refute a measure by asserting that it will have certain unwanted consequences which will outweigh any potential advantage, **S. Pragmatic**. In the example given above, the refutation is radical, the perverse effect being not a side effect, as yet unnoticed by the author of the proposition, but exactly the *reverse* of the intended effect. This is a case of pure and simple inversion of causality (see infra), which is frequent in polemical discourse.

2.7 Self-fulfilling prophecies
In the case of *self-fulfilling prophecies*, the announcement of an event is the cause of the event:

> S1_1: — *In truth, I tell you: there will be a food shortage!*

So people run into the shops and there is a food shortage.

> S1_2: — *You see, I told you so!*
> S2: — *If you hadn't have caused the people to panic, there wouldn't have been a shortage.*

Self-fulfilling prophecies are close to manipulation:

> We are certainly going to war, so we must rearm and warn the population.
> ... Now we are the strongest, and our people are behind us. We can wage war.

2.8 Conversion of cause and effect.
The reversal of cause and effect is a form of refutation common in ordinary argument. Two facts **A** and **B** vary concomitantly. To account for this concomitance, some assert that there is a causal link from **A** to **B**, others claim a link from **B** to **A**. The protagonists defend the converse propositions, "**A** is the cause of **B**" and "**B** is the cause of **A**". Do we cry because we are sad? Or are we sad because we cry? Does aggression provoke fear? Or does fear result in aggression?

> L1: — *I am afraid of dogs, they can attack and bite!*
> L2: — *No, they attack because they see that you're scared.*

> L1: — *OK, I'm aggressive, that's because they persecute me!*
> L2: — *No, they persecute you because you're aggressive.*

In the first case, the affair originates with the dog and the supposed bully, in the second case with the self-claimed victim. It is said that single people are more likely to commit suicide than people with a partner. We might therefore ask whether single people have such problems because they are single, or whether they are single because they have such problems? This form of refutation by permutation of the cause and effect is simple and radical. It is worth noting, however, that it is not always possible to apply this process, as seen above, in the case of bees and pesticides.

This causal shift is particularly popular in ordinary causal argumentation. This play on permutation of terms illustrates the pervasiveness of language-based argumentation schemes. It is easier and more exciting to argue that politics determines morality or that morality determines politics, than to argue that there is no link, or very complex ones, between morality and politics, S. **Converse**.

2.9 Causality, subjectivity, responsibilities

The causal chain might be badly cut: the expression of causality as "**A** is the cause of **B**" is a potentially misleading simplification. Every cause is itself caused, except God, who is said to be his own cause and cause of all that ensues. The phenomenon designated as the cause can itself be constructed as the effect of a deeper cause, and its effects as new causes of new effects. We are therefore not dealing with a link between two terms, but with a real causal chain of potentially infinite length.

Consider the deadly events which took place in Sheffield on Sunday, April 16, 1988. They were extensively reported and commented on in the French press. On the following day, the front page of *L'Équipe* (a sports newspaper) read as follows:

> *The Horror!*
> Eighty-four people were killed on Saturday in Sheffield stadium, where the Liverpool-Nottingham FA Cup semi-final took place.

Typically, this kind of event causes anxiety which in turn stimulates the search for causal explanations. Readers will ask themselves "*Why? How can such things be possible?*" The same day, the headlines in *Le Figaro* newspaper (news and opinions) were:

> *Football: Why so many dead?*
> Four explanations for the drama:
> • The madness of supporters   • Police negligence
> • The age of the stadium   • Inadequate relief

The answers provided in the newspaper refer to a broad causality for the first question, and to a narrow causality for the others. The same day, the newspaper *Libération* (news and opinions) asserts a broad causality:

> *94 dead in Sheffield stadium*
> Deadly stadium

Crushed to death by the throng of supporters, victims who had come to see the Liverpool-Nottingham Forest football match made a dramatic tribute to the most popular sport in Thatcher's Britain.

Still the same day, *L'Humanité* newspaper (news and opinions) combines local causes and so-called deeper causes:

After the drama of Sheffield, Liverpool in mourning
The Last Stage of Horror
90 dead and at least 170 wounded, such is the appalling toll of the Hillsborough catastrophe. The vast majority of victims are children and young people from working class backgrounds who had come to support their teams. The age of the stadiums and their segregational character, and the hold that money has on the world of football are now in the dock. The destruction of industry and the resulting disorganization of leisure activities all have their share of responsibility in the transformation of sports into high-risk activities.

Examination of the causal chain mobilizes specialists in each of the areas of responsibility mentioned. Police officers and judges investigate narrow causalities, whilst sociologists, economists, politicians and historians discuss long-term causalities and responsibilities. In short, what is the cause? The fragility of the victims' rib cage, the poor quality of care to victims, the tardy response of the emergency services, the incompetence of the police services, the poor standard of the stadium, the financial greed of the organizers, the supporters' behavior, unemployment, social exclusion, the capitalist system...? To assign a cause is to assign responsibility and apportion blame and perhaps even bring shame upon the relevant parties. This case shows that causality functions as a discursive object@, S. **Cause — Effect.**

Moreover, the causal chains intermingle and combine into a "fabric of causes". Argumentation is based on this fabric, as "causal threads" are picked up and cut at a given point. This point determines the nature of the chosen cause attached to the salient problematic event considered as an "effect". The selection of a cause, correlatively, determines the responsible agent, person or institution, to blame or to praise. All the process depends on the interests and aims of the arguing party. The speaker fully projects his own subjectivity on the causal chain he or she has selected, and on the cause he or she has isolated. It would therefore be quite illusory to consider that ordinary arguments based on causal links are *ipso facto* more rigorous and less subjective than arguments based, for example, on analogy.

Cause To Effect Argumentation

Cause to effect argumentation is based on the existence of a cause-effect@ relation and the actual finding of the cause to culminate in to effect. Such argumentation is oriented towards the future:

Argument:
>> There is a state of affairs **c**.
>> This state of affairs **c** falls into the category of facts **C**.

Cause- Effect Rule:
>> There is a known causal law linking state of affairs **C** to state of affairs **E**.

Conclusion:
>> **C** will / must have an effect **e**, of type **E**.

The causal deduction allows prediction:
> This bridge is made of metal.
> When heated, this metal expands by a certain coefficient.
> In summer the bridge will expand by such and such amount.

This causal argument can be supplemented by a pragmatic@ argument.
> Such dilatation can have dangerous consequences: Expansion can twist metal.

Which must be prevented:
> It is therefore necessary to provide sufficient space for the bridge to expand.

Circumstances

Three forms of argumentation use the notion of *circumstance*:

— The *fallacy of omission of the relevant circumstances*, a criticism addressed to an argumentation.
— The *argumentation by the circumstances*.
— In the expression "circumstantial *ad hominem*", the circumstances alluded to are the characteristics of the person implicated in the *ad hominem* argument, S. *Ad hominem*.

1. Fallacy of omission of relevant circumstances

> The fallacy of omission of circumstances is sometimes referred to by the Latin label *secundum quid* fallacy, which abbreviates the phrase *a dicto secundum quid ad dictum simpliciter*, "from a restricted affirmation to an absolute affirmation".

Aristotle classifies the fallacy of omission of relevant circumstances as a kind of fallacy "independent of language" (*Soph.* 4; 165b20; S. **Fallacy**), occurring when an expression is used "absolutely or in a certain respect" (*Soph.* 5; 166b35):

> "If < what is not is the object of an opinion >, then < what is not is >" (*ibid.*; our emphasis and parenthesis).

"*What is not is the object of an opinion*" is a semantically complete, syntactically integrated utterance, a unique and complete speech act. All its components are necessary and interdependent; none can be subtracted without altering what the speaker said and meant, and he has only said *one* thing. It is not possible to extract from this complete utterance any arbitrarily chosen segment as long as it makes some sense, and attribute the resulting utterance to the speaker of the former statement. Such considerations are crucial when it comes to determining what is an elementary well-formed linguistic formula.

Other examples: the specified expression "**A** is (Place, Time)", "*A is here now*" can be reshaped into the corresponding, non-qualified, one "**A** is (Place)", "*A is here*". Vice versa, the non-specified construction "*Peter crossed the street*" cannot be specified into "*Peter crossed the street yesterday*" (which can be non fallaciously reduced to "*Peter crossed the street*").

This kind of de-contextualization of a qualified statement may result in irony, **S. Irony**:

    S1:   — *The weather is fine!* (said in the morning, when the weather is fine).

    S2:   — *Ah hah! And you said that the weather is fine!* (said in the evening, while it is raining).

This fallacy passes over relevant contextual data, treating as an absolute assertion what has been asserted with reservation, in a particular context, with precise reference and intention. This radicalization of assertions and positions makes them very easy to refute.

To be relevant in a methodologically equipped context, the refutation must relate exactly to the expression as used, and take into account all the reservations specifically mentioned. The fallacy is particularly vicious when it pretends that the speaker had fully said and fully assumed something he or she has only said, in the flux of a dispute, as a concession to the opponent.

    Prime Minister: — *Our country cannot take in all the misery of the world* (**P1**) *but it must take its share* (**P2**).

    Opponent : — *As Mr. Prime Minister said, we cannot welcome all the misery of the world.*

In Goffman's words, in statement **P1** the Prime Minister speaks as an *Animator*, quoting an unknown *Principal*, whom he opposes; whereas he speaks as the Principal of **S2**, taking full responsibility for the content and actions, intentions and consequences of what **S2** said, **S. Roles**. The opponent forces him to speak as *Principal of P1*. While the Prime Minister advocates *receiving refugees*, the opponent, who advocates *closing the frontiers*, makes an *ally* of the Prime minister who actually *rejected* his or her position.

2. Argumentation by the circumstances

Argumentation by the circumstances establishes indirectly the existence of a fact, exploiting peripheral, unnecessary indices of an action that have no real probative value, but nevertheless point to a fact:

    Question: — *Is he corrupt?*

    Accuser: — *Certainly. He needed money; we have seen him receiving thick envelopes; and yesterday, he bought a brand new car.*

In classical terms, the argumentation by circumstances can help to solve a conjectural cause, **S. Stasis**, such as "*did he commit this crime?*" (Cicero, *Top.*, XI, 50; p. 82). To answer, one "[looks] for the circumstances that preceded the fact, that accompanied it, that followed it" (Cicero, *ibid*; XI, 51, p. 83), interpreting "an appointment [...] the shadow of a body [...] pallor... and other indications of trouble and remorse" (*id.*, XI, 53, p. 83). This is part of the investigatory technique:

> "*He went out murmuring...*: this is to argue from what precedes the action; *we saw him stealing behind a bush...*: that's what accompanies it. [...] *a malicious joy, which he endeavored to keep concealed, appeared on his face, mixed with fright:* which is what follows."
> Bossuet [1677], p. 140, **S. Collections (III)**

These observed circumstances are probable signs, **S. Sign**. Argumentation by the circumstances is a powerful instrument in the arts of suspicion and construction of the culprit.

3. Terminological delicacies

On §53 of the *Topics* Cicero deals with arguments drawn from "consequences, antecedents, contradictory things [*ex consequentibus et antecedentibus et repugnantibus*]" (Top., XI, 53: 83). This paragraph deals with *logical* antecedence and consequence, involving semantically "necessary" links (*id.*), referring to questions of *a priori* and *a posteriori* reasoning, definition, rules of implication and to the principle of contradiction, **S. *A priori*, *A posteriori*; Definition: Cause; Implication; Deduction; Principle of contradiction.**

Bossuet speaks, in connection with the argument by circumstances, of places "derived from what precedes, from what accompanies and what follows [the action], *ab antecedentibus, ab adjunctis, a consequentibus*" ([1677], p.140). Here, the links of the preceding and following events with the central event are no longer semantical or logical but purely chronological (the change of preposition - *ex antecedentibus* for the logical consequence and the necessary link vs. *ab antecedentibus* for temporal anteriority has nothing to do with this distinction).

Collections (I) and Typologies of Arguments Schemes

The tradition has bequeathed us more or less systematized inventories of argument schemes, **S. Collections,** and a series of questions about them:

— About their nature and number,
— Lists of argument schemes have been compiled, and still are; but what is the unifying factor underlying these lists? Have they a proper systematic organization? Are they amenable to some elementary headings (Blair 2012, Chap. 12 and 13)?
— Where do they come from? Are they recurring remarkable stable structures picked up in (successful) argumentative discourses of all kinds? Or are they construed from the a priori categories of the human mind?
— Are they logical, cultural or anthropological beings? Are they culture-dependent?
— What kind of historical change, if any, can affect the topics? The question arises, when the 19 "forms of reasoning" of Toulmin, Rieke & Janik are com-

pared with the Ciceronian and post-Ciceronian lists of topoi, S. **Collections (I)** and **(II)**.

1. Categorization of arguments: typologies and collections

A *class* is a set of beings; basically a *typology* is a class subdivided into various subtypes; the same class can admit organized different subtypes, S. **Taxonomies and Categories**. A *catalog* can be considered as a single-level typology.

A typology of *arguments* is a set of topics or argument schemes linking the argument to the conclusion. Typologies of arguments include from ten to several dozens argument types, S. **Topic; Enthymeme; Typologies (I) to (IV)**.

To categorize a speech segment (an individual, level 0) as a "pragmatic argument" is the process by which the characteristic features that define the pragmatic argument are recognized in this segment. This operation is itself argumentative, and obeys the rules of argumentation by definition. S. **Nomination; Definition; Scheme**.

The idea of argument type, the possibility of drawing up inventories of these types, and giving an internal structure to these inventories, in order to build a "typology of topics", is the very foundation of the theory of rhetorical argumentation. Walter Ong sees these typologies of arguments as engaged in a perpetual movement of renewal and attempt at redefining:

> As the general intellectual tradition changes, the active associative nodes for ideas change, and classification changes too. Revising the tradition has been a common phenomenon in antiquity, when Aristotle differed from the sophists in the list of topics he proposed, Cicero from Aristotle, Quintilian from Cicero, Themistius from all these, and Boethius from all of them again and from Themistius as well. The revision continues in our day with Professor Mortimer Adler's "Great ideas" (augmented beyond their original hundred), and with such articles as Père Gardeil's very helpful study of the *lieux communs* in the *Dictionnaire de théologie catholique*, where, after reporting Melchior Cano's description of the loci (which he notes are taken at times verbatim from Agricola) and Cano's organization of theological loci, Gardeil proposes, in true topical tradition, a still better classification of his own. (Ong 1958, p. 122)

There are many lessons to be learnt from this passage. First it provides us with a definition of topics as "active associative nodes for ideas", as theorized since the birth of rhetoric in the context of the theory of argumentation in discourse. Yet the particular interest of this passage lies in the description of the taxonomic trap. To bring the irritating proliferation of typologies to an end, one might be tempted to propose a new and final one, thus bringing everyone into agreement — but, in the end, it appears that an additional typology has been added to an already overloaded list, aggravating the very evil, which it claimed to remedy. This observation can be read as an ironic historical counterpoint to the works that, in that year, 1958, were reviving reflection on topics and arguments.

2. Place of typologies in the theories of the argumentation

The question of argument schemes plays a key role in some argumentation theories whilst in other schemes it is either re-defined, or plays only a marginal role.

(i) The question of argument types does not arise in Anscombre and Ducrot's theory of Argumentation within Language. The concept of topos is defined as a semantic link between predicates. It follows that the number of topoi is extremely large, uncountable even, while classical theories enumerate less than one hundred topoi, **S. Collections.**

(ii) Grize's "Natural Logic" is based on the concept of schematization@. The operations of "reasoned organization", or "shoring" amounts, in substance, to the classical concept of a conclusion supported by an argument. The types of arguments correspond to types of scaffolding. To my knowledge, this line is not further developed. Grize focuses on inference, causality, explanation.

(iii) In Toulmin's terminology, a type of *warrant* corresponds to a type of argument, as shown by Ehninger and Brockriede ([1960]). Moreover, Toulmin, Rieke and Janik (1984) proposed a brief collection of arguments, **S. Collections (iv)**. The example illustrating Toulmin's "layout@ of argument" corresponds to a very productive topic, the categorization@ of an individual in a category.

(iv) The concept is central to the *New Rhetoric* of Perelman & Olbrechts-Tyteca, as well as for Pragma-Dialectic and Informal Logic, **S. Collections (IV).**

3. Wealth of the typologies: number of argument types

Classic lists of argument schemes tend to propose a relatively large number or argument schemes. The *Rhetoric* of Aristotle offers a set of twenty-eight schemes, plus some "lines of argument that form the spurious enthymemes" (*Rhet.*, II, 24; RR, p. 379; **S. Fallacy**), plus some rules taken from the *Topics*. Cicero's *Topica* lists a dozen of schemes, and Quintilian's *Institutio Oratoria* twenty-five. Boethius passed fifteen forms on to the Middle, **S. Collections (II).**

The Dupleix's *Logic* (1607) and Bossuet's *Logic* (1677), can probably be considered as representative, in modern times, of this classic tradition. The former retains fourteen schemes and the latter twenty schemes.

Other modern typologies are quite divergent: Locke [1690] proposes a typology — if it can be considered as such — consisting of four elements to which Leibniz [1765] adds one. Locke's scientific world is, however, extremely different from, and antagonistic to the rhetorical world of the classics.

Bentham enumerates thirty-one argumentative formulas for the field of political argumentation, **S. Collections (III).**

In contemporary times, Conley counts "more than eighty different argument types" in Perelman & Olbrechts-Tyteca *Treatise* (Conley 1984, p. 180-181) **S. Collections (IV).**

4. Forms of the typologies

In the *Rhetoric*, Aristotle presents a catalogue of twenty-eight topoi listed completely randomly, **S. Collections (II)**.

Perelman & Olbrechts-Tyteca have constructed a clearly organized four-level typology of the various "techniques of argumentation"

— A speech segment (an individual, level 0) can, for example be categorized as a "pragmatic argument"; that is, this segment presents the essential features that define the pragmatic argument (level 1).

— Level 1 arguments are grouped within a super-category; for example, a "pragmatic argument" is classified as an "argument based on the structure of reality" (level 2).

— Level 2 arguments are grouped in the class of the "techniques of association", (level 3), one of the two kinds of "techniques of argumentation" (Level 4, top level), **S. Collections (IV)**.

5. Foundations of typologies

The typologies of argument schemes can be organized in different ways.

(i) From the perspective of their *contribution to the growth of scientific knowledge*, *inconclusive* arguments are opposed to *compelling* arguments. The latter are, in modern times, generally equated with mathematical demonstration and scientific evidence. In the words of Locke, they "bring true instruction with [them] and advance us on our way to knowledge" (Locke [1690], Chap. 17, § 19-22), **S. Collections (III)**. Person-centered arguments are, from this point of view, irrelevant. The same might be said of those that play only on the guiles of natural language and the nuances of interpersonal relationship.

(ii) From the perspective of their *linguistic functioning*, *metonymic* arguments based on a relationship of contiguity, can be distinguished from the *metaphoric* arguments based on a relationship of similarity. This distinction mirrors the opposition between the arguments "establishing the structure of reality" (analogy type) and those "based on the structure of reality' (causal type) (Perelman & Olbrechts-Tyteca [1958] p. 261; 350). **S. Metonymy; Metaphor; Collections (IV)**.

(iii) From the point of view of their productivity. The productivity of an argument scheme depends on the number of actual arguments (enthymemes) derived therefrom. Intuitively, some topics are very productive. One might think for example of those based on the twin argument schemes by categorization@ and definition; or arguments based on causal or analogical relations, or from the contraries, etc. Others, including the argument from sacrifice@ are less productive. Other argument schemes are apparently, no longer in use, such as the argumentative exploitation of syzygies@.

(iv) From the point of view of their legitimating power. A good example of organizing topical forms according to their strength is given by the hierarchy of legal and theological arguments in the Arab-Muslim culture and religion, such as proposed by Khallaf ([1942]). He distinguishes between ten sources, ordered according to their degree of legitimacy. The most legitimate forms are those based on the Quran and the Tradition. Those that have the weakest degree of legitimacy are, on the one hand, "the laws of monotheistic peoples", and, on the other hand, perhaps quite surprisingly given the situation in 2017, "the opinions of the Prophet's companions", in that order. In other words, the argument put forward at the time of the origin of Islam is granted the *smallest* possible weight in the hierarchy of arguments. Such was the situation in 1942; it has undergone significant change with the rise of Salafism.

Collections (II): From Aristotle to Boethius

1. Aristotle, *Rhetoric* (between 329 & 323 b.c)

1.1 The catalog and its position in the system of Aristotelian proofs

The catalog of the *Rhetoric* must be viewed within the framework of the Aristotelian typology of the different types of reasoning carried by different types of discourses. In this typology of proofs, rhetorical discourse is opposed to dialectical dialogue and to scientific (syllogistic) discourse. Tricot points out that "*syllogism* is the genre, *scientific* (producer of science) [is] the specific difference that separates the scientific demonstration from the dialectical and rhetorical syllogisms" (*S. A.*, I, 2, 15-25; p. 8, note 3). The concept of persuasion in the *Rhetoric* must be seen in this context: scientific discourse produces *apodictic* knowledge, dialectical interaction produces *probable* truth and rhetorical syllogism or enthymeme is an element of *persuasive* discourse. Thus, by its very definition, rhetorical discourse cannot be probative; in short, the phrase "rhetorical evidence persuades" is a pleonasm.

The catalog of arguments is situated as follows in the sub-typology organizing the rhetorical proofs (proof = *pistis*, "means of pressure").

1.2 Wavering distinctions

Aristotle establishes the following distinctions between the various kinds of rhetorical proof:

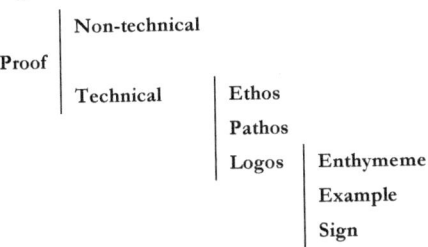

The proofs attached to the logos are *enthymemes*@, which correspond to deduction; *examples*@, which corresponds to induction; and arguments based on *natu-*

ral@ *signs*, that are probable or certain. Enthymeme and example are said to be common to the three ancient rhetorical genres (epideictic, deliberative, judicial, **S. Rhetoric**.) But the articulation of these different kinds of proofs, and the consistency of the text of the *Rhetoric* such as we read it now, is problematic (McAdon 2003, 2004). The classification of proofs attached to logos has important variants:

> (a) "I call an *enthymeme* a rhetorical syllogism, and an example rhetorical induction. Now all orators produce belief by employing as proofs either examples or enthymemes, and nothing else." (*Rhet.*, I, 2, 8; Fr., p. 19)
>
> (b) "the materials from which the enthymemes are derived […] being probabilities and signs […]." (*Ibid* I, 2, 14; p. 25)
>
> (c) "Now the material of enthymemes is derived from four sources — probabilities, examples, necessary signs and signs." (*Ibid* II, 15, 8; p. 337)

The example is placed on the same level as the enthymeme in (a), but is considered a form of enthymeme in (c); enthymemes have four sources in (c), and two in (b). Thus, it would be risky to look for a rigorous system in these presentations of rhetorical proof, and the above table must be considered as a simple reminder.

1.3 The topics of the *Rhetoric*

The *Rhetoric* enumerates twenty-eight *topics* or "lines of argument" (*Rhet*, II, 23), as listed in the following table. An *enthymeme* is an instance of a topic, **S. Enthymeme**.

They are designated by their English label, when available, or by a short description, both quoted from Freese (F) or Rhys Roberts (RR).

1. "From opposites" (F). **S. Contraries**
2. "From similar inflexions" (F). **S. Derived Words**
3. "From relative terms" (F);"upon correlative ideas" (R). **S. Correlative Terms**
4. "From the more or less" (F); *a fortiori* (R). **S.** *A fortiori*
5. "The consideration of time" (F). **S. Consistency**
6. "Turning upon the opponent what has been said against ourselves" (F). **S. Ethos;** *A fortiori.*
7. "From definition" (F). **S. Definition**
8. "Topic from the different significations of a word" (F). Aristotle explicitly refers this topic in his *Topics*. **S. Homonymy; Ambiguity.**
9. "From division" (F). **S. Case-by-case**
10. "From induction" (F). **S. Induction**
11. "From a previous judgment in regard to the same or a similar or contrary matter", this judgment having been given by one of "those whose judgment it is not possible to contradict" (F). **S. Precedent;** *Ab exemplo*; **Authority; Modesty; Politeness**
12. "From enumerating the parts" (F). **S. Case-by-case**

13. "Since in most human affairs the same thing is accompanied by some bad or good result, [...] employing the consequences to exhort or dissuade, accuse or defend, praise or blame" (F). **S. Pragmatic; Dilemma**
14. *[id. 13]*, "but there is this difference that in the former case *[i.e., 13]* things of any kind whatever, in the latter *[i.e., 13]* opposites" (F). **S. Pragmatic; Dilemma**
15. "Men do not praise the same thing in public and in secret" (F). **S. Motives**
16. "From analogy in things" (F). **S. Analogy; Opposites.**
17. "Concluding the identity of precedents from the identity of results" Instance: "There is as much impiety in asserting that the gods are born as in saying that they die; for either way the result is that at some time or other they did not exist" (F). **S. Consequence; Implication.**
18. "The same men do not always choose the same thing before and after but the contrary" (F). **S. Consistency.**
19. "Maintaining that the cause of something which is or has been is something which would generally, or possibly might be the cause of it; for example, if one were to make a present of something to another, in order to cause him pain by depriving him of it" (F). **S. Motives**
20. "Examining what is hortatory and dissuasive, and the reasons which make men act or not" (F). **S. Motives**
21. "Things which are thought to happen but are incredible" (F). **S. Probable.**
22. "Another line of argument is to refute your opponent's case by noting any contrast or contradiction of dates, acts or words that it anywhere displays" (RR). **S. Contradiction; Consistency;** *Ad hominem*.
23. "Another topic, when men or things have been attacked by slander [...] consists in stating the reason for the false opinion" (F). **S. Motives; Interpretation**
24. "Another topic is derived from the cause. If the cause exists, the effect exists; if the cause does not exist, the effect does not exist" (F). **S. Motives**
25. "Whether there was or is another better course than that which is advised, or is being, or has been carried out" (F). **S. Consistency; Motives**
26. "Another topic, when something contrary to what has already been done is on the point of being done, consists in examining them together" (F). **S. Consistency**
27. "Another topic consists in making use of errors committed for purposes of accusation or defense" (F). **S. Contradiction; Consistency.**
28. "From the meaning of a name" (F). **S. Proper Name**

Even if no clear order emerges from this enumeration, it can be noted that an important subset of topics deal basically with the world of human action and its determination, where motives have been substituted for causes, and behavioral stereotypes on human nature and human motivations have replaced strict scientific causality and taxonomies.

2. Cicero, *Topica*, "Topics" (44 b. c.)

Cicero proposes a typology of arguments in an early work, *De Inventione*, "On Invention" and in his latest book on argument, *Topica*, "Topics". Unlike the *Topics* of Aristotle, which exposes a method of finding and criticizing arguments in the context of a dialectical philosophical exchange, Cicero's observations and examples constantly refer to rhetoric as a judicial practice. In this context, Cicero proposes the following distinction:

(i) *Intrinsic* arguments, either "inherent in the very nature of the subject which is under discussion" or "closely connected with the subject which is investigated" (*Top.*, I, 8; p. 387-389).

(ii) Arguments taken "from external circumstances", or "*extrinsic* arguments" (*Top.*, II, 8; p. 388; IV, 24, p. 397), corresponding to the so-called non-technical@ arguments, mainly testimonies and their conditions of validity, and including authority (*Top.*, IV, 24; p. 397).

Objects and facts are built and discussed on the basis of arguments drawn from five main sources.

- **Definition**

    Arguments — by genus and species of the genus (*a genere*; *a forma generis*).
    — by enumeration of the parts (*partium enumeratio*)
    — from "etymology" (*ex notatione*)
    — from words of the same family (*a conjugata*)
    — "based on difference" (*a differentia*).

    S. **Categorization and Nomination; Definition; Genus; Case-by-case; Etymology; Derived Words**

- **Causal relations**, S. **Causality**

    Arguments — from efficient causes (*ab efficientibus causis*)
    — from effects (*ab effectis*).

- **Analogy** (*a similitudine*). S. **Analogy**

- **Opposites** (*ex contrario*). S. **Opposites;** *A contrario*.

- **Circumstances@**

    Arguments — from antecedents, *ab antecedentibus*,
    — from consequents, *a consequentibus*,

This brief and articulated list of arguments is all important in the Western tradition of argumentation studies. They were transmitted in the Middle Ages by Boethius (around 480-524) *On Topical Differences* (*Top.*, c. 522), and were taken up by medieval logic, dialectic and philosophy. They remained in use until well into the modern era, S. **Collections (III)**.

3. Quintilian, *Institutio Oratoria*, "The Orator's Education" (c. 95)

In Book V, Chap. 10 of the *Institutes of Oratory*, dealing with arguments, Quintilian summarizes a list of 24 argumentative lines (*IO*, V, 10, 94). A first series deals with substantial topics, S. **Substantial Topics**.

A second series is a catalog of argument types: the French translator, J. Cousin, notes that "this list-summary, which seems to be a loan, recalls previous classifications, with their elements arranged in a different order: [...] Later rhetoricians condense or develop without apparent reason" (1976, p. 240).

Collections (III): Modernity and Tradition

1. Scipion Dupleix, *Logic, or the art of speaking and Thinking* (1607) Jacques-Bénigne Bossuet, *Logic for the Dauphin* (1677)

These works most probably have no particular historical importance, yet they certainly provide an idea of seventeenth century terminology, clearly akin to the Ciceronian system, S. **Collections (II)**.

As the title suggests, Bossuet's *Logic* functions as a pedagogical guide to everyday *argumentation*: 'Dauphin' was the title given to the heir of the French Kingdom.

Table:
— First column, Bossuet, 1677
— Second column, Dupleix, 1607
The order of the lines is that of Bossuet. To facilitate reading, the order of Dupleix was changed, so that the same types of arguments are on the same line; the numbering corresponds to the order in Dupleix's typology.

Bossuet, 1677	Dupleix, 1607
1. Etymology	3. Etymology
2. Conjugates	4. *Conjugata*
3. Definition	1. Definition
4. Division	
5. Genus	5. Genus and Species
6. Species	
7. Property	
8. Accident	
9. Resemblance	6. Similitude,
10. Dissemblance	7. Dissimilitude
11. Cause	13. Cause
12. Effects	14. Effects
13. What comes before[1]	10. Antecedents[1]
14. What accompanies[1]	9. Adjuncts or conjuncts[1]
15. What follows[1]	11. Consequents[1]
16. Contraries	8. Contraries
17. *A repugnantibus*[3]	
	12. Repugnants
18. All and parts[2]	2. Enumeration of the parts[2]
19. Comparison	15. Comparison with things bigger, equal and smaller
20. Example, or Induction	

(¹) S. Circumstances

(²) Bossuet's topic n°18, "enumeration of the parts" is akin to the topic of definition. For example, what is a "good captain" is defined by enumeration of his relevant qualities: brave, wise, etc. Dupleix's topic n°2, "all and parts" relates more to composition and division

(³) Dupleix's topic n°12, from "repugnants" refers to predication: "stone" and "man" are *repugnant* because " — *be a stone*" cannot be said of *man*; whereas Bossuet's topic n°17, "*a repugnantibus*", refers to a kind of ad hominem@.

Both typologies prioritize arguments exploiting the resources contributing to the definition@ of a word or a concept, in view of their exploitation in syllogistic reasoning, **S. Categorization, Taxonomies**. This enumeration of the core set of arguments is followed by the usual enumeration of arguments schemes drawing on causality, analogy, comparison, peripheral circumstances, opposites and induction. This set will reemerge under a new re-organization in the *New Rhetoric*.

2. John Locke, *An Essay concerning Human Understanding* (1690)
Wilhelm Leibniz, *New Essays Concerning Human Understanding* (1765)

In *An Essay concerning Human Understanding* John Locke briefly mentions "four sorts of arguments, that men, in their reasoning with others, do ordinarily make use of to prevail on their assent; or at least so to awe them as to silence their opposition" (IV, 17, "Of Reason", § 19-22; p. 410). These four arguments are:

— *ad verecundiam*, **S. Ethos. S. Modesty; Authority.**
— *ad ignorantiam*, **S. Ignorance.**
— *ad hominem*, **S. *Ad hominem*.**
— *ad judicium*, **S. Matter**

In his *New Essays Concerning Human Understanding*, Leibniz comments on this list, and qualifies Locke's abrupt and general condemnation by taking into consideration the circumstances; see the above mentioned entries. In addition, Leibniz adds a new kind of argument, the argument *ad vertiginem*, **S. Vertigo.**

This brief list has nothing to do with the previous Ciceronian ones; its aim is to oppose the first three fallacious arguments to the last one, the only one to "bring true instruction with it, and advance us in our way to knowledge" (*op. cit.*, p. 411). Reasoning and the methods used in mathematics and experimental sciences are introduced under the heading *ad judicium*. Contrary to the classical typologies, these arguments are not associated to a logic itself backed by a natural ontology, but rather to the requirements of scientific method, **S. Fallacy.** We are thus entering a new argumentative world.

3. Jeremy Bentham, The Book of fallacies (1824) S. Political Arguments.

Collections (IV) : Contemporary Innovations and Structurations

1. Chaïm Perelman & Lucie Olbrechts-Tyteca, *A Treatise on Argumentation — The New Rhetoric*, 1958

In the *New Rhetoric — A Treatise on Argumentation* (1958), Perelman & Olbrechts-Tyteca propose a sophisticated typology of arguments. Some twenty years later, in *The Rhetorical Empire* [*L'Empire Rhétorique*, 1977], Perelman takes up the essential elements of the 1958 typology, making some significant simplifications. In *Juridical Logic* [*Logique Juridique*, 1979] he presents a specific set of juridical arguments, **S. Juridical Arguments**.

1.1 The typology of the *Treatise*

According to Conley, the *Treatise* contains "more than eighty different forms of argumentation, and illuminating remarks on more than sixty-five figures" (1984, p. 180-181), and contrasts these achievements with "Toulmin's renegade logic" (*ibid.*).

The "forms of argumentation" are described in the third part of the *Treatise*, entitled "Techniques of argumentation". They are presented as a set of "*association techniques*", (Chap. 1 to 3), along with two other kinds of technique, the "*dissociation technique*" (Chap. 4), and the "Interaction of arguments" (Chap. 5). This latter Chapter exposes a set of *disposition techniques*, and discusses the relative persuasive effects of the various arrangements of arguments in a speech, that is, issues in classical "*dispositio*".

1.2 The association techniques

The *association techniques* correspond to the classical argument schemes. They are classified under three categories:

    Chap. 1. Quasi-logical arguments
    Chap. 2. Arguments based on the structure of reality
    Chap. 3. The relations establishing the structure of reality

"Quasi-logical arguments" (§46-59)

This category lists arguments which "lay claim to a certain power of conviction in the degree that they claim to be similar to the formal reasoning of logic or mathematics" (p. 192); this definition should be brought closer to the definition of a fallacious argument as "one that *seems to be valid* but *is not* so." (Hamblin 1970, p. 12), **S. Fallacies**. The category covers the following argument schemes:

    §46-49 Contradiction and incompatibility
    §50 Identity and definition
    §51 Analyticity, analysis and tautology
    §52 The rule of justice
    §53 Arguments of reciprocity
    §54 Arguments by transitivity
    §55 Inclusions of the part in the whole
    §56 Division of the whole into its parts

Collections (IV) : Contemporary Innovations and Structurations

§57 Arguments by comparison
§58 Argumentation by sacrifice
§59 Probabilities

In *The Rhetorical Empire*, the Chapter on "Quasi-Logical Arguments" essentially recapitulates the class as presented in the *Treatise*.

"Arguments based on the structure of reality" (§60-77)

The broad label "argument based on the structure of reality" may be interpreted as referring to arguments which exploit *syntagmatic*, or *metonymic@* relations. This category in fact lists arguments "alleged to be in agreement with the very nature of things" (p. 191); these arguments "make use of [the structure of reality] to establish a solidarity between accepted judgments and others which one wishes to promote" (p. 261). The "causal link" and the "relation of succession" are fundamental to this category.

Arguments within this category include:

§61-63 "Causal link", "Pragmatic argument"

§63-73 discuss arguments where the person is considered to be a causal agent, such as:

§64-68 "Ends and means", among which:
§65 "Argument of waste"
§66 "The Argument of direction"
§68-73 "The Person and his acts", among which:
§70 "Argument from authority"
§73 "The Group and its members"

§74-75 extend the notion of "relation of coexistence" to:
§74 "Act and essence"
§75 "The symbolic relation"

§76-77 present "more complex", second level arguments:
§74 "Double hierarchy"
§75 "Differences of degree and of order"

The Rhetorical Empire, Chapter VIII, recapitulates the same class of *arguments based on the structure of reality* under different groupings:
— Relations of succession
— Relations of coexistence
— The Symbolic relation, the double hierarchy argument, argument about the differences of order.

"Relations establishing the structure of reality" §78-88

The inclusive label "Relations establishing the structure of reality" might be interpreted as referring to a set of arguments exploiting *paradigmatic* or *metaphoric@* relations. This category of relations is defined on the basis of two of its prototypical members, arguments from "the particular case", and "arguments by analogy". The following argument schemes come under this category:

§78 "Argumentation by example"
§79 "Illustration"
§80-81 "Model and anti-model"
§82-87, On analogy
§87-88, On metaphor.

In *Rhetorical Empire*, the title "establishing the structure of reality" is not retained; its contents are grouped under two distinct chapters:
Chap. IX, Arguments by example, illustration and model
Chap. X, Analogy and metaphor

This can be construed as a waiver of the distinction between arguments "*establishing*" the structure of reality, and those "*based on*" the structure of reality. It might, however, also be argued that this couple of concepts does not characterize causal arguments in opposition to analogical ones, but indeed applies to *both* argument schemes. The successful use of an argument "*based on*" authority, for example, presupposes that the invoked authority has been previously "*established*". This distinction is especially helpful in the case of arguments from authority@, definition@, causality@ and analogy@.

1.3 The dissociation techniques.
The basic difference between *association* and *dissociation* techniques is that the former operate on *judgments*; they "establish a solidarity between accepted judgments and others which one wishes to promote" (p. 261); they correspond to argument schemes. In contrast, *dissociation* techniques operate on "*concepts*" (p. 411; my emphasis): "[they] are mainly characterized by the modifications which they introduce into notions, since they aim less at using the accepted language than at moving towards a new formulation" (p. 191-192), **S. Dissociation,** *Distingue*; **Persuasive Definition.**
The two terms of the opposition association / dissociation are thus of a very different nature.

2. Toulmin, Rieke, Janik, *An introduction to reasoning* (1984)
Toulmin, Rieke, Janik consider nine "forms of reasoning" (1984, p. 199)

from analogy	from generalization
from sign	from cause
from authority	from dilemma
from classification	from opposites
from degree	

3. Kienpointner, *Alltagslogik* [Everyday Logic] 1992.
Kienpointner (1992, p. 231-402) synthetizes six contemporary typologies (Perelman, Olbrechts-Tyteca [1958]; Toulmin, Rieke, Janik 1984; Govier 1987; Schellens 1987; van Eemeren, Kruiger 1987; Benoit, Lindsey 1987), summarized in the following table (1992, p. 246):

3.1 Rule-using argument schemes

Classificatory Schemes	Definition
	Genus - Species
	Part - Whole
Comparison Schemes	Equivalence
	Resemblance
	Difference
	A fortiori
Opposition Schemes	Contradictories
	Contraries
	Relative terms
	Incompatibility
Causal Schemes	Cause – Effect
	Consequences
	Reason
	Means - End

3.2 Rule-establishing argument schemes

Argumentation by example
Inductive argumentation

3.3 Other schemes

Argument by example, illustrative argumentation
Arg. by analogy
Arg. by authority

4. Douglas Walton, Chris Reed, Fabrizio Macagno, *Argumentation schemes*, 2008.

Walton, Reed and Macagno present an extensive and exhaustive investigation including "a user's compendium of argumentation schemes" (2008, p. 308-346). The schemes are consistently designated as *argument* schemes, with the exception of (19), (20), (21), referred to as *argumentation* from values, from sacrifice, from the group and its members.

The following list mentions only the main schemes; they may include subtypes.

4.1 Authorities: position, expertise, testimony, number (p. 309-314)

1. Argument from position to know
2. Arg. from expert opinion
3. Arg. from witness testimony
4. Arg. from popular opinion, *ad populum*
5. Arg. from popular practice.

Arguments (4) are drawn from what people generally believe, whereas arguments (5) refer to what people generally do.

Collections (IV): Contemporary Innovations and Structurations

4.2 Example, analogy (p. 315-316)
  6. Argument from example
  7. Arg. from analogy
  8. Practical reasoning from analogy

Arguments (7) concern beliefs; arguments (8) concern ways to do things.

4.3 Composition and division (p. 316-317)
  9. Argument from composition
  10. Arg. from division

4.4 Negation, opposition (p. 317-318)
  11. Arg. from opposition (contradictory, contrary, converse, incompatible)
  12. Rhetorical argument from opposition

Negation-based argumentation schemes can be logically valid or not; they are frequently not well defined.

4.5 Alternative (p. 318-319)
  13. Arg. from alternatives

This scheme concludes with the elimination of a member of an alternative due to the requirement of the other member. It corresponds to a case-by-case argument between two cases.

4.6 Classification (p. 319-320)
  14. Arg. from verbal classification
  "for all **x**, if **x** has property **F**, then **x** can be classified as having property **G**."

Set **F** is included in set **G**.

  15. Arg. from definition to verbal classification

If an individual *a* is defined (categorized) as a **D**, and if **D**s generally have property **P**, then *a* has property **P**.

  16. Arg. from vagueness of a verbal classification
  17. Arg. from arbitrariness of a verbal classification

Schemes 16. and 17. conclude with the rejection of an argument as "too vague" or "too arbitrarily defined" in some aspects. These cases can also be seen as an application of Grice's Cooperation Principle.

4.7 Persons, values, actions and sacrifice (p. 321-327)
  18. Argument from interaction of act and person
  19. Arg. from values
  20. Arg. from sacrifice
  21. Arg. from the group and its members

These schemes consider a group whose members are supposed to share quality **Q**, and attribute this quality to any member of the group. A member of a racist association can legitimately be supposed to be racist.

Not all characteristics of its members can be composed and attributed to the group as such; a large set is not necessarily composed of large elements.

> 22. Practical reasoning
> 23. Two-person practical reasoning

If one pursues an end, then one must accept the means and steps necessary to attain it.

> 24. Argument from waste
> 25. Arg. from sunk costs

Pages 10-11 (*id.*) consider as synonyms the labels *argument from waste*, (with reference to Perelman & Olbrechts-Tyteca), and *argument from sunk costs*. Nonetheless, they are discussed here as two separate entries.

4.8 Ignorance (p. 327-328)

> 26. Arg. from ignorance
> 27. Epistemic argument from ignorance

This argument covers the case "*if it were true, the newspapers would certainly speak of it*" (*id.*, p. 99)

4.9 Cause, effect; abduction; consequence (p. 328-333)

> 28. Argument from cause to effect
> 29. Arg. from correlation to cause
> 30. Argument from sign
>
> 31. Abductive argumentation scheme
> 32. Argument from evidence to a hypothesis
>
> 33. Arg. from consequences
> 34. Pragmatic argument from alternatives

Scheme (34) is a special case of (33), the choice is between doing/not doing something and suffering/not suffering negative consequences.

4.10 Arguments from threat, fear, danger (p. 333-335)

> 35. Argument from threat
> 36. Arg. from fear appeal
> 37. Arg. from danger appeal

Schemes (35), (36), (37) schematize different strategies of fear.

> 38. Arg. from need for help
> 39. Arg. from distress

4.11 Commitments, ethos, *ad hominem* (p. 335-339)

> 40. Arg. from commitment
> 41. Ethotic argument

Collections (IV) : Contemporary Innovations and Structurations

    42. Generic *ad hominem*

    43. Pragmatic inconsistency
    44. Argument from inconsistent commitment
    45. Circumstantial *ad hominem*

Scheme (44) draws a distinction between *committed* and *not really* so. Schemes (43) and (45) express forms of contradictions between personal commitments and actions.

    46. Argument from bias
    47. Bias *ad hominem*

Schemes (46) and (47) are closely related. According to (46), argument from bias: "L *is biased, so the conclusions are suspect*". According to (47), "bias *ad hominem*": "L is biased, so I do not trust him". Biases are relative to a domain, but it is convenient to consider that the whole personality is biased; L has a "false mind".

4.12 Gradualism; slippery slope (p. 339-341)

    48. Argument from gradualism

The comments (*id.* p. 114-115), show that this scheme can be likened to the slippery slope forms, (49) to (53). It expresses the sorites@ paradox, also mentioned in (52): "*If you remove a grain from a pile of grains, you always have a heap; if you remove another grain, you still have a heap ... up to what extent?*"

    49. Slippery slope argument

    50. Precedent slippery slope argument

The slippery slope argument is used to reject an exceptional treatment, on the ground that this exception would open a line of precedents leading to something unacceptable.

    51. Sorites slippery slope argument
    52. Verbal slippery slope argument

The slippery slope principle is used to reject the assignment of a property to an object because this property is transmitted by contiguity up to an object that obviously does not or should not possess it. This is a variety of argument to the absurd, based on a demonstration by recurrence.

    53. Full slippery slope argument

4.13 Rules, exceptions, precedent (p. 342-345)

    54. Argument for constitutive-rule claim

Scheme (54) relates to rules of language (synonymy) and to principles of categorization in institutionally codified languages ("*D* counts as *W*").

    55. Arg. from rules

    56. Arg. for an exceptional case

57. Arg. from precedent
58. Arg. from plea for excuse

Confronted with an exceptional case, one can waive the usual rule (56) or change it (57). Excuses and extenuating circumstances can suspend the rule.

4.14 Perception, memory (345-346)
59. Arg. from perception
60. Arg. from memory

Scheme (59), (60) argue that one can reasonably believe in a given fact on the basis of the perception or memory of the said fact.

Common Place

The expression *commonplace* corresponds to the Latin *locus communis*, which translates the Greek *topos*.

— Often reduced to *place* (*locus*, pl. *loci*), an *inferential common place* is an inferential topic, or argument scheme@.

— A substantial common place is an endoxon, *a formulary expression of a common thought*. Traditional rhetorical invention specialized in the argumentative use of substantial common places.

1. Topical questions: An ontology for doxa-based argumentation

Everyday argumentation is based on an ontology organizing the world of events according to the following broad parameters:

Person, Action, Time, Place, Manner, Cause or Reason.

These dimensions mirror the system of sentence complementation:

Yesterday,	in Philadelphia,	with great difficulty,
Time	Place	Manner
Peter	met Paul	to settle their business
Focus person	Action	Cause, Reason

The corresponding interrogative words guide the methodical procedure to follow in order to gather and organize information about an event:

Who? What? When? Where? How? Why?

[Interrogative words] have already been recognized in various languages for different purposes: for speculative purposes, in the Latin of the scholastics: *cur?, quomodo?, quando?* [why? how? when?]; or for military purposes in German, where the tetralogy *Wer? Wo? Wann? Wie?* is taught to all military recruits as an information framework that any scout on a reconnaissance mission must be capable of providing and reporting back to his superiors. (Tesnière 1959, p. 194)

These common basic dimensions of reality are rubric or "heads of chapters", generating more or less general ideas and formulas. Their application is extremely general. They might be used to frame a description or narration of any kind, a scout report, a newspaper article, or an event-based essay. Such questions also guide moral evaluation, for example an action such as "having carnal intercourse" will be evaluated as shameful if that if was "with forbidden persons" (*With Whom?*), or "at wrong times" (*When?*) or "in wrong place" (*When?*) (Aristotle, 1383b 15-20; RR p. 279).

When attached to a particular field, these ontological parameters are expressed using words which have *a full lexical meaning*. For example, the classical guide to political decisions includes questions such as: "*Honorable? Will the proposed measure turn out honorable, or embarrassing for us?* **S. Political Arguments**.

These questions governing the quest for information about a given issue or event, form the very foundation of rhetorical argumentation. They might be answered *a posteriori*, that is after a full documented inquiry into the specificities of the case. They can also be answered *a priori*, on the basis of endoxa, that is pre-conceived ideas. The undue prominence given to stereotyped ideas in the construction of arguments, leads to the strong and indignant criticism of rhetoric as a fallacious verbiage@, **S. Ornamental**

2. The method: stereotyped portrait-based argumentation

Consider the argumentative question "*Has Mr. So and So committed this hideous murder?*"

— The question *Who?* is applied to the defendant: "*Who is this Mr. So-and-So?*". The sub-topos *Which nation?* provides the categorizing@ information: "*Mr. So-and-so is Syldavian*", and likewise for all questions parameterizing the topical person.

— Endoxon on the Syldavians: to the category Syldavian is attached a set of defining endoxical predicates such as "*the Syldavians are like that*", each having a specific argumentative orientation:

    the Syldavians are peaceful / bloodthirsty people.

These predicates provide an endoxic encyclopedic-semantic definition of the Syldavian.

— The instantiation of the endoxic definition@ backs the conclusion:

    the guilt of Mr. So-and-So is likely / unlikely.

Other topical questions regarding the same Mr. So-and-So will provide other, possibly contradictory, orientations. Such questions thereby play a role in the creation or dismissal of inculpations or exculpations, shifting the burden of proof on the whim of pre-established judgments, regardless of the outcome of any detailed investigation of the matter@.

3. Common place based portrayal in literature and argumentation

Each and every one of these questions can itself become the source of subquestions, and these can be developed considerably, to produce a detailed grid of investigation. The results yielded via this technique depend entirely on the method of investigation used to answer the question; an armchair argument for which the 'research' is based on common sense and common places will deliver commonplace conclusions.

The richest set of detailed questions concerns the key element of these rhetorical scenarios, that being the person (*Who?*). Their application produces a portrait of this person, which can be taken as a literary feat (if successful), and a base for argumentative categorizing inferences.

These commonplaces serve as ready-made arguments, from which the investigating party may select the most appropriate, depending on his or her aims.

Quintilian identified the following doxically relevant facets of a person in order to compound the *a priori* rhetorical representation of a person, independently of any concrete information about the action under discussion.

— "*Birth*, for people are mostly thought similar in character to their fathers and forefathers, and sometimes derive from their origin motives for living an honorable or dishonorable life" (*IO*, V, 10, 24).

To answer the sub-question "Birth?" the inquiry about the family collects information such as "*he is from a well-known honorable family*", or "*his father was sentenced*". The first information provides arguments allowing for example the application of the rule "*like father, like son*", "*he is a chip of the old block*", which serves inferences like:

> He made a mistake, but his family affords all the necessary guarantees; good blood cannot lie, he deserves a second chance.

The second information leads to different conclusions:

> The father was sentenced, so the son has a heavy inheritance. Bring me more information about him!

The commonplace "*the miser's son is a spendthrift*" opposes the preceding one. If the father has a *vice*, the doxa now credits the son not of the corresponding *virtue*, but either of the *same vice* or an opposite vice.

— "*Nation?*" (*ibid.*) and "*Country?*" (*id.*, 25). The answers will introduce national stereotypes: "if he is a *Spanish*, he is proud, if he is *British*, he is phlegmatic". These conclusions, "*he is proud, he is phlegmatic*", may prove useful for the discussion to come "*he is Spanish, so he is proud, so he certainly strongly reacted to this personal attack*".

— "*Sex*, for you would more readily believe a charge of robbery with regard to a man, and poisoning with regard to a woman" (*ibid.*) The prejudiced investigator will follow the commonplace suggestion: in case of poisoning, he will tend to look for a woman. A French book, "*The Famous Poisoners*" [*Les Empoisonneuses Célèbres*] is exclusively dedicated to famous *female* poisoners.

— "*Age?*", "*Education?*", "*Bodily constitution*, for beauty is often drawn into an argument for libertinism, and strength for insolence, and the contrary qualities for contrary conduct" (*id.*, 25-26). In other words, *"he is handsome, he must be a debauchee"* is more probable@ than *"he is handsome, therefore he must live an austere life"*. If **A** is stronger than **B**, then "**A** is more aggressive than **B**" is likely, and therefore, if **A** and **B** had a row, "certainly, **A** attacked **B**", in other words, **A** bears the burden of proof. These inferences can be turned around by application of the paradox of plausibility: "actually, **B** must have attacked **A**, because he knew that the appearances were against **A**".

— *"Fortune*, for the same charge is not equally credible in reference to a rich and a poor man, in reference to one who is surrounded with relations, friends and clients, and one who is destitute of all such support" (*id.*, 26). The commonplaces associated with social roles and positions come under this heading. An elderly man from the countryside, sitting on a bench in the setting sun, will certainly deliver some deep and true thought about the current state of affairs, **S. Rich and Poor.**

— *"Natural disposition*, for avarice, passionateness, sensibility, cruelty, austerity, and other similar affections of the mind, frequently either cause credit to be given to an accusation or to be withheld from it" (*id.*, 27): *"the assassination was committed in a particularly cruel manner, Peter is cruel, therefore he is the murderer"*, **S. Circumstances.**

— *"Manner of living*, for it is often a matter of inquiry whether a person is luxurious, or parsimonious, or mean" (*ibid.*).

The following questions refer to arguments based on desires and motives_ (*ibid.*):
— "What a person affects, whether he would wish to appear rich or eloquent, just or powerful" (*id.*, 28).

— "Previous doings and sayings" (*ibid.*), used to find motives@ and precedents@.

— "Commotion of the mind, [...] a temporary excitement of the feelings, as anger, or fear" (*ibid.*), **S. Emotions**

— "Designs" (*id.*, 29)

This set of commonplaces underlies portraits such as:
> A man in his thirties, Canadian, West Coast, sporty, from a well-known and respected family, has never completed his law education, very kind with his neighbors, living a conventional life, works in a pharmacy, with limited prospect for the future...

This portrait can be read as an (unsuccessful) literary attempt, a police form, etc. In all cases, it is a stock of premises. Doxa-based argumentation is based on pieces of information like *"the man is **X**"*, draws on the stereotyped categories attached to **X**s, "*the **X** are like that*", and concludes that "*the man is like that*", **S. Categorization; Definition**.

4. The literature of characters

This topology has a derived argumentative function and a direct aesthetic-cognitive function. It is linked to the question of the socio-linguistic or doxical beliefs, that is to the prejudiced identity of the person. It is antagonistic with a problematic of identity as deep being, the psychological nucleus of the person. Providing a technique for the construction of the portrait, it thus establishes a bridge between argumentation and literature through the genre of "Characters", as those of the Greek Theophrastus, and, more generally, the classical literature of portraits and mores.

We are no longer in the realm of *ethos@* as an autofiction, but in the pure world of the *ethopoeia*, that is to say, of the fictional representation of a "character", such as "the Miser" or "the Garrulous person" via his or her typical manners, discourse and actions. Such de-contextualized portraits can be used as authorized and respectable sources about the character which they are used to depict, as prolegomena to the exercise of the argumentation in situation, where they will be applied to a particular person.

Historically, this is part of a coherent educative, esthetic and cognitive process of controlled, systematic writing and thinking, the very antithesis of any uncontrolled automatic writing.

5. "This noxious fertility of common thoughts" (Port-Royal)

When based exclusively on common knowledge, that is language associations and doxa-based knowledge, this technique makes it possible to quickly compose fairly convincing, true-to-life pictures of things and events. Critically, these are justifiably very difficult to rebut, as they are the mere expression of shared preconstructed knowledge. The vicious circle between persuader and persuadee is an example of such a situation, **S. Persuasion**. Such compositions are not scientific characterizations of the individual, as can be developed in psychology or philosophy, but the perfect stronghold for all positive or negative social prejudices. Port-Royal has severely condemned this "noxious fertility of common thoughts":

> Now, so far is it from being useful to obtain this sort of abundance, that there is nothing which more depraves the judgment, nothing which more chokes up good seed, than a crowd of noxious weeds; nothing renders a mind more barren of just and weighty thoughts than this noxious fertility of common thoughts. The mind is accustomed to this facility, and no longer makes any effort to find appropriate, special and natural reasons, which can only be discovered by an attentive consideration of the subject. (Arnauld, Nicole, [1662], III, XVII; p. 235)

Common Sense ▶ Doxa; Authority; Common place

Comparison

Comparison is the process of establishing whether or not two individuals, two situations, two systems... present or not some *similarities* or *analogies*. A process of comparison is involved in many argumentative activities, such that the label *argument by comparison* (*a comparatione*) is used with different meanings. These primarily correspond with the argument *a fortiori*@, the arguments *a pari*@, by *analogy*@, by *example*@ or *exemplum*@.

Comparison and categorization@ — Comparison is the basis for the categorization-nomination process; the individual to categorize is compared either with a known individual belonging to the category, or with the prototypical member defining the category. **S. Justice**

Intra-categorical comparison — Two beings belonging to the same category are *identical* from the point of view of this category. Despite this, they can still be compared in terms of:
— their non-categorical properties; **S. Analogy (III)**.
— their position relative to a prototypical subcategory of this category. A rat and a whale, for example, are identical insofar as both are *mammals*; considering that the cow is a prototypical mammal, we can say that a rat, being nearer to a cow than to a whale, is "more" a mammal than a whale. **S. Categorization.**
— Hierarchized categories contain by definition built-in comparisons: Bachelor, Master, and Doctorate are three kinds of academic degrees, listed by ascending order. They can enter in an *a fortiori*@ argument.

Comparison and structural analogy — A process of comparison is also involved in *establishing a structural analogy*, **S. Analogy (IV)**.

Completeness

   Argument *a completudine*; Lat. *completudo*, "completeness".

The evolution of society can be manifested by the emergence of legal cases that do not find clear solutions in the existing system of laws, whether in national, international or human rights legislations (Tarello 1972, quoted in Perelman 1977, p. 55). Nonetheless, the judge is under an *obligation to judge*, that is, he or she must pass a sentence upon all the cases before him or her. That is to say, he or she cannot refuse to make a decision upon a case by arguing that there is *no law* applicable to that case, or that *no interpretation* of an existing law can settle it. In other words, the principle of completeness assumes that the existing system of law, duly interpreted, can qualify all and any human act as permitted, tolerated, or prohibited.

Meta-principles such as the following complement the system of laws:

> In civil matters, in the absence of specific law, the judge is obliged to proceed in accordance with equity. To decide according to equity, he must call on natural law and on reason, or on the usages received, when the primitive law is silent.
>
> Fortuné Anthoine de Saint Joseph, [*Concordance between the Foreign Civil Code and the Napoléon Code*], 1856.[1]

The argument of completeness is parallel to the topos of the *impotent legislator*, the nature of things rendering the application of the law impossible, S. **Force of Circumstances**.

Composition and Division

Aristotle considers *composition* or "combination of words" *and division* as verbal fallacies, that is fallacies of words, as opposed to fallacies of things or method, S. **Fallacies (II)**. They are discussed in the *Sophistical Refutations* (RS 4) and in the *Rhetoric* (II, 24, 1401a20 – 1402b5; RR p. 128).

The label *argumentation by division* is sometimes used to refer to case-by-case argumentation, S. **Case-by-Case**.

1. Grammar of composition and division

Composition and division involve the conjunction *and* that can coordinate:
— Phrases:

    (1) Peter and Paul came.   (No and N1) + Verb
    (2) Peter smoked and prayed.   No + (V1 and V2)

— Statements:

    (3) Peter came and Paul came.   (N + V1) and (N1 + V2)
    (4) Peter smoked and Peter prayed.   (N + V1) and (N1 + V2)

In Aristotelian logical-grammatical terminology:
(3) and (4) are obtained by *division* respectively from (1) and (2).
(1) and (2) are obtained by *composition* respectively from (3) and (4).

The compound and divided statements are sometimes semantically equivalent and sometimes not.

(i) Equivalent — (1) and (3) on the one hand, (2) and (4) on the other hand are roughly equivalent, although it seems that (1), not (3), implies that Peter and Paul came *together*. In this case, composition and division are possible, and the coordination is used simply to avoid repetition.

---

[1] Fortuné Anthoine de Saint Joseph, *Concordance entre les codes civils étrangers et le Code Napoléon*, 2nd ed. t. II. Paris: Cotillon, 1856. P. 460.

(ii) Not equivalent — sometimes phrase coordination (composed statement) is not equivalent to sentence coordination (divided statement). The semantic phenomena involved are of very different types.

>   Peter got married and Mary got married.
>   ≠ Peter and Mary married.

If Peter and Mary are brother and sister, the custom being what it is, the composition is unambiguous. Without such information, the composition introduces an ambiguity.

The operation of division can produce a meaningless discourse:

>   The flag is red and black.
>   * The flag is red and the flag is black.

>   **B** is between **A** and **C**.
>   * **B** is between **A** and **B** is between **C**.

Sometimes a syntactic operation applied to a statement produces a paraphrase of this statement. At other time, the same operation applied to another statement having apparently the same structure as the first one produces a statement that has no meaning, or whose meaning and truth conditions entirely differ from those of the original statement.

2. Aristotelian logic of composition and division

The study of paraphrastic systems is a classical object of *syntactic* theory. Aristotelian logic considers composition and division as a problem in *logic*. As Hintikka (1987) has repeatedly pointed out, the Aristotelian notion of fallacy is dialogical, S. **Fallacy (I)**. The fallacious maneuver throws the interlocutor into confusion, and this is precisely what happens with composition and division. The following case is one of the oldest and most famous illustrations of the fallacy of composition:

>   This dog is your dog (is yours); and this dog is a father (of several puppies).
>   So this dog is your father and you are the brother of the puppies.

The interlocutor is disoriented, and everyone finds that very funny (Plato, *Euth.*, XXIV, 298a-299d, pp. 141-142). S. **Sophism**.

Aristotle analyzes this kind of sophistical and sophisticated problem in the *Sophistical Refutations* and in the *Rhetoric* under the heading of "paralogism of composition and division". He shows that the issue extends to a variety of discursive phenomena, under what conditions can *judgments* made on the basis of isolated statements be "composed" into a *discourse* of connected statements? The discussion is illustrated by several examples showing the full scope of the interpretation issues that are raised, even if their wording may seem contrived.

(i) Consider the statement: *"it is possible to write while not writing"* (RS, 4); it can be interpreted in two ways:

— Interpretation 1 *composes* the meaning: *"one can at the same time write and not write*" (*ibid.*), in the sense of: "*one can (write and not write)*". The composition is misleading and absurd.

— Interpretation 2 *divides* the meaning; when one does not write one still retains the capacity to write, meaning: "*one can know how to write, while not writing*", which is correct. Under certain circumstances, a person who can write cannot physically do so, for example if one's hands are tied. The modal power is ambiguous between "*having the capacity to*" and "*having the possibility to exercise that capacity*".

(ii) The following example also uses the modal *can*, this time in its relation to time. Consider the statement "*if you can carry one thing, you can carry several*" (RS, 4, 166a30: 11):

(1) (I can carry the table) and (I can carry the cabinet)

Therefore, by composition of the two statements into one:

(2) I can carry (the table along with the cabinet)

Which is not necessarily the case.

(iii) The fallacy of division is illustrated by the example "*five is equal to three and two*" (after RS, 4, 166a30, p.12):

— Interpretation (1) divides meaning, that is, it decomposes the utterance into two coordinated propositions, which is both absurd and fallacious:

(Five is equal to three) and (five is equal to two)

— Interpretation (2) composes the meaning, which is correct:

Five is equal to (three and two)

In the *Rhetoric*, the notion of composition is discussed with several examples that clearly show the relevance of the issue for argumentation. The argument by composition and division "[asserts] of the whole what is true of the parts, or of the parts what is true of the whole" (*Rhet*, II, 24, 1401a20-30; RR, pp. 381), which makes it possible to present things from quite different angles. This technique of argumentation involves statements constructed around appreciative and modal predicates such as:

— *is good;* — *is just;* — *is able to* —; — *can* —;
— *knows* —; — *said*.

The following example is taken from Sophocles play, *Electra*. Clytemnestra killed her husband, Agamemnon. Then their son Orestes kills his mother to avenge his father. Was Orestes morally and legally entitled to do this?

"'T'is right that she who slays her lord should die'; 'it is right too, that the son should avenge his father'. Very good: these two things are what Orestes has done." Still, perhaps the two things, once they are put together, do not form a right act. (*Rhet.*, II. 24, 1401a35-b5, RR, 383).

Orestes justifies what he did, arguing that his two actions can be composed. His accuser rejects the composition.

This technique of decomposing a doubtful action into a series of commendable, or at least innocent, acts is arguably very productive. *Stealing* is just *taking* the bag that is there, taking it somewhere else, and failing to put it back in the same place. The division blocks the overall assessment.

A second example clearly shows that fallacy and argument are two sides of the same coin:

> If a double portion of a certain thing is harmful to health, then a single portion must not be called wholesome, since it is absurd that two good things should make one bad thing. Put thus, the enthymeme is refutative; put as follows, demonstrative "for one good thing cannot be made up of two bad things". The whole line of argument is fallacious. (*Rhet.*, Ii. 24, 1401a30, RR p.381-383)

Abstainers start from an agreement upon the fact that *"having a lot of drinks makes you sick"*, and divide: *"so having a drink makes you sick"*. Permissive people follow the other line: *"having a drink is good for health"*, and proceed by composition. Abstainers argue by division, and this is considered to be fallacious by permissive individuals. Permissive individuals argue by composition, and this is considered to be fallacious by abstainers.

3. Whole and parts argument

The two labels "composition and division" and "part and whole" are in practice considered equivalent (van Eemeren & Garssen, 2009).

3.1 Whole to parts and division

The argument based on the whole assigns to each of its parts a property evidenced on the whole:

> If the *whole* is P, then each of its *parts* must be P.
> If the *country* is rich, each of its *regions* (*inhabitants*…) must be rich.
> Americans are rich, so *he* is rich; let's ransom him!

The problem faced by whole to parts arguments mirrors that of the argument by division: can the property evidenced on the whole be transferred to each of its parts?

3.2 Parts to whole and composition

The argument based on the parts assigns to the whole they make up the properties evidenced on each of its parts:

> If every part of a whole is **P**, then the whole is **P**.
> If every player is good, then the team is good (?).

The problem faced by parts to whole arguments mirrors that of the argument by composition: is the property evidenced by each part also evidenced by the whole?

4. Complex wholes and emerging properties

Accidental or *Mechanical* wholes are composed of a set of disconnected objects in a relation of neighborhood. *Essential* or *complex* wholes are made up of the conjunction of the parts plus some *emerging* extra properties, which distinguishes them from an inert juxtaposition of components. The degree of *complexity* of the whole is superior to the simple arithmetical addition of its parts. This process is referred to as a *composition effect*. The case of the superiority of the group over the individual alleged by Aristotle is an example of such an effect, S. *Ad populum*.

This issue is also found in rhetoric, where a distinction is made between metonymy and synecdoche, the first focusing upon neighborhood relations and the second on relations between a complex whole and its parts.

Concession

Concessions may be *negotiated* in an organized discussion, or *presented as such* in a monological discourse.

1. Negotiated concession

Through negotiated concessions, the arguer modifies his or her original position by decreasing the original demand or by granting to the adversary a controversial sub-point. From a strategic point of view, this move may amount to an orderly retreat, possibly for future benefit, hoping that the opponent will do the same when it comes to another point.

Aristotelian logical-dialectical games ignore concessions, as a violation of the principle of excluded middle, things being either entirely true, or entirely false; conclusively defended or not, **S. Dialectic**. In contrast, conceding is a key moment in the negotiation process of human affairs, understood as a discussion leading to a reasonable agreement (Kerbrat-Orecchioni, 2000).

By making concessions, the arguer recognizes that the opponent's point of view is to some extent valid, whilst continuing to uphold the value of his or her own positions and conclusions. The arguing party may believe that his or her remaining arguments are:

— More compelling, or of a different type than those of the opponent.
— Not strong arguments, but nonetheless arguments grounded on personal values and deep convictions (identity-based arguments).

The original position should thus be maintained against all odds, according to the formula "*I do know, but still…*".

In everyday discussions, concessions are valued as manifestations of openness to the others, and as constitutive of a positive ethos. Nonetheless, concessions may be ironic, **S. Epitrope**.

2. Concession as a speech act

In grammar, concessive constructions "**A**(claim) + **C**(concession)" co-ordinate two statements having opposite argumentative orientations, while retaining the overarching orientation determined by the first proposition **A**:

"Although **C**, **A**"; "certainly **C**, but **A**"
"I admit, I understand **C** but I stick to **A**".

C takes up or reformulates the speech of the opponent, or evokes the speech of a fictitious opponent; **A** reaffirms the speaker's claim.

> Social relations are indeed extremely tense these days, but we must nonetheless go on restructuring the company.

Unlike negotiated concession, linguistic concession is structural. The speaker sets out:
— first, a virtual character or voice developing the argument *"social relations are extremely tense"*, oriented towards conclusions such as *"stop the restructuring of the company"*,
—followed by a second argument, putting forward the *opposite* position *"we must go on restructuring the company"*, and *identifies with this second character*. In Goffman's words, the speaker is the *animator* of **A**, and the *animator* and *principal* of **C**. In other words, the speaker recognizes the existence of arguments supporting an opposing conclusion, but at the same time refuses to conclude on this basis. The concession here is a simple acknowledgment of the fact that somebody, somewhere, says, or may say something opposite to that claimed by the speaker. This amounts to a de-activation of the argumentative strength of the aforementioned argument. This kind of concession is by no means the expression of the goodwill of a reasonable negotiator, but a mere phagocytosis and castration of the opponent's arguments.

The two forms of concession may be superimposed, by rationalizing the linguistic concession. One considers that linguistic concession occurs when the speaker has taken the opponent's arguments into consideration and confronted them with his or her own (even if this examination often leaves no discursive trace), and that, finally, in the grand scheme of things, he or she thinks that her or his arguments are better. But since language gives for real and true that which it signifies, a purely *linguistic* concession automatically produces a *negotiated* concession effect, whether or not it is really the case. This does not mean that linguistic concession is always mere lip service, but that negotiated concession can only be studied on corpora built to that effect.

Conclusion ▶ Argument — Conclusion

Conditions of Discussion

The *Treatise on Argumentation* insists on the necessity and variety of "prior agreements" between participants to develop an *argumentation* — that is, an *argument$_1$*; no previous agreements are necessary to engage in an *argument$_2$*:

> For argumentation to exist, an effective community of minds must be realized at a given moment. There must first of all be agreement, in principle, on the formation of this intellectual community, and, after that, on the fact of debating a specific question together: now, this does not come about automatically. (Perelman & Olbrechts-Tyteca [1958], p. 14)

Two different kinds of agreements are mentioned here, and, as the text points out, neither of them can be taken for granted.

1. Formation of speech communities

This first kind of agreement deals with the realization of an "effective community of minds", constituted upon the free decision taken by the participants. It may be considered as an ideal form of argumentative communication. Its nearest approximation may be philosophical or scientific friendly encounters.

Not all argumentative practices depend on the production of such a community. The court is a prototypical argumentative place, and no prior voluntary agreement must be made with criminals to assure their timely appearance; when necessary, legal coercion may be used. Institutions defining specific forums, problems and rules of interaction determine the social and legal conventions ruling argumentative communities. The existence of these social infrastructures makes it possible to avoid previous cumbersome negotiations among speech communities.

2. Agreement about the issue

To discuss an issue, must we first "agree to discuss this issue together"? As was the case for the kind of agreements described immediately above, the different legal systems establish who has the legal right to determine the charges leading to the appearance of a given party; the defendant does not necessarily agree to discuss the matter, but is summoned by the judge.

Prior discussions may be useful in institutionally structured communities in order to establish the points that will be discussed at a particular meeting. But the agenda is not necessarily decided upon by mutual agreement among the future participants in the discussion; it may be the prerogative of an individual in charge of the organization. On the other hand, the issue itself, may be reframed during the encounter.

Intellectual communities are also social communities, even when they address questions concerning the human condition in general. The *disputability* of an issue is itself an argumentative exercise, in the same way as the process of *discussing* the issue itself. Two quite distinct subquestions must be envisioned, first, a *central* one, the conditions on the "disputability" of the issue properly said, and second, if all the potential partners agree to discuss such and such issue, a *practical* issue must be settled, the material conditions on the discussion itself — where, when, who will chair the discussion, etc. — not to mention the shape of the table.

The dispute about the maximization vs. minimization of the right to discuss define what may be called *the stasis@ of stasis*.

2.1 Maximizing the right to discuss

Concerning the substantial issue, one can either stress the principle of radical free expression according to which any point of view can be affirmed and challenged, or emphasize the pragmatic conditions of such discussion. The first of the "Ten Commandments for Reasonable Discussants" posits that:

> Commandment 1, *Freedom rule*: Discussants may not prevent each other from advancing standpoints or calling standpoints into question. (van Eemeren, Grootendorst, 2004, p. 190) **S. Rules.**

This is also the position taken by Stuart Mill:

> If all mankind minus one, were of one opinion, and only one person were of the contrary opinion, mankind would be no more justified in silencing that one person, than he, if he had the power, would be justified in silencing mankind. (Mill, [1859], p. 76)

2.2 Conditioning the rights to discussion

Absolute liberty of expression would give free rein to racist speech, hate speech, collective verbal and non-verbal persecution of the individual chosen as a scapegoat a group, types of speech which many would find unacceptable. If individuals are free to privately discuss anything, provided they can find a partner willing to do so, actual speech communities put conditions on social discussions. For example, the *res judicata* principle prevents the reopening of an issue which has already been judged, unless a new fact is to be considered.

Moreover, the proper functioning of a speech community must take into consideration the fact that it is not possible to discuss *anything* (condition on the subject, on the agenda), with *anyone* (condition on the participants), *anywhere* and *anytime* (material conditions on place and time), *no matter how* (according to what procedure), **S. Manipulation:**

> Some Truths Are Not for Common Ears. It is lawful to speak the truth; it is not expedient to speak the truth to everybody at every time and in every way.
> Erasmus, [1524], *On the Freedom of the Will.* (no pag.)[1]

The *Treatise* is very sensitive to the "anyone" condition:

> There are beings with whom any contact may seem superfluous or undesirable. There are some one cannot be bothered to talk to. There are also others

---

[1] Quoted after Desiderius Erasmus, *On the Freedom of the Will*. Trans. by E. Gordon Rupp (no pag., no date). www.sjsu.edu/people/james.lindahl/courses/Hum1B/s3/Erasmus-and-Luther-on-Free-Will-and-Salvation.pdf (05-23-17).

with whom one does not wish to discuss things, but to whom one merely gives orders. (Perelman & Olbrechts-Tyteca [1958], p. 15)

Aristotle limits topics of legitimate discussion to the *endoxa*, and rejects debates questioning "anything", that is to say, affirmations which in practice nobody doubts:

> Not every problem, nor every thesis, should be examined, but only one which might puzzle one of those who need argument, not punishment or perception. For people who are puzzled to know whether one ought to honor the gods and love one's parents or not need punishment, while those who are puzzled to know whether snow is white or not need perception. (*Top.*, 11)

The undisputable refers to three kinds of evidence: *sense data* evidence, *"snow is white"*; *religious* evidence, *"we must honor the gods"*; and the *social* evidence *"we must love our parents"*; these statements are uncontroversial because it is unconceivable that anyone would argue otherwise — in Aristotle's Athenian society of course. In order for an opinion to be worthy of doubt, it must, on the one hand, fall within the scope of the doxa. That is, it must be part of the defining beliefs of the community, or seriously claimed by some of its honorable members or a subgroup, **S. Doxa.**

On the other hand, the doubt must be serious, that is motivated. Arguing being a costly activity, one must have a good reason to doubt. In other words, the person who wants to challenge an accepted statement bears the burden@ of proof.

In the same spirit, the theory of stasis categorizes as uncontroversial (a-stasic) misplaced, badly worded or intractable questions, or, conversely, questions whose answer is obvious, **S. Dialectics; Self evidence; Stasis; Argumentative question.**

On the legitimizing effects of debate, **S. Paradoxes.**

3. Agreement on what counts as an argument

Agreements on the community of speech and on the issue must be supplemented by agreements on beings, facts, rules and values (Perelman, Olbrechts-Tyteca [1958], II, 1). Agreements here should establish what counts as an argument: condition of truth; of relevance of the true statement for the defended conclusion; of relevance of the conclusion (defended by a true and relevant statement) for the debate itself, **S. Relevance.**

When it is impossible to determine whether a statement is true, relevant to a conclusion itself relevant to a debate, a general system of acceptance or tacit agreement is invoked. In serious global disagreements, sub-agreements are difficult to reach; the disputants anticipate their opponent's conclusion, and know very well that once the argument is accepted, the conclusion will quickly follow, hence the tendency to postulate disagreement as a ruling principle, including upon what should be considered as facts, **S. Politeness; Dissensus; Disagreement.**

This "appeal to agreement" is actually grounded on an argument by perverse effects, considering that the absence of agreement would condemn the debate to an undesired state of deepening disagreement, that can indeed lead to a collapse of the discussion (Doury 1997). In practice, two facts must be taken into account. Firstly, points of agreement and disagreement can be negotiated on the spot, during the discussion. Secondly, the lack of agreement does not preclude argumentation, it suffices that *third parties* take the reins of the discussion. The decision of the judge, and more generally that of the third party, is commonly made on the basis of an argument rejected or ignored by one party, or by both, **S. Roles**. Judicial organizations intervene precisely when no agreement can be passed between the parties; as representing the ruling power, they dispense with agreements — not with arguments.

In general, if one agrees on the data and rules, the conclusion automatically follows; argumentation becomes demonstrative. But argumentation is a linguistic way of dealing with the different in a system of generalized disagreement and uncertainty. There is a decisive incompatibility between the material interests at stake: one can indeed divide the pie, but what is eaten by any one person cannot be shared with the other. Serious, deep, intractable… disagreement *between the parties*, proponent and opponent, should be considered to be the basic condition of argumentation; that is why *third parties* have a key role to play in argumentative devices.

Conductive Argument

Conductive arguments are defined by Wellman as third kind of argument, parallel to deduction and induction. In view of examples such as those below (my numbering, CP), he notes that, "it is tempting, therefore, to define a conductive argument as any argument that is neither deductive nor inductive" (1971, p. 51):

    (1) You have to take your son to the circus because you promised.

    (2) This is a good book because it is interesting and thought provoking.

    (3) Although he is tactless and nonconformist, he is still a morally good man because of his underlying kindness and real integrity. (*Ibid.*)

Wellman distinguishes between three types of conductive arguments

(i) "A single reason is given for the conclusion" (*id.* p. 55), as in
    (4) You ought to help him because he has been very kind to you.
    (5) That was a good play because the characters were so well drawn. (*Ibid.*)

(ii) "In the second pattern of conduction, several reasons are given for the conclusion" (*id.*, p. 56), as in:
    (6) You ought to take your son to the movie, because you promised to do so, it is a good movie, and you have nothing better to do this afternoon.

(7) This is not a good book, because it fails to hold one's interest, is full of vague description, and has a very implausible plot. (*Ibid.*)

(iii) "The third pattern of conduction is that form of argument in which some conclusion is drawn from both positive and negative considerations. In this pattern, reasons against the conclusion are included as well as reasons for it" (*id.*, p. 57), as in

(8) In spite of a certain dissonance, that piece of music is beautiful because of its dynamic quality and its final conclusion.

(9) Although your lawn needs cutting, you ought to take your son to the movie because the picture is ideal for children and will be gone by tomorrow. (*Ibid.*)

The key characteristic of conductive reasoning appears to be condition (3), where, depending on the speakers, and with the same reasons, the pros can outweigh the cons or vice versa (Blair 2011). From the same data, another speaker might draw the opposite conclusion.

(8.1) In spite of a certain dynamic quality and its final conclusion, that piece of music is ugly because of its dissonance.

The adjective *certain* seems to be attached to the connective *in spite of*, indicating that the speaker will not argue on the basis of this argument (will not identify with this voice), **S. Interaction, Dialogue, Polyphony.**

A conductive argument does not seem amenable to default reasoning. Their conditions of refutation are different. Default reasoning might be updated or changed when new information is accessed, while conductive reasoning does not depend on information as such. A conductive argument typically deals with value@, either moral or aesthetic. The specific issue of conduction is the hierarchization, or balance, of values. Whilst some pairs of values will be very difficult, if not impossible, to balance, others will be quite plausibly balanced. So, sentence (8) for example can be plausibly converted as (8.1), because the three implied values cannot, in my view, be hierarchized, whilst (9) invokes values which seem easier to balance:

(9.1) I know, the movie is ideal for children and won't be showing in the cinema after tomorrow, but you ought to cut your lawn.

Cutting the lawn seems to be a task which is easy to postpone, in view of the children's education and their legitimate satisfaction, which might be prioritized. So, in the case of (9), the consensus would be that pros clearly outweigh the cons.

In any case, more complex interactional data could provide some clue as to how dissenting speakers fare when dealing with competing values.

Connective

A *connective* word is a *function word* that combines several propositions, simple or complex, into a new, integrated, (more) complex proposition.

1. Connectives in propositional calculus

Logical connectives articulate simple or complex well-formed propositions so as to construct well-formed complex propositions, or *formulas*. *Propositional calculus* studies *logical syntax*, that is the rules of construction of well-formed *formulas*. It determines, among these formulas, which are *valid formulas* (*logical laws, tautologies*).

Propositions@ are denoted by the capital letters **P, Q, R**... They are said to be *unanalyzed*, that is, taken as a whole, in opposition with *analyzed* propositions "[Subject] *is* [Predicate]" considered in the *predicate calculus*.

A *binary logical connective* combines two propositions (simple or complex) **P** and **Q** into a new complex proposition "**P** [connective] **Q**". Logical connectives (or connectors) are also called *functors, function words* or *logical operators*

The most used connectives are denoted and read as follows:

↔	equivalence, "P *is equivalent to* Q",
→	implication, "*if* P *then* Q"
&	conjunction, "P *and* Q"
∨	disjunction, "P *or* Q"
W	exclusive disjunction, "*either* P *or* Q (not both)"

Logical connectives are defined on the basis of the possible truth-values given to the propositions they combine. A specific logical connective is defined by the kind of combination it accepts between the truth-values of the component proposition.

1.1 The truth tables approach to binary connectives

A logical connective is defined by its associated *truth table*. The truth table of a "**P** *connec* **Q**" binary connective is a three-column, five-line table.

— The letters **P, Q** ... denote the propositions; the letters **T** and **F** denote their truth-values: true (**T**) or false (**F**). **P** and **Q** *are* propositions, while truth and falsehood are *said of* propositions, "P is **T**rue", "P is **F**alse"; so the corresponding abbreviating letters use a different typographic character.

P	Q	P connec Q
T	T	(depends on the connective)
T	F	(depends on the connective)
F	T	(depends on the connective)
F	F	(depends on the connective)

— Columns:

> The truth-values of the proposition **P** are expressed in the first column
> The truth-values of the proposition **Q** are expressed in the second column
> The corresponding truth-values of the complex formula "**P** *connec* **Q**" are expressed in the third column.

— Lines:

> The first line mentions all the propositions to take into account, **P**, **Q** and "**P** connec **Q**".
>
> The four following lines express the truth-values of these propositions. As each proposition can be T or F, four combinations must be considered, each corresponding to one line.

Conjunction "&"

By definition, the conjunction "**P & Q**"
> — is true when P and Q are simultaneously true: *line 2*
> — is false when • one of the two is false: *line 3 and 4*
> • both are false: *line 5*.

This is expressed in the following truth table:

P	Q	P & Q
T	T	T
T	F	F
F	T	F
F	F	F

This truth table reads:
> line 1: "when **P** is true and **Q** is true, then '**P & Q**' is true"
> line 2: "when **P** is true and **Q** is false, then '**P & Q**' is false"
> line 3: "when **P** is false and **Q** is true, then '**P & Q**' is false"
> line 4: "when **P** is false and **Q** is false, then '**P & Q**' is false"

Equivalence, " ↔ "

The logical equivalence "**P ↔ Q**" reads "P is equivalent to Q". This resulting proposition is true if and only if the original propositions have the same truth-values.

Truth table of logical equivalence:

P	Q	P ↔ Q
T	T	T
T	F	F
F	T	F
F	F	T

Under this definition, *all true propositions* are mutually equivalent, *all false propositions* are mutually equivalent, regardless of their meaning.

Disjunctions: Inclusive "∨"; Exclusive, "W"

The *inclusive* disjunction "**P** ∨ **Q**" is false if and only if **P** and **Q** are simultaneously false; in all other cases, it is true.

Truth table of the inclusive disjunction:

P	Q	P ∨ Q
T	T	T
T	F	T
F	T	T
F	F	F

The *exclusive* disjunction <**P** W **Q**> is true if and only if only one of the two propositions it conjoins is true. In all other cases it is false.

Truth table of the exclusive disjunction:

P	Q	P W Q
T	T	F
T	F	T
F	T	T
F	F	F

Implication "→"

The logical implication symbol "→" reads "**P** implies **Q**". **P** is the *antecedent* of the implication and **Q**, its *consequent*.

Truth table of logical implication:

P	Q	P → Q
T	T	T
T	F	F
F	T	T
F	F	T

This table reads:
- line 2: The true implies the true
- line 3: The true does not imply the false
- line 4: The false implies the true
- line 5: The false implies the false

Only truth can be logically derived from truth (line 1), whereas, anything can follow from a false assertion, a truth as well as a falsehood.

The equivalence, conjunction, inclusive disjunction and exclusive disjunction connectives are symmetrical, that is, for these connectives, "**P** connective **Q**" and "**Q** connective **P**" are *equivalent (convertible)*:

P ↔ Q	↔	Q ↔ P
P & Q	↔	Q & P
P ∨ Q	↔	Q ∨ P
P W Q	↔	Q W P

The implication connective is *not* convertible ; that is "**P** → **Q**" and "**Q** → **P**" have different truth tables.

The laws of implication express the notions of necessary and sufficient condition:

    A → **B** (is true)
    A is a *sufficient* condition for **B**
    B is a *necessary* condition for **A**

Causal relation may be expressed as an implication. To say that if it rains, the road is wet, means that rain is a *sufficient* condition for the road to be wet, and that, *necessarily*, the road is wet when it rains.

The implication thus defined is called *material* implication; it has nothing to do with the *substantial* logic of Toulmin.

The implication "**P** → **Q**" is false only when **P** is true and **Q** false (line 2). In other words, "**P** → **Q**" is true if and only if "**not-(P & not-Q)**" is true.

Line (3) asserts the truth of the implication "*If the moon is a soft cheese* (false proposition), then *Napoleon died in St. Helena* (true proposition)". Like the other logical connectives, the implication is indifferent to the meaning of the propositions it connects. It takes into consideration only their truth-values. The *strict implication* of Lewis tries to erase this paradox by requiring that for "**P** → **Q**" to be true, **Q** must be deducible from **P**. This new definition introduces semantic conditions, in addition to the truth-values. This explains why the word "implication" is sometimes taken in the sense of "deductive inference".

Systems of *natural deduction* are defined in logic (Vax 1982, *Deduction*). They have nothing to do with Grize's *Natural Logic*.

1.2 Logical laws

Using connectors and simple or complex propositions, one is able to construct complex propositional expressions, for example "**(P & Q)** → **R**". The truth-value of such a complex expression is only a function of the truth of its component propositions. Truth tables can be used to evaluate these expressions. Some of them are *always true*, they correspond to *logical laws*.

"Laws of thought"

Binary connectors combine in equivalences known as De Morgan's laws, considered to be *laws of thought*. For example, the connectives "&" and "V" enter in the following equivalences:

> *The negation of an inclusive disjunction is equivalent to the conjunction of the negations of its components:*
> $\neg (P \vee Q) \leftrightarrow (\neg P \,\&\, \neg Q)$

> *The negation of a conjunction is equivalent to the disjunction of the negations of its components:*
> $\neg (P \,\&\, Q) \leftrightarrow (\neg P \vee \neg Q)$

Case-by-case@ argumentation is based upon inclusive disjunction.

Hypothetical (or conditional) syllogism, S. Deduction

Conjunctive syllogism

The following statement expresses a logical law:

> *"If a conjunction is false and one of its components true, then the other component is false"*
> $[\neg(P \,\&\, Q) \,\&\, P] \rightarrow \neg Q$

The corresponding three-steps deduction is known as a *conjunctive syllogism*:

$\neg(P \,\&\, Q)$	the major proposition denies a conjunction
P	the minor affirms one of the two propositions
$\neg Q$	the conclusion excludes the other

An adaptation to ordinary reasoning:

> Nobody can be in two places at the same time
> Peter was seen in Bordeaux yesterday at 6:30pm (UT)
> So, he was not in London yesterday at 6:30 pm. (UT)

Knowing that Peter is suspect; that his interest is to hide that he was actually in Bordeaux, and that the witness is more reliable than the suspect, we may conclude that Peter lied when he pretended to be in London yesterday at 6:30pm.

In the following example, the major of the disjunctive syllogism is the negation of an exclusive disjunction:

$\neg(P \,W\, Q)$	a candidate cannot be admitted and rejected
$\neg P$	my name is not on the list of successful candidates
$\neg Q$	I am rejected

All these deductions are common in ordinary speech, where their self-evidence ensures that they go unnoticed. It would be a mistake not to take them into account on the pretext that, since these arguments are *valid*, they are *not* arguments.

2. Connectives in logic and in language

Introductory logic courses make a consistent use of ordinary language to illustrate both the capacities and specificities of logical languages. Generally speaking, logic can be "applied to the usual language" (Kleene 1967: p. 67-73) as an instrument for expressing, analyzing and evaluating ordinary arguments as valid or invalid reasoning. These translation exercises run as follows (*id.* p. 59):

> I will only pay you for your
> TV installation only if it works   translated as $P \rightarrow W$
> Your installation does not work   translated as $\neg W$
> So I will not pay you   translated as $\neg P$

Using the truth table method for example, this reasoning is then tested for validity, and declared valid.

In order to identify similarities and differences, natural language components and properties can be compared with their counterpart in a logical language. This enables us to better understand both kinds of languages. Such exercises are helpful when it comes to gaining a better understanding of logical or linguistic systems, and may also be of benefit when it comes to argumentation education. Nonetheless, there are some additional facts which should be taken into one consideration when using this methodology.

(i) The preceding exercise did not focus on the correct combination of the truth-values of *semantically independent* propositions such as in the logical talk about the moon and Napoleon (cf. supra §1.4). The exercise introduces a strong condition on semantic coherence between the linked propositions, which belong to the same domain of action, in this case, TV installation.

(ii) Natural language connectives do not connect propositions in the way logical connectives do. The former can be said to be *between* the two propositions, whereas the latter are syntactically *attached to the second proposition*. Logical connectives and natural language connectives have two different syntaxes.

As a consequence, the right-scope of a linguistic connective is essentially defined by the sentence to which it belongs, whereas its left-scope can be much larger, and may include a whole narration, with various twists and turns:

> *Thus*, the prince married the princess — The End —

Connectors are classically considered as connecting two statements in a complete discourse, such as *yet* in:

> the path was dark, *yet* I slowly found my way (google)

Nonetheless, in:

> It is good, *yet* it could be improved (d.c, *Yet*)

yet introduces a more complex scenario, and the preceding example is not a complete discourse. *Yet* announces that more indications are to come specifying the weak points of the assessed task.

(iii) In many cases, the logical reconstruction of ordinary reasoning must introduce new propositions which are said to be present but are left implicit in the considered discursive string. This string is then said to contain an "incomplete argument", **S. Enthymeme**

(iv) Logical reasoning does not cover all ordinary reasoning:
> I have eaten three apples and two oranges, so I have had my five fruits diet today

First, this apparently crystal clear reasoning is loaded with implicit knowledge, such as *"apples are fruits"*, *"oranges are fruits"* and that *"no orange is an apple"*: *"three citrus fruits and two oranges"* sum up as five fruits only if none of the mentioned three citrus fruits is an orange.

Second, the critical fact here is that the conclusion is based upon an *addition* that is easier to solve in arithmetic than in a logical language. Toulmin's layout@ would meet this condition by adding a warrant-backing system referring to the laws of arithmetic.

(v) Logical connectives capture only a small part of the linguistic role played by natural language connectives. The connector "&" requires only that the conjoined clauses are true. This property is common to many ordinary words, *and, but, yet* ... and to all concessive words:
> The circumstances which render the compound true are always the same, viz. joint truth of the two components, regardless of whether 'and', 'but' or 'although' is used. Use of one of these words rather than another may make a difference in naturalness of idiom and may also provide some incidental evidence to what is going on in the speaker's mind, but it is incapable of making the difference between truth and falsehood of the compound. The difference in meaning between 'and', 'but', and 'although' is rhetorical, not logical. Logical notation, unconcerned with rhetorical distinctions, expresses conjunction uniformly. (Quine 1959, p. 40-41)

In other words, classical logical theory does not have adequate concepts to deal with phenomena of argumentative orientation, and imposes no obligation in this respect. Quine's argumentative strategy consists in minimizing the problem and delegating it to rhetoric, seen as a refuse site for problems left unsolved by logical analysis.

And carries with it subtle semantic conditions, for example, a sensibility to temporal succession. If "P & Q" is true, then "Q & P" is true. But these two statements do not contain the same information, and this is no longer a matter of rhetoric, whatever the meaning given to this word:
> They married and had many children.
> They had many children and were married.

One might consider that, under certain conditions, this logical analysis introduces a third proposition "events succeeded in this order". For other conditions influencing the use of *and*, S. Composition and division.

3. No subordination, but bilateral relations

There is no ideal way to envision the relation between logical and natural language; everything depends on the theoretical and practical objectives of the researcher, whether building a conversational robot, developing a formal syntax for ordinary language, or teaching second-level argumentation courses.

Logic is an autonomous mathematical language, that can be constructed from the suggestions of some chosen segments of ordinary language. From the very beginning, the teaching of *logic* may draw more or less heavily on the resources of *ordinary language*. The same applies to the teaching of *everyday argument* in relation to the resources provided by *logical language*. The teacher is free to make pedagogical choices, and possible alternative approaches should be judged by their results, according to the standard methods used for the evaluation of educational methods.

Consensus

1. Consensus as agreement, S. Agreement; Persuasion

2 Argument from consensus

The label *argument from consensus, appeal to consensus*, covers a family of arguments claiming that a belief is true or that things must be done in such and such a way on the basis that everybody thinks or does this, and that other proposals should be rejected. It implies that by flouting the existing consensus, the proponent of a new measure, that is the opponent to consensus, is on the verge of being excluded from this community, S. Burden of proof. These arguments have the general form:

> We always thought, desired, did ... like that; so buy (please, do...) like that.
> Everybody loves the product *So-and-So*.
> Everybody puts *Such and Such* ketchup on their burger!

The ***universal consensus*** **argument** claims that "*all men in all times have thought so and things have always been done that way*". The existence of God has been argued upon the universal consensus argument.

The argument from the *relative (partial) consensus* covers the argument from *majority*, the argument from *number* (Lat. *ad numerum; numerus*, "number") and related expressions:

> The majority / many people ... think, desire, do ... X.
> Three million Syldavians have already adopted it!
> My book sold better than yours.
> He is a well-known actor.

Common Sense — The argument of consensus includes the kind of authority generously granted to *traditional wisdom* or to *common sense*, **S. Authority; Matter**.
    I know that all true Syldavians approve of this decision
    Only extremes attack me, all people of common sense will agree with me.

Populist argument is based on a kind of consensus among the people (or attributed to it), **S. *Ad Populum*.**

***Bandwagon* argument and fallacy** — The bandwagon argument is a special case of the argument from consensus about an action. The bandwagon being the decorated wagon that leads the orchestra through the city, the bandwagon argument adds joy and enthusiasm to the dry argument from consensus. To climb on the bandwagon is to follow the popular movement, to share in a popular "emotion" in the etymological sense, "a public upheaval". Joining a party to have fun and sing should not be condemned as systematically fallacious; but, seen by any opposing party, climbing on the bandwagon can be considered as fallacious, as a follow-the-group or follow-my-leader attitude, sheepish behavior, uncritically adopting the views of the most vocal or visible group.

Consequence and Effect

    Ad consequentiam, lat. *consequentia*, "continuation, succession"

The word *consequence* can mean:
— *Effect*, referring to a *causal*, cause / effect relationship **S. Causation (I).**
— *Consequent*, referring to a *logical*, antecedent / consequent relationship.
The Latin word *consequens* can mean "what temporally follows". **S. Circumstances.**

1. Effect to cause argumentation
The effect to cause argument goes back from the consequences to the cause. Given data is considered the effect of a hypothetical cause that can be reconstructed on the basis of this data combined with a known causal relationship between these type of facts and their cause. Other expressions can also be used, such as argument *by the effect*, or *from the effect to the cause*.
    You have a temperature, so you have an infection

— *Argument*: A confirmed fact **t**, the patient's temperature. This fact **t** belongs to the category of facts or events **T**, *"having a temperature"*, as defined by medicine. This is a categorization process.
— *Causal Law*: There is a causal law linking **I** facts *"having an infection"* to **T** facts, *"having a temperature"*
— *Conclusion*: **t** has a type **T** cause, an infection, and the patient should be treated accordingly.
This corresponds to the diagnostic process; one could speak of *diagnostic reasoning*.

Here, the effect (the temperature) is the *natural sign@* of the cause. Such natural, palpable, effects provide endless basis for argument by natural signs:
> See! The cinders are still hot, there was a recent fire (... they cannot be very far)

In the area of socio-political decision, the argument by the consequences corresponds to the pragmatic argument, transferring upon the measure itself the positive or negative evaluation of the effects of a proposed measure.

The argument *by the consequences* is sometimes referred to in Latin as argumentation *quia* "because" in opposition to the arguments *by the cause* or *propter quid* "because of which".

S. Pragmatic; S. Causation; A priori, a posteriori.

2. Arguments by the identity of the consequences

The same kind of argumentation applies to deductions made from the implied meaning of words, as an appeal to the sense of semantic coherence or logical consecution:
> *Scheme*: "Another topic consists in concluding the identity of precedents from the identity of results"
>
> *Instance*: "There is as much impiety in asserting that the gods are born as in saying that they die; for either way the result is that at some time or other they did not exist" (Aristotle, *Rhet*. II, 23, 1399b5; F. p. 313-315).

If something is condemned because it forcibly involves mechanically something negative, then it automatically creates a category of causes *"having that kind of negative consequences"*, which must also be condemned. If the reason given for banning the consumption of marijuana is that it causes a loss of control, then all substances that cause a loss of control must also be banned, including for example alcohol.

3. Refutation by contradictory consequences

The refutation by contradictory consequences is a kind of *ad hominem**, used in dialectic:
> Peter says "**S** is **P**".
> **S** has the consequence **Q**: the fact is known and accepted by the opponent.
> **P** and **Q** are incompatible
> So Peter says incompatible things about **S**.

Example:
> Pierre says that *power is good*.
> Yet, everyone agrees that *power corrupts* (consequence)
> Corruption is an *evil*.
> The *good* is incompatible with the *evil*; to be good, power should *exclude* corruption.
> Peter says contradictory things.

Consistency

The fundamental expression of argumentative *coherence* or *consistency* is *non-contradiction*, S. Non-contradiction; Absurd; *Ad hominem*. The consistency requirement is of special importance in systems of regulations of human behavior, religion, law, as well as ordinary institutional or familial rulings.

The consistency requirement is expressed *a contrario* in the refutation strategy mentioned in Aristotle's *Rhetoric*, topic n° 22:

> Another line of argument is to refute your opponent's case by noting any contrast or contradiction of dates, acts or words that it anywhere displays. (1400a15; RR p. 373).

1. After the event as before

The topic n° 5, "on consideration of time" appeals to consistency. This topic is not explicitly stated, but presented through two examples:

> If before doing the deed I had bargained that, if I did it, I should have a statue, you would have given me one. Will you not give me one now that I have done the deed? (*Rhet*, II, 23, 5; RR, p. 361).

The situation is this:
1. **X** (asks nothing and) accomplishes a feat (maybe an impulsive heroic act)
2. After this, he asks for a reward.
3. Argument: if he had asked before, they would have agreed on a reward.

The hero considers that all feats must be paid for as such. It is as if the definition of the word *feat* includes the characteristic *"deserves a reward"*:

L1: — *If you do, you'll receive…*
L2: — *I have done and done well, so give me…*

This topic explains the disappointment of one who reports the found wallet and receives no reward.

2. Human (in)consistency

Consistency may be the rule, but inconsistency is a fact of life. This is what topic n° 18 expresses:

> Men do not always make the same choice on a later and on an earlier occasion, but reverse their previous choice. (*Rhet*, II, 23, 18; RR, p. 371)

This topic materializes in the following enthymeme:

> When we were exiles, we fought in order to return; now we have returned, it would be strange to choose exile in order not to have to fight. (*ibid.*)

The enthymeme seems to assume the following situation. In the past, exiles fought to return home, and they returned; in the current situation, they are suspected of refusing to fight, and preferring exile. They deny the charge by this enthymeme, which is a claim of consistency, as in:

> You fought for this position, now you can't accept being thrown out like that!

This is a kind of positive *ad hominem* argument; it may presuppose an *a fortiori*: "*We fought to return to our homeland, a fortiori we will fight to not be chased out of it!*" Those accusing them reply that "*Men do not always make the same choice, etc.*"

The opposing party argues from an opposing vision of human nature; the two opinions "men are constant / inconstant", are equally probable (see *ibid* I, 2, 14; p. 25). They can thus be the basis for two antagonistic conclusions.
S. *Ad hominem*; Consistency; *A fortiori*.

3. Consistency of the system of laws and stability of the objects of the law
Lat. arg. *a cohærentia*, de *cohærentia*, "formation into a compact whole".

3.1 Principle of coherence of laws, *a cohærentia*
This principle requires that, within a legal system, one norm cannot conflict with another; the system does not allow *antinomies*. An argumentative line can therefore be rejected if it leads to the view that two laws are contradictory; this is a form of argumentation from the absurd.
In practice, this principle excludes the possibility of the same case being settled in two different ways by the courts.
By applying this principle, if two laws contradict one another, they are said to do so only *in appearance*, and, as a consequence, they must be *interpreted* so as to eliminate the contradiction. If one of these laws is obscure, it must be clarified by reference to a less doubtful one.
The argument *a cohærentia* is used to solve conflicts of standards. To prevent this kind of conflict, the legal system provides for adages, which are interpretative meta-principles, such as "the most recent law takes precedence over the oldest". These adages are interpretative meta-principles, coming from Roman law and sometimes expressed in Latin: "*lex posterior derogat legi priori*".

3.2 Principle of stability of the object of the law, *in pari materia*
Lat. *in pari materia*: lat. *par*, "equal, like"; *materia*, "topic, subject" argument "in a similar case, on the same subject".

The argument *a cohærentia* deals with the formal non-contradiction of laws in a legal system. The argument *in pari materia*, or argument "on the same subject", expresses a substantial form of consistency. It requires that a law be understood in the context of other laws having the same goal or relating to the same subjects, that is to say the same beings (persons, things, acts) or the same topic.
The given definition of the subject of the law must be stable and consistent. The application of the argumentation *a pari* presupposes the stability of the legal categories. S. *A pari*; **Taxonomy and categories**.

This principle of consistency prompts the legislator to harmonize the system of laws on the same subject. What constitutes the same subject and the set of laws on the same subject might be questioned. Anti-terrorism laws, for example, are a package of different statutory provisions, for which it is necessary to ensure that the definition of "terrorism" remains the same in each of the passages that uses the term. If this is not the case, these laws need to be made consistent, which implies that they themselves must be underpinned by consistent policy.

The two topoi discussed in the two following paragraphs are taken from Aristotle's *Rhetoric*. They are based on the two incompatible, but equally recognized substantial topoi, "*human conduct is, or should be consistent*" and "*human conduct is inconsistent*".

4. Argument from narrative inconsistency

As a particular case of *ad hominem* argumentation, showing inconsistencies in the accusatory narrative can rebut a charge:
    S1:   — *you are the heir, you benefit from the crime, you killed to inherit!*
    S2:   — *if so, I should have murdered the other legatees too.*

The prosecution will have to prove that **S2** also intended to murder the other heir, or otherwise find an alternative motive. The defense starts from the hypothesis put forward by the prosecution to show that the actions of the suspect do not fit in the proposed scenario; the accusatory narrative contains flaws or contradictions.

The argument of incoherent accusation exploits a basic principle of practical rationality: the actions of the suspect must be consistent with his or her claimed goal. The accused can refute the accusatory narrative by showing that, according to this narrative, he acted inconsistently:

> You say I'm the murderer. But it has been proven that just before the crime, I spent an hour at the cafe in front of the victim's home, everyone saw me. It is not coherent conduct on the part of a murderer to show himself at the scene of his crime.

Any weakness identified in the prosecution scenario can then be used to clear the defendant.

The principle of consistency of laws and the principle of stability of the subject of the law concern the coherence of the legal system. The argument from the inconsistency of the narrative exploits the resources of narrative rationality: all the narratives offered as excuses, all the narratives mingled with argumentation are vulnerable to this type of refutation.

Conversely, the argument seems plausible and reasonable because the story is so, and because the speaker knows how to tell it.

The strategies described in the topoi n° 22, 25 and 27 and probably 18 (cf. supra) of the *Rhetoric* are relevant to this discussion (Aristotle, *Rhet.*, II, 23), S. **Collections (II)**.

Contradiction

1. In dialogue, a contradiction *emerges* when a first speech turn is not ratified by the partner's following turn. The contradiction is *open* when the two parties produce anti-oriented speech turns. When the opposition is thematized and ratified by both participants, it gives rise to an argumentative situation. **S. Disagreement; Question; Stasis; Denial; Refutation; Counter-argumentation.**
Contradictions can be solved on the spot by a series of adjustments and arrangements, by playing on the margins of indeterminacy and windows of opportunity left by ordinary language and actions.

2. Non-Contradiction@ principle, S. *Ad hominem*; **Consistency.**

3. Contradiction as a relation between opposite@ terms, S. **Contrary and contradictory.**

4. Contradiction as a relation between propositions: S. **Contrary and contradictory; Absurd.**

Contrary and Contradictory

1. Definition
In logic, the square of oppositions links the affirmative and negative propositions, universal and particular, according to a set of immediate inferences, among them the relations of contradiction and contrariety, S. **Classical Logic (II).**
Two propositions **P** and **Q** are *contradictory* when they cannot be simultaneously true or simultaneously false; that is, one of them is true, and the other is false, as shown in the truth-table below.
Two propositions **P** and **Q** are *contrary* when they cannot be simultaneously true, but can be simultaneously false, S. **Logic (IV).**
These terms can be easily mixed up. The easiest way to avoid confusion is to refer the relations of *contrariety* and *contradiction* to two kinds of universes, defining two kinds of opposites. Let **U** be a universe including a series of individuals.

(i) Contradictories — In the case of *contradiction*, the opposition is within a *bi-dimensional* universe, such as the traditional system of genre. "— *is a man*" and "— *is a woman*" are *contradictory* predicates in this system. In a non-traditional system of genres, they are *contrary* propositions.
U is a two dimensional universe; two properties P_1 and P_2 are defined upon this universe, such as:
— Any members of this universe possess *either* the property P_1 *or* the property P_2: $(P_1 \vee P_2)$
— None possess both properties P_1 and P_2: neither is both $(P_1 \& P_2)$. This is noted $(P_1 W P_2)$, with the symbol 'W' for "disjunctive *or*".

P_1 and P_2 are *complementary* properties; they divide the universe **U** into two complementary (non-overlapping) sets.
— P_1 and P_2 are *contradictories (opposites)*; they are in a relation of *contradiction*.

(ii) **Contraries** — In the case of *contrariety*, the opposition is within a *multi-dimensional universe* such as the universe of colors. "— *has white hair*" and "— *has red hair*" are *contrary* predicates: one person cannot have both white and red hair (notwithstanding the case of badly dyed hair roots); and he or she may have brown hair.
U is a **n**-dimensional (more than two dimensions) universe: $P_1, \ldots P_i, \ldots P_n$.
— Any members of this universe possess one of these properties, P_j; that is, is either a $P_1, \ldots$ or a $P_i, \ldots$ or a P_n.
— None possess two or more properties $P_1, \ldots P_i, \ldots P_n$, that is, none is both (P_k & P_l).
— $P_1, \ldots P_i, \ldots P_n$ are *contraries*; they are in a relation of *contrariety*.

To sum up, semantically connected predicates, or properties, are *opposite* if they divide exhaustively their universe of reference into a series of non-overlapping sets. If there are just two such properties, they are said to be *contradictory properties*; if there are more than two, they are said to be *contrary properties*. So, contradictories are the limit case of contraries.

Opposites	Two-dimensions opposition: the two opposite properties are *contradictories*
	More than two-dimensions opposition: the more-than-two opposite properties are *contraries*

2. Refutation by substitution of contrariety to contradiction

It follows that an assertion based on a contradiction can be refuted by showing that the universe under discussion should not be considered as two-dimensional, but multi-dimensional. This seems to be the case in the following example.

> *In 1864 Pope Pius IX published the* Syllabus, *that is, a collection or a catalog summarizing the positions of the Vatican about "modernist" ideas. Considered as retrograde, the Syllabus is strongly attacked by "the modernists". In 1865, Mgr. Dupanloup, defended the Syllabus in the following terms; "they" refers to the modernists.*

> It is an elementary rule of interpretation that the condemnation of a proposal, condemned as false, erroneous and even heretical, does not necessarily imply the assertion of the contrary, that could be another mistake, but only its contradictory. The contradictory proposition is the one that simply excludes the condemned proposition. The contrary is the one that goes beyond the simple exclusion.

> Well! It is this common rule that they apparently have not even suspected in the inconceivable interpretation of the *Encyclical* and the *Syllabus* they have been giving us for the past three weeks. The Pope condemns this proposition: "*It is permitted to refuse obedience to legitimate princes*" (Prop. 63).
> They claim that, according to the Pope, disobedience is never permitted, and that it is always necessary to bend under the will of princes. This is jumping to the last end of the contrary, and attributing to the vicar of Jesus Christ, the most brutal despotism, and slavish obedience to all the whims of the kings. This is the extinction of the noblest of liberties, the holy freedom of souls. And that's what they claim the Pope said!
>
> Félix Dupanloup, Bishop of Orleans,
> [*The September 15th Convention, and the December 8th [1864] Encyclical*], 1865[1].

Is the universe of the *Encyclical* binary or multidimensional? Let's consider a position **X**.

— If it comes in a binary opposition, "*allowed* vs. *forbidden*", then the proposals "*it is permitted (to refuse obedience)*" / "*it is forbidden (to refuse obedience)*" are *contradictory contraries*: only one of these propositions is true. If we condemn the proposition "*it is permitted to refuse obedience to legitimate princes*", then we have to conclude that the contradictory is true, that is to say, "*it is forbidden to refuse obedience to legitimate princes*", otherwise said: "*we must always bow our heads under the will of the princes.*"

Thus, for Dupanloup, the malevolent "modernists" substitute contradictories for contraries, what he describes as "jumping to the last end of the contrary", which is a proper designation of the contradictories.

He accuses his opponents of reframing the Pope's position, using a strategy of absurdification (an exaggeration up to the absurd, **S Exaggeration.**

— If the position **X** enters a three dimensional universe, as "prescribed / permitted (indifferent) / forbidden" then the proposals "it is allowed / it is forbidden" (to refuse obedience) are not contradictories but contraries: they are not simultaneously true, but they can be simultaneously false, e.g. if **X** is indifferent. The inference "If **X** is not fought, X is required" is not valid. If we condemn "*it is permitted to refuse obedience to legitimate princes*" then we can only conclude one or the other of these opposites:

> *It is prescribed to refuse obedience to legitimate princes.*
> *It is forbidden to refuse obedience to legitimate princes.*

As it would be difficult to admit that Pius IX, or anyone else, prescribes a systematic duty of disobedience to the legitimate rulers, we are left with the other member of the disjunction, that is, "X is forbidden."

— If two or more additional options, "encouraged" and "discouraged" are introduced, we get a five dimensional universe "prescribed / advised / permit-

---

[1] Quoted after Félix Dupanloup, *La Convention du 15 Septembre et l'Encyclique du 8 décembre [1864]*. In Pius IX, *Quanta Cura and the Syllabus*. Paris: Pauvert, 1967. P. 104-105.

ted (indifferent) / recommended / forbidden". The interpretation "encouraged" is hardly possible, for reasons previously seen; "discouraged" could correspond to the intention of the *Syllabus*, such as interpreted by Dupanloup. One then wonders why this sentence seems so solemn : if we admit that something which is not recommended is something that we do not do without good reason, it is obvious that one does not disobey the legitimate prince without some good reason.

Convergent

Convergence is a basic mode of organization of complex discourse supporting a conclusion, S. Convergent, Linked, Serial.

Two or more arguments are *convergent* when they independently support the same conclusion. The arguments are said to be *convergent* or *co-oriented*, and the argumentation is called *convergent* or *multiple*.

"Two reasons are better than one": in a *convergent argumentation*, a claim is defended on the basis of several arguments which, considered separately, can be relatively inconclusive, but, considered as a whole, combine to make an stronger case: "*My computer is beginning to age, there are discounts on the price of my favorite brand, I've just got a bonus, I will buy one!*".

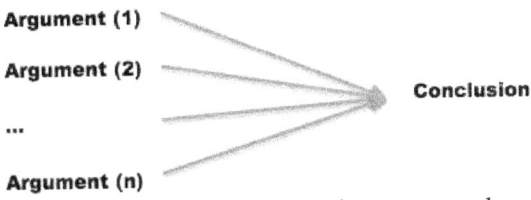

In the above diagram, each argument is represented as a whole. The following diagram spells out the transition laws according to Toulmin's proposal, S. Layout; compare with linked@ argumentation:

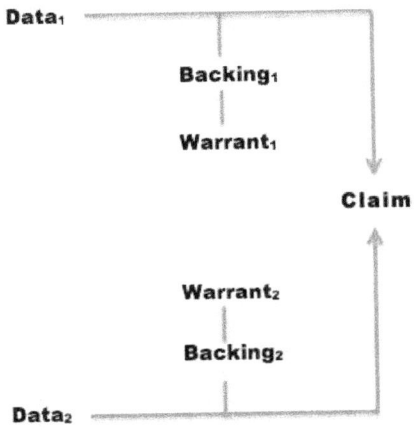

As well as pro-arguments, counter-arguments can converge to refute a claim. S. **Script**.

This open structure defines the argumentative *net*, as opposed to the demonstrative *chain*. In the demonstrative chain, each step is necessary and sufficient; if one step is invalid, the constituent parts, and, in turn, the whole construction collapses. In the case of the argumentative net, if one link in the mesh breaks, the net can still be used to catch fish, at least the biggest ones.

In a convergent argumentation, the organization of the sequence of arguments is relevant. If the arguments are of a very different strength, a weak argument alongside a strong argument risks damaging the whole argumentation, especially if this argument ends the enumeration:
> He's a great hunter, he killed two deer, three wild boars and a rabbit.

In classical rhetoric, the theory of discourse general organization (Lat. *dispositio*) discussed the supposed different persuasive effects of the various possible textual arrangements of converging arguments of different strength, S. **Rhetoric**.

Convergent arguments can be merely listed (paratactic disposition):
> Arg, Arg *and* Arg, *so* Concl

The argument can be connected by any listing or additive connective:
> *first*, Arg1; *second*, Arg2; *third*, *Arg*3; so Concl.
> *Additionally, also, in addition, let alone, moreover, not only, besides*

Connectives such as *besides, not only, in addition, let alone, not to mention...* not only add argument upon argument(s), they present them as if each one was actually sufficient for the conclusion, and are adduced just "for good measure" (Ducrot *& al.* 1980, pp. 193-232):
> No, Peter will not come on Sunday, he has work, as usual, besides his car broke down.

The additive approach considers that each argument brings in a part of truth, and that these parts can be arithmetically added to create one big decisive discourse. Speech activity theory considers that by nature, an argument is presented as sufficient, and that their addition actually obeys the logic of commercial display for consumers (the audience), that is to say the speaker offers the audience a range of equally satisfying and self-sufficient arguments.

Case-by-case@ — To refute the conclusion of a convergent argument, each of the arguments supporting this conclusion must be discarded. Thus, a convergent argument is countered by a case-by-case refutation, limited to cases that have been advanced by the proponent.

Convergent — Linked — Serial

The conclusion of an argumentation is usually expressed in a single statement, possibly expanded in a brief conclusive speech, S. **Argument - Conclusion**. The argument part, supporting and sometimes surrounding the conclusion, can be considerably developed along quite different lines, referred to as:

— *Convergent@ argumentation*, also called *multiple argumentation*, combines several co-oriented arguments, S. **Convergence**.
— *Linked@ argumentation*, also called *coordinate argumentation* is composed of several statements combining into an argument, S. **Linked**.
— *Serial@ argumentation*, also called *subordinate argumentation* is composed of a succession of argumentations, such as the *conclusion* of the first one is taken as *argument* to support a second one and so on, S. **Sorite**.

Conversion

1. Logic
In logic, two propositions are *converse* (in a relation of *conversion*) if they swap their subjects and their predicates. "**As** *are* **Bs**" and "**Bs** *are* **As**" are converse propositions. The converse proposition of a true proposition is not necessarily true, S. **Proposition**.

2. Grammar and argumentation
In grammar, conversion can apply to any binary structures. Restructuring an expression of the opponent, that is, playing with his or her words, can be instrumental in reversing the global orientation of his or her discourse, according to the mechanisms of the *antimetabole*, S. **Orientation Reversal**

> Well, you know, this talk about the so-called *pleasures of retirement* is just empty talk to mask the *retirement of pleasures*.

> Personally, I'd prefer a *frightful end* to this *endless fright*.

> González, on Kohl '*He fought for a* European Germany, *never again a* German Europe."(*El País*, 07-01-2017)

One can radically counter-argue a proposition by emphatically supporting its converse, S. **Causality (II); Analogy**:

S1 — **A** is the cause of **B**;
  A is like **B**; **A** mimics, copies **B**.

S2 — Not at all! <u>**B**</u> is the cause **A**!
  <u>**B**</u> is like A; <u>**B**</u> copies A.

In the same way, a sweeping defense strategy consists in converting the roles of accuser and accused, first by applying the reciprocity principle, "*it takes one to know one*":

You blame me (for **X**), I blame you (for **Y**)
You filed a complaint against me (for **X**), I file a complaint against you (for **Y**).

and, second, by converting the position about the *same* criminal offense:
<u>You</u> are the culprit, *you* did it, <u>you</u>, who accuse me!

The child's reply *"he who says it did it"* converts the accusation:
S1 — <u>You</u> *stole the orange!*
S2 — *No,* <u>you</u> *stole it, who says it did it!*

The fact that **S1** accuses **S2** is used by **S2** as an argument to accuse **S1**. **S. Reciprocity; Stasis.**

Cooperative Principle

According to H. P. Grice, the intelligibility of the conversation is ruled by "a rough general principle which participants will be expected (ceteris paribus) to observe", namely:

> 'Make your conversational contribution such as is required, at the stage at which it occurs, by the accepted purpose or direction of the talk exchange in which you are engaged'. One might label this the COOPERATIVE PRINCIPLE. (1975, p. 45; capitalized in the text).

This "Principle of Cooperation", is specified under four forms, "Quantity, Quality, Relationship and Manner" (*ibid.*).

— *Quantity*: "I expect your contribution to be neither more nor less than is required" (*ibid.*).

— *Quality*: "I expect your contribution to be genuine and not spurious" (*ibid.*). This can be compared to the requirement of accuracy mentioned in the pragma-dialectical rule 8; the same concern is also found in Hedge's Rule 1 "For an honorable controversy", **S. Rules.**

— *Relation*: "I expect a partner's contribution to be appropriate to immediate needs at each stage of the transaction" (*ibid.*). This concerns in particular the relevance of the turn in relation to the present topic of dialogue and action. Grice recognizes the difficulty of identifying what is relevant in an exchange. The pragma-dialectical "Relevance rule" deals with this same requirement (van Eemeren, Grootendorst (2004, p. 192). **S. Relevance; Rules.**

— *Manner*: "I expect a partner to make it clear what contribution he is making" (*ibid.*). This entry can cover the refusal of the obscurity of expression and action; of ambiguity (the first of the Aristotelian fallacies); of the unnecessary prolixity, corresponding to the fallacy of *verbiage@*.

Grice holds that his principles capture the rational character of conversation:

> One of my avowed aims is to see talking as a special case or variety of purposive, indeed rational behavior. (*Id.*, p. 47)

as well as its reasonable character: Respecting these principles is not merely "something that all or most do IN FACT follow, but as something that it is REASONABLE for us to follow, that we should not abandon" (*id.*, p. 48; capitalized in the text).

These four principles can be compared with those advanced by normative theories of argument**, S. Rules.**

A statement violating Grice's principles is not eliminated as fallacious, but is understood as an *indirect speech act*. When a participant notes that something is not in conformity with a conversational rule, the reaction is not to accuse the partner of making an irrelevant or irrational contribution, but to engage in an interpretive process to identify why he or she has flouted the conversational rule. The analysis of fallacies reverts to this interpretive orientation whenever it adds to its logic pragmatic considerations taking into account the contextual conditions of the exchange.

In an argumentative situation, the concept of cooperation is a strategic issue redefined by the participants, who are not necessarily willing to cooperate, for example in their own refutation. There is nothing scandalous or irrational about this, insofar as partners are aware of being in such an intentionally opaque context**, S. Politeness**. *Rational, reasonable*, as well as *honorable* rules for discussion are intended to reintroduce or strengthen cooperation in such antagonistic contexts.

Coordinate Argumentation ▶ Linked

Correlative Terms

Correlative terms are also called *relative* or *reciprocal* terms, and may be considered as opposite@ terms. *Mother* and *child* are correlative terms, that is, they are linked by the immediate inference:

    if **A** is the mother of **B**, then **B** is the child of **A**

Correlative terms are defined by reference to one another; *mother* is defined as "woman with children"; *child* as "son or daughter of **M**".

The following terms are correlatives:

    cause / effect   double / half   master / slave
    action / passion   sell / buy

Generally speaking, two predicates **R1** and **R2** are in a correlation relation when
    A_**R1**_B <=> B_**R2**_A
    A_*mother*_B <=> B_*child*_A

"By definition, correlatives are opposites"; they are "ontologically simultaneous" (Hamelin [1905], p. 133).

The topic of the correlative is n°3 on Aristotle's list:
> Another line of proof is based upon correlative ideas (*Rhet*, II, 23, 3; RR, p. 357)

The topic is exemplified by the enthymemes:
> Where it is right to command obedience, it must have been right to obey the command.
> The tax-farmer: "if it is no disgrace for you to sell it, it is no disgrace for us to buy it" (*ibid*.).

These inferences have limitations:
> If it is legal/tolerated to buy 2 g of marijuana, then one may sell 2 g of marijuana.

But what about "possessing" and "buying"?
> if it is legal/tolerated to possess 2g of marijuana,
> then it is legal/tolerated to buy 2 g,
> then it is legal/tolerated to sell it

given that for me, the only way to get marijuana is to buy it. But the law can make a distinction between two kinds of "possession": the possession of drugs *for private consumption* is not an offence, while possession *for trafficking* is.

The following case deals with two pairs of correlatives, *know / learn*, and *order / obey*, articulated by the topic of the opposites:
> If you want to command, you must first learn to obey.
> The executive, when he was on his way up, had to learn to obey so that he should know how to command (quoted in *Linguee*).

Counter-Argumentation

The expression *counter-argumentation* can be used to refer to any kind of discourse, argued refutation or objection, going openly against an argumentation. A mere "*No!*" can be considered as a counter-argumentative move, even a non-verbal expression of rejection clearly interpretable as such.

Unlike direct refutation, a specific "argumentation *vs.* counter-argumentation" situation occurs when the refutation is reciprocal and indirect:
— Speaker **S1** argues for proposition **M**.
— Speaker **S2** counter-argues for proposition **R**, incompatible with **M**:
> S1 — *Let's built the new school here, the land is cheaper.*
> S2 — *Let's built the new school there, the students will waste less time commuting*

S2 makes a *counter-proposition* **R**, providing an alternative to **M**.
Argumentation and counter-argumentation play a reciprocal role in refutation. In such a polarized situation, the fact of providing a reason for doing **R**, incompatible with **M**, serves as a reason for *not* doing **M**. Any good reason for supporting **R** is seen as a *counter-argument* to **M**.

Counter-Discourse ▶ *Counter-Argumentation*

The argumentation / counter-argumentation structure may correspond to an emerging argumentative situation, or to the moments when the participants present and argue their position without considering the antagonist's proposal, which can occur at any time in a concrete argumentative situation.

An argued position can be presented in isolation in an autonomous text without refuting or even mentioning any existing counter-argumentation. Adopting such a strongly assertive strategy avoids the paradoxes of refutation, but can be seen as a kind of contempt for the argument put forward by an opposing party. **S. Question; Contradiction; Antithesis; Dismissal.**

As is the case with weak refutations, a weak counter-argumentation will reinforce the attacked position. In the following passage, Noam Chomsky considers that his opponent, the philosopher Hillary Putnam, has failed to develop a counter-argumentation, even a counter-proposal, and argues that this shows that he, Chomsky, must be right:

> So far, in my view, not only [Putnam] has not justified his positions, but he has not been able to clarify what these positions are. The fact that even such an outstanding philosopher fails to do so, may allow us to conclude that…
> Noam Chomsky, [Discussion on Putnam's Comments], 1979.[1]

The praise of the adversary as "an outstanding philosopher", is a characteristic move in this kind of refutative strategy, **S. Politeness; Ignorance; Paradoxes.**

Counter-Discourse ▶ Counter-Argumentation

Counter-Proposition ▶ Counter-Argumentation

Criticism — Rationalities — Rationalizations

1. Rationalities

In the modern and contemporary world, *scientific rationality*, based on experience and shaped by mathematics has taken the upper hand upon the current vision of rationality. Scientific discourse is taken as the prototype of rational discourse, while argumentation is seen as the instrument of reason as *reasonableness* in human affairs. This position has been strongly reasserted by Perelman & Olbrechts-Tyteca ([1958]), **S. To Persuade; Persuasion.**

Ordinary discourse in action embodies different kinds of rationalities.

Rationality as common sense — Rationality as common sense can be defined as the art of thinking complying with the rules and intuitions embodied in

---

[1] Noam Chomsky, Discussion sur les Commentaires de Putnam. In Piattelli-Palmarini M. (ed.). *Théorie du Langage, Théorie de l'Apprentissage.* Paris: Le Seuil. 1979. P. 461.

traditional logic and adapted to social necessities by rhetorical argumentation. As a scientific concept, this vision of rationality has been shaken to its foundations by the development of axiomatic thinking, as exemplified by non-Euclidian geometries or by the invention of the imaginary unit **i**, such as $i^2 = -1$. In human sciences, the Freudian invention of the unconscious and the development of studies on ideologies and social determinisms, have most certainly challenged the vision of a sovereign subject transparent to him/herself, consciously mastering his or her calculus, intentions, discourses and actions. This double crisis directly affects the classical vision of the rational well-intentioned rhetorical orator.

Rationality as adaptation of a conduct to a goal — Rationality as adaptation of conduct directed towards a goal covers all forms of action guided by a script, recipe or pre-established conventional plan. To make good custard, for example, it is more rational to pour the hot milk on the eggs than to put the eggs in the hot milk, so that the cream will be more homogeneous. This principle of rationality merges with the consistency requirement between *conduct* and *objective*. It is exploited by all forms of refutation revealing a contradiction in the opponent's conceptions and actions S. *Ad hominem*; **Consistency**. Since it is human to pursue several objectives at the same time, the resulting practical rationality is perpetually destabilized.

Rationality as adaptation of a conduct to a goal is compatible with crime. The Marquis de Sade is an outstanding arguer. Hence the possibility of *delirious and despotic* rationalities serving equally perverse goals.

Rationality related to a domain — Rationality depends on domains. A given behavior (with or without a linguistic component) is said to be rational if it conforms to recognized practices in the relevant domain, technical field, scientific paradigm or tradition of thought, S. **Rules**.

Democratic rationality — Democratic rationality is a quality of societies and groups where information is accessible; where free and contradictory examination of political positions and oppositions may develop with a view to effective decision-making; where there is a right of reply; and where the safety of the opponents is ensured. It is a form of society in which the holders of legal power and violence are brought to account for their use.

Rationality is sometimes thought to be governed by rules; if one tries to express the preceding conditions as a set of rules, they will have to be hierarchized and context-sensitive in order to integrate various genres and practices.

2. Discursive and argumentative rationality

Language rationality — From a linguistic point of view, a discourse is deemed rational if it is well built, if it is understandable, if the speaker can account for it and if it makes sense in relation to the problem discussed or the task under way.

The reasonableness paradox created in an argumentative situation driven by a question is that each of the competing discourses taken in isolation makes sense, but, taken together, they become contradictory. To discriminate between these answers, theorists of argumentation need a criterion, which would be stronger than meaning, and, to that effect, introduce the notion of *rational* or *reasonable discourse* into their models. The different families of theories of argumentation can be related to different visions of rationality.

Discourse rationality and discourse types — Argumentative discourse is *not* the unique receptacle of discourse rationality. There is not one, but several discursive rationalities: *argumentative* rationality, *narrative* rationality, *descriptive* rationality, and so on. Irrationality is manifested in incoherent and delirious narrations, descriptions or prescriptions; any ill-conceived installation diagram which can be called irrational, because it is useless.

Rational discourse and effective rhetoric — Effective rhetoric, focused on the persuasion of an actual, relevant audience is a case of goal adaptive rationality. It is compatible with verbal and nonverbal manipulation.

Rational discourse as justified and rectified discourse — The definition of rational discourse as a justified discourse develops the idea that a discourse is reasonable insofar as its claim is not asserted on the basis of individual certainty, but openly supported by other propositions, exploiting some kind of public data connected to the claim by some recognized rule, albeit fragile. Its rationality increases if it exhibits its weak points, suggesting the directions that must be taken to improve it; as Bachelard says, there is no truth, only rectified errors. The Toulminian layout@ meets these requirements: the *Claim* is based on *Data*, according to a *Warrant*, itself supported by a *Backing*, and duly *Qualified*. The critical instance is represented by its trace, the *Rebuttal*, indicating the potential point of refutation.

The practice of dialogue, whether remote or face-to-face, can be considered to be the exercise of the *critical function of language*. A speech is more rational if it has been duly criticized, that is, if it has survived a number of contradictory encounters. Criticizing does not mean "denigrating" or "rejecting", but "passing a judgment", positive or negative, on an activity. The observation of the data shows that the partners involved in an argument spend much time evaluating their partner's arguments (Finocchiaro 1994, p. 21). Argumentative speech is evaluated in a meta-discourse, produced under any conditions, face-to-face or at a distance in space and in time. Any approach to argumentative discourse concerned with empirical adequacy must take this critical dimension into account.

For the New Rhetoric, arguments are assessed by the participants in the rhetorical event; the rationality of an argument increases with the number and quality of the interested and competent audiences who accept it. The progression towards human rationality is seen as an evolution from a particular to a universal audience, **S. Persuasion.**

The dialogue models of argumentation put the critical activity at the center of their concerns. Pragma-Dialectic and Informal Logic develop a critique of argument based on the notion of fallacy. To detect fallacies, pragma-dialectics uses a system of rules, while informal logicians use the technique of critical questions. **S. Paralogism; Sophism; Fallacy; Norm; Rules; Evaluation**

3. Rational argumentation, as a "dream of language"

The Argumentation within Language theory of Anscombre and Ducrot and the Natural Logic of Grize make no commitment to rationality; they are *not irrational* but *a-rational*. Any discourse being argumentative, the idea of rectifying a discourse in order to improve its argumentativity or its rationality does not make sense. These theories are just concerned with the fact that to be rational a discourse must first be meaningful, **S. Schematization; Orientation.**

The Argumentation within Language theory proposes a radical criticism of the capacity of discourse to achieve any kind of rationality. Conclusions are seen as mere semantic developments of the arguments, the argumentation process being driven by the linguistic orientations of the utterances; the discourse develops according to the orientations of natural language, which are denounced as biases by fallacies theories, in search of a referenced, neutral, objective language. Rephrased in the language of fallacies, this amounts to the claim that argumentation in natural language is circular, so fallacious. It results that argumentation as a rational process is a "dream of discourse" (Ducrot 1993, p. 234). Following this metaphor, the rational pretension of argument (as found in Perelman, for example) will be seen as a "rationalization of the dream", and the criticism of the arguments, as a "criticism of the dream", whereas dreams can only be exposed and interpreted as such. **S. Demonstration.**

4. Rationality and rationalization

Psychoanalysis uses the terms *rationalization* or *intellectualization* to refer to discursive constructions claimed to be *rational* by the subject who tries to *account for* his or her actions, representations, feelings, symptoms or delirium. Psychoanalysis objects to such reconstructions that the subject has no conscious intellectual access to their true source (Laplanche and Pontalis, 1967, art. [*Rationalization*]):

> Whenever possible, [the ego] tries to remain in good terms with the *id*; it clothes the *id*'s unconscious commands with its preconscious rationalizations [...] In its position midway between the *id* and reality, it only too often yields to the temptation to become sycophantic, opportunist and lying, like a politician who sees the truth but wants to keep his place in popular favor.
>
> (Freud [1923], p. 55).

D

Debate

Typically Western debates and discussions implement all the facets of argumentative activity: constructing points of view, producing good reasons; interacting with different people and points of view, building more or less ephemeral alliances, integrating / refuting / destroying the positions of others, backing arguments by drawing on personal involvement in the debated issues. Sometimes the two terms *arguing* and *debating* are assimilated, with TV debates implicitly considered as the prototypical argumentative genre.

This vision of argumentation has major limitations, such as the automatic association of argumentation with polemical debate, which is a non-cooperative and non-conclusive form of argumentation. TV debates may try to influence the decision, but they have no decision-making power. Work meetings, family discussions are certainly more representative of the complexity of argumentation. In a work meeting where issues are debated with both short term and long term implications, different kinds of sequences must be managed in different episodes: new participants are introduced; the agenda is read; relevant information is given (to all, to less informed participants), conclusions are written down — not to mention the episodes devoted to interaction management, including digression and jokes. The level and kind of argumentativity of these episodes can be extremely varied.

The form and efficiency of the arguments put forward in a debate depend on the relative power of the participants in the relevant sphere. If taken on a majority basis, the decision compels the minority, whether or not persuaded, and regardless of whether or not the winning argument is the strongest from the point of view of an external evaluator.

1. The informed and properly argued debate as a source of legitimacy

From a *foundational perspective*, a political decision may be considered legitimate if conforms with, or is derived from an original pact, a social contract that the ancestors, or ideal representatives of the community, freely convened in a mythical original time, or in an ideal rational space.

Democracy values debate. A decision is considered legitimate only if the issue has been publicly argued pro and contra, in a safe, open, free and contradictory space. In principle, the decision should take the results of debate into account; whether or not this decision is really supported by the best argument, is another issue; authority and power play a role. Debate as a form of argument is at the heart of democratic life. At school, it is considered to be the key instrument of "democratic learning", be it in Citizenship education, in History, or in Science education.

2. Criticism of debate

Debate, however, is not an innocent and miraculous practice which can solve all issues in education, society and uneven development. Debate, particularly debate in the media, or in any public space, is the target of a critical argument that includes the following points.

— Resorting to debate may be merely an artifice of presentation. The topic is framed as an issue, as being the focal point of two antagonistic discourses, as if things were "interesting" only insofar as they radiate some polemical heat.

— Paradoxically, *"the debate is open"* can be a convenient conclusive formula, when listeners in both camps have got their share of good reasons, as if the main virtue of a debate is furthering the debate, and justifying further debates.

— A dubious and interrogative posture can be very comfortable. Debate merges the variety of positions in one unique global voice saying everything and the opposite; *but* articulates such unresolved contradictions very well. Correlatively, debate is a fertile field for argumentative personalities to flourish.

— Becoming an end in itself, debate becomes a *performance*, and loses all connection with the search for truth, clarification of the issues and positions, agreement or exploring and deepening the differences. This is the sophistical *ad ludicrum* tendency rightly and abundantly condemned as *playing to the gallery*; a delighted audience consents to its own manipulation, **S. Laughter and Seriousness**.

— From an educational point of view, debate can promote *confrontational* forms of argumentation. In fact, debate does not systematically break with symbolic violence, but can simply displace it. Some cultures find open interpersonal confrontation repugnant, or at least rude and counterproductive. Pressing students into a debate can be an educational blunder. Moreover, debates on serious issues divide groups, and can put at risk the reputation and even the security of the individual summoned to expose his or her creeds, networks and communities. Such self-exposure cannot be an option in some communities and cultures.

— Even coming from the best-organized public socio-political forum, the argument deemed the best might differ according to the parties. What is more, once taken, the decision can necessitate a new discussion about how it should be implemented, this being a regulatory or legal issue, in the hands of the current regime. There is a broad open and opaque space between argumentation and decision, and another one between decision and implementation.

— The ideal space in which the debate is held is framed as egalitarian and free. It denies any imbalance of power, at least it puts power relations between parenthesis. But every place has its own rules that impose formal and substantive standards. Such *rules of the place* apply to all participants. Debate *presupposes* democracy, as well as it *promotes* democracy.

Debate is a powerful resource, which must be used with care. Debate alone will not resolve all social and individual ills, nor global hardships.

Deduction

1. In ordinary language

In ordinary language, the word *deduction* is homonymous. As a derivative of *to deduct*, deduction means "subtraction", and does not directly concern argumentation. As a derivative of *to deduce*, it can be used as an umbrella term, to refer to any kind of argument, that is of derivation of a conclusion from a set of data taken as premises. Deductions are given as *valid* and *sound* by the arguer to the other participants.

The well-known Holmesian "deductive method" proceeds as follows:

> *Watson visits Sherlock Holmes. Opening sequence:*
> 'In practice again, I observe. You did not tell me you intended to go into harness.'
> 'Then how do you know?'
> 'I see it, I deduce it. How do I know that you have been getting yourself very wet lately, and that you have a most clumsy and careless servant girl?'
> 'My dear Holmes, this is too much. You would certainly have been burned, had you lived a few centuries ago. It is true that I had a country walk on Thursday and came home in a dreadful mess, but I have changed my clothes I can't imagine how you deduce it. As to Mary Jane, she is incorrigible, and my wife has given her notice; but there again, I fail to see how you work it out.'
> He chuckled to himself and rubbed his long, nervous hands together.
> 'It is simplicity itself," said he, "my eyes tell me that on the inside of your left shoe, just where the firelight strikes it, the leather is scored by six almost parallel cuts. Obviously they have been caused by someone who has very carelessly

scraped round the edges of the sole in order to remove crusted mud from it. Hence, you see my double deduction that you had been out in vile weather, and that you had a particularly malignant boot-slitting specimen of the London slavey.'

Arthur Conan Doyle, *Adventures of Sherlock Holmes — Scandal in Bohemia*, 1891[1].

The reasoning seems to correspond to an argument from natural@ sign, or if considered as the derivation of an explanatory hypotheses, to an abductive@ argument, more than to a logical deduction.

2. In Cartesian philosophy

A deduction is a series of operations linking, according to *valid* rules, a set of *true* premises (axioms, true propositions) to a conclusion:

> Many things are known although not self-evident, so long as they are deduced from principles known to be true by a continuous and uninterrupted movement of thought, with clear intuition of each point. (Descartes [1628], Rule III).

In this sense, a well-led deduction is a *demonstration*, producing *apodictic* (incontestable) knowledge, defined as "any necessary conclusion from other things known with certainty" (*ibid.*).

Valid and sound syllogistic reasoning is a kind of deductive reasoning, sometimes taken as the reference for valid argumentation. Argumentation developing the definition@ of a word and its implications, or the various forms of argumentation from the absurd@, are examples of deductions in natural language.

3. In logic

According to Kleene, a *proof* is based on *axioms*, while a *deduction* is based on *hypothesis*:

> The proof of theorems, or the deduction of consequences of assumptions, in mathematics typically proceeds à la Euclid, by putting sentences in a list called a "proof" or "deduction". We use the word "proof" (and call the assumptions "axioms") when the assumptions have a permanent status for a theory under consideration, "deduction" when we are not thinking of them as permanent" (1967, §9, Proof theory: provability and deducibility, p. 33)

In logic, "a (formal) proof (in the propositional calculus)" is defined as "a finite list of (occurrences of) formulas $B_1 \ldots \ldots B_l$ such as each of which is an axiom of the propositional calculus, or comes by the $\supset$–rule from a pair of formulas preceding in the list" (*id.* p. 34).

The $\supset$–rule is "the modus ponens or rule of detachment", defined as "the operation of passing from two formulas of the respective form A and $A \supset B$ to

---

[1] Quoted after Arthur Conan Doyle, *The Penguin Complete Sherlock Holmes*. London: Penguin Books, 1981. P. 162.

the formula **B**, for any choice of **A** and **B** [...]. In an inference by this rule, the formulas **A** and **A ⊃ B** are the premises and **B** is the conclusion" (*ibid.*).

3.1 Validity and Soundness

Under such a definition, deduction is taken as a *valid* and *sound* deduction. Now, a string of propositions can be advanced by as speaker as a valid and sound deduction without being really so.

To be valid, the deduction has to be led according to the laws of (a well-defined system of) logic. For example, the inference from a false proposition to a true one "**P(F) → Q(T)**" is valid, but not sound: to be sound, the reasoning has to start from axioms or, generally speaking, from true propositions.

The *implication* (conditional) is a binary logical connective@. A *deduction* is a chain of operations linking well-formed expressions by means of a rule. For example, the rule of *modus ponens* (⊃–rule, cf. supra) makes it possible to deduce "**B**" from the two premises "**A → B**" and "**A**" (hypothetical syllogism), by a three-step deduction:

$$\frac{A \rightarrow B \quad A}{B}$$

The same reasoning can be expressed as an implication expressing a logical law,

S. Connective:
"*If the implication is true and the antecedent true, then the consequent is true*"
[(A → B) & A] → B

Let's consider a true conditional "**R → W**", "*If it rains, the lawn is wet*". **W** is a *necessary* condition for **R**; **R** is a *sufficient* condition for **W**.

3.2 If a sufficient condition for W is met, then W

If the antecedent of a true conditional is true, then its consequent is true.

R → W	R is a sufficient condition for W	*If it rains, the grass is wet*
R	this sufficient condition is met	*It is raining*
so W	so W is met	so *the grass is wet*

This rule proceeds from the *affirmation of the antecedent* of a true implication. It is also known as the *modus (ponendo) ponens* rule: the deduction poses (*ponendo*) the truth of the antecedent **R,** in order to affirm (*ponens*) the truth of the consequent **W**.

The idea of sufficient condition is also expressed as:
    not-(A & not-B)

In the ordinary world and natural language, a situation in which it might rain without the grass becoming wet is unthinkable.

3.3 If a necessary condition for R is not met, then R is not met

If the consequent of a true conditional is not true, then its antecedent is not true.

R → W	W is a necessary condition for R	If it rains, the grass is wet
not-W	this sufficient condition is not met	The grass is not wet
so not-R	so R is not met	So it is not raining

This rule proceeds from the *negation of the consequent* of a true implication, also known as the *modus (tollendo) tollens* rule, the mode that, by denying (the consequent), denies (the antecedent).

All reasoning from natural@ signs involves this kind of deduction.

4. Paralogisms of deduction

4.1 Denying the antecedent

It is not possible to deny the existence of a phenomenon on the basis of the absence of a sufficient condition for the given phenomenon. The following deduction is invalid:

R → W	R is a sufficient condition for W	If it rains, the lawn is wet
not-R	this sufficient condition is not met	It does not rain
*so not-W	*so W is not met	*So the lawn is not wet

Raining, a sufficient condition for the grass to be wet, has been incorrectly considered as necessary.

4.2 Affirming the consequent

It is not possible to infer the existence a phenomenon in view of the prevalence of a necessary condition of this phenomenon. The following deduction is invalid:

R → W	W is a necessary condition for R	If it rains, the lawn is wet
W	this necessary condition is met	The lawn is wet
*so R	*so R is met	*So it is raining

To find that the grass is wet is not a sufficient basis to conclude that it is raining.

5. Pragmatic of deduction

The rules of deduction are defined within the framework of a logical system in which all the components of reasoning are explicit and well defined.

Ordinary situations are different; in particular, and ordinary reasoning only makes relevant knowledge explicit. Let us suppose that the lawn could be wet because it has rained, because the lawn has been watered, because a pipe has leaked, or due simply to a heavy dew. If it is contextually evident that the lawn has not been watered (I know what I have done), that there is no water leaking (for the simple reason that there is no water pipe in the garden), and there is no dew (at that time of the day), then I can safely say that if the grass is wet, it is because it rained, or is raining.

Only the superficial form of reasoning is fallacious. Full evaluation must take the context into account and re-build the argument explicitly, on a case-by-case basis thereby eliminating the other sufficient conditions, transforming the latter into a necessary and sufficient condition. This is a simple application of Grice's cooperation@ principle.

Default Reasoning

Researchers in artificial intelligence have developed the formal study of argumentation as *defeasible reasoning* in a logical, computational, and epistemological perspective.

1. Default reasoning

From the logical point of view, defeasible reasoning is studied within non-monotonic logic. Unlike conventional ("monotonic") logic, non-monotonic logic admits the possibility that a conclusion can be deductible from a set of premises {P1} and not from {P1} plus new premises. In terms of belief, the challenge is to formalize the basic idea that the provision of new information may lead to revision of the belief derived from a formerly limited set of data.

From an epistemological perspective, the theory of defeasible reasoning (Koons 2005) concerns beliefs that permit exceptions: in general, birds fly; but penguins (*Sphenisciformes, Spheniscidae*) are birds and do not fly. As a consequence, if the only thing one knows about Tweety is that Tweety is a bird, it is not possible, strictly speaking, to infer that Tweety can or cannot fly. Nonetheless, in the absence of any information suggesting that Tweety is a penguin (or some other flightless bird), the theory of revisable reasoning admits the conclusion "*Tweety flies*". It validates exception-conditioned inferences:

Since **A** (*Tweety is a bird*), normally **B** (*Tweety flies*).

The premise does support the conclusion, but it may nonetheless be true and the conclusion false. A conclusion considered to be correct on the basis of the knowledge which has now become available, may later turn out to be false if further knowledge is gained.

The theory of defeasible reasoning also addresses more complex issues such as the following. We know that:
  (1) Birds fly
  (2) Tweety is a bird
  (3) Tweety does not fly
  (4) Birds have highly developed wings muscles

In these conditions, can we deduce (5) from (1) - (4)?
(5) Tweety has highly developed wings muscles

The property of having highly developed wing muscles is linked to having the capacity to fly, which, according to the available information (3), is not true in Tweety's case. The inference from (1) and (4) to (5) is therefore invalidated. In other words, the conclusion "*Tweety has highly developed wings muscles*" is deducible not from "*Tweety is a bird*" but from "*Tweety is a flying bird*".

A conclusion **C** asserted through defeasible reasoning can be rebutted in two ways:

— On the one hand, upon the existence of good arguments for a conclusion inconsistent with **C** ("rebutting defeater", Koons 2005), that is to say upon the existence of a strong counter-argumentation.

— On the other hand, upon the existence of good reasons to think that the transition principles usually invoked in the argument do not apply in the case considered ("undercutting defeaters", *ibid*), **S. Refutation**.

2. Representation of default reasoning

The default inference is represented as a default rule:
> If Tweety is a bird,
>> in the absence of information suggesting that Tweety may be a penguin (etc.),
>> it is legitimate to conclude that Tweety flies.

The sequence is represented as:

$$\frac{\text{Tweety is a bird: Tweety is not a penguin (etc.)}}{\text{Tweety flies}}$$

$$\frac{\zeta : \eta}{\theta}$$

ζ: Prerequisite: we know that ζ
η: justification: η is compatible with available information
θ: conclusion

The historical origins of the theory of revisable reasoning are sought in dialectical reasoning and the *Topics* of Aristotle. The restriction "*in the absence of information*" corresponds exactly to the "modal" component of Toulmin's layout@ of argument; the basic intuitions and concepts are the same. Toulmin layout can be schematized as:

$$\frac{\mathbf{D}\ (Data) : \mathbf{R}\ (Rebuttal)}{\mathbf{C}\ (Claim)}$$

D, Data: Prerequisites, we know that **D**;

R, Justification: The inference from **D** to **C** could be rebutted under the conditions **R1… Rn**; but we have no information leading us to believe that these rebuttal conditions are actually true.
C, Claim: So, the conclusion **C** can be accepted; one can work on the basis that **C**.

Gabbay & Woods (2003) develops a study of practical reasoning combining the insights of and relevance theory and default reasoning theory, **S. Relevance**.

Definition (I): Definition and Argument

All typologies have an entry "definition", frequently the first on the list. Issues about definition, that is *questions of definitions* (issues focused on definition of terms, S. **Stasis**) arise in highly productive forms of argumentation. We shall distinguish between two cases:
— argumentation *justifying* a definition
— argumentation *using* a definition.

1. Argumentations justifying a definition, S. Definition (II)

Argumentations justifying or building a definition appear when a conflict of definition occurs, that is, when the dispute is structured by a family of questions such as:
    What is a terrorist, a democracy, a spin doctor?
    What is the correct definition of the word "terrorist", "democracy", "spin doctor"?
    What do you precisely call a terrorist, a democracy, a spin doctor?

Argumentations justifying or rebutting definitions support claims like:
    [This discourse] is a good / bad definition [of the word **W**].

2. Argumentation using a definition

In the second case, the definition of a word is used as a stock of arguments.

2.1 Definition used to enrich the description of an individual

In this form of argumentation, the speaker allocates to an individual any feature mentioned in the definition of the category to which this individual belongs and bears the name. Any quality, property, right, duty, values, commonplaces, knowledge attached to the definition can be safely attributed to the individual. If Syldavia is a *democracy* (category), and that *"having fixed elections dates"* is a characteristic of democracy, then he or she might infer that there will be elections in Syldavia in the not too distant future. S. **Definition (III)**.

2.2 Definition used to categorize an individual

The argumentation categorizing an individual attaches this individual to a category designated by a word **W**, accompanied by the definitional discourse. The structuring question is of the form "*is this being an W?*", or "*Is this bird a goldfinch?*". In order to answer such a question, the speaker will look for a definition and description of **X** in the relevant dictionary or encyclopedia. He or she might, for example, look at the entry "goldfinch", and so find a definition which includes the descriptive features of a goldfinch, and, if the bird considered fits the definitional features, he can securely claim that "*this is certainly a goldfinch!*". S. **Categorization and Definition**.

3. Persuasive definitions@

Definition and categorization work together in a coordinated way: 1) establishing a definition of the term **W**, then 2) categorizing an individual **w** as "a **W**" then 3) enriching the description of **w** by attaching to it any feature taken from the definition of **W**s. In order that this 'assembly line' functions correctly, the definition must *avoid circularity*, that is to have been established without consideration of such and such individual that one would possibly like to include or exclude from the category. In other words, a definition becomes persuasive when operations (1) and (2) collapse; that is, when the definition is locally and opportunistically re-framed to include a predetermined individual.

4. Others

Other forms of argumentation draw on categorization and definition; S. *A pari*; *A fortiori*; **Argumentative scale**.

Definition (II): Argumentation Justifying a Definition

The meaning of a word in ordinary language is not a backstage spirit animating the word. To define a word (or a phrase) is to associate to this word a discourse "equivalent"; "having the same meaning":

  uncle   = "brother of the mother or the father"
  [*definiendum*]   = [*definiens*]

The definition establishes a semantic equivalence between a term, the *definiendum*, "what is to define", the dictionary entry, and a discourse, the *definiens* "what defines" (sometimes called "definition" by metonymy).

The *definiens* is a discourse answering questions like "*what does the word X mean?*" "*What is X?*"…

From a logical perspective, the equivalence *definiendum* / *definiens* meets two requirements, one semantic and one formal.

— In semantic terms (intension) *definiens* and *definiendum* must have the same meaning.

— In formal terms (extension), the *definiens* and *definiendum* must be intersubstitutable in all contexts, the global meaning of the passage remaining the same.

Definition (II): Argumentation Justifying a Definition

The definition is substituted to the word defined, when the discourse containing this word has to be clarified; the word is substituted for its definition when the discourse has to be abridged.

Not only *words* but also *phrases* are in need of definition:
> —What is a single parent? An educated person? What constitutes an emergency situation? An urgent case?

Depending on the nature of the word and the circumstances of the questioning, these questions ask about the meaning of the word, or about information about the kind of object to which the word refers, or about the conventional circumstances in which it is possible to use the word.

The definition of "fish" as a species of animal draws on the field of natural sciences. The definition of "democracy", "citizen" and "citizenship", combines political sciences and political and ideological ideals. The definition of "single parent" refers to laws and ordinances. The vague concept of a "cultivated person" will combine a little of all the arts and letters. Advances in knowledge, history, and usages will change the meaning of the words and the kind of beings and objects they refers to.

Argumentative situations de-stabilize the meaning of words, and the definition of commonly used words may require revision and further precization.

The same overall *methodological* concern governs the system of rules for the construction of a proper definition and the rules for building a good causality, a good authority, and a good analogy. **S. Causation (I); Authority; Analogy.** The following paragraph presents a sample of the very distinct methods used not only to build a definition but also to de-stabilize an unsatisfactory definition, or justify a challenged one.

1. Techniques of definition
There different techniques of definition and kinds of definition:
- def. by ostention
- def. by exemplification
- operational def.
- functional def.
- stipulative def.
- def. by enumeration (in extension)
- essentialist def. (in intension).

However, actual definitions tend to combine these various techniques, or otherwise favor one, according to the need the definition is intended to satisfy, and the interests at stake in the discussion. When *a question of definition* arises, the arguers may play one kind of definition against others (cf. infra § 3).

Definition (II): Argumentation Justifying a Definition

1.1 Definition by ostention

Ostension is a gesture, the act of showing to somebody a concrete object. Defining a concrete noun by ostension is to show a sample of the objects or beings referred to:

> Want to know what a duck is? Well, look at that one just flying by!

Ostensive definitions can be applied to concrete beings only. Such definitions are based on initial contact or experience with a given being or object named by the word. Ostension is fundamentally ambiguous: the same gesture shows the chestnut *horse* and its chestnut *color*, but it is disambiguated by the context.

To the extent that the definition concerns the meaning of the word, ostension is not really a definition in itself, as it bypasses meaning, relying on the object being pointed to. It thus lacks the discursive element considered essential for a proper definition. Ostension nevertheless provides a good introduction to the adequate use of a term, by seeing ducks we do learn what ducks look like and are.

The pictures illustrating dictionary definitions introduce an element of ostension into a dictionary. Ostension is a key auxiliary for the definition of concrete things. The more closely the concrete object or being resembles the prototype of its category, the more effective ostension will prove to be.

Ostension underlies the famous argument

> I cannot explain how, but I do recognize a *boletus badius* when I see one!

1.2 Definition by exemplification

The definition by exemplification approaches the meaning of a word by giving contextualized examples of its use:

> What is a canard? Well, that is, for example, remember when the media announced the partition of Syldavia into two independent states?

The example given, if prototypical, provides a basis from which the meaning can be reconstructed by generalization. If the answer enumerates large enough number of cases, the examples can serve as a basis for an inductive construction of a good definition.

The consideration of a variety of cases is crucial for the criticism of definition: Does the definition under scrutiny permit to correctly refer to all the beings or cases currently referred to by the corresponding name?

1.3 Operational definition

Operational definition associates a term **X** with a set of operations permitting to determine whether or not that individual is an **X**. Operational definitions do not say what an **X** *is*; they simply indicate how the signifier **X** is correctly used.

The expression *"prime number"* is defined as *"a number that is only divisible by itself and by the unit"*. For any number, this definition allows one to unambiguously determine if it is or not a prime number.

Definition (II): Argumentation Justifying a Definition

1.4 Functional definitions
Functional definitions do not consider the essence, or the technical design of the instrument named. The definition is expressed in terms of functions, goals, objectives. To know the meaning of the word *compass* is to know what it is used for, "*it points north (magnetic)*".

1.5 Stipulative definition, neology and baptism
Stipulative definitions are also called "definition of name":
> The only definitions recognized in geometry are what the logicians calls *definitions of name*, that is, the arbitrary application of names to things which are clearly designated by terms perfectly known. (Pascal *Geom.*, p. 525)

They play a key role in the scientific creation of words. When a new class of phenomena or beings has been identified and characterized, they must be given a name. While in the general case, the definition begins with a given term and looks to its pre-established definition, stipulative definitions start with a clear and well-established meaning (the *definiens*), and seek a word to refer to this content; it is a baptism. To this end, one might choose a usual word emptied of its ordinary meaning. By convention, physicists use the word *charm* to speak of a particular particle, the *charm quark*. The equivalence condition between the word and its definition is fully satisfied.

In other cases, the word chosen to name the new phenomenon retains something of its ordinary meaning, and it is arguable that "*my word fits better than yours the nature of the phenomenon*". As each and every person has a preferred terminology, the relatively arbitrary nature of the stipulative neologism can lead to terminological inflation and a "war of words", which can be calmed or perhaps even overcome by invoking the primacy of the reality of things. Should we call such argumentative patterns:
> *serial reasoning* or *subordinate argumentation*?
> *linked reasoning* or *coordinate argumentation*?
> *convergent reasoning* or *multiple argumentation*?

If no agreement can be reached, the issue can be radically settled, "You may even call it '*Ivan Ivanovich*' as long as we all know what you mean." (Jakobson 1971, p. 557).

1.6 Definitions in extension (by enumeration)
Definitions in extension proceed through the enumeration of all the (classes of) objects the word or expression refers to. Thus, the expression "*conventional binary logic connector*" is defined in extension as a member of the set {~, &, V, W}; a democracy is a state mentioned in the list of democracies established in the Democracy Dictionary:
> Syldavia is a democracy since it is on the "Democracy List".

Definition by extension provides the basis for case-by-case arguments. If "*honestly acquired money*" is defined as acquired "*either through work, inheritance, financial investment, or winning the lottery*", then it can be indirectly proved that a sum of money was ill-gotten by showing that it has not been acquired by work, nor by inheritance, nor is the legitimate product of a financial investment, S. **Case-by-case.**

1.7 Essentialist definition (definition in intension)

Essentialist definitions require that the definition "focus on the essence (and not the accident), and proceed by next genus and specific difference" (Chenique 1975, p. 117). A unique being may be referred to through a series of specific features providing an unequivocal designation. A dictionary of Syldavian institutions would include an entry "President of the Syldavian Republic (SR)", defined through the modes of election, the constitutional role etc. These core elements are often accompanied by anecdotal characteristics, such as "lives in the Parnassus Palace", "the spouse is called 'the first lady or man of Syldavia'", etc. The latter information *refers* unambiguously to the President (they apply to him or her and only to him or her, the substitutability condition is fulfilled), but doesn't contribute to clarifying the meaning of "President of the SR". In Aristotelian terms, free accommodation at the Parnassus Palace is not an essential property attached to the office of President of the SR.

Essentialist definitions seek to express, beyond the linguistic knowledge of the word (*lexical definition*), and even beyond the knowledge of the thing defined (*encyclopedic definition*), always reflecting an imperfect state of knowledge, the *true sense* of the word, expressing the very *nature* of the thing it designates, that is, its permanent *essence*. In Platonic terms, an essentialist definition claims to retain the *idea* of the thing: *"what is virtue?"*. In theory, the essentialist definition is ruled by a methodology, based on an "intuition of the essence of the thing", S. **Taxonomies and Categories.** *Ancient dialectic* was the instrument used to build correct essentialist definitions.

While a pragmatic definition of the word *democracy* is based on the many sociohistorical uses of the word, an essentialist definition tries to establish the ideal, essential characteristics of democracy, sometimes to condemn the current uses of the word on behalf of *"true democracy"*, S. **True meaning.** It is possible that no functioning democracy really expresses the essence of democracy. The essentialist definition is used as an important critical tool in idealist or conservative argumentation (Weaver 1953).

1.8 Scientific definition

Encyclopedias collect only conceptual terms. Encyclopedic definitions summarize the state of knowledge about things and concepts referred to by the term. A good definition of a thing stabilizes a well-constructed knowledge.

The scientific definition can use a re-defined common term. The *mass* of the physicist is not the *mass* of the language dictionary:

> In physics, *mass* is a property of a physical body. It is the measure of an object's resistance to acceleration (a change in its state of motion) when a force is

applied. It also determines the strength of its mutual gravitational attraction to other bodies. In the theory of relativity a related concept is the mass-energy content of a system. The SI unit of mass is the kilogram (kg). (Wikipedia, *Mass*).

Whereas the word *mass* is commonly defined and illustrated as
> 1 a: a quantity or aggregate of matter usually of considerable size
> b (1): expanse, bulk — (2): massive quality or effect —(3): the main part or body <*the great* mass *of the continent is buried under an ice cap* (...) (4): aggregate, whole <*men in the* mass>
> c: the property of a body that is a measure of its inertia and that is commonly taken as a measure of the amount of material it contains and causes it to have weight in a gravitational field
> 2 : a large quantity, amount, or number <a *mass* of material>
> 3 a: a large body of persons in a group <a *mass* of spectators> b: the great body of the people as contrasted with the elite —often used in plural <the underprivileged and disadvantaged *masses* (...) (MW, *Mass*)

Arguments establishing a definition of things are domain-dependent. An astronomy conference was necessary to redefine the term *planet*, and end the controversy over the status of Pluto. The usual definition can be hardly recognizable under the technical definition. The following definition correspond to an everyday experience:
> 1. A blocking of the alpha activity preceded by a transitional element that is expressed in the cortex region (a temporal tip-cortex);
> 2. A more or less pronounced muscle jerk (a start);
> 3. Neuro-vegetative events, such as tachycardia and decreased skin resistance.
> So I was referring to the "classical" reaction of surprise that you all know.
>
> Henri Gastaud, [Discussion], 1974[1]

This is a scientific definition of *surprise*, "in the sense of 'surprise reaction' that is to say the set of phenomena observed by the neurophysiologist, when a sudden unexpected stimulus occurs." (*Ibid.*)

1.9 Lexicographical definition
Lexicographical definitions are found in *language* dictionaries, as opposed to *encyclopedic* dictionaries. Language dictionaries must meet multiple conditions:
— Collect all the words and idioms of a language (or the vocabulary used at a particular period).
— Provide a description of their various meanings, their uses in speech, and their stereotypical figurative uses.
— Give the typical contexts of use associated with these meanings.
— Specify the syntactic constructions corresponding to these meanings.

---
[1] Gastaud H. (1974) "Discussion". In Morin E. & Piattelli-Palmarini M. (eds). (1974). *L'Unité humaine*. Paris: Le Seuil. P. 183.

— Locate them in the various semantic fields to which they belong, that is, specify their relationships with their (quasi-) synonyms and antonyms, and their position in their derivational families.

The dictionary is a highly legitimized and legitimizing institution. From the perspective of argumentation studies, lexical meaning being inferential, the dictionary should be seen first of all, as a huge stock of "inferring principles", S. **Definition (III)**.

Linguistic definitions simultaneously draw on different kinds of definition. Knowledge of words (lexical definitions) and knowledge of things (encyclopedic scientific definitions) are theoretically clearly separated. They are, however, inextricably linked for current terms having an encyclopedic definition. "*When the barometer falls, the weather turns bad*": is the deduction backed by a meteorological physical law expressing knowledge about the variations in atmospheric pressure? Or is it included in the linguistic meaning of the word? Knowing the operative meaning of the word "*barometer*" is to know that "*when it falls, the weather turns bad.*"

All words are worthy of a lexical definition, but only those having "plenty of being" are worthy of scientific knowledge, and are registered in the encyclopedia. The border between the two categories is unstable and dependent on the state of research; conversation, once considered a futile and elusive thing, was conceptualized fruitfully by conversation analysis and ethnomethodology. These sciences have given "more being" to their object.

2. Issues and critical questions about definitions

When a definition is at issue, one technique can be played against another. Typically, definitions based on *common usage*, on *true meaning* of the word, on *the scientific meaning* of the word can be opposed to each other.

Just as there are rules for arguments establishing a correct causal relationship or a correct analogy, there are rules for *arguments establishing a correct* definition. The methodology of definition specifies the rules for constructing, and therefore for evaluating, definitions. These rules depend on the social or scientific fields to which the defined beings belong, and adapt to the various definition types. The more general ones are as follows.

(i) Does the definition correctly *disambiguate* the term according to its meanings (homonymy) and acceptances (polysemy)? S. **Ambiguity**.

(ii) Does the definition *avoid circularity*? If not, it enters a vicious@ circle. Words being defined with words, the whole dictionary is actually circular. As explanations or arguments in general, definitions should try to defer circularity as far as possible; that is, the definition (*definiens*) cannot use the word to define (*definiendum*), nor a (near) synonym of the word. Nonetheless, definition through synonyms or the simple negation of an antonym does help if one of these defining words is better known than the *definiendum*.

(iii) Does the definition *cover all the uses* of the word? Does the meaning of a passage remain the same when the *definiens* is replaced by the *definiendum*? If not, the definition should be amended or rejected.

(iv) Does the definition make it possible *to sort out the beings* that are called by that name from those that are not? A definition might be criticized because it is *too broad* (it applies to heterogeneous objects or beings) or because it is *too narrow* (it excludes objects or beings it would be desirable to integrate). See supra the role of ostension and exemplification.

(v) *Does it help?* That is, does it provide sufficient *information* to clarify the meaning of the word, and, if need be, does it give some functional indications, or point to the scientific or specialized uses of the word?

(vi) Is the definition *brief, clear, and simple*? Does it use unknown words to define unknown ones, ambiguous words to define ambiguous ones?

(vii) Is the definition *objective*? Does it exclude the judgments of value and ideological preferences of the author towards the beings or properties defined? S. **Bias; Persuasive definitions**.

Methods and rules such as those mentioned above serve as a guide for the establishment of definitions and, consequently, for their criticism.
— They are mobilized in debates on definitions or involving definitions (Schiappa 1993; 2000), that is, when there is a *stasis of definition*.
— They are fundamental to the criticism of argumentations *that use a definition*, showing for example that the underlying definitions are poorly constructed and do not comply with such and such a rule.

3. Questions of definition

A *stasis* of definition, or *question of definition*, occurs when it appears that discourse and counter-discourse are based on incompatible definitions of the same object:

> S1: — *The rights of free speech and demonstration are fundamental to democracy.*
> S2: — *What is fundamental in a democracy, is the right to have enough to eat and an iphone.*

A definitional question ensues: which features are essential (central) features and which ones are accidental (peripheral) to characterize a democratic state?

Incompatible categorizations result in a question of definition:

> S1: — *A Syldavian Diplomat killed in an accident*
> S2: — *Murder of a Syldavian Diplomat*
>
> *Confidential information was disclosed:*
> S1: — *A new manifestation of the malfunctioning of Syldavian Services*
> S2: — *There are traitors in our services.*

The investigator, in the role@ of the third party, transforms the two conflicting

Definition (II): Argumentation Justifying a Definition

discourses into an argumentative question, and initiates an investigation to clarify what happened, on the basis of the legal definitions:

What is murder? What is an accident?
What are the crucial differences between carelessness and betrayal?

The stasis of definition can develop as follows:

S1_1: — *Syldavia is now a true democracy!*
S2_1: — *How dare you talk about democracy in a country that does not recognize the rights of minorities?*
S1_2: — *According to the dictionary, democracy is ...; nothing in this definition mentions the rights of minorities; so, Syldavia is for sure a true democracy*
S2_2: — *This definition is poor and ideologically biased.*

— The confrontation of the positions **S1_1** and **S2_1** produces a question of *categorization*.

— **S1_2** rejects the objection of **S2_1** by referring to the dictionary; he or she might as well have quoted the recognized conventions, international law, consensus, etc.

— **S2_2** ratifies the question of *definition*.

According to Humpty Dumpty, the best way to resolve of a stasis of definition is to appeal to power:

> [Humpty Dumpty] [...] — and that shows that there are three hundred and sixty-four days when you might get un-birthday presents—"
> "Certainly," said Alice.
> "And only *one* for birthday presents, you know. There's glory for you!"
> "I don't know what you mean by 'glory,' " Alice said.
> Humpty Dumpty smiled contemptuously. "Of course you don't—till I tell you. I meant 'there's a nice knock-down argument for you!'"
> "But 'glory' doesn't mean 'a nice knock-down argument,'" Alice objected.
> "When *I* use a word," Humpty Dumpty said in rather a scornful tone, "it means just what I choose it to mean—neither more nor less."
> "The question is," said Alice, "whether you *can* make words mean so many different things."
> "The question is," said Humpty Dumpty, "which is to be master—that's all."
> Alice was too much puzzled to say anything, so after a minute Humpty Dumpty began again. [...]
>
> Lewis Carroll, *Through the Looking Glass*, 1872[1]

4. A Discursive ploy: Requiring a definition

The request for a definition might be made with the intention of blocking the development of the opponent's argumentative line. The following exchange takes place in a discussion about various personalities competing for a scientific distinction:

---

[1] Quoted after Lewis Carroll, *Through the Looking Glass*. Chapter 6, Humpty-Dumpty. 2016. No pag. http://www.gutenberg.org/files/12/12-h/12-h.htm#link2HCH0006. (11-08-2017)

S1: — *Doe has a lot of prestige.*
S2: — *What do you call prestige?*

This inevitably leads to a stasis of definition, into which participants may not want to enter. The internal magazine of a research institution objects to a traditional claim from laboratories:

"[Lack of technical staff] would lead to a lack of "optimum efficiency" in laboratories. First, how do we define the optimal efficiency of a laboratory?

Definition (III): Argumentations Based on a Definition

1. Argumentation by definition

The definition (the *definiens*) of a word or an expression (*boy, scotch bonnet, democracy, single parent, educated person, British citizen, natural disaster*...) provides a stock of definitional features applicable to all the beings, individuals, institutions, events... designated by the *definiendum* (belonging to the category named by the word). Argumentation by definition applies the definition of the name to a being designated by that name. It operates as follows:

1. *An argument*: a statement of the form "**I** is a **D**": **I** is an individual (identified, categorized, perceived, named... as) "a **D**".
2. *A license to infer*, found in the definition of **D** considered as authoritative.
3. *A conclusion*: everything said of the **D** or with them can be truly said of **I**.

A definition (a *definiens*) is a rich set of proposals about "what that kind of being is". It includes *doxical assertions* based on common knowledge about these beings, to be found in the *examples* illustrating the *definiens* as well as in the *definiens* properly said. To call a being "a **D**" is to allocate to this being all the properties defining the name "**D**", as well as scripts, duties and obligations attached "to **D**s". In other words, the definition (the *definiens*) of "what is a **D**" is a stock of inference licenses applicable to all the persons and objects called **D**.

Using the definition allows inferences of the following type, **S. Common Place**.
— "*Harry is a British citizen*": this claim expresses a categorization@ of the person Harry, derived from the information that he was born in Bermuda, **S. Layout**. The categorization ("— *is a British Citizen*") corresponds to a local modeling of the person "*as-a-British-citizen*", which makes it accessible to the inferential definitional machine. Armed with this information, we can draw from the knowledge stock that defines "what it is to be an Englishman", and conclude, according to the needs of the moment, that:

He takes tea at five
He will need a drop of milk
We can certainly address him in English
If he has committed a crime abroad, his judicial treatment will be led according the relevant international convention

— "*My dear, you're a little girl!*" Common knowledge says that girls are like this, should do this and that, etc. So, my daughter, you're like that, and you must behave accordingly:

— "*This is a Scotch bonnet*" so, it is "*very aromatic, it is delicious prepared in an omelet*"; better yet, you can "*dry it out, and use it as an aromatic*"[1], **S. Categorization.**

— "*Now you are undoubtedly one of the great democracies*" so we can re-establish diplomatic relations and encourage our citizens to spend their vacation on your beaches.

— "*Mrs. Doe is a mother who lives alone*" so, on the basis of such and such administrative and financial provisions, she is entitled to a single parent allowance of a certain amount.

— "*Mrs. Smith is registered for a doctorate*" so she enjoys certain rights and must fulfill certain duties defined by the PhD Charter in force at the university where she is enrolled.

— "*He is a bastard*" so I do not trust him.

Argumentation by definition ascribes to a definite being a feature actually found in the definition of its name, as found in a dictionary or an encyclopedia. More broadly, it attaches to a being any feature borrowed from the stereotyped notion of the kind of beings bearing that name.

Argumentation by definition is the epitome of what Billig calls "bureaucratic thinking", which is fundamental in everyday life (Billig [1987], p. 124).

If the criteria used for categorization are defined within a rigorous scientific framework, then argument by definition will be an essential scientific tool. Similarly, in the legal domain, the criteria qualifying an act make it possible to apply the *legal syllogism*, which delivers routine legal decisions.

2. Lexical definitions as inferential resources for categorization and argumentation from the definition

Some basic argumentative inferences embedded in a word are made explicit in its lexical definition and are suggested in the examples of its usage. Language dictionaries are stocks of accepted ideas and accepted connections between ideas; as such, they provide legitimate inferences from and to a word in the language and culture to which they belong (Raccah 2014) **S. Orientation.** These inferences are considered *rational* and *convincing* insofar as they are the expression of a shared *semantic* heritage, the treasure trove of discursive *rationality*. Let us consider the word *rich*. By merging the definitions of some current dictionaries, we are able to gain some insight into the elementary "licenses to infer", diversions, or "drifts" from and to this word, that is, the semantic inferences which characterize a basic understanding of the word "rich". The following information comes from definitions from MW, tfd; CD).

(i) ... *SO he is rich*. This claim is justified:

---

[1] Entry *Mousseron* in J. and J. Manuel Montegut (1975). *Atlas des Champignons* [*Atlas of Mushrooms*] Paris: Globus, 1975.

— On an analytical basis: ... *(he has) a lot of money; of valuable assets,* SO *he is rich*

— On the basis of signs: ... *(he owns) expensive materials, workmanship (such as mahogany furniture),* SO *he is rich*

— On the basis of his or her moral character and motives@:
*he is determined to **get** rich quickly,* SO *he will probably become rich*

(ii) *He is rich,* SO ...
On the same analytical basis, or from signs, one can deduce:
... *(he has) a lot of money; of valuable assets*
... *(he owns) expensive materials, workmanship (such as mahogany furniture)*
... *he does not have to work*
... *he has forgotten his humble background* —
This last conclusion admits of exceptions: He is rich, BUT...
... *Even when he became rich and famous, he never forgot his humble background.*

(iii) An implicit principle, *"everybody can get rich"* eliminates two rebuttals:
Having an humble background
Even when he became rich and famous, he never forgot her humble background
Lacking of formal education
A lack of formal education is no bar to becoming rich.

(iv) A main opposition: *the rich* vs. *the poor,* allows the application of the topic from opposites@:
there's one law for the rich and another for the poor.

Definition (IV): Persuasive Definition

Stevenson ([1938]) introduced the concept of persuasive definition in the following terms:

> In any "persuasive definition" the term defined is a familiar one, whose meaning is both descriptive and strongly emotive. The purport of the definition is to alter the descriptive meaning of the term, usually by giving it greater precision within the boundaries of its customary vagueness; but the definition does not make any substantial change in the term's emotive meaning. And the definition is used, consciously or unconsciously, in an effort to secure, by this interplay between emotive and descriptive meaning, a redirection of people's attitudes. (Stevenson [1938], p. 210-211)

To make a definition persuasive, within Stevenson's meaning, its descriptive content must be redefined, whilst its "emotional force" must be kept intact so as to be applied to the redefined content. Stevenson gives the following example; **A** and **B** are "discussing a mutual friend" (*id*, p. 211.)

— **A** points out a number of shortcomings of that person (education, conversation, literary references, subtlety of spirit) and concludes that *"he is definitely lacking in culture."*

Definition (IV): Persuasive Definition

— **B** describes this friend under a number of favorable lines (imagination, sensibility, originality) and concludes that "*he is a man of far deeper culture than many of us who have had superior advantages in education*".

First, both **A** and **B** value culture, and are willing to give the word *culture* and the judgment "X is a cultured person" a positive emotional orientation. Moreover, the word *culture* has a vague descriptive sense; **B** carves out of this descriptive sense a new definition, and shows that it fits their mutual friend. Stevenson analyses **B**'s argumentative move as follows

> "His purpose was to redirect **A**'s attitudes, feeling that **A** was insufficiently appreciative of their friend's merits" (*id.*, p. 211).

The argumentative trick is located at point (b), that is:
> (a) **B** wishes to value his friend.
> (b) He redefines culture "within the boundaries of its customary vagueness" according to qualities possessed by his friend;
> (c) and he concludes that his friend is cultured;
> (d) and the friend benefits from the positive opinion associated with the idea of culture and cultured person.

Thus, a persuasive definition redefines the descriptive contents of a term not on the basis of context-free, objective general considerations, but with a view to applying this term to a pre-determined person, a singular case. This is what would make it deceptive. It should be noted however that Stevenson attributes a persuasive definition to **B** only. Yet it might be argued that **A** also carves his definition out of the vague meaning of "culture", "within the boundaries of its customary vagueness". **A** thus has a persuasive definition, in much the same way as **B**, of "culture" as literary references, etc., allowing him to exclude the common friend from the circle of the cultivated. **A** seeks to influence **B** just as much as **B** tries to influence **A**.

Point (d) implies that the argumentative orientation (called here "the emotive content") is independent of the cognitive content, and not affected by the redefinition. Thus, this orientation has to be attached directly to the signifier.

As it operates a redistribution of meaning, persuasive definition exploits the processes of *distinguo*@, and dissociation@.

A persuasive definition is a definition that does not meet the condition of *separability* between, on the one hand, the construction process of the definition, and, on the other hand, the use of the definitional features to include an individual or a special case in the category it determines and calls it by the category's name. In other words, a persuasive definition is a definition which is conditioned by the intention of including a specific object, that is, an *ad hoc* definition, imagined or altered on the spot, for the purpose at hand, **S. Bias**.

The criteria of what is "good school task" must be applied regardless of the categorization of such work in or out of that category. A skewed definition does not meet this criterion:

> A good school task is a task on which students worked hard and invested heavily. My son spent his weekend on his history course. Thus he handed in an excellent piece of work, and deserves a good grade.

The category "is a good school task" has been redefined so that it can apply to Mr. Doe's son, leaving aside the contents of the work, traditionally regarded as the decisive factor. The target has been re-designed to fit the arrow, and the limited capacities of the archer.

Demonstration and Argumentation

To demonstrate comes from the Latin *demonstrare* "to show, to point out". To demonstrate and to show verbs are synonymous in some contexts: *"in what follows, I'll show (= demonstrate)* that…".

In ordinary life, people engage in *demonstrations of friendship, solidarity, affection*… making *an exhibition, a show* of their sentiments, as they give *proofs of love*.

The word *demonstration*, even in its most abstract uses, keeps a link with the visual and pictorial; if a proof involves *touching* with the finger, a demonstration *shows*. *Argumentation* has no such metaphorical backgrounds; it originates and deploys within language.

In rhetoric, besides the meaning of "proof", the word *demonstration* is used with two totally different meanings.

— A *demonstration* is a vivid representation of an event or a state of affairs as a picture, for an audience or a reader, put in the position of witness of the represented event. This figure is also called *evidence* or *hypotyposis* (Lausberg [1960], § 810).

—The *demonstrative genre* is another name for the epideictic or panegyric or laudatory genre, next to the deliberative and judicial genres (Lausberg [1960], § 239).

Demonstration is often opposed to argumentation as belonging to two different cultures without contact and communication, the world of *science* vs. the world of *human affairs*, the world of *truth* vs. that of *opinion*. This popular opposition is often considered a definitional characteristic of argumentation. Nonetheless, its substance and actual scope, the precise relations between argumentation and demonstration, should be considered as an essential issue for the development of argumentation studies.

1. The hypothetico-deductive demonstration, ideal of proof?

In logic, a demonstration is a discourse proceeding from *axioms* to *theorem*, according to specified deduction rules. The construction of a demonstrative sequence is guided by intentionality, since it aims at a stopping point, a remarkable, detachable result, the theorem.

A proof has been formalized if it can be presented as a mathematical demonstration. Formal proof is seen as characteristic of science as pure *calculus*, and is sometimes considered as the ideal of proof. This vision is contrasted with science as a *description of reality* (geography, zoology), or as a combination of *calculation, observation and experimentation* (physics, chemistry).

In the sciences, a demonstration is a discourse, bearing on true propositions: (true by hypothesis; or as a result of observations or experiments carried out according to a validated protocol; or as results obtained from previous demonstrations), and leading to a new, stable, true, proposition. Such a proposition marks a step forward in the field, and is likely to guide further developments in research.

Scientific practice presupposes many non-formal linguistic, cognitive or material operations, other than demonstration. Such operations might include grasping a situation, formulating the problem, conceiving a hypothesis, defining an experience, realizing an experimental setup, manipulating the objects and instruments, selecting, observing and describing the relevant data, making quantitative measurements and the relevant calculation, checking the results, imagining new experiences, drawing conclusions, editing the results for oral communication and publication, answering the colleagues' objections, revising the claims, etc. We might add to this all the professional argumentative situations in which researchers must apply for new funding, write or evaluate a research project or to employ a new colleague. These argumentative operations require the coordinated management of technical, mathematical and natural languages, including a variety of semiotic media, figures, tables, schemes and diagrams. Argumentation in natural language plays a key role in all these mixed activities.

2. Two distinct fields: What we know, what we do

Argumentation deals with what is to be believed. Argumentation concerns the question of proof and demonstration, but goes beyond this. The exploratory function of argumentation extends beyond its epistemic role, to *practical* discussion (internal or external) of what, in view of one's current interests, would be the most sensible next move. So for example, one might ask, *"should I apply for this or that position, buy this or that car, ignore or accept offers of negotiation"*. And human affairs extend still further, beyond the realm of practical decisions; generally speaking, argumentative situations emerge as soon as any kind of choice is possible. Argumentative situations can thus arise in regard to antagonistic *feelings*, what is really worthy of admiration or love, S. **Emotions**. In these areas the language of proof and demonstration does not makes sense, whereas the language of argument does.

One might think that in the case of certain issues concerning true beliefs and accurate scientific predictions, doubt is *provisional*, and that any doubt will be removed in view of scientific progress. When considering situations involving human agents, however, doubt is an *essential* component. In such situations, it is often impossible to completely dispel doubt, and one can legitimately ask what would have happened if ...

We turn to argumentation when the data is incomplete or of poor quality and the assumptions and laws imperfectly defined; the deductions are, therefore, subject to a continuous principle of revision. As the last resort, we are referred to the question of time: an argued claim is a bet. Linked with urgency and occasion, argumentation is a time-limited process, different from the unlimited time afforded to the philosophical or scientific demonstration. There are essential differences in the modus operandi of argumentation and demonstration, their fields of application and the kind of problem they can apply to.

When operating in the field of knowledge, argumentation has an *exploratory* and *creative* function which goes beyond its *demonstrative* and *critical* function, **S. Abduction**. Argumentation produces hypotheses, opens up discussions and triggers the critical process of verification and revision.

Demonstration is by definition related to a domain; argumentation may combine *heterogeneous evidence*. Argumentation is the art of hierarchizing not only *values*@ (**S. Values**), but also proofs and demonstrations. If one wants to explore the possibilities and economic interests of a major environmental management works, constructing a channel between the Green and the Yellow Oceans, for example, then the technical proofs, solutions and objections offered by geologists, economists and ecologists must be articulated and confronted with those of neighbors, citizens, investors and politicians. The negotiation will take place in view of calculations and technical proofs each as unique as the others, and argumentation in natural language will have to fully exert its synthetizing function.

3. Argumentation-proof and argumentation-demonstration: The heritage

Several theories of otherwise very different orientations come together in order to oppose argumentation to demonstration. Historically, the notions of demonstration and argumentation inherited through Western tradition were developed in ancient Greece. *Demonstration* in science and mathematics (Archimedes, Euclid) was built without relation to *argumentation* in social affairs. According to Lloyd, Aristotle elucidated "the explicit concept of rigorous proof" ([1990], p. 77) in a scientific context where four types of argument were currently used:

> The first of these is arguments in the legal and political domains, the second those in early Greek cosmology and medicine, the third mathematics in pre-Aristotelian period, and the fourth deductive arguments in philosophy. The first two relate primarily, to informal, the second pair to rigorous proof. (*Ibid.*)

The unity of the disciplines of proof can be shown by the examination of their vocabulary:

> The same vocabulary, not only of evidence, examination, judgment, but also proof, appears also outside the specifically legal or political domain, notably in a variety of contexts in early Greek speculative thought. Both cosmology and medicine, and some extended passages from the Hippocratic Corpus merit particular attention. (*Id.*, p. 78)

In Aristotle's work, *convincing* rhetorical argumentation is characterized by its differences with *valid* logical demonstration (and *probable* dialectical deduction). Since then, argumentation has been conventionally referred to logical demonstration, to *argumentation-demonstration*, and not to *argumentation-proof* such as exemplified in the practices of scientists, practitioners, historians, police investigators, etc. Argumentation is most strongly linked to these practices, in virtue of its substantial nature and its relationship to practical action. For example, the essential concept of argumentative question does not derive from a *logical* concept, but from the *medical* descriptive concept of stasis@, that is a state in which physiological fluids are blocked, and, metaphorically collaborative speech and actions are suspended.

This non-operational opposition between demonstration and argumentation, which now functions as a commonplace, has been considerably restated and strengthened by the New Rhetoric, as well as by the non-referentialist positions of the theory of Argumentation within Language.

4. Demonstration against argumentation?

Demonstration and logical proof are classically opposed to argumentation on the basis of their premises and modes of inference. Things go much deeper than that, however. Natural language and discourse are inherently *subjective*, that is self- and we-related, focused on the "here" and "now", **S. Subjectivity**. Words allow synonymy, homonymy and polysemy; their meaning is context-sensitive. Syntactic constructions must be interpreted. Discourse is figurative. Meaning and reference are negotiated and managed by relevance principles. These processes are stigmatized as being inherently "vague and elastic"; but the polymorphism of language should simultaneously be praised for its adaptability to new situations and its rule-changing capacity.

At a general level, it should be noted, firstly, that there is no reason to favor elementary logical demonstration over other scientific activities, of which it is, unquestionably a distinguished member. Secondly, oppositions make sense only if the opposed domains are comparable. Experimentation, mathematics, computerization, have taken the techno-sciences worlds apart, and it does not make much sense to compare a paper in a scientific journal publishing cutting edge research with a column in a newspaper.

4.1 The New Rhetoric

Perelman & Olbrechts-Tyteca's *Treatise* has constructed a powerful, autonomous concept of argument by rejecting emotions out of the field of argument on the one hand, and by setting argumentation against demonstration on the other. The purpose of the *Treatise* is to circumscribe an autonomous discursive domain, where speech develops cut off from demonstration and emotion. In the very words of the *Treatise*, the couple argumentation / demonstration functions as an "antithetical pair" (Perelman & Olbrechts-Tyteca [1958], p. 422.), whose terms are the subject of a genuine "breaking of connecting links" or "dissociation" (*id.* p. 411 sq.). Systematically, the *Treatise* opposes argumentation to demonstration, as can be checked on every occurrence of the word *demonstration* mentioned in the index. This strategy, constitutes one of the building blocks of the *Treatise*.

The fundamental question of the difference of languages between argumentation and demonstration, is not addressed. In the *Treatise*, the form of demonstration opposed to argumentation is taken in a particular discipline, formal logic, which would be prototypical of demonstration as the inaccessible ideal of argumentation. This hardened and simplified image of demonstration promotes the antagonism argumentation / demonstration. This results in the exclusion of anything concerning sciences from the *Treatise*:

> We seek here to construct such a theory [of argumentation] by analyzing the methods of proof used in the human sciences, law and philosophy. We shall examine arguments put forwards by advertisers in their newspapers, politicians in speeches, lawyers in pleadings, judges in decisions, and philosophers in treatises ([1958], p. 10)

No reference is made to any type of scientific activity. Argumentation addresses human affairs only, and demonstration concerns mathematics and science. exclusively. The gap between "the two cultures" (Snow, 1961) is thus effective at the very foundation of argumentation as a discipline.

4.2 The Argumentation within Language theory

This theory considers that argumentative orientation@ is an essential characteristic of the semantic level of language, and concludes to the impossibility of developing argumentation as good reasons in discourse and interaction. Consider the following passage:

> It has often been remarked that discourses concerning everyday life cannot achieve "demonstrations" in the logical sense of the word. Aristotle already noted that, by opposing to the necessary demonstration of the syllogism the incomplete and only probable argumentation of the enthymeme. Perelman, Grize, Eggs insisted on this idea. At first I thought I was merely following this tradition, my only originality being to refer to the nature of language the necessity of substituting argumentation for demonstration. I thought that the words of language were the cause or the sign of the fundamentally rhetorical, or, as we said, the "argumentative", character of discourse. But I am now led to say much more. Not only do words not allow demonstration, but likewise they do not allow that degraded form of demonstration that would be argumentation. Argumentation is only a dream of discourse, and our theory should rather be called "the theory of non-argumentation" (Ducrot 1993, p. 234).

As Ducrot's structuralist framework reduces the order of speech to that of language (Saussurian *langue*), it is quite coherent to deny any principle of intelligibility to argumentation in discourse.

5. Arguing the non-demonstrative character of argumentation

The refutation of the possibly demonstrative nature of ordinary discourse is threatened by skeptical paradoxes and exposed to self-refutation. It is difficult to argue about the argumentative or non-argumentative character of natural language discourse, whilst using natural language discourse.

Interaction studies have taught us a great deal about what everyday life discourses can achieve. Brief, local reasoning is accomplished in sequences in which language combines with action to achieve operational conclusions. We define, categorize, articulate causes and effects, and make analogies, which are all more or less insufficient, but which are all susceptible to criticism and rectification. Sometimes, these result in the satisfaction of all parties concerned.

By means of some conventions and adjustments, more sophisticated reasoning episodes can be developed in ordinary language. If the syllogism constitutes an example of a necessary demonstration, since the syllogism consists of a sequence of utterances in natural language, words allow at least syllogistic demonstration. All the same, figures and calculus are not entirely foreign to natural language, which also allows for some correct geometrical conclusions, so that the tenon exactly fits the mortise. Not only a logic, but also an everyday geometry, arithmetic, physics... underpin linguistic practices, and no metaphysical lack stops them from concluding properly, as shown by the following little calculation:

> *The Abbé du Chaila is one of the essential architects of the repression of the Protestants of the Cevennes, in Southern France. His murder is the origin of the Camisards' war, in the 18th century.*
>
> The date of birth of the future abbot of Chaila remains a mystery, due to the disappearance of parish registers. It should be at the beginning of the year 1648. Indeed, François's parents, Balthazar de Langlade and Francoise d'Apchier, were married on the 9th of April 1643 and had successively eight boys and two girls in ten years, at a rate of a child a year. François being the fifth child of the family was thus born in 1648, the four previous brothers being born in 1644, 1645, 1646 and 1647.
>
> Robert Poujol, [*The Abbé du Chaila (1648-1702)*], 2001[1].

Any assertion about the demonstrative character of argumentation in general is hard to assess, regardless of the prestige of the authority supporting it. Arguments from natural signs, case-by-case arguments cannot be treated as an appeal to authority or arguments by analogy. Ordinary argumentative discourse might combine entirely heterogeneous types of arguments and fields of evidence, including rather technical and scientific episodes. One might argue correctly in natural language; sometimes, some truth emerges from judicial and historical debates when properly framed and managed; and argumentation plays a role in science acquisition.

---

[1] Poujol R. *L'Abbé du Chaila (1648-1702)*. Montpellier: Les Presses du Languedoc, 2001. P. 31.

6. Argumentation in science education

Other connections have to be found between argumentation and scientific activities. The great rule followed by Quine to construct his formal logic shows the way:

> This course is prompted by an inclination to work directly with ordinary language until there is a clear gain in departing from it. (1980, p. x).

Mutatis mutandis, we will say that the teaching and learning of scientific method are necessarily anchored in natural language and everyday argumentation, and that they depart from them only when they find a decisive gain in doing so. This leads to a focus on argumentation as an instrument for knowledge acquisition.

Demonstrative-scientific proof can be considered, on the one hand, as a *finished* product, impeccably exposed in published papers and textbooks; and, on the other hand, as a *process*, which can leave room for dialogue, arguments and rectification and progression. Argumentation being on the side of the *process*, its claims are in the making. It might therefore be an interesting option to orient the arguer's capacity towards the exploration of scientific domains, monitored by competent advisers.

Finally, we must consider the question as to whether there is a break or a continuity between argumentation and demonstration. This is certainly a very important philosophical and epistemological issue, yet it is quite different from the empirical issue as to how best to construct scientific capacities. The teacher might consider that the gap is a fact (Duval 1992-1993, 1995) and choose to break with ordinary language and practices, as did teachers of "modern mathematics" in the seventies in France. They might otherwise try to use everyday capacities to build knowledge. Gap and continuity are pedagogical choices and constructs.

Science acquisition, scientific "enculturation" and education are key situations for the development of argumentation studies (Erduran & Jiménez-Aleixandre 2007). The humanities remain largely trapped in a conception of argumentation based on logocentric discourses, in which all and everything can be claimed. From this conception, a comfortable antagonism has been developed, with "logical demonstration" serving as a convenient antagonist. The repositioning of argumentation as a complex, combinatorial activity which seeks to manage heterogeneous evidence in possibly complex material contexts enables us to distance ourselves from this traditional logocentric vision. Discussions between two mechanics disagreeing on how to repair a failing engine, or two students disagreeing on the shape of the beams coming out of the lens are as prototypical of what is an argumentative situation as an ideological debate where the language is perpetually referred to itself.

The research program on argumentation in science education emerged in the late 1990s and early 2000s. It now represents a key field of development for argumentation in the near future (Baker 1996, De Vries, Lund, Baker 2002; Buty & Plantin 2009; Erduran & Jiménez-Aleixandre 2008).

Denying

The act of denying operates on words and on sentences.

1. Word negation
The lexical relation of opposition@ can connect morphologically different words, or pairs of words produced by prefixation.
The attachment of a *prefix* to a base word or to a root morpheme produces a new word, belonging to the same grammatical category. *Negative prefixes* produce derived negative terms. The base term and the derived terms are antonyms, that is opposites.
Frequently, the derived negative word serves to add a "not" to the whole semantic content of the positive word:
> agree, agreement => don't agree / dis-agree / dis-agreement.

Negative derived words do, however, tend to become independent:
> they made, reached an agreement / they made, reached ≠ a disagreement

The specific nature of the opposition between base and derived words is idiosyncratic, that is, it is not possible to attach a semantic-lexical rule to the negative prefix in order to pinpoint the meaning of the derived word from the meaning of the base word.
Various negative prefixes can operate on the same basis:
> social => *un*social, *a*social, *anti*social, *non*social (after WCD)

Some *dis-* words do not have a positive counterpart, but are clearly negative, for example *to discard*: "to get rid of… useless, unwanted." (MW, art. *Discard*).

Argumentation based on derived words is characterized by the fact that it *leaves aside* the variation of meaning between the base word and its derivative, in particular negative derivatives, S. **Derived words.**

2. Sentence negation
A negative statement **E1** can be analyzed as **not-E°** (but cf. 2.3). *Total* negation rejects **E°** as globally untrue, incorrect, inadequate; it dismisses, turns down, refutes, rebuts, rectifies, … the primitive statement **E°**. *Partial* negation rectifies a segment or a feature of **E°**.
From the point of view of practical argumentation analysis, and following Ducrot (1972), there are three main types of sentence negation.

2.1 Dialogic negation
E° corresponds to an existing statement previously produced by another participant in the same linguistic action. This "confrontational metalinguistic nega-

tion" (Ducrot 1972, p. 38) is basic for refutation. Examples (after Ducrot, s. d).
— Rejection of a claim:
    L0: — *The next presidential election will be held in two years.*
    L1: — *No, it will take place* next year.
— Invalidation of a presupposition:
    L0: — *Peter stopped smoking*
    L1: — *Peter* never *smoked.*
— Rectification of the degree:
    L0: — *Flood damages are substantial*
    L1: — *They are not substantial, they are indeed* negligible / catastrophic.
— Correction of a linguistic rule:
    L0: — *Look at the childs*
    L1: — *Not* the childs, the children.
— Correction of a contextual mismatch:
    Student to teacher: — *Wyhh, it's 3.30!* (end of class; in a whining and demanding tone)
    Teacher to student: — *No, it's not* 3.30! (said in the same tone), *it is* 3.30 (said in a factual and positive tone)

When working on a corpus of texts or argumentative interactions, the practical rule for the analysis of a negative statement **E1** = "not **E°**" is to browse through the previous context for an addressed statement **E°** (or something close to the semantic content of **E°**). If there is one then, **E1** rectifies **E°**, and the precise nature of the rectification can be specified, in the broad context of the argumentative question structuring the exchange.

E° may be in the "short" or "long" memory of the interaction. When dealing with a complex argumentative situation, that is to say with a question debated at different times, on various places according to several genres and formats, the discursive distance to retrieve **E°** may be rather long.

2.2 Polyphonic negation: E° is not recoverable in the context
There is no actual statement or semantic content corresponding to **E°**, for example when the speaker of **E1** anticipates a foreseeable objection, **S. Prolepsis**. In that case, according to Ducrot's original and robust version of the polyphonic nature of language, we can consider that the negative utterance articulates two voices, that of the *rectifier* and that of the *rectified*. As in the preceding case, the speaker adopts the position of the *rectifier*. Ducrot speaks in such cases of "conflictual polemical negation" (*ibid.*).

The two uses of negation, according to whether **E°** is or is not recoverable in context, are perfectly continuous. If the **E°** statement cannot be recovered in the immediate context, one will opt for a polyphonic analysis, referring the contents to voices, and not to actual participants, **S. Interaction, Dialogue, Polyphony**. There will however remain some doubt as to the precise scope of the rectification operated by the negation.

2.3 Descriptive negation

Ducrot mentions the case of a "descriptive negation", which could not be split into two antagonistic voices:

> Some uses of a syntactically negative sentence have neither conflictual nor opposing character. Negation is used without paying attention to its negative character, without, therefore, introducing into it any idea of dispute or doubt. Thus, to point out that today the weather is perfectly fine, I can use a negative sentence *"not a cloud in the sky"* as well as a positive sentence *"the sky is perfectly blue and clear"*. (Ibid.)

Such negative sentences have an autonomous meaning. This analysis is suitable for negative polarity statements, from which it is impossible to retrieve an underlying positive statement:

> You can't hold a candle to him.

It is also appropriate for negative prefix words without corresponding positive terms (see above).

3. Denying

The dialogic character of negation is systematically exploited in psychoanalysis, in which the negative utterance is considered to be the result of a negotiation between the conscious and the unconscious:

> The manner in which our patients bring forward their associations during the work of analysis gives us an opportunity for making some interesting observations. *"Now you'll think I mean to say something insulting, but really I've no such intention."* We realize that this is repudiation, by projection, of an idea that has just come up. Or: *"You ask who this person in the dream can be. It's not my mother."* We amend this to: *"So it is his mother."* In our interpretation we take the liberty of disregarding the negation and of picking out the subject matter alone of the association. It is as though the patient had said: *"it's true that my mother came into my mind as I thought of this person, but I don't feel inclined to let the association work."*
> Thus the content of a repressed image or idea can make its way into consciousness, on condition that it is negated. Negation is a way of taking cognizance of what is repressed; indeed it is already a lifting of the repression, though not, of course, an acceptance of what is repressed. We can see how in this the intellectual function is separated from the affective process.
> (Freud, [1925], p. 235)

4. Argumentative strategies using various forms of negation

The relation between discourse and counter-discourse is fundamental for the definition of an argumentative situation, negation and denial are therefore at the very foundation of argumentation studies. **S. Contradiction; Opposition; Opposite; Destruction; Objection; Refutation; Counter-argumentation.**

Denying the Antecedent ▶ Deduction

Derived Words

1. Argumentations based on word derivation

A *derived word* is a word formed on a base or a stem word combined with a prefix or a suffix. A *derivational family* is made up of all the words derived from the same root or base word.

The argument based on derivatives uses this mechanism of morphological derivation. As the *signifier* of the root word is found in the derived word, one may think that the *meaning* of the root word is also transferred to the derivative, which is not necessarily the case. The president of a rather powerless *commission of conciliation* addressed his fellow members of this commission as *commissioners*; this clever label gives him and his colleagues the authority associated with the word *(police) commissioner* and some superiority over the people who appeal to the *commission*.

The argumentation by derivative exploits a sense of semantic obviousness arising from the morphological similarity between words belonging to the same derivational family, which produces a statement apparently impossible to deny, because true by virtue of its seeming analytical form, "**A** is **A**":

> I am human, nothing human is foreign to me.

This famous speech made by General de Gaulle uses such self-argued statements, **S. Self-argued**:

> As for the legislative elections, they will take place within the period established by the Constitution, unless the whole French people are to be gagged, preventing them from speaking as they are prevented from living, by the same means that prevent students from studying, teachers from teaching and workers from working. (Charles de Gaulle, *Speech on May 30, 1968*[1])

In a well-made world, "students study, teachers teach and workers work" if not, the semantic disorder argues the abnormality of beings who don't act according to their essential principle.

These self-evident arguments are based on a license to infer according to which the derivational families are semantically consistent. The morphological similarity may obscure deep semantic differences between the root word and the derived word, which meaning may range from the conservation of the root meaning, to opposition between their connotations or argumentative orientations, to the complete independence of meanings in synchrony. By a kind of antanaclasis **S. Orientation**, the following exchange plays on the opposite argumentative orientations of words belonging to the same lexical family, *politic*:

> S1 — *By signing this compromise at a convenient moment, the president made a highly political decision.*
>
> S2 — *We are just witnessing a new example of the President's usual politicking*

---

[1] Quoted after http://archives.charles-de-gaulle.org/pages/espace-pedagogique/le-point-sur/les-textes-a-connaitre/discours-du-30-mai-1968.php (11-08-2017)

The French present participle-adjective *aliénant*, "alienating", and the past participle-adjective *aliéné*, "alienated", derive morphologically from the verb *aliéner*, "alienate", but have two different meanings. *Aliénant* refers to *socio-political* conditions whilst *aliéné* refers to severe *mental* conditions. In the following case, the speaker rebuts a social claim by aligning the former on the latter:

> If you find your work *alienating* [Fr. *aliénant*], then we will direct you to an *asylum* [Fr. *asile d'aliénés*].

It must be emphasized that argumentation based on word derivations are strictly dependent on the linguistic structure of the specific language considered.

Rebuttal — The argument by derivation can be rejected as a fallacy of *form of expression*. The identity of the visible forms of the derivative word with its base word suggests that their meanings are the same; but this supposition is misleading, **S. Expression**. They are therefore refuted as "plays on words" by highlighting the differences in meaning between root word and derived word. In turn, this rebuttal will be rejected as "semantic nitpicking".

2. Other designations and related forms

Topic of related words — Cicero considers the same argumentative device under the label topic of *related* terms (*coniugata*), that is "arguments based on words of the same family"; related terms are terms such as *"wise, wisely, wisdom"* (*Top.*, III, 12, p. 391):

> If a field is "common" (*compascuus*), it is legal to use it as a common pasture (*compascere*). (*Ibid.*)

Since it is a *common field*, the sheep can graze there in *common*. But does that mean that all herds can graze there *simultaneously* or *successively*?

Topic of the derivative — Topic n° 2 of Aristotle defines the "topic of derivative" as follows:

> Another topic is derived from similar inflexions, for in like manner the derivative must either be predicable of the subject or not; for instance, that the just is not entirely good, for in that case good would be predicable of anything that happens justly; but to be justly put to death is not desirable. (*Rhet.*, II, 23, 2; Freese, p. 297)

This is a dialectical exercise. Problem: *"Is the just desirable?"*, that is to say, is the predicate *"— is good, desirable"* part of the essential definition of the word *just*? The answer is no, because "If you find that the *just* is desirable, then you find that being *justly* put to death just is desirable", which is rarely the case.

Etymology, *notatio nominis*, *conjugata* — For Bossuet there are two kinds of topics drawn from the noun.

— On the one side, the topic "drawn from etymology, in Latin *notatio nominis*, that is from the root the words originate from, like 'to be a master, you have to master the masters'." (after Reverso; Fr. "if you are king [*roi*], then reign! [*régnez*]". The example corresponds to Cicero *conjugata*.

— On the other side, the scheme "taken from words that have all the same origin, called *conjugata*", giving as an example of this relationship the pair *homo / hominis*, two inflected forms of the same word.

The terminology might seem a little confusing, but the bottom line is clear, whenever two terms are linked by morphology, lexicon or etymology, the conclusions established for one of the two can be transferred to the other.

Destruction of Speech

The argumentative forms of rebuttal are based upon what is *said*, that is to say upon a critical examination of the content of the rejected speech, of its relevance to the current issue, or upon considerations related to the person who holds it. Good or bad, the refutation is explicitly argued, S. **Refutation**.

Argumentative discourse, as any discourse, can be put under attack, either by such an argued refutative discourse or through more radical, linguistic or non-linguistic coups. *Speech destruction* tries to impair, cancel, exclude, the targeted speech; to make nonsense of what it says, leaving it devoid of substance and import; to make it unbearable, untenable, repulsive — and, first of all, to make it innocuous, to ensure that it will have no practical impact upon the group.

1. Discourse destruction and freedom of expression

In view of their material exclusion from the public sphere, argued beliefs and proposals can be neutralized by the legal prohibition of their expression, and the imprisonment of the opponents. This can be seen as attacks on freedom of expression; nonetheless, many democratic countries agree to prohibit by law hate speech as an incitement to crime.

Free expression can also be hindered by popular demonstrations, thus making public expression inaudible, by means of shouting, blowing horns, etc.

2. Destruction through interactional behavior

In ordinary face-to-face situations, discourse can be destroyed by non-verbal interactional maneuvers, the most radical being the refusal to listen, and let the others listen, the discourse of the other. Agreement is manifested by various phenomena of ratification, and, conversely, a simple lack of ratification, the inertia of the partner, may induce the speaker to withdraw the speech, S. **Disagreement**.

The following interaction takes place in a high school physics lab. The lesson is on the notion of force, and exploits a small device, a stone suspended from a gallows.[1] The two male students **F** and **G** are working in pair. The question asked by the teacher is:

> What are the objects that act on the stone?

---

[1] Example taken from the VISA database: https://visa-video.ens-lyon.fr/visa-web/ (09-20-2017).

Puzzled, the two students look at the teacher. Then, still addressing the class, she adds:
> Well, I took an object in the most general sense that is to say, all that can act on the stone er: visibly or invisibly if— well\

Then, student **F** immediately answers the teacher's question, addressing his partner:
> Well the air/ the air/ ... the air it acts the air when you do that the air\

After an interruption, **F** resumes his argumentation, waving vigorously his arm up – down – up, intensely addressing his partner (simplified transcription):
> When you do that there will be air afterwards since y'know when you make a fast movement like that\ it is the same there is the air\ I'M SURE\ but here for now we do not answer that yet but/

Then student **G**, playing with the stone, says:
> There is the attraction\

F's argument is perfectly in line with Toulmin's model of argument. The claim is *"the air [acts upon the stone]"*. It is supported by an appeal to analogy, *"it's the same"*, referring to an arguing ad hoc gestures, mimicking and emphasizing some self-evident fact. The conclusion is duly emphatically modalized, *"I'm sure"* — and immediately withdrawn: *"but for now we do not answer that yet"*. In view of this strongly asserted argument, this withdrawal is quite unexpected. It is understandable only in view of the interactional behavior of the conversation partner **G**, who stares at the stone and gives no sign of ratification throughout, not even signaling that he is listening to **F**'s argument (with whom he gets on very well, as shown by their following fully collaborative exchanges).

3. Rejecting the expression

An embarrassing discourse can be destroyed through a criticism focusing upon the *style and expression* of the opponent without taking into consideration the argument itself. The reply *"I don't agree"* actually demonstrates a high level of cooperation.

Ancient rhetoric enumerates a trio of major linguistic qualities of discourse, *quality* of language, *clarity* and *vivacity* of expression (respectively *latinitas, perspicuitas* and *ornatus*). Destruction strategies can develop out of any of these points.

Quality of the language — *"You are hardly understandable, you don't even know the language you pretend to speak, you use dialect expressions you should try to speak classical Syldavian"*. In a polemical situation, the opponent can reject *a priori* a discourse arguing from its grammatical defects. It would be wrong to think that these strategies are marginal or ineffective:
> In an uncertain spelling, Mrs. X challenges the evaluation of her language skills by the jury of the competition.

Mrs. X failed her exam about her language skills. Now, she disputes the jury's decision, and the jury answers mentioning the "uncertain spelling" of her complaint letter. Stricto sensu, these misspellings do not prove that her exam paper was also misspelled, but can certainly be used as a suggestion to that effect. In any case it justifies a charge for neglect, showing a disregard for the jury, which is enough to devaluate the significance of her complaint.

Clarity and vivacity of the expression — Similar devastating strategies appeal to the lack of clarity of expression: *"the presentation was unclear and confusing"*, or vivacity *"so boring!"*.
It is of course better for an argumentative speech to be grammatically correct, clear and interesting. On the other hand, it is human nature to consider correct, clear, and interesting the speeches with which one agrees. This is not just a psychological or bad faith issue; it has a cognitive relevance. The discourse with which one agrees is better known; its deep principles being well accepted, it is easier to recover the ellipsed contents and the missing links; its variations are better tolerated; it is better memorized, etc. When it comes to an opponent's discourse, it is relatively natural to translate as speech defects the corresponding difficulties, and to conclude by denying that the minimum conditions of mutual comprehension are satisfied.

Making fun and puns out of the opponent's discourse, is a popular way to get rid of the problems and arguments defended S. **Laughter and Seriousness; Orientation Reversal**

4. Leaving aside the argumentative details
A class of refutative maneuvers refers to the opponent's discourse without considering its argumentative details, for example:
— *Declaring the discourse sub-argumentative*, unworthy of a refutation, S. **Dismissal**.
— *Misrepresenting the argument*, S. **Straw Man**.

5. Disqualifying the arguer
Personal@ attacks against the speaker set aside the argument and try to disqualify the arguer.

For other forms on the verge of *destruction* and propositional *refutation*, S. **Refutation**.

Dialectic

> *Dialectics* and *dialogue* have the same Greek etymology *dia-* + *legein*, *dia-* "through", *legein* "say". The prefix *dia-* is different from the prefix *di-* meaning "two". Etymologically, a dialogue is not a two-person conversation (which could be referred to as a *dilogue*). The condition is not on the number of participants, but on discourse circulation. However, the historical notion of *dialectic* does refer to a two-partner dialogue.

1. The ancient dialectical method
Aristotelian dialectic is a dialogical method used to solve questions of the form "**P** or **not P**?", such as *"is being rich a good thing or not?"*, by eliminating one of the options, in a standardized question-answer interaction using dialectical syllogisms.

Dialectic is a philosophical instrument used in the a priori search for the definition of fundamental concepts. In this function of clarification of the first principles, it has been replaced by axiomatization.

1.1 Dialectical reasoning
As "mathematical science" and "rhetorical argument", "dialectical reasoning" proceeds by syllogism and induction (Aristotle, *Post. An.*, I, 1). While scientific syllogistic deduction proceeds from "true and primary" premises, dialectic uses generally accepted premises (*Top.* I, 1), or simple "opinions", *endoxon*:

> Our treatise proposes to find a line of inquiry whereby we shall be able to reason from opinions that are generally accepted about every problem propounded to us, and also shall ourselves, when standing up to an argument, avoid saying anything that will obstruct us. (*Ibid.*)

The word *endoxa* translates as "probable premises" or as "accepted ideas". The strict deduction rules of the syllogism are replaced by argument schemes.

1.2 Dialectical game
The dialectical game is played by two partners, the *Respondent* and the *Questioner* (Brunschwig 1967, p. 29). It is a bounded interaction governed by strict rules, proceeding by questions and answers, with a winner and a loser. The Respondent first chooses to assert either **P** or **not P**. The Questioner must refute the proposition that the Respondent has chosen to support, by means of total questions (*yes* or *no* questions). On the basis of these answers, the Questioner attempts to make the Respondent admit a statement which contradicts the original assertion. If the Questioner succeeds, then he or she will win the dialectical game; if he or she fails, the Respondent will win.

The terms *Proponent* and *Opponent* used to refer to the core partners of an argumentative situation, are borrowed from this dialectical theory. Unlike the Proponent of a substantial proposition in an argumentative situation, the Respondent in the dialectical game does not have to build a positive proof of the proposition put forward, but must simply avoid being led into a self-contradiction.

1.3 Dialectical authority
To be worthy of a dialectical debate, the proposition must be an *endoxon*, that is to say, it must be endorsed by some social or intellectual authority:

> Now a dialectical proposition consists in asking something that is held by all men or by most men or by philosophers, i.e., either by all, or by most, or by the most notable of these. (*Top.*, 10)

The Aristotelian continuum values different orders of endoxa. We are far from the vision of the doxa as cliché or stereotype as "ready-to-think", or, just as mechanically, "ready-to-denounce".

Endoxa are opinions worthy of discussion. They define a contrario what a *thesis* as "a supposition of some eminent philosopher that conflicts with the general opinion"; the philosopher must be eminent, "for to take notice when any ordinary person expresses views contrary to men's usual opinions would be silly" (Aristotle, *Top*., I, 11). In other words, "if it were the first comer who emitted paradoxes, it would be absurd to pay attention to it" (Aristotle, *Top*., Brunschwig, I, 1, 100b20, p.17). The authority entering the debate is clearly socially referenced as such.

It is remarkable to see that it is the *plurality* and *competition* between authorities — rather than the *call* to authority — which is placed at the core of intellectual debate. Authority is not invoked in order to close the discussion but rather to open it. To say that a proposal is supported by an authority is not to say that it is true, but to say that it deserves discussion.

2. The scholastic dispute

The scholastic dispute (*disputatio*) corresponds to the medieval practice of a dialectical game. It is an instrument of research and teaching, based upon a specific substantial question, as proposed by a master. At the end of the discussion, the master proposes a solution and refutes the arguments against it (Weijers 1999).

3. The revival of dialectic

The ancient dialectical method, which had been declining since the Renaissance (Ong 1958), was reconstructed in the second half of the twentieth century within the framework of logical dialogue games. It has been put at the forefront of argumentation studies by the Pragma-Dialectic and by the Informal Logic programs. The Pragma-Dialectic program of Frans van Eemeren and Rob Grootendorst (1996, etc.) is a "New Dialectic", a counterpart of Perelman's "New Rhetoric" (van Eemeren, Grootendorst, 1996 "La Nouvelle Dialectique" ["The New Dialectic"]). In the Informal Logic framework, the study of "logical dialogue games" has been developed by Douglas Walton (Walton 1984; Walton 1998, *The New Dialectic*).

In a continuation of a general definition of dialectic as, "the practice of reasoned dialogue, [the art] of arguing by questions and answers" (Brunschwig 1967, p. 10), one can consider that the conversational process is "dialectized" insofar as 1) it relates to a specific and mutually agreed problem; 2) it is played out between equal partners, 3) driven by the search for the truth, the just or the common good; 4) between which the speech circulates freely, but nonetheless 5) respects explicitly established rules.

4. Aristotelian dialectic and Hegelian dialectic

Unlike Aristotelian dialectic, Hegelian dialectic does not proceed by the *elimination* of the false, but by *synthesis* of the antagonistic positions. The original opposition is not resolved but abolished and transcended. Aristotelian dialectic is founded on the principle of non-contradiction, whereas Hegelian dialectic tends towards something "beyond" contradiction, S. **Contradiction**.

Nonetheless, going beyond contradiction should not imply that a speaker may hold an inconsistent discourse:

> [HL] claims that "since the world is torn by contradictions, only dialectic (which admits the contradiction) makes it possible to consider it as a whole and to find out its meaning and direction". In other words, since the world is contradiction, the idea of the world must be contradictory. The idea of a thing must be of the same nature as this thing: The idea of blue must be blue.
>
> Julien Benda, *The Betrayal of the Intellectuals*, [1927][1]

Conversational dialectic, made up of negotiations and adjustments, enables the opponents to save face, whereas Aristotelian dialectic does not take into account the questions of faces and politeness.

5. Rhetoric and dialectic

According to their ancient definitions, dialectic and rhetoric are the two arts of discourse. Argumentative rhetoric is "the counterpart of dialectic" (Aristotle, *Rhet*, I). Rhetoric is to public speech what dialectic is to private, conversational speech. Rhetoric concerns long and continuous discourse, whilst dialectic is a technique of discussion between two partners, proceeding by (brief) questions and answers. Fundamentally, dialectic is *legislative*, it serves the discussion of the *a priori* foundations which will serve as premises for scientific deduction. Rhetoric has an *executive* function: it deals with current, public, legal and political affairs, and, with the development of Christianity, religious belief; it strengthens the principles that govern these practices, via epideictic means.

Diallel ▶ Vicious Circle

Dilemma

A *dilemma* is a schematization of a situation as an alternative whose terms are equally undesirable. Used as an argumentative strategy, the dilemma corresponds to a case-by-case refutation, consisting in cornering one's opponent by showing that all his or her lines of defense lead to the same negative conclusion:

---

[1] Quoted after Julien Benda, *La Trahison des Clercs*. Excerpt from the Preface to the 1946 edition. Paris: Grasset, 1975. P. 63.

Direction ▶ *Gradualism; Slippery Slope*

— Either you were aware of what was going on in your services, and you are an accomplice, at least passively, of what has happened, and you must resign.
— Or you were not aware, then you do not control your services, and you must resign.
Either way, you will have to resign.

A dilemma can be rejected as *poorly built*, as a *false dilemma*, an artificial radicalization of a more complex opposition, which can be reconstructed in order to show that there is a third way out of the dilemma, **S. Case-by-case**.

> If I have clear and strong support from the citizens to remain in office, the future of the new Republic will be secured. If not, there can be no doubt that it [*the new Republic*] will immediately collapse and that France will have to endure, this time without remedy, a confusion of the State even more disastrous than that which it once knew.
> Charles de Gaulle, 4 Nov. 1965 Speech, announcing his candidacy for the December 1965 presidential election[1]

This relatively common practice of framing the political situation can be rephrased as the slogan *"it's either me or chaos"*. A supporter of the speaker will take this statement as offering a realistic clear-cut choice between good and evil. An opponent will reject it as an arrogant and inadequate means of pressure. Undecided citizens may see it as the expression of a real dilemma, a choice to make between two equally undesirable options.

Direction ▶ Gradualism; Slippery Slope

Disagreement

1. Preference for agreement

Argument is a means of building a new consensus from an established consensus, **S. Agreement; Persuasion**. Such a construction can be seen as the "macro" expression of a trend observable at the "micro" level of the interactional sequence, *preference for agreement*. This concept is fundamental for the organization of turns of speech in interaction. In an adjacency pair, the first turn "prefers", i.e., is *oriented* towards a specific kind of second turn. The preferred response to an invitation is acceptance, rather than refusal; proposals are made to be accepted and not rejected; affirmations are put forward to be ratified, not to be rejected, etc.

The *preferred sequence* is unmarked; the second speaker aligns with the first; agreement is a given, a minimal linguistic mark may suffice: (*yes, OK, let's go...*), or a quasi-verbal ratification (*mm hm*) or a minimal bodily action (*nodding*). The

---

[1] http://fresques.ina.fr/jalons/fiche-media/InaEdu00101/de-gaulle-Fact-de-candidature-en-1965.html] (11-08-2017). The last phrase alludes to the June 1940 military rout.

preference for agreement is also reflected in practices such as the avoidance of frontal opposition, the absence of ratification of emerging disagreements and the preference for micro-adjustments to reach an agreement without explicitly bringing up the disagreement for an overt discussion.

The *dispreferred sequence* is marked, that is to say, it contains specific features such as hesitation, presence of pre-turns (underlined in **S2_2**) and justifications (bold characters in **S2_2**):

S1_1 — *What are you doing tonight?*
S2_1 — *Well I don't know ...*
S1_2 — *Come for a drink!*
S2_2 — (silence) *hmm, well, you know,* *I'd prefer not to,* **I have got a little work to do**.

Giving reasons for accepting an invitation is almost an offense:

S1 — *Come to dinner tomorrow night!*
S2 — *With pleasure, it'll mean I won't have to cook, and I will take down the trash.*

This preference for agreement is not a psychological fact, but an observational conversational regularity. It can be compared with Grice's principle of co-operation, or with Ducrot's observations on the polemical effect produced by second turns which do not accept the presuppositions of the first turn, **S. Presupposition.**

2. Conversational divergences and overt arguments

Face to face disagreement is expressed by a series of specific coordinated behaviors, either verbal "*I don't agree*", or paraverbal: fights for the floor; interruptions; non collaborative overlappings; accelerated speech flows; raised voices; negative regulators, heads shaking, sighs, agitation — or ironic excesses of signs of approval; non-addressed partner behavior, etc.

Sequences of *conversational divergence* appear randomly; they follow unforeseen patterns; they have a potentially negative impact on the goals of the overall interaction; they introduce a delicate balance between somehow sacrificing one specific vision of things to maintain good relations with the other party; or taking the risk of damaging the relationship to maintain and sharpen extreme difference of opinion. In the majority of cases, conversational disagreements are resolved immediately, through step-by-step micro-adjustments and negotiation, to be forgotten.

At other times, conversational divergences serves to deepen differences. When *conversational divergences* are explained and disagreement ratified, each position backed by arguments and counter-arguments, the interaction becomes strongly argumentative. Such interactions can be momentous, kept in mind, ruminated upon and elaborated. They may generate new interactions, referring to the root disagreement, where the parties will develop planned interventions. The treatment of what has become an issue is now the rationale of these interactions.

3. *Enantiosis*: emerging argumentation

The argumentative role of an opponent may develop from his or her interactional role as a listener, ratifying the existence of an argumentative situation, where two discourses concerning the same topic are in explicit competition.

> During a friendly conversation at a party, between people who barely know each other:
> S1 — *if we watch the TV candidates debate together tonight, maybe we should know something about each other, personally I vote for candidate Smith.*
> S2 — *oh, well, for me it's not quite so...*

Before this exchange, **S2** is simply the interlocutor of **S1**. During the exchange, a political divergence emerges, which initiates a restructuring of the interaction, that can lead to a re-framing of the interlocutors as political antagonists. A full-blown argumentative situation can develop from that point, depending on whether or not the subsequent turns will thematize the emerging opposition.

The figure of rhetoric called *enantiosis* seems particularly well suited to designate this transitional moment, where opposition is looming large without yet being ratified by the participants. The Greek adjective [*enantios*] can mean:

> 1. Being in front of, such as shores that face each other; things that are offered to the gaze of somebody.
> 2. With an orientation towards hostility, *which stands in front of*: "those in front of us", that is *the enemy*; in general, *the opposing party*, the adversary.
> 3. Opposed, contrary to: *the opposite party*, *the opponent* (after Bailly, [*enantios*]).

According to this development of meaning, in a dialogue, the adjective *enantios* refers first to the person standing here, in front of you, for example, in the *interlocutor's position*. The idea of *hostility* appears in a second instance, and then the *interlocutor* becomes *the opponent* (the "*adversarius*" in a rhetorical encounter, Lausberg [1960], §274).

The word *enantiosis* is also used as a synonym of "antithesis", and can refer to oppositions such as "*good* vs. *bad*; *even* vs. *odd*"; *one* vs. *multiple*" (Dupriez 1984, *Énantiose*). This kind of binary opposition is characteristic of the sometimes Manichaean diptych corresponding to antagonistic argumentation. The semantic palette of enantiosis covers the dynamics of this emergence and the initial stabilization of the argumentative situation:

> The person facing you > with hostility: the opponent
> > the argumentative antithesis, discourse vs. counter-discourse.

4. "Deep disagreement", S. Dissensus

Dismissal

Dismissal is a method of processing the opponent's discourse on the brink of refutation@ and destruction@. The standard forms of refutation are based on a substantial examination of the content of the rejected speech, or on more or less relevant considerations about the person holding it. Even in the latter case, the rejection is, however badly argued, at least backed by some justificatory discourse.

The opponent can *dismiss* a discourse simply by declaring that the bad quality of the proposed argument is self-evident and self-denouncing:

> No comment.
> Your arguments are shabby, insufficient, miserable, distressing
> I will not give your statement the honor of a refutation.
> What you say is not even false.

This brings to mind Uncle Toby's reaction, "whistling half a dozen bars of *Lillabullero.*", S. *Ab –, ad –, ex –*. In ancient rhetoric, this move declaring the argument to be "childish" or "obviously absurd or practically null", is called *apodioxis*, (Dupriez 1984, *Apodioxis*; Molinié 1992, *Apodioxis*), S. **Pathetic**.

The opponent who uses it can speak in perfectly good faith, which can lead to paradoxical situations. If the discourse of Big Jones is really self-denouncing, then:

> One should give Big Jones a greater say, the more he speaks, the more foolish he appears, the fewer votes he will get.

But this is a perilous strategy, inspired more by the arguer's self-confidence than by any self-evidence about the discourse.

To top it all off, the opponent might adopt a strategy of irony@, and contribute to the dissemination of the opponent's speech. This is the extraordinary case reported by Wayne Booth, when talking about events taking place in his university, where students were clashing with their University administration:

> At one point, things got so bad that each side found itself reduplicating broadsides produced by the other side, and distributing them, in thousands of copies, without comment; to each side it seemed as if the other side's rhetoric was self-damning, so absurd had it become. (Booth 1974, p. 8-9)

Obviously, the other side cannot even hear such a disqualification, which targets third parties. Used in particularly contentious argumentative situations, such a maneuver makes any deal between the discussants impossible, S. **Conditions of discussion**.

From the ethotic perspective, such a (non-) arguer displays a kind of moral *indignation*, whereas the opponent can make an accusation of *arrogance* and contempt.

Ad lapidem argument (Lat. *lapis*, "stone")

The name of this fallacy is derived from a famous incident in which Dr. Samuel Johnson claimed to disprove Bishop Berkeley's immaterialist philosophy (that there are no material objects, only minds and ideas in those minds) by kicking a large stone and asserting 'I refute it thus' (after Wikipedia, *Ad lapidem*).

This clear *contempt of verbal argument* is akin to "the proof of the pudding is in the eating", a proof by the *facts* or the *act*.

Dissensus

Rhetorical argumentation focuses on persuasion, adherence, communion, consensus, co-construction... These terms sound much like moral incitements, *"don't be different, be the same"*; and it's difficult to disagree with the principle of agreement. The emphasis on persuasion and consensus suggests that unanimity would be the normal, healthy state of society, as opposed to the pathological state of controversy, or dissensus.

1. The passion for dissensus as sin and fallacy

The passion for dissensus characterizes *polemical* exchanges; verbal violence is not associated with controversies as it is with polemics. Emotional dramatization and personal involvement are expressed in the speech acts opening the debate: *to rise up against, to be outraged, to protest...* When it comes to emotional repercussions, controversy and polemic might hurt the feelings of the parties.

The polemicist refuses to close the debate, and allow the other party's argument to prevail, even if it is the stronger argument. This refusal to defer to the arguments of the other is a paralogism of obstinacy, stigmatized by Rule 9 of the critical discussion, that asks the proponent to bow before a conclusive argument (van Eemeren & Grootendorst 2004, p. 195; **S. Rules**). But who says that the point of view has been conclusively defended? The polemicist refuses to admit that the point of view of his or her opponent has been defended conclusively, and posits that the veracity of his or her viewpoint is beyond reasonable doubt. As a last resort, he or she might appeal to intimate conviction, as a way of preserving a jeopardized identity.

The condemnation of argumentativeness and polemic has deep historical roots. The Middle Ages considered *contentio*, that is contentiousness, as a sin of the tongue, **S. Fallacies and Sins**.

> *Contentio* is a war of words. It may be a defensive war waged by the stubborn individual, who refuses without reason alter his position. But *contentio* is most often manifested as a display of aggression in one of many forms. This might be an unnecessary verbal attack against one's neighbor, an aim not to seek the truth, but to simply manifest aggression (Aymon); a quarrel which, abandoning any quest for truth, gives rise to dispute and goes as far as blasphemy (Isidore); a refined and malevolent argumentation that opposes the truth to satisfy an irresistible desire for victory (*Glossa ordinaria*); a wicked, contentious and violent altercation (Vincent of Beauvais); an attack against the truth led by the strength of the *clamor* ["public outcry", CP] (*Glossa ordinaria*, Peter Lombard). Often, however, the *contentio* appears in texts without ever being defined, as if the connotation of violent verbal antagonism attached to the term is sufficient to indicate that it should be avoided and condemned as a sin.
>
> Casagrande & Vecchio ([1987], p. 213-214.

Contentio is a second level sin, derived from first level sins such as envy, vainglory and pride. There is one reservation to be mentioned here, namely that such definitions restrict the sin of *contentio* to violent attacks against *religious truth*. It is not, however, a sin to violently and continuously attack *error and sin*; anger becomes a *holy anger*.

2. Polemics and "deep disagreement"

The concept of *deep disagreement* was introduced by Fogelin (1985). Deep disagreement involves incompatible values or metaphysical principles, rather than empirically testable epistemic issues. The solution of scientific conflicts, including in mathematics and logic, call for technical treatment (Woods 2003), while deep disagreement is more akin to polemics, involving intense personal commitment on the part of the participants. Nonetheless, polemics seems to prefer (face-to-face) confrontation, while deeply disagreeing position can be developed in parallel and in mutual ignorance, thus appearing beyond the field of argued dialogue.

In human affairs, the existence of such intractable divergences may be considered as a "radically shocking" challenge (Turner & Campolo 2005, p. 1) to the argumentative enterprise itself. "if [Fogelin] was right, what would become of the field? Even more important, arguably, what could be done about deep disagreements themselves? The field and all of the good it meant to accomplish seemed to be threatened all at once" (*ibid.*).

3. The post-persuasion era and the normality of dissensus

Any serious argumentative debate contains an element of radicalism, which calls for a de-demonization of dissensus, and, as a consequence, for a re-evaluation of the role of the ratified third parties, who have the power to make a decision. As Willard, who has written extensively on this subject, states:

> To prize dissensus goes against an older tradition in argumentation, that values opposition less than the rules that constrain it. (Willard 1989, p. 149)

The preference for consensus does not exclude the reality of dissensus. Argumentation studies must confront situations in which differences of opinion are produced, managed, solved, amplified or transformed through their discursive confrontation. Determining which differences of opinion should be reduced and how, and which ones should rather be encouraged and deepened is a major social and scientific issue, having critical educational implications.

Argumentation can be used to divide opinion; this is what the discourse of Christ achieves in the Christian vision of the world:

> Do not suppose that I have come to bring peace to the earth. I did not come to bring peace, but a sword. 35 For I have come to turn a man against his father, a daughter against her mother, a daughter-in-law against her mother-in-law, a man's enemies will be the members of his own household.
>
> Matthew 10: 34-36[1]

---

[1] Matthew 10:34-36. Quoted after *The Bible*, New International Version (NIV),

The first virtue of argumentation is not that it solves the conflicts, but that it is able *to give words to conflicts*; it is a precious method of managing differences, sometimes reducing them, sometimes increasing them and causing them to multiply. In an over-consensual context, it may be the noble task of argumentation to bring about relevant dissensual discourses, and to value and stimulate the emergence of *differences of opinion*.

The majority rule does not imply that the majority is the holder of the truth, and is entitled to enforce its rule over a disgraced minority who spuriously resist the persuasive power of the orator, or refuse to acknowledge the defeat inflicted upon them by the dialectician. One can hypothesize that, in our terrestrial world, the coexistence of contradictory opinions represents the normal state, neither pathological nor transitory, of the socio-political ideological field; deep disagreement is the routine and rule. Hegelians would add that contradiction is the dialectical engine of history.

In any case, democracy does not eliminate differences, and voting does not eliminate minorities and their opinions. In such conditions *"No se trata de convencer sino de convivir"*[1], the objective is not to convince others, but to enable groups to coexist. Argument is a way of managing these differences, sometimes eliminating them, sometimes promoting them for the common good.

Dissociation

The concept of *dissociation* was introduced by Perelman & Olbrechts-Tyteca. According to the *Treatise*, the techniques of argumentation are of two kinds "association and dissociation" ([1958], p. 190). The former of these concerns two or more *propositions*, making up an argumentation, while the latter operates on a single *concept*. The *dissociation* technique is thus placed on a par with *association* techniques as a whole, that is, the large set of argument schemes.

Dissociation is defined as the splitting of the meaning of a word or a concept, to avoid a contradiction. The meaning of the problematic term T is re-framed as containing an internal contradiction, "an incompatibility", "an antinomy", and dissociation is the mechanism by which it can be solved ([1958], 550-609). T is split into a term T_1 and a term T_2, this operation coming with a negative evaluation of T_1 and a positive evaluation of T_2. Dissociation appears as a kind of "semantic cleansing", through which an unwanted content or connotation, T_1, can be disposed of. The word *reality* can thus be divided, "dissociated", into the pair T_1 = *appearance* vs. T_2= *reality*, the latter being "the true reality".

> While the primitive status of what is given as the starting point of dissociation is undecided and indeterminate, the dissociation in terms I and II will value the aspects corresponding to term II and will devalue the aspects that oppose

---

www.biblegateway.com/passage/?search=Matthew%2010:34-36 (11-08-2017)
[1] A. Ortega, "La razón razonable", *El País*, 25-09-2006.

it. Term I, the appearance, in the narrow sense of this word, is only illusion and error. (Perelman 1977, p. 141)

Dissociation can operate in a monologue or a dialogue:
X: — *Well old chap, that's democracy!*
Y: — *There is democracy and democracy.*

According to Perelman, the dissociation technique is, "hardly mentioned by traditional rhetoric, for it is especially important for the analysis of systematic philosophical thought as systematic" (1977, p. 139). An example is taken from Kant, for whom natural sciences postulate a universal *determinism* while morality postulates the *liberty* of the individual; hence the necessity of dissociating the Term *reality*, a confused notion, into a *phenomenal* reality, in which determinism reigns, and a *noumenal* reality where the individual can freely choose and act upon his or her decision. In that case, dissociation is equivalent to *distinguo®*, but without a preferred term.

It seems to follow from the examples given above that the same notion can be dissociated according to the arguer's objectives, dissociation being the key operation to derive a concept from the ordinary meaning of a word.

1. Linguistic aspects of dissociation

Reasoning through dissociation is characterized first of all by the opposition between *appearance* and *reality*. This can be applied to any notion, as soon as one makes use of the adjectives such as *apparent, illusory* on the one hand, *real, true* on the other. To use an expression such as *apparent peace* or *genuine democracy* is to indicate the absence of *genuine peace*, or the presence of an *apparent democracy*: one of these adjectives refers to the other. (*Id.*, p. 147)

The linguistic markers of dissociations are very diverse:

A prefix such as *pseudo-* (*pseudo-atheist*), *quasi- not-* the adjective *alleged*, the use of quotes indicate that we are dealing with the term I, while the capital letter (*Being*), the definite article (*the solution*), the adjective *unique* or *true* denote a term II. (*Id.*, p. 148)

Other dissociations are stabilized as pairs of antithetical terms or "philosophical pairs" such as "opinion / science; sense knowledge / rational knowledge; body / soul; just / legal, etc." (Perelman [1958], 563). Some of these dissociated pairs are traditional and constitute the oppositions generating foundational ideological discourses. As for all antonymic pairs, one term is linguistically preferred to the other, and this preference can be reversed. The T_1 *vs.* T_2 opposition "superficial *vs.* deep" can be reversed through a praise of the superficial — "the skin is the deepest thing there is" (Paul Valéry). The dissociated pair, "rhetoric *vs.* argumentation" is engaged in a permanently revolving evaluation.

2. Dissociation as a shielding operation

Dissociation has a concessive facet. For example, one might assume that some intellectuals would make good businessmen, while conceding that they are only a tiny minority. Dissociation does the same, but via an outright exclusion of this sub-category from the general category, "intellectuals":

(1)   S1   — *When it comes to business, intellectuals are hopeless*
       S2   — *Or they are not true intellectuals.*

(2)   S1_1   — *Germans drink beer.*
       S2   — *Not Hans!*
       S1_2   — *Normal, Hans is not a true German.*

In (2) S2 refutes $S1_1$ by the production of a contrary case. $S1_2$ recognizes that Hans is German and does not drink beer, and maintains his original claim by splitting the category "German" into "true Germans *vs.* not true Germans". This amendment to the argument may or may not be substantiated; S1 might have replied:

       S1_3   — *But Hans is not a real German, he was brought up in the United States*

— Assuming that Americans drink less beer than Germans do. S1_3 introduces a justificatory line showing that Hans departs from the stereotype of the true German; the category created by S1_3 is based upon an explicit criterion, independent of the current discussion. In the original dialogue, the only criterion contextually available is "beer-drinking". The word *Germans* in S1 refers to all German people; if Germans are re-defined as *true Germans* on the basis of the criterion, "*Germans who drink beer*", the statement $S1_1$ is indeed compelling, since "*Germans who drink beer*" do drink beer.

The category rectification serves to exclude individuals from the category under re-analysis. In politics, this strategy opposes the, "true Syldavian" as good citizens to exclude other citizens as, "bad citizens". In practice, dissociation transforms a formerly necessary and sufficient condition (*to be a Syldavian one must be a Syldavian citizen*) into a necessary one, "*to be a true Syldavian, one must have Syldavian nationality and share our ideology*".

The following case opposes "La Réunion"[1], that is "the people living in La Réunion", to "the true Réunion", an *ad hoc* subcategory of this group.

Roland Sicard (RS) is the host of the TV program. Marine Le Pen is the candidate for the National Front ("Front National", a far right party) in the 2012 French presidential election. Gilbert Collard (GC) is a lawyer, chairman of her Supporting Committee.

       RS   — good morning Gibert Collard [...] er- a word about Marine Le Pen's trip to La Réunion\ she has been heckled, one feels that the candidates of the National Front is still in a lot of trouble overseas/

       GC   — listen I know La Réunion very well since I went there as a lawyer very often and then in particularly sensitive cases and— there are: er two Réunions eh there is a Réunion which is instrumentalized which organizes the usual reception committee for Marine Le Pen they are quite unsignificant eh\ well and then there is the true Réunion made of men with di-

---

[1] The Réunion Island is an overseas French department, East of Madagascar.

vergent views of— women with opi— but that is no more difficult in the overseas departements than in metropolitan France anyway\ no I do not think what makes it difficult is the instrumentalization of the media hmm [...]

TV program [*Home Truths*] France 2, 08 Feb., 2012.[1]

S. Opposite; Categorization; *Distinguo*; Orientation.

Distinguo

Lat. *distinguo*, 1st person singular present indicative of the Latin verb *distinguere*, "to separate; to distinguish".

Distinguo is a strategy developed in order to avoid a *terminological difficulty* or *confusion*, either perceived in the discourse of an opponent or envisioned in a polyphonical space as a possible mistake.

The word *distinguo* is also used as a synonym for *paradiastole*, **S. Orientation Reversal**.

1. *Distinguo* used as an analytical tool

Distinguos are useful for clarifying *definitions*@ of complex realities. In current language, to make a *distinguo* is to draw distinctions in order to clarify a complex notion.

> The system of 'territorial development' is based on the interaction between its two components: the local economic system on the one hand, and the so-called 'territorial' system on the other.
> The *distinguo* between the latter two systems stems from oppositions relating to the underlying logics that bear them. The economic system obeys principles that are recognized and exposed in economics. [...] The territorial system, for its part, covers all the human, social, economic and urban functions of the place.
> G. Loinger & J.-C. Nemery, [*Recomposition and development of territories*], 1998.[2]

2. *Distinguo* used to rebut a reasoning

The *distinguo* is an instrument used to reduce ambiguity@: "*do not mix everything up!*". It can be used for example to detect a four terms paralogism@, or, generally a shift of meaning in a reasoning. It is justified when it is based on socially recognized distinctions, independently established in a language dictionary or an encyclopedia, for example to eliminate the confusion created by the use of the word *metal* to refer to *a chemical element*as well as to *an alloy*.

In a second instance, *distinguo* is used to re-establish a blurred distinction (Mackenzie 1988). To make a *distinguo* is to say, "*I distinguish [in your speech] some truth*

---

[1] TV program *Les Quatre Vérités* France 2. Feb. 8, 2012.
[2] Loinger G. & Nemery J.-C.. *Recomposition et Développement des Territoires*, Paris: L'Harmattan, 1998. P. 126.

and some errors, and I'm going to rectify the mistakes". Consider the following theological syllogism (after Chenique, 1975, p. 9):

>Every man is a sinner.
>No sinner will enter heaven.
>No man will enter heaven.

The opponent says:
(i) I agree with the minor proposition *"every man is a sinner"*.
(ii) In the major, *"no sinner shall enter heaven"*, *distinguo*, I distinguish two different statements:

— "(No sinner) *as a sinner* shall enter heaven", I agree: *"no man in a state of sin will enter heaven"*;

— "(No sinner) *as a forgiven* sinner shall enter heaven": I deny this proposition. The *distinguo* does not bear upon the meaning of the word *sinner*, but two categories of sinners.

(iii) Therefore, I deny your conclusion.

The opponent therefore objects that the syllogism is fallacious, for the minor is true in one sense, and false in another.

It should be emphasized at this point that this is not a case of a four terms syllogism fallacious by homonymy, **S. Syllogistic Paralogism**. *Sinner* is not ambiguous by homonymy, but because, in discourse it can be construed in two different ways.

Distinguo is a figure traditionally dismissed as being "scholastic", and used to draw spurious oppositions. Thomas Diafoirus courts Angélique, who in fact hates him:

>*Angélique:* — But the greatest mark of love is to submit to the will of her who is loved.
>*Thomas Diafoirus:* — *Distinguo*, madam; in what does not have to do with possessing her, *concedo*; but in what does have to do with it, *nego*.
>
>Molière, [*The Imaginary Invalid*], [1673][1]

Thomas Diafoirus is brutal and pedantic; he claims his right to *possess* Angélique, against her will; apart from this, however, he is ready to submit to her will. The *distinguo* is an instrument which prevents or rectifies ambiguities, but when it introduces distinctions into a perfectly clear expression, it can itself cause confusion.

In these cases, *distinguo* may or may not be accepted according to the value of the distinction operated. In the case of the sinner, the *distinguo* might be justified by the parallel case of the criminal: a criminal having served his or her sentence cannot be called a criminal without qualification, one cannot say, "*he is a crimi-*

---

[1] Molière, *Le Malade imaginaire* [1673], act II, scene 6. Quoted after Ch. Franks, D. Lettau, http://www.gutenberg.org/files/9070/9070-h/9070-h.htm (11-08-2017)

nal, *let's call the police!*", a *distinguo* is clearly necessary.
In the case of Angélique, the *distinguo* invokes an arbitrary, ad hoc distinction. In this case it can be countered by a third round of speech such as "*stop it now!, enough with your scholastic distinguos!*", "*stop quibbling please!*".

Division ▶ Case-by-Case; Composition

Doubt

Doubt is a mental state and a behavior typically attached to an argumentative situation.

— As a *psychological* state, doubt means *discomfort and apprehension*, S. **Emotion**. Argumentation is a costly and time-consuming activity, from the cognitive, emotional and interactional points of view. Non-argumentative individuals are reluctant to engage in an argumentative situation, where they will have to face the resistance of the other party.

— At the *cognitive* level, to doubt is to be in a state of *suspended assent*@ of a proposition, or a state of indecision about what to do.

— From a *linguistic* point of view, doubtful propositions are worded by the speaker, without these being *asserted or rejected*. In Goffman's words, the speaker is, at most, the "Author" of the proposition, not its "Principal"; he or she is not committed to the statement, S. **Roles**.

— From an *interactional* point of view, doubt is cast upon a turn of speech when this turn is not ratified or overtly rejected by the interlocutor, S. **Disagreement; Question**. Such rejection cannot remain unfounded and reservations must be justified, either in the addition of arguments supporting another point of view, or by refuting the reasons given in support of the original proposal.

— In a full-blown *argumentative situation*, one or the other party does not necessarily assume doubt. A party may be absolutely and entirely confident of the truth of his or her argument, and argue that **P** is the case or the right thing to do in perfectly good faith, whilst the other party will have no doubt that it is *not* the case. Doubt is systematically taken in charge by the third party, S. **Roles**. The dialogue outsources these different operations by giving them specific linguistic shapes and micro-social configurations.

Argumentative doubt, Cartesian doubt, skeptical doubt

Argumentative doubt is opposed to *Cartesian doubt*. Descartes rejects "all such merely probable knowledge and makes it a rule to trust only what is completely known and incapable of being doubted" ([1628], Rule II; Geach). He reconstructs a system of certain beliefs on the basis of the only absolute certainty, that of the *cogito*: "*I think, therefore I am*". This kind of doubt is opposed to skeptical doubt:

> Cartesian doubt does not consist in floating, uncertain, between affirmation and negation. On the contrary, it clearly demonstrates that what is in doubt is false, or insufficiently self-evident, and so cannot be asserted to be true. Skeptic doubt considers uncertainty to be the normal state of thought, whereas Descartes regards it as a disease he wants to cure. Even when he takes up the Skeptics' arguments, it is in a spirit quite opposite to theirs. (Gilson, Note 1, p. 85 to Descartes [1637])

Argumentative doubt is opposed to *skeptical* doubt in that it does not privilege the indefinite suspension of assent@ over resolution of dispute.

Doxa

The word *doxa* is modeled on an ancient Greek word, meaning "opinion, reputation, what is said of things or people". The doxa corresponds to a set of socially predominant, fuzzy, sometimes contradictory, representations, considered in their current linguistic formulation. The word shares the deprecating meaning of *cliché* or *commonplace*; it may be given the meaning of "ideology" or "dogma", particularly when it is called into question (Amossy 1991, Nicolas 2007). Its derived adjective is *doxic* (or *doxical*).

Aristotle defines the *endoxa* (sg. *endoxon*) as the common opinions of a community, used in dialectical and rhetorical reasoning:
> Those opinions are 'generally accepted' which are accepted by everyone or by the majority or by the philosophers, i.e., by all, or by the majority, or by the most notable and illustrious of them. (Aristotle, *Top.*, I, 1)

An *endoxic* idea is therefore an idea based on a form of social authority, ranging from the authority of common people as a group, to that of the wise (S. **Dialectic**), according to a gradation ranging from the purely quantitative to the qualitative, from the opinion of the human (the universal consensus) to the authority of the enlightened opinion.

Thus, *endoxic* is an antonym of *paradoxic, paradoxical*. Latin translates the adjective derived *endoxos* "endoxic" by *probabilis*, "probable".

The endoxa is the target of philosophical criticism addressed to common sense and common opinion. This criticism extends to conclusions based on the *endoxon - inferential topic* system, used in dialectic and rhetoric. Yet, to say of a proposition that it is endoxic, is not pejorative:
> It is well known that Aristotle confides, under conditions of scrutiny, in the collective representations and the natural vocation of mankind toward truth. (Brunschwig, *Preface* to Aristotle, *Top.*, p. xxv)

Rhetoric and dialectic are based on *endoxa*: dialectical arguments test the endoxa, and rhetorical argument exploits them, pro and contra, in the context of a particular conflict.

In a judicial situation, the salient doxic elements, without being taken as true, may determine who bears the burden of proof, in other words, it determines who is at first sight, the object of suspicion, who is accused by the rumor, S. **Common places.**

Many forms of argument rest on the authority of the doxa:
— Appeal to common belief, S. **Authority; Matter.**
— Appeal to the feeling of the crowd, S. *Ad populum*.

E

Ecthesis ▶ Example

Effect to Cause Argument ▶ Consequence and Effect

Emotion

1. Emotion

1.1 Psychology
From a psychological point of view, emotion is a *syndrome*, a temporary synthesis of different states:

— A *psychic* state of consciousness.

— A *neurophysiological* state, perceptible or not to the subject, such as goose bumps associated with emotions such as fear or pleasure; or the adrenaline rush accompanying rage.

— An *altered self-presentation*: transformation of facial expression; of body posture; specific attitudes and emergence of actions, such as the flight reaction, characteristic of fear.

— A *cognitive state*, including a *structured representation of reality*.

The direction of causality between these components is discussed: common sense considers that the psychic state determines the neurophysiological and attitudinal changes, *"he cries because he is sad"*, but it can be shown that, if one puts a subject in the physical state corresponding to a particular emotion, he or she experiences this emotion, so, literally *"he is sad because he cries"* (James, 1884).

1.2 Basic emotions

The emotions listed by Aristotle in the *Rhetoric* and taken up by the Latin rhetoricians can be considered as the very first set of *basic social emotions* in the Western world S. **Pathos**. Modern philosophers propose their own lists of emotions; for example, Descartes holds that there are only six "simple and primitive" passions, "wonder, love, hatred, desire, joy, and sadness. [...] all the others are composed of some of these six or are species of them" ([1649], §69).

Psychologists define basic emotions as universal, independent of languages and cultures. The lists are variable and more or less developed; they generally include *fear, anger, disgust, sadness, joy, surprise*. Ekman (1999) counts fifteen basic emotions: *amusement, anger, contempt, satisfaction, disgust, embarrassment, excitement, fear, guilt, pride in success, relief, sadness - distress, satisfaction, sensory pleasure*, and *shame*. In theology, the capital sins — *pride, envy, anger, sadness (acedia: sloth, depression), avarice, gluttony, lust*, can be seen as emotional leakages, considered as sins insofar as they are left uncontrolled.

1.2 Emotions and mood: phasic and thymic

Moods are defined as stable or *thymic* affective *states*, whereas emotions are *phasic*, that is developing in an *event* structure, according to a bell-shaped curve pattern, S. **Calm**

1.3 Emotion and situation

An emotion is related to a situation. Causal theories of emotion analyze the situation as a *stimulus* mechanically inducing a *response*, which is the corresponding emotion. This view does not, however, explain the possibility of emotional injunctions and disagreements about emotion (see below). In fact, emotion is linked not to a kind of objective situation, but to a subjective perception of the situation; the stimulus is a *situation under a certain description*. In other words, the situation perceived as emotional is part of the emotion.

A distinction can be made between *experienced* emotion and *spoken* emotion. The relation between these two modes of emotion is analogous to that between *time* as an extra-linguistic reality, and *tense*, the shape language gives to time. Rhetorical-argumentative emotions concern emotion-tense, whereas psychology is interested in emotion-time.

2. Arguing emotions

Serious argumentative situations are *intrinsically emotional*. Contradiction, conflictive or not, disrupts routine beliefs and action plans. So, for example, having to decide what to do introduces tension on the social, cognitive, and emotional levels. The arguer must confront an uncomfortable situation; relations with the other, as well as social statuses are potentially threatened; representations of the world are destabilized, as are the personal identities based on these representations.

2.1 Emotions as issues in argumentative discourse

The situation related to emotion is not a causal source of emotion; when it rains there is no argument as to whether or not one will get wet. This *negotiability of emotions* is evidenced by the existence of emotional injunctions, such as,

"*Time for Outrage!*"(Stéphane Hessel)
"*A Call to Outrage*" (Ignacio Ramonet)
"*Indignant? We should be*" (Simon Kuper).[1]

In one given situation, there can be huge discrepancies between the emotional states of different people:

S1 — *Let us cry, the father of the Nation is dead!*
S2 — *Let's rejoice, the tyrant is dead!*

S1 — *I'm not afraid!*
S2 — *You should be.*

In the second example, by refusing to align with **S1**, **S2** opens a debate, and must explain why he or she does not agree. **S2** must reveal *his or her reasons for being afraid*, and argue his or her emotion. Reciprocally, **S1** is now at risk of being *refuted* by **S2**, and left with an inadequate emotion. An emotion is a point of view.

As for argumentation in general, we can distinguish cases where emotion is explicitly argued, and those where the argument is left implicit, and where we are dealing with an orientation towards a particular, unnamed, emotion. In both cases, the point of departure of the emotion is in the participants' perception of the situation. Ultimately, formatted situation and experienced emotion form a compact whole. Therefore, in order to justify an emotion, one has to give a detailed account of the situation including objective specifications about what happened and subjective emotional appraisal of the latter. This formatting obeys a relatively simple system of "emotional parameters", which determine the nature and intensity of the emotion, depending on the more or less predictable and pleasant character of the situation, its origin, its distance, control, norms and values of the experiencer, etc. (Scherer [1984a], p. 107; 1984b).

Aristotle's *Rhetoric* presents an excellent description of the thematic structure of speech constructing specific emotions, that is, of the *topics* of emotion. The

---

[1] Stéphane Hessel (2011). *Time for Outrage!* London: Charles Glass Books.
Ignacio Ramonet (2011). Quoted after http://www.ipsnews.net/2011/02/a-call-to-outrage/ (11-08-2017)
Simon Kuper (2011). Quoted after https://www.ft.com/content/280c9816-192c-11e0-9311-00144feab49a?mhq5j=e1 (11-08-2017)

book is not about the psychology of emotion but rather a treatise on what discourse can do with emotions and how an emotional social thrust can be monitored, constructed or refuted. The question is not what anger or calm *are*, but how discourses which are likely *to prompt* or *to calm* anger are constructed. That is why, from an argumentative perspective, action predicates should be preferred to substantives where we refer to emotions:

— To anger vs. to cool down the anger;
— To inspire friendship vs. to break the bonds of friendship fueling anger and hatred;
— To frighten vs. to buoy up;
— To shame vs. to despise other's opinion;
— To be grateful vs. to feel no obligation;
— To pity vs. to be indifferent;
— To kindle rivalry, jealousy, envy *vs.* to instigate a spirit of competition.

Emotions belong in the field of discursive action. In the *Rhetoric*, they are defined on the basis of typical scenarios, activated and developed by the speaker. This description of discursive strategies which generate emotion is one of the major achievements of rhetorical argumentative theory.

Anti-oriented discourses construct anti-oriented emotions. Speech alters representations, thus arousing or appeasing, counterbalancing emotions, just as any viewpoint can be fought, turned back, or circumvented. The examples of pity and anger can give an idea of these basic argumentative techniques.

2.2 Pitying and pitiless

Moving to pity — **A** pities **B** if he considers that **B** is the victim of an evil that is not deserved; and if **A** is well aware that one day he or she may suffer from the same evil (*id.*, 1385b10-15, RR, p. 291). For **A** to pity **B**, the *distance* between **A** and **B** must have been calibrated correctly; one feels pity towards people who are similar and close to us. Generally speaking, the "distance" dimension plays an essential role in the construction of emotion, not as an objective metric, but as a cultural, language-built notion. It follows from this description that pity should not be considered an automatic feeling. In particular, those who have nothing to fear for themselves would be insensitive to pity. According to the theory of the moral characters (mores) of the audience, the locally relevant construction of an emotion depends on a good analysis of the audience, S. **Ethos**. It follows that to directly induce pity, **B** must show that he or she is suffering, and that he or she does not deserve to be, that the same thing could happen to you, his or her interlocutor, etc., and then, of course, amplify these substantial common places. If pity is constructed according to such parameters, it is *justified* and judged to be *decent* and *reasonable*.

Refuting misplaced pity — Walton has shown along which lines the target can resist pity, in other words, he has shown how to build a discourse *against pity*, which allows the target to remain *calm*, *insensitive*, not to yield to a movement of pity. This discourse is constructed first along a specific "information line", about the situation. It is then subjected to a relevance condition, pity can only be appealed to if the domain admits of personal involvement. For example, scientific discourse, excluding ordinary subjectivity, does not allow appeals to pity, which will then be deemed "irrelevant" (Walton 1992, p. 27)

When relevant, the appeal to pity routinely functions in the general conflict of pro and contra arguments, concerning personal involvement. In the case of dismissed workers, for example, the appeal to pity (*ad misericordiam*) confronts the need to preserve the interests of shareholders (*ad pecuniam* vs. *ad misericordiam*), to place the company in a good position in the market (*ad rivalitatem* vs. *ad misericordiam*), or to preserve the jobs of other employees of the company (*ad misericordiam* vs. *ad misericordiam*).

2.3 Anger: getting angry and calming down

Argumentation theory has glorified appeal to pity with a Latin name, *ad misericordiam*. Actually, from an argumentative point of view, there is no reason to set pity apart from other emotions. All should be given the same lexical consideration, particularly the appeal to anger, *ad iram*, a highly argued and argumentative emotion.

Causing anger — Anger is a basic rhetorical emotion. When willing to cause the public anger, the speaker will develop his or her righteous indignation or holy anger, and will adopt a virtuous ethos. From the same virtuous posture, the opponent will denounce rage, fury and hatred, **S. Pathos**.

Discursive representations play an essential role in these oppositions. To make **A** angry with **B**, the speaker has to show to his or her interlocutor **A** that:

— **B** *baffles, offends, mocks* **A**; **B** makes an obstacle to **A**'s plans, wishes, and takes *pleasure* in it.
— **A** *suffers* and seeks revenge by harming **B**.
— **A** fantasizes a vengeance, and *enjoys* it.

These are the basic lines of inflammatory speeches. It should be noted that anger is not an atomic emotion, a crude response to the bite of a stimulus, but the complex result of an aggregate of emotions such as *humiliation*, *contempt* and even *pleasure*. The rationality or morality of anger depends on the proper construction of this feeling of *injustice*. Anger can be fully virtuous, rational and emotional, if these distinctions have any meaning here.

Anger triggers the mechanisms of revenge. In a typical serial episode, *anger* constructed and justified in the first sequence, is turned, into an argument for subsequent action.

Anger is not hatred; anger can be rationally justified, hatred cannot; there is no acceptable reason for hate. From a religious point of view, hate speech is a sin "*love, at least patiently bear one another!*". The status of hate speech serves as an example of how social evaluation@ is achieved. Any citizen can legitimately comment on and take stock of anger speech, outrage speech and hate speech.

Politicians and judges have the authority to make judgments about such issues. They may of course also draw on the assistance of other parties that they consider helpful, anthropologists, moralists, and, of course, linguists and logicians.

From anger back to serenity — In order to calm down an enraged **A**, **B**'s advocate will develop a soothing discourse concluding that **A**'s expression of anger is poorly worded and badly constructed, and that this anger is unreasonable. He or she will argue that **B**'s behavior was not contemptuous, mocking, abusive, or outrageous. It rather the case that **B** has been misunderstood; he or she was joking; the intention was not hostile; **B** behaves like this with people that he or she loves; **B** repents, and offers excuses, compensations; **B** has already been punished, etc. — and the soothing discourse will conclude that all that happened a long time ago, and that the situation has changed, S. **Kettle**.

Enantiosis ▶ Disagreement

Enthymeme

1. The Greek word

The Greek word corresponding to the English word *enthymeme* (adjective *enthymematic*) means (after Bailly, [*enthymema*]):
   1. Thought, reflection.
   2. Invention, particularly war stratagem.
   3. Reasoning, counseling, warning.
   4. A reason, a motive.

The general meaning of "thought, reflection" is present in all ancient rhetoric: "Any expression of thought is properly called an [*enthymema*]. (Cicero, *Top.*, XIII, 55; p. 423). Quintilian also alludes the meaning "everything that is conceived in the mind", to put it aside (*IO*, V, 10, 1).

2. An instance of an argument scheme

In rhetorical argumentation, an enthymeme is essentially an instance of a topic, an argument scheme@. An argument scheme is a general formula having an inferential (associative) form; an enthymeme is the application of such a formula to a specific case. This general definition combines with the following orientations.

(i) In relation to logic, the enthymeme is:
   — A form of *syllogism*:
      • A syllogism based on *plausibility* or on *sign*.
      • A *truncated* syllogism.
   — The *counterpart* of the syllogism.

(iii) Functionally, the enthymeme is seen as a manifestation of *cooperation* with the audience.

(iv) Marginally, the enthymeme has also been defined as a *concluding formula*.

3. A special kind of syllogism

3.1 The enthymeme, a syllogism based on "a probability" or "a sign"
In the *Prior Analytics*, Aristotle defines the enthymeme as "a syllogism starting from probabilities or signs." (*P. A.*, II, 27)
An enthymeme is a probable reasoning such as:
> Peter is tired, he must have worked hard.

The arguer can be charged with mistaking necessary and sufficient conditions, or trusted as knowing for sure that Peter did not spend the whole night celebrating, according to the context span taken into consideration by the analyst.

A *natural@ sign* is a proposition expressing a natural connection between two states of things. The connection can be probable (to be *red* is a *sign* or a *symptom* of fever) or necessary (as *smoke* to *fire*).
A *probability@* is a proposition expressing either a probable natural relation or a social agreement:
> A probability is a generally approved proposition: what men know to happen or not to happen, to be or not to be, for the most part thus and thus, is a probability, e.g. *"the envious hate"*, *"the beloved show affection"*. (Aristotle, *PA*, II, 27)

These are excellent examples of associative semantic inferences (+ *envious*, + *hate*); (+ *love*, + *show love*), **S. Orientation; Topos**. Such substantial probabilities are based on common sense views of basic human tendencies. The corresponding topics underlie the current production of arguments; **S. Common place**.
The big strong man will prevail over the small weak one, and mothers love their children. Sometimes, however, this is not the case. A characteristic of reasoning from social probabilities is that it can be reversed, as expressed in the key Aristotelian topic n°21, "incredible things do happen" (*Rhet*, n°22, 1400a5; RR p. 373).
Consistency is generally a source of probabilities. Humans are rational and intentional beings; they make plans and are expected to act according to these plans, to remain true to their words and intentions. Their behavior is assumed to be *probably* consistent. Inconsistency is the sign of a defective personality, or of a basic mistake, **S. Consistency; *Ad hominem***. Showing that the opponent is incoherent is a key tool for claims or narratives to be rejected. But consistency is only a probability, as noted in topic n°21, and probabilities cannot hold against hard evidence; they are default qualifications. Other topics are based on inconsistent behavior, people change their minds and criminal actions might be badly planned, **S. Mobiles**.

3.2 The enthymeme as a truncated syllogism
The enthymeme is also defined as a categorical syllogism where a premise is omitted:

> Men are fallible, you are fallible.
> You are a man, you are fallible.

Or the conclusion:
> Men are fallible, after all you are a man!

The *Logic* of Port-Royal defines the enthymeme as:
> A syllogism perfect in the mind, but imperfect in the expression, since one of the propositions is suppressed as too clear and too well known, and as being easily supplied by the mind of those to whom we speak. (Arnauld, Nicole, [1662], p. 224)

> No enthymeme is conclusive, save in virtue of a proposition understood, which, consequently, has to be in the mind though it be not expressed. (*Id.*, p. 207)

The example in the preceding paragraph can therefore be called an enthymeme for two reasons: on the one hand because it is based on *probable* indices and on the other hand because it is an *incomplete* syllogism. The definition of an enthymeme as a truncated syllogism is often not considered to be Aristotelian: "It is not of the essence of the enthymeme to be incomplete" (Tricot's Note to Aristotle, *PA*, II, 27, 10, p. 323). Moreover, according to Conley, this conception of the enthymeme as a truncated syllogism is not widely used in ancient rhetoric. He finds it only in a passage by Quintilian (Conley 1984, p.174). However, the *First Analytics* does consider the case of the truncated syllogism, "Men do not say the latter [*Pittacus is wise*] because they know it" (*PA*, II, 27, 10). On the other hand, we read in the *Rhetoric* that:
> If any of these propositions is a familiar fact, there is no need even to mention it; the hearer adds it himself. Thus, to show that Dorieus has been victor in a contest for which the price is a crown, it is enough to say '*For he has been victor in the Olympic games*', without adding '*And in the Olympic games the prize is a crown*', a fact which everybody knows. (*Rhet.*, I, 2, 1357a15; RR, p. 113).

Under this definition, the enthymeme can be considered as a *figure of speech* by ellipsis, precisely *a figure of thought*.

4. The rhetorical counterpart of the syllogism

In the Aristotelian systematic, the proof is obtained by inference, whether scientific (logical), dialectical, or rhetorical. Aristotle considers that there are two kinds of scientific inferences, syllogistic deduction and induction. In rhetoric, scientific inference is replaced by "rhetorical inference" or enthymeme, the requirements of rhetorical discourse not being compatible with the exercise of scientific inference:
> I call the enthymeme a rhetorical syllogism, and the example the rhetorical induction. (*Rhet.*, I, 2, 1356b5, RR, p. 109)

The syllogism (scientific inference) and the enthymeme (rhetorical inference) are defined in a strictly parallel way:

> When it is shown that certain propositions being true, a further and quite distinct proposition must also be true in consequence, whether invariably or usually, this is called syllogism in dialectic, enthymeme in rhetoric. (*Rhet.*, I, 2, 1356b15; RR, p. 109)

But, unlike the syllogism, derived from true propositions, the enthymeme is drawn from "probabilities and signs" (*Rhet.*, I, 2, 1357a30; RR, p. 113), see supra § 3.1.

The enthymeme is "the substance of persuasion", "a sort of demonstration" (*Rhet.*, I, 1, 1354a10, RR p. 95; 1355a5, RR p. 99). It deals centrally with the issue, the substance of the debate, "the fact" (*Rhet.*, I, 1, 1354a25, RR p. 97. As such, the enthymeme is opposed to the reckless use of ethos and pathos, S. **Emotion; Pathos; Ethos.**

The enthymeme is also called a *rhetorical syllogism*, considered as an *imperfect* syllogism. These labels refer rhetoric to syllogistic. However, the scientific / dialectical / rhetorical parallelism, however attractive it may be, is problematic. If one accepts this opposition, one enters a very uncomfortable and empirically inadequate notional grid. On the one hand, the distinction between the three types of reasoning creates a divide between *categorical* scientific syllogism and *probable* dialectic syllogism, versus *persuasive* rhetorical enthymeme, the socially relevant discourse being posited as inherently unable to deal with well-grounded truth. On the other hand, argumentative rhetoric is straightjacketed in the opposition between *technical@ evidence*, rhetorical evidence proper, and non-technical proof, which obviously do not fit into the previous notional framework. Common judicial discourse routinely combines the two types of proof, in perfectly syllogistic forms of reasoning, S. **Layout; Demonstration**.

The reasons given for binding the enthymeme to syllogistic discourse are somewhat paradoxical. The enthymeme as a truncated syllogism is supposed to suit rhetoric because it would be less pedantic than the complete syllogism; this assumes that the missing premise is easy to retrieve. Another reason put forward is that one would use an enthymeme because the ordinary audience is composed of people of a mediocre intelligence, unable to follow a rigorous syllogistic chain. This second justification supposes that the missing premise is too difficult to recover: these two justifications are not immediately compatible.

5. Enthymeme and interpretative cooperation

From the point of view of argumentative communication, the enthymeme exploits what is implicit to achieve persuasion:

> Everyone who effects persuasion through proof does in fact use either enthymeme or examples; there is no other way. (*Rhet.*, I, 2, 1356b5, RR p. 109)

As Bitzer notes (1959, p. 408), the enthymematic form is a way of connecting speaker and audience in a process of co-construction of the meaning of discourse, "the enthymeme is satisfied if merely what is stated in it be understood", (Quintilian, *IO*, V, 14, 24). Building a common speech space, implicitness produces intersubjectivity. The orator frames the audience as good listeners, and thus creates a "good intelligence" and an atmosphere of complicity. Communicative fusion thus contributes to the formation of an ethos: *"you understand me; you can read my mind, I am like you, we are together"*.

In Jakobson's words, the enthymematic formulation of reasoning has a *phatic* function, that is to say, it keeps the communicative channel open. The effect of surprise associated with the ellipsis is supposed to wake up somnolent audiences: *"Something missing!"* (see supra § 3.2).

6. The enthymeme as a conclusive formula
The ancient rhetorical practice accorded a superior efficiency to the enthymeme founded on opposites. As the paragon, this specific enthymeme has appropriated the name of the class:

> Although every expression of thought may be called enthymeme, the one which is based on contraries has, for it seems the most pointed form of argument, appropriated the common name for its sole possession. (Cicero, *Top.*, XIII, 55; 423)

And gives as an example:

> What you know is of no use; is what you do not know hindrance? (Cicero, *Top.*, XIII, 55: 425).

The second sentence is a rhetorical question *"so, what you do not know should be useful"*.

Epicheirema

> The word *epicheirema* comes from the Greek *"epicheirein,* to endeavor, attempt to prove"* (Webster, *epicheirema*). It translates into Latin as *ratiocinatio* (Cicero), "reasoning", or as *argumentatio* (*Ad. Her.*)

The term *epicheirema* is used in ancient argumentation theory with three distinct definitions.

1. Epicheirema as dialectical reasoning
The Aristotelian theory of syllogistic reasoning opposes *philosopheme* to *epicheirema*. Philosopheme is another name for the analytical or scientific syllogism, where the premises are true and the rule of deduction is valid (*Top.*, VIII, 11; p. 156). In contrast "epicheirema is a dialectical inference" (*ibid.*), that is, a syllogism founded on premises taken from the doxa@, hence only probable; this inference concludes to a probability.

2. Epicheirema as an argumentation whose premises are supported
In rhetorical argumentation, the word *epicheirema* is a synonym of *probable (rhetorical) syllogism, enthymeme* and *argumentation*. A well-built, convincing, rhetorical

proof is defined as an argumentation (*ratiocinatio*) whose premises are only probable, and, consequently, should be explicitly backed by their proofs (Cicero, *Inv.* I, 34; Hubbell, p. 98-99). In short, a probable premise accompanied by its proof becomes certain. Cicero discusses the following rhetorical syllogism (*id.*, 101-103)

— *Premise 1 + Proof of Premise 1:*
- *Premise 1:* "Things that are governed by design are managed better than those that are governed without design"
- *Proof of Premise 1:* "The house that is managed in accordance with a reasoned plan is better managed that those that are governed without design. The army […] The ship […]"

— *Premise 2 + Proof of Premise 2:*
- *Premise 2:* "Of all things, nothing is better governed than the universe"
- *Proof of Premise 2 (our numbering and presentation)*
  (a) "the rising and the setting of the constellations keep a fixed order"
  (b) "and the changes of the seasons not only (b1) proceed in the same way by a fixed law but (b2) are also adapted to the advantage of all nature,"
  (c) "and the alternation of night and day has never through any variation done any harm."

— *Conclusion:* "Therefore, the universe is governed by design."

Premise 1 is the conclusion of an *induction*@, that is an enumeration of examples, sharing the same structure and orientation. In premise 2, case (b), the element (b2) argues not only for a design but also for a *benevolent design*, as does case (c).

Structure of an epicheirema

The question as to whether an epicheirema includes five or three components is disputed (Solmsen 1941, p. 170). On the surface level, an epicheirema is indeed a sequence consisting of five components:

Premise 1 + Proof of Premise 1 + Premise 2 + Proof of Premise 2 + Conclusion

This corresponds to a three-element deep structure:

(Premise 1 and its Proof) — (Premise 2 and its Proof) — Conclusion

This is Quintilian's position: "To me, as well as to the greater number of authors, there appears to be not more than three [parts]" (*IO*, V, 14, 6).

The epicheirema corresponds to a *linked*@ argumentation, represented as follows:

Proof_1 => Premise_1
Proof_2 => Premise_2   ] Argument => Conclusion

3. Epicheirema, as a communicated argument

The *Rhetoric to Herennius* defines "the most complete and perfect argument [*argumentatio*]" as "that which consists of five parts: the Proposition, the Reason, the Proof of the Reason, the Embellishment and the Résumé" (*Ad Her.*, II, 28). This perfect rhetorical argument is described as a sequence consisting of five components, like a logical epicheirema, but with a quite different organization. The first three elements correspond to the *logical* component, establishing the Proposition:

Reason 1 + Proof of the Reason + Proposition

The proof of the *Reason*, "corroborates by means of additional arguments, the briefly presented Reason" (*Id.*, p. 107). The argumentation must now be seen as *serial*@:

[Argument$_1$ =>   {(Conclusion] = Argument$_2$)   => Conclusion}

Proof of the Reason   *Reason*   *Proposition*

The *Embellishment* is a reformulation that "adorn[s] and enrich[s] the argument (*argumentatio*)". The Résumé is not the conclusion; its "[brevity]" contrasts with the preceding amplification episode, creating a kind of hot / cold contrast. This second component of the argumentation articulates two elements that clearly have a *communicative* function.

Epitrope

An *epitrope* is defined as, "a figure of rhetoric, consisting in granting something that one can dispute, in order to give more authority to what one wants to persuade" (Littré, Epitrope), **S. Concession.**
Under ordinary conditions, as described by Grice's principles, the arguer refutes everything possible, and concedes anything else. So, "*Peter concedes* **P**" pragmatically implies that Peter is not able to refute **P**. If the arguer concedes a doubtful proposition, he or she will be considered a bad arguer; if he or she concedes something that could obviously be refuted, the speech will be interpreted as being ironic:

P is obviously wrong:
L: — *P, okay, but / yet Q*

Embedding **P** in a concessive structure, assigns **P** to the opponent, whether or not he or she wishes to endorse it:

About a writer whose qualities of style have just been discussed in a rather negative way:

I am ready to consider him a good stylist, but he doesn't know what a plot is.

Irony may also arise from the exaggeration given to the granted position:
I may have visions, but I also have some hard evidence.

S. Concession; Irony; Exaggeration.

Ethos

1. The word *ethos*
The word *ethos* is borrowed from the ancient Greek word ἦθος (ēthos), having two meanings:
> "I. *In pl.* Usual stay, familiar places, dwelling. *Speaking of animals:* cowshed, stable, den, nest. [...]
> II. Usual character, *hence* custom, usage; the manner of being or habit of a person, his character; [...] *by extension*, mores"
>
> (Bailly, [*ethos*])

In rhetoric, *ethos* refers to "the moral impression (produced by an orator)" (*ibid.*).

In Latin rhetoric, *ethos* is translated as *mores*, "manners", or *sensus* "common sense". Quintilian considers that *ethos* "manners" and *pathos* "passions" are subcategories of *feeling* [*adfectus*]:
> Of feelings [*adfectus*] as we are taught by the old writers, there are two kinds, the first of which the Greeks included under the term πάθος (pathos), which we translate rightly and literally by the word "passion" [*adfectus*]. The other, to which they give the appellation ἦθος (ēthos), for which, as I consider, the Roman language has no equivalent term, is rendered however, by mores, "manners"; whence that part of philosophy, which the Greeks call ἠθική (ēthikē), is called *moralis*. (IO, VI, 2, 8)

The same opposition ethos / pathos can also be translated in Latin as *sensus* / *dolor*:
> *Sensus* is one of those vague terms by which Latin tries to express what Greek rhetoric designates by [ethos]. [...] It is distinct from *dolor*, which responds to [pathos] (Cicero, *De Or.* III, 25, 96). (Courbaud, note 2 to Cicero, *De Or.*, II, XLIII, 184; p. 80)

The noun *sensus* basically refers to *physical* perception, also to "an *intellectual* way of seeing things", and a *moral* perception of the situation in terms of right and wrong, a "moral sense" (after Gaffiot, *Sensus*). To display *sensus* is therefore to jointly display good perceptual, analytic and moral skills.

Sensus also points toward *sensus communis*, "common sense", as a synthesis capacity in agreement with what people consider to be "[soundness and prudence]" (MW, *Common sense*). The good orator is a man of common sense with the ability to achieve synthesis.

The English words *ethos, ethics, ethopoeia, ethology* are borrowed and adapted from the Greek.

Ethology is the science of the behavior of animal species in their natural environment, cf. supra, meaning (I).

The noun *ethos* is used in rhetoric, up to the present time. *Mores* is borrowed from the Latin *mores*, which itself translates the Greek *ethos*.

The noun *ethopoeia* is used in rhetoric, and literary theory, referring to a "moral and psychological portrait".

Ethics is the part of philosophy dealing with morality and values.

The rhetorical notion of *ethos* refers to the fact that the speaker is projected into discourse and holds part control over this projection. The *ethics* of discourse refers to an inner moral authority controlling discourse. The *ethotic* dimension of rhetorical discourse can be seen as a discursive projection of ego ideals, whereas its *ethical* dimension would be a discursive projection of the superego imperatives.

Such moral control is central to the rhetorical definition of an orator as a *vir bonus dicendi peritus* "a good man having public speaking skills". In contemporary argumentative theory, the criticism of discourse is referred to a *rational* control, whereas classical rhetoric refers discourse to moral control as well.

2. The arguer's ethos

Ethotic strategies deal with the social "presentation of self" (Goffman [1956]). The ethos of the orator is a professional ethos. All professions have their ethos, for example, beyond its strictly professional capacities, the traditional waiter embodies a set of peripheral professional virtues: may be finding the cocktail best adapted to the customer's mood, having the art to deftly drop into and out of the conversation, etc.

Aristotle deals with ethos in two passages of *Rhetoric*. It describes on the one hand *ethos proper*, the auto-fiction that constitutes the construction of the face that the orator intends to present to the public; and, on the other, the *ethos of the audience*, the synthesis of information which enables him or her to adequately orient his or her argument.

2.1 Aristotle: The combined effect of discourse and reputation

In the Aristotelian system, *ethos* is one of the main leverages for persuasion, the other two being *logos*@ and *pathos*@. The *Rhetoric* poses the primacy of ethos over logo-ic proofs, "[the speaker's] character may almost be called the most effective means of persuasion" (*Rhet.*, I, 2, 1356a10; RR, p. 106). The concept of *ethos* is introduced as follows:

> Persuasion is achieved by the speaker's personal character when the speech is so spoken as to make us think him to be credible. We believe good men more fully and more readily than others: this is true generally whatever the question is, and absolutely true where exact certainty is impossible and opinions are di-

vided. This kind of persuasion, like the others, should be achieved by what the speaker says, not by what people think of his character before he begins to speak. It is not true, as some writers assume in their treatises on rhetoric, that the personal goodness revealed by the speakers contributes nothing to his power of persuasion; on the contrary his character may almost be called the most effective means of persuasion he possesses. (*Rhet.*, I, 2, 1356a1-15; RR, p. 107)

The speaker's ethos is the product of a discursive strategy that builds a complex authority based on three components:

There are three things which inspire confidence in the orator's own character — the three namely, that induce us to believe a thing, apart from any proof of it: good sense, good moral character, and goodwill. (*Rhet.*, II, 1, 1378a; RR, p. 245).

Good sense is *phronesis*, that is to say, "prudence"; good moral character, *arete*, "virtue"; and good will is *eunoia*, or "goodwill". The arguer has persuasive authority because he or she is (or appears to be) clever, honest, and on our side. To no lesser extent than pathos, ethos has a pathemic structure; ethotic authority combines expertise, morality and benevolence into a unique feeling of *trust*, the perfect persuasive cocktail:

These qualities are all that is necessary, so that the speaker who appears to possess all three will necessarily convince his hearers. (*Rhet.*, II, 1, 1378a15; Freese, p. 171)

The verb *to appear* (and not *to be*), will seem suspicious. Rhetoric is always suspected to give to the incompetents, vicious and crooks, the means to deceive their partners. As Groucho Marx says or repeats, "*sincerity — If you can fake that, you've got it made*". But the ablest and truest arguer remains subject to the "paradox of the actor", that is to say, he or she can be suspected of feigning the skills, virtues, and intentions he claims and shows, and therefore must not only *be* but *appear* sincere and true. The arts of appearance are no less necessary to honest people than to scoundrels.

Under this definition, the Aristotelian ethos attracts identification on the basis of a shared community feeling. Disruptive rhetoric implements another ethotic positioning, as an influential minority group: "*we are different from all of you ... I bring a new world ... yes your wise men call it madness.*"

The text of the *Rhetoric* is somewhat puzzling. On the one side, ethotic persuasion "should be achieved by what the speaker says, not by what people think of his or her character before he or she begins to speak". In line with the classical doctrine of technical and non-technical proofs, this amounts to an outright rejection of *non-technical ethos* (not the speaker's character *before* the speech) in favor of *technical ethos* (what the speaker says). Nonetheless, the following sentence seems to prioritize the former over the latter, probably because both play their part in actual discourse, as suggested in Ruelle's translation: "It is necessary, moreover, that this result should be obtained by the force of discourse and *not merely by* a preference favorable to the speaker." (Aristotle, *Rhet.* Ruelle, emphasis added).

2.3 Challenging the ethos
Ethos can be seen as a public exhibition of one's best possible self, in view of influencing the recipient. Critical theories of argumentation focus on the subject matter of the debate, protect the participants by keeping at least some part of their personalities distanced from the dispute, when they have nothing to do with it. They make a crucial distinction between the *charisma* of the speaker, which is rejected by principle as exerting an irrational influence, and the exercise of the *authority*@ legitimately attached to his or her specific competences.

Ethos and *personal*@ *attacks* on the opponent are the obverse and reverse of the same discursive coin, as theoretically shown by *politeness*@ *theory*. Exhibiting ethos, the speaker exploits his own person as a resource to accredit his point of view, whilst when attacking personally the opponent, the speaker exploits the person of the opponent to refute or discredit his point of view. In both cases, the discourse eludes the substance of the issue and turns to a discussion about the participants, either to discredit or to accredit their positions.

From a critical perspective which postulates that only explicit arguments about the *matter*@ itself are relevant, and potentially valid, there cannot be something like an ethotic argument simply because the propositional requirement is not met. Due to its implicit and global positioning, ethotic authority cannot be challenged by any refutation on the *matter*@; accordingly, the opponent will be tempted by an *ad personam* counter-attack.

In a face-to-face situation, the ethotic grip seeks to establish an asymmetrical relationship framing the interactional relation on a high / low opposition, humbling the opponent into the low position, in order to inhibit free criticism.
S. Modesty. So, from a critical point of view, the ethotic yoke must be shaken off, as a preliminary of any constructive discussion. The charismatic facets of ethos are first rejected outright, as irrelevant and fallacious. Second, an explicit component is extracted from a synthetic form of ethos, the *argument from authority*@, which satisfies the propositionality condition and is accessible to criticism. This authority is integrated as peripheral evidence, to be dealt with within the appropriate critical framework.

3. Ethos and discursive identities
Contemporary and ancient discussions about ethos deal with a broadly recognized fact, language splits the speaker into several discursive roles@. *Ethos* is a hub concept, connecting argumentation studies with linguistic studies on subjectivity in language (Benveniste 1958) and with literary studies in narratology, confronting *author* and *narrator*, *real* and *implicit* readers.

Argumentative discourse, as any discourse, articulates three identity-building elements, *ethos* strictly speaking, *reputation*, *self-portraying*. The ethotic impact of discourse is the result of these three forces:

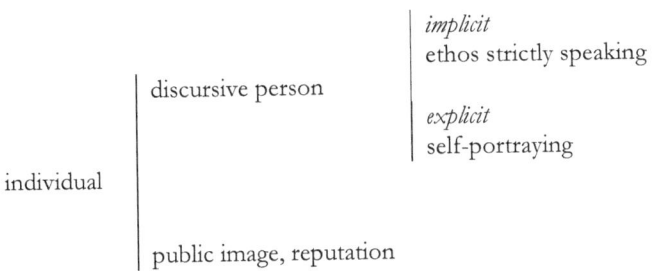

Ethos Itself — Ducrot integrates the notion of ethos into the general theory of polyphonic discourse: "Ethos is attached to the speaker as such: the character attributed to him or her as the source of the utterance, make this utterance acceptable or not" (Ducrot 1984, p. 200). In Goffman's terminology, ethos is attributed to the *Figure*, **S. Roles**.

Explicit Self-Portrayal — Ducrot introduces as a second, intra-discursive element "what the orator could explicitly say about himself" (1984, p. 201). The arguer can be the author of her own portrait: "*I raised my three children myself*" but these self-accounts are quite distinct from what can be indirectly revealed through the discourse. Having an accent is not the same thing as saying "*Yes, I have an accent and I am proud of it*". In an argumentative situation, participants systematically value their persons and actions in order to legitimize themselves. The requirements of this situation prevail over the principles of linguistic *politeness@*.

Reputation — Some social actors are well-known people, that is, they have a reputation, prestige, and perhaps even charisma, positive or negative. This established image is called the "prior", or the "preliminary" ethos by Amossy:

> We shall therefore call preliminary ethos or preliminary image, the image that the audience can have of the speaker before the speech, as opposed to ethos (or oratory ethos, which is fully discursive). [...] Preliminary ethos is developed on the basis of the role played by the speaker in society (its institutional functions, status and power) but also on the basis of the collective representation or stereotype of this person [...]. Indeed, the image projected by the speaker integrates prior social and individual data, which necessarily plays a role in interaction and contributes significantly to the power of his speech. (Amossy 1999b, p. 70; Maingueneau 1999)

"Pre-discursive" does not mean "language-free". Reputations are based on discourse as well as upon actions. Ethos can be said to be pre-discursive only in the sense of "preceding a particular speech act".

Public relations agencies can construct, manage and repair the image of human beings and commercial products (Benoit 1995).

The operating and control systems of these different identity layers are very different, and each layer can conflict with the two others.

Reputation is a socio-historical construct, which can be socially managed and controlled. Reputation can be inconsistent; the self-representation that the arguer has of his reputation can be different from the representation his audience has of him.

Self-portrayal is an explicit, declarative, controlled activity, an "argumentation of the self" as it is properly termed.

Ethos building is an on-going speech activity. All speech, spontaneous or elaborated, contains subjective features. This fact is transparent for the participants. The speaker knows that his or her conversation partners know (that he or she knows, etc.) that at least some of these subjective features will be elaborated and interpreted as clues to the speaker's identity, through standard argumentations from natural@ signs. The arguer might therefore intentionally arrange these subjective features in order to channel these interpretations according to his or her intended aims and perspectives.

The concept of ethos can be used as a descriptive category, relevant to the analysis of any form of ordinary discourse (Kallmeyer 1996). This trend towards generalization, coming with the naturalization of ethos, is typical of modern theories of argumentation such as that of Argumentation within Language or Natural Logic. Argumentative ethos is specifically a category of rhetorical action, a *strategic resource* available to the arguer, a functional element, intentionally elaborated or distorted.

Generally speaking, inferences to the speaker's (deep) identity(ies) are based on inferences from linguistic and encyclopedic clues. Like all interpretation processes, such inferences are open-ended, the only restrictions are those of the imagination of the interpreting party: the identity of the speaker is in the eye and ear of the receiver. When it comes to the specificity of ethos, argumentative analysis focuses on the strategic dimension of the presentation of self in argumentation. Its reconstruction program, distinct from the psychoanalytic approach, dovetails with the semiotic and stylistics program.

4. Ethos as a stylistic category

"Style is the man", and ethos is the style. When looking for a systematic method to study ethos, we come across the stylistic tradition. For example, Quintilian thus emphasizes the effectiveness of a style linked to the choice of vocabulary having a "majestic" ethotic effect:

> Words derived from antiquity have not only illustrious patrons, but also confer on style a certain majesty [not without charm], for they have the authority of age and, as they have been disused for a time, bring with them a charm similar to that of novelty. (Quintilian, *IO*, I, 6, 39, slightly modified)

The authority of the uttered word is constitutive of the ethos of the speaker. The ethos is constructed from features belonging to any linguistic level, beginning with the voice — a powerful vector of attraction or repulsion — the art of hesitating, repeating, faltering, and so on. Ethotic inferences can be drawn from any feature of the argument. He or she who:
— makes concessions is moderate / weak.
— does not make concessions is straight / sectarian.
— appeals to the authorities is conservative / dogmatic.
— uses pragmatic arguments upon causes and consequences is sensible and realistic, pragmatic / opportunist.
— refers his arguments to the nature of things and their definition is a man of conviction / conservative.
Other argumentation lines (by absurdity, by analogy...) do not have such clearly associated ethos.

The link of ethos with style is explicitly made in the *Rhetorical Art* of Hermogenes of Tarsus (160-ca. 225 CE). Hermogenes considers that discourse can be evaluated along seven stylistic categories:

> Clarity, grandeur, beauty, vivacity, ethos, sincerity and skill (Hermogene, AR, 217, 20 - 218, 05; Patillon, 1988, p. 213).

Ethos is one of these categories of discourse; in any given speech, there may be *a little* or *a lot of ethos*.

Ethos has four components, *simplicity; moderation; sincerity; severity*. These qualities compare with the qualities of *wisdom, expertise* and *benevolence* that make up Aristotelian ethos. Each of these components is characterized by specific *thoughts, methods, words, figures of speech*, and *rhythms*.

As strange as this might sound, *sincerity*, the key ethotic element is a *style*. Sincerity is a linguistic condition attached to:
— *Emotions@*, and particularly a feeling, *indignation*.
— *Severity* in the accusation of others or oneself is shown by using *harsh* and *vehement* words.
— A method of discourse management, in particular the balance achieved between what is *openly* discussed and what is left *suggested*.
— The use of derogatory demonstrative pronouns; of figures: apostrophe, and particularly figures of embarrassment (reticence, doubt, hesitation, corrections, interrogations).
— Personal comments suspending the speech (after Patillon 1988, pp. 259; p. 261 et seq.)

Thus, a *sincere character* is not an extra-linguistic supplement that would be introduced into the discourse from outside, by a moral exhortation. It is the product of a discursive strategy@. Any ethics of discourse should take this into account. In particular, figures of speech serve the construction of ethos, and they therefore are instrumental in argumentation in general. We are very far from post-Ramusian rhetoric where invention is divorced from elocution.

5. Character of the audiences

After having defined the ethos of the orator in a brief passage of the *Rhetoric*, Aristotle takes a very different perspective to deal with the characters of the audiences:

> Let us now consider the various types of human character, in relation to the emotions and moral qualities, showing how they correspond to our various age and fortune. (*Rhet*, II, 12, 1388b31, RR p. 311).

This section describes a set of "ideal-types", that is, human characters classified and characterized according to their social condition, wealth and power (noble, rich, powerful, and lucky) and age (youthful, mature, old). These "elements of sociology for the rhetoricians" conclude with a practical remark:

> Such are the characters of Young men and Elderly Men. People always think well of speeches adapted to, and reflecting, their own character: and we can now see how to compose our speeches so as to adapt both them and ourselves to our audience. (RT, II, 13, 1390a20-29, RR p. 319)

Such a passage clearly shows that the adaptation-identification to the audience is the key to persuasion@. It will be regarded as fallacious by the normative theories of argumentation requiring that one speaks the truth, not upon the basis of the specific beliefs@ of a particular audience (*ex datis*);

Compared to the three statuses distinguished for the ethos of the speaker (ethos strictly speaking, self-portraying, reputation), we see that the character of the audience is entirely of the latter kind, that is *reputation*, not that of a person but of a group: "young people are like that". Strictly speaking, however, any audience is able to express its rhetorical ethos by means of its spontaneous reactions to speech.

Etymology ▶ True Meaning of the Word

Evaluation and Evaluators

In general, to evaluate an argumentative discourse is to pass upon this discourse a justified positive or negative "value judgment", **S. Value**. The assessment activity is one argumentative activity among others, which may be misleading or well founded, regardless of the quality of the discourse it approves or condemns. In order to avoid arbitrariness, the evaluation must specify its method, criteria and reference assessment scale, and remain open to criticism — as is the case for any other argumentation.

1. Dimensions of evaluation

1.1 Assessment scales
Arguments may be evaluated on the basis of different kinds of scales, such as:

An efficiency scale — The best argument is the one that best orients its target towards the thesis it defends, or the action it advocates.

A scale of logical-scientific validity — Good arguments are valid deductions, starting from true premises and conveying their truth to their conclusion according to valid rules and methods.

Invalid arguments are misleading, and effective argumentation may be so; in fact, effective arguments are systematically suspected of being misleading. Conversely, a valid argument may be totally ineffective: for example, "**P**, so **P**" is a valid deductive inference but it has no persuasive power. It could be argued that ordinary arguments often simply just camouflage this kind of truism by using two distinct formulations of the same proposition **P** "**P**, therefore (paraphrase of **P**)". Since trains never leave before the scheduled time, the following account is not a real justification:

> Due to the delay, the train will not start on time.

1.2 Binary and gradual evaluation

— *Binary evaluation* classifies arguments as valid (good, accepted) and invalid (bad, rejected). This evaluation follows the rules and criteria of formal logic. It requires the translation of the argumentation produced in ordinary language into a logical language. The evaluation bears on this logical characterization, taken as expressing the essence of the argumentation, and this logical evaluation is then transferred to the original discourse, **S. Connectives**.

— *Gradual evaluation* positions the argument on a gradual scale, as more or less good or bad. In practice, the evaluation criteria depend on the argument scheme considered, and on the availability of a relevant set of critical questions, which can be rather heterogeneous. The argument under consideration is then checked for each condition, and its global evaluation may be only a precarious synthesis of the results of these different operations.

2. The diagnosis of fallacy
The imputation of fallacy is an adversarial procedure condemning, rejecting, or disqualifying a discourse. The accused arguer has the right to defend his or her argument, in view of the principle of "no execution without representation". Discussions about the fallacious nature of argumentation are, in principle, open-ended and their conclusions are defeasible and adjustable. They are arguments like any others, possibly themselves fallacious. At any rate, meta-argumentative disputes about argument evaluation provide interesting data for argumentative analysis.

Evaluators

Hamblin gives a clear answer to this question: the logician is *not* the arbiter of the argument or dispute (Hamblin 1970, p. 244-245):

> Consider, now, the position of the onlooker and, particularly, that of the logician, who is interested in analyzing and, perhaps, passing judgment on what transpires. If he says "Smith's premises are true" or "Jones argument is invalid", he is taking part in the dialogue exactly as if he were a participant in it; but, unless he is in fact engaged in a second-order dialogue with other onlookers, his formulation says no more than the formulation "I accept Smith's premises" or "I disapprove of Jones's argument". Logicians are, of course, allowed to express their sentiments but there is something repugnant about the idea that Logic is a vehicle for the expression of the logician's own judgments of acceptance and rejection of statements and arguments. The logician does not stand above and outside practical argumentation or, necessarily, pass judgment on it. He is not a judge or a court of appeal, and there is no such judge or court: he is, at best, a trained advocate. It follows that it is not the logician's particular job to declare the truth of any statement or the validity of any argument.
>
> While we are using a legal metaphor it might be worthwhile to draw an analogy from legal precedent. If a complaint is made by a member of some civil association such as a club or a public company, that the officials or management have failed to observe some of the association's rules or some part of its constitution, the courts will, in general, refuse to handle it. In effect the plaintiff will be told: "Take your complaint back to the association itself. You have all the powers you need to call public meetings, move rescission motions, vote the managers out of office. We shall intervene on your behalf only if there is an offence such as a fraud." The logician's attitude to actual argument should be something like this.

The diagnosis of fallacious speech operates on a meta-argumentative level, but this second level does not transcend the dialogue under scrutiny, it remains an integral part of the argumentative game. In other words, the judgment *"this argument is fallacious"* works in the same way as any other ordinary refutation, whether carried by a participant (ordinary use of the word *fallacy*) or by an analyst, who then behaves as a participant. One must then speak of an *ad fallaciam* argument.

In a letter to Scherer, the economist Leon Walras refers to a controversy between Scherer himself and Guéroult:

> I take [...] your [= Scherer's] study of December 30 to the point where you [...] clearly and plainly address the more general considerations about the divergence between his [*Guéroult's*] opinions and yours.
> "*Perfectibility*, you say *is a modern idea, one of those that best indicate the distance between the old world and the new world. It bears within itself its own self-evidence, so that its adversaries are only a few sophists or some misanthropes. It has passed into the common law of intelligence.* Yet, as M. Guéroult seems sometimes to do, perfectibility should not be confused with the possibility of perfection. This confusion is not merely a matter of words; for those

> *who understand the scope of the questions, it marks the dividing point between two systems, liberalism and socialism. Socialism, reduced to its principle, is nothing other than the belief in the possible perfection of society and the effort to realize this state."*
> This is clear and precise. M. Guéroult and you agree up to a point: for both of you, humanity advances and does not retreat; the law of the development and organization of society is a law of progress and not of decadence. Beyond these limits, you separate, you think that society is only perfectible, while M. Guéroult, on his part, thinks that society, sooner or later, will be perfect; you are a liberal, M. Guéroult is a socialist. Perfectibility or perfection, liberalism or socialism, such is the alternative and the question that is raised.
> <div align="right">Léon Walras, [<i>"Socialism and liberalism"</i>], [1863][1].</div>

Schérer argues that Guéroult concludes from the possibility of the *perfectible* (point upon which they agree) the possibility of the *perfect* (point upon which they disagree). This is a typical argument built upon derived@ words. Schérer does not consider this inference to be a sophism (he does not attribute to Guéroult the intention to mislead his readers) not even to be a mistake, a fallacy of "confusion", simply a criticism. The analysis is not made from an *external logical* point of view; it comes from a *political opponent*, making this point as a sub-issue of a greater debate "*Liberalism or Socialism?*".

3. For a laissez-faire in argumentation

Ordinary argumentation is carried out in specific fields by speech communities corresponding to what Hamblin calls "civil association[s]", having their, interests, programs, ways of thinking and rules for deliberation and action. In these fields, the logician does not, *as such*, have the substantial specialist skills required. This remark is at the heart of "critical liberalism", advocating a laissez-faire attitude with argumentation.

From such a viewpoint, what becomes of evaluation? Hamblin's objection is taken into account, and the evaluation of the arguments is entrusted to the "civil association" with which the arguing party, interested in the outcome of the issue, is affiliated. The data to be considered for the evaluation is not limited to the one isolated argument under scrutiny, but consists in a well-defined selection of contradictory discourses developed around the same issue.

As a result, the evaluation process may be empirically documented and criticized on three levels:

— *Non-thematized criticism*: description of the practices of evaluation in action, such as concessions, objections, refutations and counter-discourses in general.

— *Emergence of a specific ordinary critical metalanguage*: charges of fallacy, misplaced authority, irrelevance, emotionality, amalgam and impugned motives, etc. (Doury 2000).

---

[1] Quoted after Léon Walras L. (1896). "Socialisme et libéralisme". In *Études d'économie sociale – Théorie de la répartition de la richesse sociale* Lausanne: Rouge & Paris: Pichon. P. 4.

— *Evaluations carried out by the specialists of the field.* This level, which includes scientific expertise, is the ultimate level of evaluation. Scientists routinely evaluate the discourses and fallacies of their colleagues; historians and social scientists evaluate the fallacies of historians (Fisher 1970), and the teachers and pupils evaluate the pupils' and teachers' arguments.

All these activities are "[meta-argumentative]", as opposed to "ground level argumentations" (Finocchiaro, 2013, p. 1). Provided that the intervention is useful and desired, the specialist in argument analysis can intervene at all levels. As Hamblin has explained, his or her function and deontological position are those of a "well-trained advocate". As such, the specialist can evaluate all the arguments of the world, the posture being that of the participant analyst and evaluator, working under a double constraint of externality / internality well known in ethnomethodology. He or she may meaningfully intervene in court as a *jurilogician* or a *jurilinguist*, that is as a *counsel*, not a *substitute* for the judge.

Argumentative discourse is in itself evaluative and critical; scholarly evaluation is a process of argumentative expansion and deepening of the issue itself. There is no super-evaluator capable of putting an end to the critical process by providing a final, conclusive evaluation to silence all other participants.

Evidentiality

Evidentiality is a set of grammatical or linguistic phenomena indicating how the information conveyed in a statement has been obtained by the speaker. Evidential systems commonly signal that the information 1) comes from sensory *experience* (auditory, visual); 2) that is has been *inferred* from something else; 3) that it reproduces a *hearsay*. Other evidential systems are much more complex. In evidential languages, the speaker must explicitly mention *the basis* on which he or she says what he or she says, that is, the kind of argument supporting the utterance.

In some languages, evidentiality is *grammaticalized*. That is, it corresponds to a specific grammatical category. In English, for example, the reported event is necessarily referred to by its temporal-aspectual coordinates. In evidential languages, the speaker is obliged to stipulate how the information he relays has been obtained (via the senses, hearsay, inference, etc.). The sub-system of the grammatical marks of evidentiality is distinct from the modal system as well as from the temporal-aspectual system.

Evidentiality can be considered as a linguistically embedded argumentation, as an "argumentation within grammar". It leads to the conception of the argumentation as a continuum, sometimes related to the *grammar and semantics of the language* and at other times, to the *grammar and semantics of discourse*.

In English, where evidentiality is not grammaticalized, evidential markers or phrases remain optional. The evidential sources can be discursively expressed as coordinated sentences or as the head of the sentence:

Peter is at home; one can hear him	I hear that Peter is at home
one can see him	I see that —
they told me	They told me that —
I read it	I read that —
I guess	I guess that —

Evidentiality may be expressed by modals. So, for example, in the statement *"Peter is at home"*, the information about that Peter's whereabouts is given in the categorical mode, and is endorsed by the highest degree of speaker certainty on an epistemic scale ranging from doubt to certitude. From an evidential perspective, the statement implies that the speaker has some direct evidence to back the speech, for example *"I just left him"*, etc.

"Peter should be at home now": this sentence communicates the same information on a lower position on the epistemic scale. From an evidential perspective, the statement implies, for example, *"I have no direct and categorical evidence of what I say, but on the basis of Peter's usual habits, I infer that he is at home"*.

The following examples are taken from Ducrot (1975). In *"Peter must/should have received my letter"*, the information *"Peter has received my letter"* is backed only by common knowledge of the usual delivery deadlines. The following case is different:

Well, I believe that Peter has received my letter!

There is now an implication that the same information has been inferred from a quite different source, that is some natural sign taken in Peter's behavior which can be explained only by referring to the letter's content; for example, the letter informed Peter of a disciplinary warning, and Peter clearly changed his behavior.

Ex — Arguments (*Ex Concessis*...)

Some argument schemes are designated by Latin labels, S. *Ab* —; *Ad* —; *Ex* —. This entry lists the labels using the Latin preposition *ex* (rarely *e, and* never *e* before a vowel).

E/ex means "taken from"; in the construction "arguments *e/ex* **N**" the Latin noun **N** refers to the substance, from which the argument is drawn.

List of the "*ex* + N" Arguments

Latin name of the argument	• Meaning of the Latin word(s)Latin • (When necessary a word-for-word translation) • English equivalent(s) • Reference to the corresponding entry/ies
ex concessis (sg. *ex concesso*) *e concessu gentium*	Lat. *concedere*, "admit; agree with sb" — *arg. from the consensus of nations ; from traditional wisdom; from what is admitted by the audience or the opponent* — S. **Consensus; Authority;** *Ex concessis*; *Ex datis*; **Beliefs of the Audience; Concession;** *Ad hominem.*
e contrario [generally *a contrario*]	Lat. *contrarius*, "contrary; opposite" — S. **Opposite**
ex datis	Lat. *datum*, "gift" — *arg. from the facts as such; from what is accepted by the audience* — S. **Ex datis**
ex notatione	Lat. *notatio*, from *notare* "stamp with a mark" — *arg. from "what the word (truly) says"; argument from the meaning or of a word.* S. **True Meaning of the Word; Derived Words**
ex silentio	Lat. *silentium*, "silence" — S. **Silence**

As the *ab* and *ad* arguments, the *ex* arguments do not refer to a unified category of arguments, or to a common semantic family, nor to a formal type.

Ex Concessis

>Lat. *ex concesso*, pl. *ex concessis*; *concessus*, "concession, consent".
>Lat. *ex concessu gentium*: *gentium*, from *gens* "race, nation, people" translated as "argument from the consensus of nations".

The Latin label *ex concessis* refers to two kinds of argumentation.
— Argumentation based upon the (alleged) *universal consensus* (*ex concessu gentium*), that is to say, from general agreement, S. **Consensus; Authority.**
— Argumentation based upon a *local consensus*, limited to the *beliefs of the audience*. The orator may or may not share these beliefs. With this meaning, the argument *ex concessis* is also called *ex datis*@.

Ex Datis

>Lat. *ex datis*, *datum*, "gift, present".
>The label "*ad auditorem* argument", lat. *auditor* "listener" is used by Schopenhauer ([1864], p. 43).

The *ex datis* argumentation is not based on facts or experience, but on what has been admitted, "given", or conceded by the interlocutor, the audience or the adversary. The arguer reasons, "from what has been granted" (Chenique 1975, p. 322). The *ex datis* argumentation is sometimes called *ex concessis*@ (*ibid.*).

Like the *ad hominem*@ argument, the *ex datis* argument is based on the beliefs of the audience. While the *ad hominem* argument exploits these beliefs to rebut the whole system, the *ex datis* argument exploits them for confirmation purposes only. Good knowledge of the audience's character is important to argumentative rhetoric, because it provides the orator with a great reserve of such *ex datis* premises, S. Ethos.

If the interactional framework does not allow the revision of beliefs, this data cannot be questioned, and the conclusions based upon it are irrefutable in this context. From this data, the arguer concludes positively, "*besides, you yourself sayit!*". Consider the issue, "*Should we take military action in Syldavia?*":

> You admit that the Syldavian troops are poorly trained and that the situation will soon be out of control;
> and that the unrest in Syldavia may extend to the region;
> we agree that this extension would threaten our security;
> and no one denies that we must intervene if our security is threatened.
> So, you should join us, the camp of the people who are in favor of a military intervention in Syldavia.

This strategy of argumentation has something to do with religious confession and philosophical maieutic. The listener is invited to assume the truth of his or her beliefs, that is the conclusion that he or she does not dare to draw, or is unable to express because of some intellectual or moral inhibition.

The *ex datis* argument scheme calls for a foundational and an ethical criticism. According to the *foundational* principles, to be valid, an inference must be based on true universal premises, whereas any *ex datis* argumentation is misleading because its premises are based upon local beliefs only. As the *ex datis* argument is the typical form of rhetorical argumentation, all such rhetorical arguments must be condemned.

From an *ethical* point of view, and unlike the logician arguing for what is, or what he truly believes to be true, the party putting forward an *ex datis* argument does not necessarily endorse and support the premises and rules borrowed from the audience. This is why the *ex concessis* argument may end up as a trap. People are generally expected to take responsibility for what they say, so the audience in good faith will normally allocate to the arguer the beliefs he or she argues from, even if the speaker only advances them *ex concessis*. Yet, if this arguer is better informed than the audience and knows that **P** is true (or false), whereas the listeners believe that **P** is false (or true); or if he or she has reliable information unknown from the audience, and takes into account only what the audience believes and knows, then, to say that the argument is ex *datis*, ex *concessis*, *ad auditorem*... is simply to say that the orator lies and manipulates the audience. S. Refutation; Manipulation.

In philosophy, Kant draws a distinction between *ex datis* knowledge based on experience, and *ex principiis* knowledge deduced from the first principles. History is the prototype of knowledge *ex datis*, philosophy and mathematics prototypes of knowledge *ex principiis*. Mere *ex datis* knowledge would be only a compilation of data. Extending the Kantian meaning, one might think that the *ex datis* argumentation is based on experience, "on the substance, on things themselves". This interpretation would make *ex datis* an equivalent of the *ad rem*, to the matter@ argument, and this does not seem to be the case. The use of the expression *ex datis* in argumentation is distinct from its use in philosophy.

Exaggeration and Euphemization

1. Exaggeration as amplification

Aristotle defines exaggeration as the use of "indignant language [...] painting a highly colored picture of the situation" (*Rhet*, II, 24, 1401b1-10, RR, p. 383), and notes its spectacular and curious effect in a judicial situation: "if the defendant does so, he produces an impression of his innocence; and if the prosecutor goes into a passion, he produces an impression of the defendant's guilt" (*ibid*).

2. Exaggeration to absurdity

Exaggeration to absurdity is a technique of refutation known under the name of *adynaton*: "the arguer uses both hyperbole and apodioxis to establish a position by the exaggeration of the absurd of the opposite position" (Molinié 1992, *Adynaton*; for *apodioxis*, S. **Contempt**)

This is a variant on refutation from the absurd, taken to the ridiculous:
>To prevent accidents, leave your car at home!
>To avoid recidivism, let us execute all offenders!

The mechanisms of argumentation are the same as those of the slippery slope argument, an invitation *"don't stop now, the path is so good"*, S. **Direction; Slippery Slope; Laughter.**
>You want to be vegetarian, no problem, eat salad, go and graze on the lawn.

In the following passage the position "criminally insane people must be judged as everyone else" is rejected by showing that if intentionality is not taken into account, the very idea of criminal behavior becomes meaningless:
>Let us judge all criminal acts. Whatever the level of consciousness of their perpetrators. And why not a dog? The news provides a tragic opportunity to further advance justice. [...] And why does the cyclone that recently ravaged the West Indies, causing several victims and immense material damage, escape the wrath of justice?
>
>L. M. Horeau, [*Obvious Delirium*]. *Le Canard Enchaîné*, 2007[1]

---

[1] L. M. Horeau, "Flagrants délires". *Le Canard Enchaîné*, (a satirical newspaper) August 29, 2007. P. 1

3. Minimization, or euphemization

Minimization strategies are used to deflect an accusation, when bad behavior is acknowledged as such, and its material significance is reduced to nothing. If I'm accused of having stolen a bicycle, for example, I might defend my actions thus: *"Oh yes, but it's just an old broken worthless bike."*
The associated feeling is *indifference*, and the accuser is encouraged to *cool down*, S. Calm. Anything can be euphemized, even torture:

> *30-7-84 Christian Von Wernich (chaplain [capellán] of the Police of Buenos Aires, currently priest in Bragado) (statement to the magazine* Siete Días*):*
> Tell me that Camps has tortured a poor guy whom nobody knows, good, okay then. But how could he have tortured Jacobo Timermann, a journalist about whom there was constant and decisive global pressure, if only for that!
> Carlos Santibáñez & Mónica Acosta, [*The Two Churches*], [1996].[1]

Example

The word *example* has two main meanings:
1. Way of being or doing worthy of imitation: *setting an example, leading by example, being an example for the community*.
2. Any item in a series of equivalent elements, one case among others. If the series is composed of different elements, a *typical example* is the most characteristic individual, central to the series.

Besides the specific forms of argumentation described below, the following forms of argumentation are related with the example: S. *Exemplum*; Imitation, paragons and models; *Ab exemplo*.

1. The example in the Aristotelian rhetorical system

In a version of the Aristotelian rhetorical system, the *induction@* and the *syllogism@* are the instruments of scientific discourse, whereas the *example* and the *enthymeme@* are their counterparts in rhetorical discourse (*Rhet*, II, 20, 1393a20-25, RR p. 335). There are different kinds of examples:

> [Argument by example] has two varieties; one consisting in the mention of actual past facts, the other in the invention of facts by the speaker. Of the latter, again, there are two varieties, the illustrative parallel and the fable. (*Id.*, 1393a25-30; RR p. 357-358)

A table of rhetorical instruments:

---

[1] Carlos Santibáñez & Mónica Acosta, *Las dos Iglesias*. Report commemorating the 20th anniversary of the assassination of Bishop Angelelli.
www.desaparecidos.org/nuncamas/web/investig/dosigles/02.htm (11-08-2017).

Example

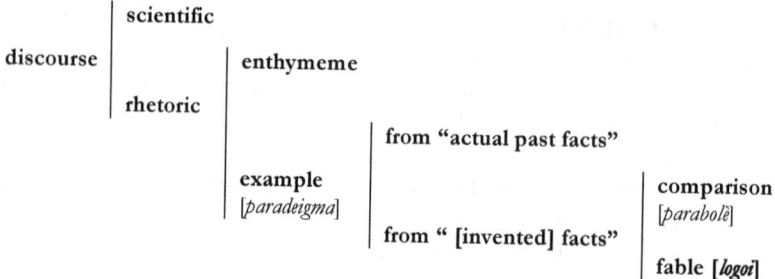

An argument drawn from an example based on past, real facts is illustrated by a form of induction leading to the conclusion, "we must prepare for war against the King of Persia and not let Egypt be subdued", in view of two past experiences which were detrimental to the Greeks:

> For Darius of old did not cross the Aegean until he had seized Egypt; but once he had seized it, he did cross. And Xerxes again did not attack us until he had seized Egypt. but once he had seized it, he did cross. (*Rhet.*, II, 20, 1393a30-b5, RR p. 335)

The reasoning can be seen as an induction@, aimed at establishing as a law that "*the conquerors who seize Egypt then cross the sea to Europe*", or as a direct stimulation to wake up bad memories. In that case, the argument by example would function as a kind of two-term@ reasoning.

Comparison — Aristotle gives as an example of a "parable", an *analogy@* drawn from the speeches of Socrates. This parable condemns the practice of drawing lots for magistrates, since one does not "use the lot to select a steersman from among a ship's crew" (*Rhet.*, II, 20, 1393b5, RR, p. 335); S. **Metaphor**.

Fable — Aristotle gives as an example of a fable of the horse that wanted revenge on the stag, and in so doing becomes a slave to man, with an application to the saviors of the fatherland who quickly became tyrants (*Rhet*, II, 20, 1393a5-25, RR p. 337). As portraits (S. **Ethos**), fables are a fully argumentative and literary genre, from Aesop (620 - 564 BCE) to modern times, S. *Exemplum*.

2. Argument by example

As a generalization (induction) based on a single specific case, the argument from example draws on an observation made on one individual, and categorically generalizes it to all individuals of the same class or of the same name:

> This butterfly is blue, so (all) butterflies are blue.

In reality it is only possible to conclude "*some* Bs are **P**" from "*this* **B** is **P**". The generalization on the basis of one single specific case corresponds to the converse of the instantiation of a universal proposition, which is valid; if "*all* Is are **P**" then "*this* **I** is **P**".

> This swan is white, it's okay, since (all) swans are black.

Argument by example is a kind of hasty generalization or induction@ on the basis of a single case, or a relatively small number of cases. It may also be a case of two-term@ reasoning.

The inductive narrative proceeds from an anecdote: *"the owners of iPhones are unbearable. Recently I was camping…"* and the anecdote develops, highlighting the terrible behavior of *one* iPhone user and generalizes this case to all iPhone users. In Aristotelian terms, the process is an inductive generalization, based on a real past fact, which is then elaborated as a truth revealing fable.

3. Argumentation from a generic example, or *ecthesis*

A *generic example* is a being in which all the properties of the genus to which it belongs are clearly manifested. It is a prototype of the class, its best incarnation, **S. Category; Analogy**. The argument from the generic example is based on such a specimen and results in conclusions being made about a given genus (about all the individuals belonging to that genus):

> The generic example consists in explaining the reasons for the validity of an assertion by performing operations or transformations on a given concrete object, considered not for itself but as a characteristic representative of a class. (Balacheff 1999, p. 207).

The process is also known as *ecthesis*, defined as "[a] technique of demonstration used especially in Euclidean geometry: to establish a theorem, you reason on a singular figure. Your inference is correct if it does not mention the characteristics peculiar to the drawn figure but only those which it shares with all the figures of its species." (Vax 1982, *Ecthèse*)

4. Exemplification of a generic or accidental feature?

The argument by example is a legitimate extrapolation if it is founded on a generic feature. If one asks for example how many wings birds may have, observation of any bird will lead the observer to discover the correct answer. On the other hand, if one asks about the average weight of a pigeon, the same procedure is absurd: *"this pigeon taken at random weighs 322 g. So the average weight of a pigeon is 322 g."*

As in many cases, it is not previously known whether the investigated feature is essential or accidental, this distinction is exploited as an argumentative resource. The proponent considers that generalization is valid because it is based on an essential trait, and the opponent argues that it is accidental and cannot be generalized. **S. Taxonomy and category; Accident.**

The remains of a single animal belonging to an unknown disappeared species provides a wealth of knowledge about this species, but its specific conditions must be duly acknowledged, as shown by the case of the Neanderthal man.

> 1. The views the scientists hold about the Neanderthals have changed over time. (after G. Burenhult, "[Towards Homo Sapiens]", 1994[1])

---

[1] G. Burenhult, Vers Homo Sapiens. In *Le Premier homme*. Preface by Y. Coppens, Paris, Bordas, 1994, p. 67.

Example

More precisely: Is the Neanderthal man our ancestor or a species different from our own?

2. First answer: The Neanderthal man belongs to our species. "It has long seemed obvious that the physical appearance of the Neanderthal man — and especially those living in Europe — was very different from ours". However, "in spite of these physical differences, Neanderthals have long been regarded as direct ancestors of the present man" (*id.*, p. 66).

3. Second answer: The Neanderthal man belongs to a different species. "Following the work of the French paleontologist Marcellin Boule these differences were judged too great" (*id.*, p. 67), and the Neanderthal man was considered to belong to a different species.
The Neanderthal of Marcellin Boule: "From 1911, the paleoanthropologist Marcellin Boule published a detailed study of the skeleton. He built an image that has conditioned the popular perception of Neanderthal man for more than thirty years. His interpretations are strongly influenced by the ideas of his time concerning this extinct hominid. He describes him as a kind of savage and brutal caveman, dragging his feet and not able to walk upright."
"Marcellin Boule describes a Neanderthal with a flattened skull, a curved vertebral column (much like gorilla), semi-flexed lower limbs and large divergent big toes. This description is in keeping with the ideas of the time on human evolution" (Wikipedia, *Marcellin Boule*).

4. But this Neanderthal was seriously handicapped: "In 1913, Marcellin Boule exaggerated the differences with us, not realizing that the skeleton he was studying — the "Old Man of the Chapelle aux Saints" (Corrèze, France) — was deformed by arthritis, as demonstrated by W. Strauss and A. J. Cave in 1952." (*id.*, p. 67)
"J.-L. Heim describes the subject as badly disabled; he suffered a deformity of the left hip (epiphysiolysis or rather trauma), a crushing of the finger of the foot, severe arthritis in the cervical vertebrae, a broken rib, and a narrowing of the channels of the spinal nerves." (Wikipedia, *id.*)

5. Conclusion: Our cousin, the Neanderthals: "Today Neanderthal men are seen as our cousins rather than as our ancestors, although they look like us in many respects" (*ibid.*).

5. Exemplification as illustration and test case example
The *generic* example functions as a basis for an abductive *generalization*, resulting in a rule or regularity about a class of cases or individuals. Specific cases can be introduced in relation with such a general discourse.
— The *illustrative* example facilitates the understanding of a concept or a law, by introducing a (typical) instantiation of the concept or the law:
A migratory bird is a bird that ... So the swallow...

Moreover, if the example chosen is (presented as) typical of the phenomenon, it renders the time-consuming and precarious work of checking a large number of cases unnecessary. In this sense, to give an argument in defense of a general statement is simply to find a case to which it applies correctly. If the general statement is the result of an a priori argumentation or illumination, the illustrative example will at least show that the conclusion is not undermined by the first example that comes to mind (see infra, § 6).

The *illustrative* example can also be used as an epideictic amplification technique:
> Whereas an example is designed to establish a rule, the role of illustration is to strengthen adherence to a known and accepted rule, by providing particular instances which clarify the general statement, show the import of this statement by calling attention to its various possible applications, and increase its presence to the consciousness. (Perelman & Olbrechts-Tyteca [1958], p. 357)

— The *test case* example is different. It may be introduced as an objection to the theory, and the speaker must show that the general principle he or she favors can be successfully applied to this case, that it *accounts for* this case.

6. Refutation by the counter-example (arg. *in contrarium*)

An example does not establish a law, but is sufficient to refute a generalization. Argument by the counterexample is the standard method of refutation of general propositions "*all A are B*": this assertion is refuted by showing an **A** which is not **B**. This strategy is perfectly operative in ordinary argument, S. **Opposite**.

Exemplum

1. The predicative rhetorical genre

The classical rhetorical genres, the deliberative, the judicial, the epidictic, all relate to civil life. Christian religious rhetoric has developed a new genre, *preaching*, where persuasion is put to the service of religious faith. *Predication* is the action name associated with the verb *to preach*, and the noun *preacher*. It has not been affected by the derogatory orientations sometimes associated with these two words in contemporary usage. It is homonymous with the word *predication* as used in grammar and logic to designate the operation by which a *predicate* (a verbal group) is associated with a subject in a sentence; and with the word *to predicate something upon*, that is to base an action or a saying upon:
> I predicated my argument on the facts. (tfd, *Predicate*)

Preaching as an argumentative genre fully complies with the definition of argumentation provided by Perelman & Olbrechts-Tyteca as a discursive effort "to induce or to increase the mind's adherence to the theses presented for its assent" ([1958]/1969, p. 4). The *theses* referred to in this case are religious *beliefs*, that are *articles of faith* from the point of view of the preacher. Assuming that the audience is composed of *believers*, by preaching to them, the pastor assures their ongoing training and increases their degree of belief, in other words, "the soul's adherence" to their creed (after Perelman & Olbrechts-Tyteca, [1958], p. 4).

If the audience is composed of *unbelievers*, the missionary might preach to them in order to instigate these same beliefs. If the audience is composed of *heretics* in a position of strength, rhetoric must give way to dialectic.

The tenants of the Catholic faith are given in the Holy Scriptures, and are commented on by the authorities, the Fathers of the Church. These contents are articulated and applied in sermons by means of various speech techniques, which have established themselves in a sometimes polemical tension between dialectical appeals to reason and rhetorical enthusiasm for faith, S. **Faith**.

2. The *exemplum*

The *exemplum* (plural *exempla*) is an instrument of preaching which has been particularly developed by the Dominican and Franciscan mendicant orders, from the beginning of the thirteenth century. Structurally, the *exemplum* is a narrative, exploiting the resources of the fable. The genus is legitimated by the very example of Christ who preached by parables. The *exempla* present models of action to be followed or avoided.

The *exemplum* is "a brief narrative given as truthful and intended to be inserted into a discourse (usually a sermon) to convince an audience by a salutary lesson" (Brémond *& al.* 1982, pp. 37-38). Brémond distinguishes metaphorical and metonymic *exempla*.

2.1 Metonymic *exemplum*

In such *exempla*, the fact is presented as being likely. There is then a certain identity of status between the heroes of the anecdote and the recipients of the exhortation. The parable of the evil rich is told to the rich, and the logicians are told the tale of one of their colleagues, who is tormented in hell for his sins, that is to say, his sophisms.

> *The following* exemplum *deals with the fate of souls after death, and especially with purgatory. The lesson it contains is a "Christian denunciation of vain pagan erudition" (Boureau, p. 94), and a call to the logicians to convert to a religious life.*
>
> For our edification, it may be useful to know that a harsh sentence is inflicted upon sinners at the end of their lives.
>
> This is what happened in Paris, according to the Parisian Cantor (= Peter the Chanter, *Petrus Cantor*). Master Silo urged one of his colleagues, who was very ill, to come and visit him after his death and to inform him of his fate. The man appeared before him a few days later, wearing a cloak of parchment covered with sophistic inscriptions and full of flames. The master asked him who he was. He replied, "*I am the one who promised you that he would visit.*" When asked what his fate was, he said, "*This cloak weighs me down and oppresses me more than a tower. They make me bear it for the vainglory which I have derived from the sophisms. The flames with which it is filled represent the delicious and varied furs I wore, and this flame tortures me and burns me*". And as the master found this slight penalty, the deceased told him to stretch out his hand to test the lightness of punishment. On his outstretched hand, the man dropped a bead of sweat, which drilled the hand of the master as fast as an arrow. The Master experienced an extraordinary agony, and the man said to him, "*so it is with all my being*". Afraid

of the harshness of this chastisement, the master decided to leave the world and enter religion. And in the morning, facing his gathered students, he composed these verses:
> *To the frogs, I give up croaking / To the ravens, cawing, / To the vain, vanity. I attach my fate / To a logic that does not fear the conclusive 'therefore' of death.*

And, abandoning the world, he took refuge in religion.

Jacobus da Varagine, The Golden Legend, written around 1260[1]

The practice of *exemplum* goes beyond the strictly religious domain. Fontenelle's *"Golden Tooth"* is actually a lay metonymic exemplum illustrating the fallacy of finding the cause of a fact that does not exist, **S. Cause - Effect**.

2.2 Metaphorical *exemplum*

In such *exempla*, "the narrative no longer quotes a sample of the rule, but a fact that resembles it" (*ibid.*):

> The hedgehog, it is said, when he enters a garden, takes on a load of apples which he fixes on his prickles. When the gardener arrives, the hedgehog wants to run away, but his load prevents him doing so, and thus he is caught with his apples. [...] This is what happens to the unfortunate sinner who is taken, when he dies, with the burden of his sins.
> Humbert from Romans, [*The Gift of Fear or the Abundance of the Examples*], written between 1263 and 1277.[2]

Explanation

In common language, the words *explain* and *explanation* refer to different scenarios, discourse genres and interactions. *Ethnomethodology* proposes to grasp the ongoing intelligibility of ordinary actions and interactions through the concept of "accounts" (justifications, explanations). *Text linguistics* considers the explanatory sequence as one of the basic sequence types (Adam 1996, p. 33), along with narration, description and argumentation. The relations between text types are complex: a *justificatory* (vs. *deliberative*) argument *explains*, or *accounts* for a decision by enumerating the good reasons having motivated the decision made in the past.

1. Structure of explanatory discourse

From the conceptual point of view, explanatory discourse connects a *less well-known, local* phenomenon, something to be explained (*explanandum*) to a *better known and complex* explanatory domain (*explanans*). *Explanation* promotes *understanding*. An explanation is an *abduction*@. One can distinguish between different

---

[1] Quoted after Jacques de Voragine, *La Légende Dorée*. Text presented by A. Boureau. In J.-C. Schmitt (ed.), Prêcher d'exemples [*Preaching* Exempla]. Paris: Stock, 1985. P. 7.
[2] Humbert from Romans, *Le Don de Crainte ou l'Abondance des Exemples*. Trans. from Lat. to French by Chr. Boyer. Lyon: PUL. 2003. P. 116.

kinds of explanation according to the kind of field-related principles invoked to connect the *explanandum* to the *explanans*:

— *Causal* explanations, allowing prediction and action, as in the following explanatory definition, **S. Causality:**

> Rainbow: A luminous meteorological phenomenon [...] produced by the refraction, reflection and dispersion of the colored radiations composing the white light (of the sun) by drops of water. (PR, Art. *Rainbow*).

— *Functional* explanations:

> Why does the heart beat? — *To circulate the blood*
> Why religion? — *To strengthen social cohesion*
> Why do oranges have slices and chocolate bars have squares? — *So they can be more easily divided among children.*

— *Analogical* explanations, **S. Analogy:**

> The atom is like the solar system

— *Intentional* explanations, **S. Motives:** "*He killed to steal*".

— *Interpretive* explanation; when it comes to an obscure text, the explanation provided is an *interpretation*@ of the text.

The specific conceptual structure of explanatory discourse in science depends on the definitions and operations governing the field considered: one can explain in history, in linguistics, in physics, in mathematics. As it relies upon a less known / better known differential, explanation also depends on the previous knowledge of the person to whom it is addressed. A good explanation must "reach home"; the explanation provided to someone having no knowledge of the given field, will not be the same as that given in a research paper in that same field.

2. Ordinary explanations

1.1 *Explain*: The word and its usages

The actors of the verb *to explain* are human (**S1, S2** ...). Explanatory discourse connects the *explanandum* to a possible *explanans*.

— Explanation typically bears upon an external phenomenon which one wishes to better understand:

> • In "**S1** explains **M** to **S2**" the explanation is a conceptual interactional sequence.
>
> • In "**E** explains **M**", the explanation is phrased as an objective conceptual monologue, containing no reference to an interactive event.

— **S1** can summon another person **S2** to explain his or her (= **S2**'s) behavior. Then, **S1** wants to clarify an interpersonal misunderstanding, or something that could be taken as an offense **O**, committed by **S2** against **S1**:

> **S1** and **S2** had an explanation about **O**
> *You owe me an explanation*! (1)

Explanation

The so-called "explanation" required is actually a justification. (1) constitutes a rather threatening opening, said in an angry tone, and anticipating an animated, even violent discussion. The "explanatory" interaction to follow will probably be an *argument$_2$* (**S. Argument – Conclusion**), made with the aim of either restoring the relationship between the two individuals, or redefining it.

In everyday usage, the word *explanation* refers to segments of speech or to interactive sequences opened by a speaker who:
— does not understand:
> "(Explain to me) *what does 'zoon politikon' mean?*": a request for a definition, a paraphrase, a translation or an interpretation.
> "(Explain to me) *what really happened?*": a request for a convincing narrative.
> "(Explain to me) *why does the shape of the moon change?*": a request for a theory, diagrams and images.

— does not know how to do something:
> "(Explain to me) *how does it works?*": a request for directions for use, a leaflet, a manual, a practical demonstration.

The structure of the explanation provided will be as diverse as the kind of activity involved.

The question of the unicity of the *concept of explanation* thus arises, as well as that of the varieties of interactional explanatory discourses. At the most general level, the need for explanation comes from the feeling of *surprise* (novelty, anomaly) before something *astonishing*. Any answer that can satisfy this astonishment and rid the speaker of any sense of surprise may be considered to be a satisfactory explanation.

1.2 In ethnomethodology

Ethnomethodology (Garfinkel 1967) attaches central importance to accounts in everyday interactions, that is to ordinary explanations, justifications or good reasons given by the participant in regard to the meaning of what they are doing and expecting. Accounts are given at two levels; firstly, as *explicit* explanations "in which social actors give an explanation for what they are doing in terms of reasons, motives or causes" (Heritage 1987, p. 26). Secondly, *implicit accounts* are provided as explanations inscribed in the ongoing flow of actions and social interactions (*ibid.*). Such implicit accounts are intended to ensure the mutual intelligibility of "what is going on", on the basis of action scripts, social expectations or practical moral standards. These explanations are said to be *situated*, i.e., *context bound*.

When it comes to conversation, explicit explanations often manifest themselves as *repairs*, when an initial turn is followed by a non-preferred sequence, for example if an invitation is rejected, the refusal will often be accompanied by a justification: "*I'm afraid I can't come with you, I have to work*". This kind of explanation or reason is required in view of a social norm, as can be seen in the conflictive turn taken by the interaction when explanations are not provided (Pomerantz 1984).

1.3 Explanatory sequences

Beyond the question "*why are things so?*", the quest for an explanation is defined as a cognitive, linguistic, interactional activity, triggered by the feeling or expression of doubt, ignorance, by a disturbance in the normal course of action, or a mere "mental discomfort" (Wittgenstein 1974, p. 26). Explanations seek to satisfy such a cognitive need, to appease doubt and so produce a sense of understanding and (inter)comprehension.

The explanatory interaction between an "explainer" and an "explainee" can be schematized as a succession of stages. The first stage is a *demand* for an explanation addressed to an *explainer* by an *explainee*, and the last one a *ratification* of the explanation by the *explainee*:

(i) **Ee** has a curiosity, a doubt, concern, a mental block ... about **M**.
(ii) **Ee** looks for an explanation from **Er**
(iii) **Er** provides an explanation
(iv) **Ee** ratifies this explanation, or not.

According to this scheme, the explanation is an answer to a request. As an epistemic-interactional act, an explanation is satisfactory if it appeases **Ee**'s "mental discomfort". This means that, if not based upon **Ee's** interrogations, the most sophisticated and true explanation, will be satisfactory, at best, for the explainer **Er**.

3. Explanation and argumentation

3.1 Explanation and justificatory argumentation

Explanations are on the side of the justificatory arguments, **S. Justification:**
— Explanation and argumentation both originate in a state of doubt about a statement which does not fit with the individual's stock of beliefs and knowledge.
— Explanation and argumentation develop from an interrogation.
— Both are connecting processes which develop a given stock of beliefs. Explanation integrates an unquestionable fact, the *explanandum* into the *explanans* system. Deliberative argumentation develops *arguments* taken in this stock of beliefs towards a *conclusion*, which will be integrated in this same stock of beliefs. Justificatory argumentation integrates a challenged known fact into an established coherent system of representation.
In *deliberative* argumentation, the argument is given as assured, doubt is attached to the consequent, the conclusion. In *justificatory* argumentation, the search for argument goes the opposite way:

My client is entirely innocent, how can I prove this to the jury?

as in explanation, where the *explanandum* is an established fact, and the *explanans* must be identified:

No doubt, the face of the moon change; how can I make sense of that?

The same laws of passage can make the connection; causal links, for example, are exploited both in explanation and in argumentation, S. **Pragmatic; Motives.**

3.2 Explanation as argumentative move
The opposition between argumentation and explanation may have an argumentative import. Explanation projects unequal interaction roles: the explainee is the ignorant profane in a *low* position, whilst the explainer is the expert in a *high* position. In argumentative situations, the roles of proponent and opponent are more equal; one "explains *something to somebody*" vs. "argues *with* or *against somebody about something*".

The question "why?", which typically introduces a request for explanation, may also be used to call into question an opinion or a behavior. In the latter case, it opens an argumentative, egalitarian, discussion. But the recipient of this question may re-frame the argumentative situation as an explanatory situation, "*Wait, let me explain!*", whereby the relations becomes asymmetric, the explainer trying to have the upper hand over the explainee.

Expression

The term *expression* is used in Aristotelian rhetorical theory and critical theory with three quite distinct meanings.

1. Language-related paralogisms
In the *Sophistical Refutations*, the label "paralogisms of expression" covers the six paralogisms "related to language": 1. Homonymy; 2. Amphiboly; 3. Composition, and 4. Division; 5. Accent; 6. Expression; S. **Fallacy (II).**
This label can also be used to specifically refer to the paralogism of homonymy.

2. Pseudo-deduction
A speech is said to be fallacious by expression when although expressed formally as a demonstration, it has no demonstrative content. The speech may take the form of a demonstration, if, for example the speaker introduces a high number of argumentative indicators. When there is no semantic connection between the connected propositions **A** and **B**, the argument "**A**, therefore **B**" is said to be fallacious due to the "*form of the expression*". The conclusion is drawn "although there has been no syllogistic process" (*Rhet.*, II, 24, 1401a1; Freese, p. 325), that is without any real argumentation.

Such examples can sometimes be found in academic essays overloaded with argument indicators, hoping that they will end up producing an argument. The discourse of Pangloss, railed at by Voltaire in *Candide*, is of that kind:

[After the earthquake that ravaged Lisbon]
Some [citizens] whom they had succored, gave them as good a dinner as they could in such disastrous circumstances; true, the repast was mournful, and the company moistened their bread with tears; but Pangloss consoled them, assuring them that things could not be otherwise. "For," said he, "all that is for the best. If there is a volcano in Lisbon it cannot be elsewhere. It is impossible that things should be other than they are; for everything is right."

Voltaire, *Candide, or The Optimism.* [1759].[1]

3. Misleading expressions

In Aristotle *Sophistical Refutations*, the fallacy of "form of expression" is also called the fallacy of "form of discourse", as well as a "figure of discourse", a label likely to introduce formidable confusions. The fallacy of form of expression corresponds exactly to the phenomenon that analytic philosophers discuss under the heading of *misleading expressions*.

For example, according to Ryle, a statement such as *"Jones hates the thought of going to hospital"* (1932, p. 161) suggests that the phrase *"the thought of going to hospital"* refers to some existing object, its reference; this expression induces a belief in the existence of *"'ideas', 'conceptions', 'thoughts' or 'judgments'"* (ibid.). Ryle considers that to eliminate such non-existing entities, the statement must be rewritten in the form corresponding to its semantic-ontological reality: *"Jones feels distressed when he thinks of what he will undergo if he goes to hospital"* (ibid.). This new formulation is not supposed to contain any reference to deceptive entities such as *"the idea of going to hospital"* (ibid.).

Analytical philosophy has devoted substantial efforts to the study of misleading expressions as expressions that generate non-existent problems, as seen in the previous case, or expressions which are superficially similar but whose semantic structure is very different, as shown by the following examples.

— According to Austin's analysis ([1962]), *descriptive* statements and *performative* statements have the same superficial grammatical structure, whilst their meanings and references are very different. The former refer to states of the world, whereas the latter produce the reality they formulate.

— The words *"the path is stony and steep"* and *"the flag is red and black"* are syntactically analogous, yet one can infer from the first that *"the path is stony"* and that *"the path is steep"* whilst it cannot be inferred from the second that *"the flag is red"* and that *"the flag is black"*. Fallacies of composition@ and division can be considered as a particular case of fallacious expression by the form of the expression.

— The similarity of superficial linguistic forms, can lead us to attribute to a word an erroneous semantic characterization. For example, *suffering* and *running*

---

[1] Quoted after Voltaire, *Candide*, Chap. V. New York, Boni and Liveright, 1918. No pag. https://archive.org/stream/candide19942gut/19942.txt (11-08-2017)

are syntactically, intransitive verbs, and, from this analogy, one might think that, like *running*, *suffering* expresses an action.

— The arguments drawn from *derivative words* might also be criticized as cases of fallacies of expression, **S. Derived Words.**

F

Faith

Lat. *ad fidem* argument, *fides*, "faith".

Revealed truth can be used either as arguments, or disputed as claims.

1. Revealed truths as arguments
Revealed truths can be used as arguments justifying some conduct; we follow the Law because our God has given it to us; because He will reward His Followers, the Good, and punish the Wicked. Appeals to religious beliefs may be dismissed as appeals to superstition, **S. Threat.**

2. Revealed truths as claims
Faith and religious mysteries can be opposed to reason and argument. Thomas Aquinas discusses *"whether sacred doctrine is a matter of argument?"* and quotes St. Ambrose's categorically negative response: *"Put arguments aside where faith is sought"* (*ST*, Part 1, Quest.1, Art. 8)[1].

For a believer, revealed truths have precedence over all other forms of truth; trying to demonstrate a revealed truth would degrade it. It should be emphasized that, for a believer, *renouncing to argue* does not imply *submitting to the argu-*

---

[1] Quoted after Thomas Aquinas, *The Summa Theologica*. Benziger Bros, 1947. Translated by Fathers of the English Dominican Province.
http://dhspriory.org/thomas/summa/FP/FP001.html#FPQ1OUTP1 (11-08-2017)

ment from authority, since he or she considers that *authority* has a human origin, while *faith* has a divine origin. Whether religious tradition is of human or divine origin is a controversial issue among theologians.

But the precedence of faith does not invalidate the necessity of argument. Thomas Aquinas distinguishes three kinds of situations, depending on whether one addresses Christians, heretics or unbelievers.

— Where a speaker is addressing a *Christian* audience, argumentation will have two significant uses. The first use is to connect two articles of faith, to show that one can be logically deduced from the other. For example, if somebody believes in the resurrection of Christ, then he or she must believe in the resurrection of the dead. In addition, arguments may be used to extend the domain of faith to deeper truths, derived from the elementary ones.

— When arguing with *heretics* who agree on some point of the dogma, an argument will be built upon this point to show that they must also accept the validity of other connected points. The technique is basically the same as in the previous case.

In both cases, argumentation about matters of faith is based on arguments postulated as true because they are taken from the corpus of revealed truths.

— Where a speaker confronts *unbelievers*, the argument will essentially be *ad hominem*@, showing that their beliefs are contradictory (after Trottman 1999, p. 148-151).

As can be seen, the Angelic Doctor does not exclude situations of deep disagreement from the field of argumentation, **S. Disagreement**.

Fallacies (I): Contemporary Approaches

1. *Fallacy*: The word

1.1 The Latin *fallacia*

Etymologically, the noun *fallacy* and the adjective *fallacious* come from the Latin *fallacia*, which means "deceit", referring to a "trick", or even a "spell". This deceit can be defined as a verbal deceit, as expressed by the adjective *fallaciloquus*, "[he] who deceives by words, astute" (Gaffiot [1934], *Fallaciloquus*). The corresponding verb *fallo, fallere* means "to deceive someone", and according to the contexts, "to disappoint the expectations of someone, to betray the word given to the enemy, to break his promises" (*id.*, *Fallo*). These meanings show that etymologically the word *fallacy* does not refer to a logical or dialectical mistake but to an interactional manipulation.

1.2 *Paralogism, sophism, fallacy*

Fallacy — The word *fallacy* has at least two meanings. On the one hand, the very general meaning of "erroneous belief, false idea" (Webster, Art. *Fallacy*).

On the other hand, it refers to an "invalid" argumentation or reasoning, the conclusion of which does not follow from the premises, and which may therefore be misleading or deceptive (*ibid.*).

Being an ordinary word, there is no guarantee that *fallacy* refers to a unique stable, highly connected domain of reality that can be systematized. It is not a priori obvious that fallacies can be theorized more coherently than, for example, errors, deceptions, blunders or carelessness, just to mention some relatively close terms.

Paralogism@ has a precise and restricted technical use, in which it refers to a formally invalid syllogism. This term is of little use outside this specialized field.

Sophism@ refers to a deliberately misleading discourse, using paralogism or any other maneuver. This imputation of bad intention is not necessarily present when one speaks of paralogism or fallacious discourse.

2. Hamblin, *Fallacies*, 1970

Hamblin revived the theory of fallacies in his book, *Fallacies* (1970). As Perelman revived ancient rhetoric, or rhetorical argumentation, from Aristotle's *Rhetoric*, Hamblin reactivated the other Aristotelian source of argumentation as a critical theory from the *Topics* and the *Sophistical Refutations*. Following Hamblin, the study of argumentation developed as a critique of bad reasoning, fallacious and specious arguments.

The Argumentation within Language or the Natural Logic theories do not approach the critical question. The *New Rhetoric* proposes an ideal critical instance, *the universal audience*, in a different perspective from that generally implemented in fallacy theories.

Hamblin gives the following definitions of *fallacy*. It should be noted that this conceptual definitions is parallel to the lexicographical definition given above.

Fallacy$_1$ — The ordinary meaning of "erroneous belief" has been dismissed by Hamblin: "a fallacy is a fallacious *argument*. [...] In one of its ordinary uses, of course, the word 'fallacy' means little more than 'false belief'; but this use does not concern us." (1970, p. 224; italics in the text).

Hamblin adds that, "there are several varieties of fallacies, or particular fallacies which have received special names, but which are not really logical fallacies at all, but merely false beliefs" (*id.*, p. 48; capital in the text). In this sense, the word corresponds to a "false concept", which may clearly be itself deceitful, S. **Expression**.

Fallacy$_2$ — In this second sense, the word *fallacy* designates the *counterfeit of argument*[1]:

---

[1] To use a title of W. Ward Fearnside & William B. Holther (1959). *Fallacy: The Counterfeit of Argument*, quoted in Hamblin 1970. P. 11.

> A fallacious argument, as almost every account from Aristotle onwards tells you, is one that *seems to be valid* but *is not* so. (*Id.*, p. 12)

This definition brings up some questions, the first one being:
> What it is for an argument to *seem valid?* The term 'seems' looks like a psychological one, and has often been passed over by logicians, confirmed in the belief that the study of fallacies does not concern them. (*Id.*, p. 253)

Following Frege, mathematicians have de-psychologized logic@. Axiomatized logic is no longer a theory of thought, **S. Logic** (I). From this point of view, truth is one, and if error is multiple, it is precisely because it is related to psychology. There is no logical theory of error. In short, a fallacious argument is an argument or argument that seems valid to *a negligent or untrained* reader; it is the reader who has a problem.

In the definition of a "fallacious argument" given above, Hamblin refers to a fallacious *argumentation*, since he speaks of validity. In English however, the word *argument* can also denote an argumentation. A *fallacy₁* is an "erroneous belief" which can obviously serve as a premise for an argumentation. Since ordinary *argumentation* demands the truth of the *arguments*, an argumentation based on a false premise is legitimately deemed to be fallacious; this is an authentic *fallacy₂*. In other words, from this *fallacious argument* (erroneous belief), derives a *fallacious₂ argumentation*, a *fallacy₂*. "To appear to be true or valid", "to look honest, solid, admissible, credible" is a property common to arguments and argumentations. There is no difference between the first former and the latter which would enable us to reject one without forcibly rejecting the other. Like argumentation, fallacy is a unitary phenomenon, both substantial and formal.

The lexical / conceptual distinction between substantial fallacies (fallacy₁) and formal fallacies (fallacies₂) is generally taken up in the theory of argumentation, as in the following text:
> Assumptions, principles, and ways of looking at things are sometimes called fallacies. Philosophers have spoken of the naturalistic fallacy, the genetic fallacy, the pathetic fallacy, the fallacy of misplaced concreteness, the descriptive fallacy, the intentional fallacy, the affective fallacy, and many more. And outside of philosophy, we also hear sophisticated people using the term 'fallacy' to characterize things which are neither arguments nor substitutes for arguments. For example, the China expert Philip Kuhn speaks of the *hardware fallacy*. This, according to him, is the mistaken assumption common among Chinese intellectuals that China can import Western science and technology without importing with it Western (i.e., decadent) values as well. (Fogelin, Duggan 1987, p. 255-256)

The distinction between form and substance is not easy to maintain. For example, the genetic fallacy, given here as an example of "a way of looking at things", that is, a *substantial* fallacy (fallacy₁) can be seen as referring to an argumentation (fallacy₂) which evaluates beings and things according to their origin, and which Hamblin admits in his list of authentic *formal* fallacies.

3. Lists of fallacies

In the chapter entitled "Standard Treatment", Hamblin proposes four lists of fallacies.

(i) The list of Aristotle in the *Sophistical Refutations*, S. **Fallacy (III)**.
(ii) The fallacies or arguments *ad* —, a list of modern fallacies, designated by Latin labels of this form, S. *Ad* — **Arguments**.
(iii) The syllogistic paralogisms, S. **Classical logic (III); Paralogisms**.
(iv) The fallacies of scientific method.

Under this final heading Hamblin proposes the following six cases:

— Fallacy of *simplism* or *pseudo-simplicity*, (*id.*, p. 45), according to which the simplest explanation is necessarily the best.

— The fallacy of *exclusive linearity* (*ibid.*), assumes that a series of factors is ordered according to a strictly linear progression. The fallacy of linearity neglects the existence of thresholds and ruptures in the development of phenomena. This is an extrapolation fallacy: for example, the conductivity of a metal or a solution decreases steadily and then drops abruptly when approaching absolute zero temperature.

— The *genetic fallacy* (*ibid.*), ostracizes an idea or a practice on the basis of their source or origin: "*This is exactly what the Bad Guys Group says*", S. **Authority**.

— Fallacy of *invalid induction* (*id.*, p. 46), S. **Induction; Example**.

— Fallacy of *insufficient statistics* (*ibid.*).

— *Hasty generalization* (*ibid.*), which may correspond to the fallacy of accident or induction, S. **Accident; Induction**.

— The *naturalistic* fallacy (*id.*, p. 48). Moore defines this fallacy of valuing the "natural" as follows:

> To argue that a thing is good *because* it is "natural," or bad *because* it is "unnatural," in these common senses of the term, is therefore certainly fallacious; and yet such arguments are very frequently used. (Moore, 1903, §29; italics in the original)

This amounts to saying that the word *natural* has a generally positive argumentative orientation, but not for the author's group. The naturalistic fallacy goes hand in hand with a range of reciprocal fallacies, named after the antonyms of "natural": *culturalist* fallacy, etc. S. **Orientation**.

Fogelin (see above) adds:

— The *descriptive* fallacy, a form of fallacy of expression, S. **Expression**.

— The fallacy of *misplaced concreteness*. Whitehead introduced this expression in the field of the philosophy of science, to denote the error of forgetting

the distinction between the model and reality, and more generally between words and things.
— *The intentional* fallacy, is invoked in literary analysis, to condemn the interpretation of a work based on intentions attributed to the author. It should be noted that, conversely, in the field of law, the argument based on the *intentions@ of the legislator* is recognized as being entirely valid.
— The *emotional* and *pathetic@* fallacies, **S. Emotion; Pathos**.

Many of these so-called fallacies view scientific language as the norm of ordinary language, and represent ordinary arguments as unsatisfactory scientific arguments.

4. Informal Logic and Pragma-Dialectic

From the 1970s onwards, following Hamblin, the literature on fallacies underwent considerable developments, particularly within the theoretical frameworks of *Informal Logic* and *Pragma-Dialectic*. These works have clearly highlighted the necessity of systematically taking the pragmatic conditions under which ordinary language reasoning operates into account.

In the Informal Logic framework, Woods and Walton represent the first generation to follow on from Hamblin. They questioned the logical and pragmatic conditions of validity ordinary arguments (Woods and Walton 1989, 1992). Woods (2013) focuses on "errors of reasoning", insisting on the necessity of formalism (Woods 2004). Walton has in particular developed and systematized a new vision of argument schemes including their "rebuttal factors" (Walton & *al.*, 2008). Argumentation is consequently defined as a default reasoning@, which is both consistent with, and goes beyond Toulmin's approach, **S. Layout**.

This development of a *counter-discourse based criticism* of argument is different from the *rule-based criticism* of argument developed by the pragma-dialectical school. The Pragma-Dialectic orientation can be read as follows, "*if you want your discussion to progress towards a decent resolution, you had better follow such and such a procedure and avoid such and such counter-productive, that is, fallacious, maneuvers*". The felicity conditions of the argumentative exchange are dependent upon the observation of ten rules@

> In principle, each of these ten discussion rules constitutes a separate and different standard or norm for critical discussion. Any infringement of one or more of the rules, whichever party commits it and at whatever stage in the discussion, is a possible threat to the resolution of a difference of opinion and must therefore be regarded as an incorrect discussion move. In the pragma-dialectic approach, fallacies are analyzed as such incorrect discussion moves in which a discussion rule has been violated. A fallacy is then defined as a speech act that prejudices or frustrates efforts to resolve a difference of opinion and the use of the term "fallacy" is thus systematically connected with the rules for critical discussion. (van Eemeren and Grootendorst, 1995, no pag.)

5. Methodological remarks

Natural argumentation develops in contexts where the question of truth is suspended. It might also arise when a decision has to be taken as a matter of urgency, even when all necessary information is not available.

Wanting to *solve a dispute rationally* is the manifestation of a specific and legitimate desire, which is obviously not a prerequisite for arguing. One can also argue to solve the dispute to one's own advantage, at all cost, to end this affair; or to uphold the truth, or to protect one's interests; to spread one's emotions, to satisfy one's ego, to fill time, for enjoyment... One might also be interested not in *solving* but rather in *deepening* the difference of opinion. If a new issue has just arisen, for example, it may be more productive and more rational, to properly articulate the problem, rather than to prematurely seek to eliminate it. One might also be interested not in *solving*, but rather in *increasing* the difference of opinion. If a new issue has just arisen, for example, it may be more productive, even more rational, to properly articulate the problem and the dispute, rather than seeking to eradicate any discussion.

There are *interesting* arguments, which contain a portion of truth, the whole truth being unknown and not entirely in a single camp. On the other hand, a speaker can put forward a weak or even doubtful argument, in an exploratory way, while explicitly emphasizing its uncertain character. It is therefore impossible to introduce a definition of fallacies based on truth and validity as a single regulatory ideal in all argumentative situations.

Discursive atomism —To criticize an argument, the analyst must first delineate the discursive passage in which this argument is intuitively seen. This basic operation must itself be technically justified, **S. Tagging; Indicators**. On the other hand, the quality of the argument must be assessed in relation to the argumentative question on which it depends, including the replicas introduced by opponents, **S. Stasis; Question; Relevance**.

The arbitrator is also a player — The diagnosis of fallacy is supposedly made by the logician who has the role of fulfilling the evaluator's "meta" function in a neutral and objective way. That is to say that he or she must fulfill this role as if he or she had no interest in the controverted issue, but only an interest in the correction of discourse evaluated according to a priori rules and principles. As Hamblin points out, this position is untenable in the case of "actual practical argument', (1970, p. 244), **S. Norms; Rules; Evaluation**. The evaluators of social arguments are by no means excluded from the argument; they are also *participants* like any others.

Natural language eliminated — These elements — an atomistic approach, an unbiased arbitrator, augmented by a strong reductionist tendency —, all feature in the practical advice by which the *Encyclopedia of Philosophy* concludes the entry on fallacies:

> As Richard Whately remarked "...a very *long* discussion is one of the most effective veils of Fallacy: ... a Fallacy which when stated barely... would not deceive a child, may deceive half the world if *diluted* in a quarto volume." (*Elements of Logic*, p. 151). Consequently, an important weapon against fallacy is condensation, extracting the substance of the argument from a mass of verbiage. But this device too has its dangers; it may produce oversimplification, that is, the fallacy *a dicto secundum quid*, of dropping relevant qualification. When we suspect a fallacy, our aim must be to discover exactly what the argument is; and, in general the way to do this is first to pick out its main outlines, and then to take into account any relevant subtleties or qualifications. (Mackie 1967, p. 179; italics in the original).

Even if one were to agree with the method, the problem of implementing the proposed solution would remain unsolved, nothing being said about how to deal with natural language and speech, seen somewhat contradictorily as an insubstantial and vicious medium.

Natural language, the common vehicle of argument, is accused of dissolving logic in an insignificant verbiage@ which serves to mask unsavory human interests. Thus, a sustained war against language would be the price to pay for a correct determination of sound arguments, that is, for eliminating fallacies. Nonetheless, it may be noted that natural language is to natural argument what air resistance is to the flight of the "light dove":

> The light dove, in free flight cutting through the air the resistance of which it feels, could get the idea that it could do even better in airless space. Likewise, Plato abandoned the world of the senses because it posed so many hindrances for understanding, and dared to go beyond it on the wings of ideas, in the empty space of pure understanding. (Kant, [1781], p. 129).

Natural language is not an *obstacle*, but the *condition* of ordinary argumentation.

The diagnosis of fallacy as an argumentative issue — Criticism of argument does not escape argumentation. First, it has to be justified. This justified diagnosis is just a move in a longer game, not the final one, not the terminal charge. This justified diagnosis is just one move in a longer game, it is by no means a final, conclusive or terminal act. In a subsequent move, the so-called "fallacious arguer" can exercise his or her right of reply, and try to rebut the accusation of fallacy. This reply can itself be challenged, and there is no rule as to who closes the game, **S. Rules; Norms; Evaluation**.

Fallacies (II): Aristotle's Foundational List

Argumentation studies are related to two Aristotelian sources, on the one hand, the rhetorical and dialectical theories of the *Rhetoric* and the *Topics*, and on the other hand, the critical analysis of fallacious sequences (fallacies, apparent enthymemes) in the *Prior Analytics*, the *Rhetoric* and mainly in the *Sophistical Refuta-*

tions (Woods 2014). This last line is the basis of the "standard treatment of the fallacies" as reconstructed by Hamblin (*Fallacies*, 1970).

The definitions from the *Sophistical Refutations* are taken up in all works dealing with fallacious arguments. The title, *Sophistical Refutations*, is ambiguous. Firstly, according to the classic joke, it is not 'an adequate description of the contents of the book', that is to say, a set of *refutations* (concerning well defined theses) which would be *sophistical*, but a refutation of the *Sophists'* arguments. The book analyses and rejects the refutations as practiced by the sophists, or "how the sophists refute".

Aristotle draws a broad distinction between two sets of paralogisms. He defines, on the one hand, paralogisms that "depend on the language used", and on the other, paralogisms which are "independent of language" (SR, 4). The "language" referred to is the language used in a dialogue, as practiced by the dialecticians or the sophists.

The *Rhetoric* lists ten "lines of argument that form the spurious enthymemes" (*Rhet*, ii, 24, 1400b35-01a5, RR 379), clearly related to language. Note that this parallelism enthymeme / spurious enthymeme may lead us to believe that the preceding enthymemes, as enumerated in *Rhet.*, II, 23 are valid, which is not the case. **S. Collections (II); S. Expression**.

1. The fallacies in the *Sophistical Refutations*

The six Aristotelian linguistic fallacies are listed in the first column of the following table:

Six fallacies "*dependent on language*" or "*verbal fallacies*" (lat. *in dictione*) (*RS 4* (=165b-167a))	
1. Homonymy	Lat. *æquivocatio; ambiguity, equivocation* — **S. Ambiguity**
2. Amphiboly	Gr. [*amphibolia*] — S. **Ambiguity**
3. Combination	lat. *fallacia compositionis, composition of words* — **S. Composition and Division**
4. Division of words	lat. *fallacia divisionis*, **S. Composition and Division**
5. Accent	lat. *fallacia accentis* ; *wrong accent* — **S. Ambiguity**
6. Form of expression	lat. *fallacia figuræ dictionis, misleading expression* — **S. Expression**

This terminology may seem obscure, but its purpose is perfectly clear; it serves to establish, through a critique of language and discourse, the basic principles of a "logical grammar for argumentation", supporting the production of reasoned texts and speeches anticipating their criticism.

The seven fallacies considered "independent from language", are listed in the first column of the following table

Seven fallacies "independent of language", RS 4 (=166b-168b) (Lat. *extra dictionem*)	
1. "Accident"	Lat. *fallacia accidentis* — S. **Accident; Definition; Categorization.**
2. "The use of an expression absolutely or not absolutely but with some qualification of respect or place, or time, or relation"	Lat. *a dicto secundum quid ad dictum simpliciter* — S. **Circumstances;** *Distinguo.*
3. "That which depends upon ignorance of what 'refutation' is"	Lat. *ignoratio elenchi*; *misconception of refutation*; *evading the question* — S. **Question; Relevance; Red Herring**
4. "That which depends upon the consequent"	Lat. *fallacia consequentis* — S. **Implication; Causality.**
5. "That which depends upon assuming the original conclusion"	Lat. *petitio principii*; *assumption of the original point*, *begging the question* — S. **Vicious Circle**
6. "Stating as cause what is not the cause"	Lat. *non causa pro causa*, *non cause as cause* — S. **Cause – Effect**
7. "The making of more than one question into one"	Lat. *fallacia quæstionis multiplicis*, *many questions*; *complex question* — S. **Many questions**

These fallacies are actually methodological mistakes.

2. Fallacies, inferences and dialectical games

In contemporary terminology, an invalid inference is often referred to as a fallacy. According to Hintikka, the Aristotelian concept of fallacy refers to something invalid, but not to an invalid inference:

> The error in thinking that the traditional fallacies are faulty inferences is what I propose to dub "the fallacy of fallacies". It is the fallacy whose recognition will, I hope, put a stop to the traditional literature on so-called fallacies. (1987, p. 211)

In other words, a fallacy cannot be simply defined as, "a fallacious argument"; just some, but not all fallacies can be "thought of as mistaken logical or conceptual inferences" (*ibid.*). Hintikka considers that a fallacy is essentially a move which transgresses a rule in a dialectical game, dialectical games being defined as "information-seeking questioning processes (interrogative games)" (*ibid.*). It is in this sense that the concept of fallacy has been taken up in the pragma-dialectical theory.

Linguistic fallacies examine the conditions a proposition must fulfill in order to qualify as a premise in a correct syllogistic inference. The fallacy of *accident* is the consequence of an error in the methodology of definition. *Misconception of refutation* reflects a poor understanding of the issues involved in the discussion and the problem. *Many questions* is also a forbidden move in dialectical games, where problems must be serialized to avoid implicit agreements. These different cases clearly demonstrate the non-inferential nature of fallacies, and, for the latter two, their links to rules-based dialogue games.

Fallacies (III): From Logic and Dialectic to Science

1. Francis Bacon, *Novum Organum*, 1620

Hamblin considers Francis Bacon's *New Organon* as a psychological turning point in the conception of fallacies (Hamblin 1970, p. 146; Walton, 1999). Bacon presents his concept of "idol" as the scientific counterpart of logical or dialectical fallacies. An idol is an obstacle to the (inductive) edification of scientific knowledge.

The word *idol* comes from a Greek term meaning "simulacrum, phantom" (Bailly, [*eidolon*]). According to Bacon, a fallacy is a simulacrum, a phantom of argument, produced under the influence of towering idols, defined as false Gods altering human reasoning:

> XXXIX. Four species of idols beset the human mind, to which (for distinction's sake) we have assigned names, calling the first Idols of the Tribe, the second Idols of the Den, the third Idols of the Market, the fourth Idols of the Theater. ([1620], p. 20)

The Idols of *the Tribe*, that is of the whole of humanity. These idols are the deformations imposed upon reality by the innate structure of the human mind, which is not a *tabula rasa* but an "uneven mirror" (*id.*). Its *a priori* categories distort reality.

The Idols of *the Den* are the product of the education and history of each individual, that is to say, prejudices or other evidences, exerting their powers through "Authority" (*id.*, p. 21).

The Idols of *the Market place* are the words themselves, which "still manifestly force understanding, throw everything into confusion, and lead mankind into vain and innumerable controversies and fallacies" (*id.*, p. 21).

The Idols of the *Theater* correspond to "the various dogmas of peculiar systems of philosophy, and also from the perverted rules of demonstration" (*id.*, p. 22). These Idols include fallacious inferences as well as substantial fallacies.

2. John Locke, *An Essay Concerning Human Understanding*, 1690

In a brief section of his *Essay*, Locke reflects "on *four sorts of arguments*, that men in their reasonings with others, do ordinarily make use of to prevail on their assent or at least so to awe them as to silence their opposition" ([1690], p. 410). This definition of an argument perfectly suits what is a rhetorical argument as pressure exerted on the audience, S. **Logos – Ethos – Pathos**. These four sorts of arguments are *(id.,* p. 410-412):

> "The *argumentum ad verecundiam*", S. **Modesty**
> "The *argumentum ad ignorantiam*", S. **Ignorance**
> "The *argumentum ad hominem*", S. ***Ad hominem***
> "The *argumentum ad judicium*", S. **Matter.**

Locke rejects the first three arguments on the ground that, at best, they "may dispose me, perhaps, for the reception of truth, but help me not to it":

> For, 1. It [*ad verecundiam*] argues not another man's opinion to be right because I, out of respect, or any other consideration but that of conviction, will not contradict him. 2. It [*ad ignorantiam*] proves not another man to be in the right way, nor that I ought to take the same way, because I know not a better. 3. Nor does it follow that another man is in the right way because he has shown me that I am in the wrong. I may be modest, and therefore not oppose another man's persuasion; I may be ignorant, and not be able to produce a better; I may be in error, and another may show me that I am so. This may dispose me, perhaps, for the reception of truth, but helps me not to it *(id.,* 411).

The concept of fallacy is redefined independently of any Aristotelian consideration. The only valid arguments are arguments *ad judicium*, that is to say "proofs drawn from any of the foundations of knowledge or probability" *(ibid.)*; truth "must come from proofs and arguments and light arising from the nature of things themselves" *(id.,* 412). Note that whilst the fallacious arguments correspond to argument schemes, the argument *ad judicium* does not correspond to just one argument scheme but to any kind of argument recognized as scientifically valid.

Leibniz ([1765]) nuanced this strict vision of fallacious arguments (see the above mentioned entries).

Fallacies (IV): A Moral and Anthropological Perspective

Antoine Arnauld and Pierre Nicole conclude the third part of their *Logic, or the Art of Thinking* (1662) with two chapters devoted to sophisms and bad reasoning. Chapter XIX, "*Of the different ways of reasoning which are called Sophisms*", takes up the Aristotelian fallacies; Chap. XX, "*Of the bad reasonings which are common in Civil Life and in Ordinary Discourse*" repositions the concept of *fallacious reasoning* an anthropological and moral issue about *fallacious discourse and discussion.*

1. The Aristotelian fallacies

The list of "ways of evil reasoning that we call sophisms" merges the Aristoteli-

an linguistic and non-linguistic fallacies, S. **Fallacies (III)**
The linguistic fallacies are grouped under two headings. The list does not mention the fallacy of many questions, and adds two new types of fallacies independent of language, "incomplete enumeration", and "defective induction".

2. On the bad reasonings in civil life

Chapter XX "*Of the bad reasonings which are common in Civil Life and in Ordinary discourse*" is much more original. Its consists of two parts:
— Of the sophisms of self-love, of interest, and of passion.
— Of the false reasonings which arise from objects themselves.
These sophisms and bad reasoning no longer reflect logical or scientific concerns, and have no connection with dialectics. On the basis of a thorough and hypercritical description of the discussant's concrete behavior, they emphasize the difficulties in bringing a debate to a successful completion and show how deceitful and useless dispute can be when truth is at stake. More than an appeal to follow rules for discussion, the conclusion is an ascetic appeal to moral reformation of the disputants. It should be kept in mind that the religious and philosophical disputes over Jansenism and Cartesianism form the background of the disillusioned discussions mentioned in this chapter.

In the following, the various sophisms and bad reasoning are designated by an expression extracted from their definition.

2.1 "Of the sophisms of self-love, of interest, and of passion"
(1) "To take our interest as the motive for believing a thing" — The first of the causes which determine belief is the spirit of belonging to "some nation, or profession, or institution" (*Id.*, p. 268). Beliefs are not determined by truth and reality, but by the social position of the believer. The disputant borrows his beliefs from the group in which he finds "his interest" and his identity.

(2) "[The] delusions of the heart" (*Id.*, p. 269) — This sophism expresses the *ad passiones* fallacies of love and hate (*ad amicitiam, ad amorem, ad odium*), it is a variant of pathetic@ argumentation:

> 'I love him, therefore, he is the cleverest man in the world; I hate him; therefore, he is nobody'. (*Ibid.*)

(3) Those "who never distinguish their authority from reason", and
> decide everything by a very general and convenient principle, which is, that they are right, that they know the truth; from which it is not difficult to infer that those who are of their opinion are deceived, — in fact, the conclusion is necessary. (*Ibid.*)

The claim to the truth of the self-centered person comes from immediate certainty (in the profane as in the sacred domain), whereas it would require an argument, S. **Authority; Humility**. This can be read as a criticism of the Cartesian's criterion of truth, as clear and distinct ideas. Interest and self-love better determine clarity and distinctness than truth does.

(4) "The clever man['s]" sophism is related to the preceding one:
> *If this were so, I should not be a clever man; now, I am a clever man; therefore, it is not so.'* (*Id.*, p. 270)

Enthymemes:
> '*What*,' said they, '*if the blood circulates, [...] if nature does not abhor a vacuum [...] — I have been ignorant of many important things in anatomy and in physics. These things, therefore, cannot be*'. (*Ibid*).

This is another fallacy *ad passiones*, the fallacy of pride, *ad superbiam*.

These first four "sophisms" are not precisely sophisms insofar as they are self-deceiving as well as other-deceiving. Nor are they correctly called fallacies insofar as they are not public reasoning, propositionally expressed. Their premises remain unsaid, perhaps unconscious:
> I'm a Syldavian, Syldavians are always right, therefore, I'm right.
> I'm right, therefore my opponent is wrong.
> I hate him; therefore, he is a nobody.
> I know everything, thus what I don't know is false.

Interests, inflated egos and passions, are epistemological obstacles ingrained in human nature.

Chapter XIX reiterates the classical belief that education about argument requires thorough knowledge of language and a good training in logic. Chapter XX adds that first of all, the arguer has to work on himself (sophisms (1)-(4)) and avoid the pitfalls of argumentative interactions (sophisms (5)-(9)): This is the substantial content of the following subset, which complements the first moral and psychological subset with factual observation of the interactional behavior of seasoned arguers.

(5) "Those who are in the right, and those who are in the wrong, with almost the same language make the same complaints and attribute to each other the same vices" (*Id.*, p. 271). From this empirical observation follows a recommendation to the wise and thoughtful, about how to properly advocate truth in a controversy.

> *First Recommendation* to the arguers: don't start a debate before having "[thoroughly establish] the truth and justice of the cause which they maintain".

Only when these rules have been correctly applied can one shift to a meta-discussion about the bad argumentative manners of the opponent. This of course presupposes that one can decide that the rules have been correctly applied.

(6) "The spirit of contradiction", is a "malignant and envious disposition":

Fallacies (IV): A Moral and Anthropological Perspective

> "*Someone else said such a thing; it is therefore false. I did not write that book; it is, therefore, a bad one*". This is the source of the spirit of contradiction so common amongst men, and which leads them, when they hear or read anything of another, to pay but little attention to the reasons which might have persuaded them, and to think only of those which they think may be offered against it. (p. 272)

(7) "The spirit of debate"
> Thus, unless at least we have been accustomed by long discipline to retain the perfect mastery over ourselves, it is very difficult not to lose sight of truth in debates, since there are scarcely any exercises which so much arouse our passions. (p. 277),

Observations (6) and (7) have a clear link with the sin of *contentio*, **S. Fallacies as Sins of the Tongue**.

From the observation that "speaking of ourselves, and the things which concern us" can "excite envy and jealousy" comes a new recommendation: when advocating truth, self-exposure should be minimized, and the arguers should rather "seek, by hiding in the crowd, to escape observation, in order that the truth which they propose may be seen alone in their discourse" (p. 273).

(8) "The Complaisant"
> For as the controversial hold as true the contrary of what is said to them, the complaisant appear to take as true everything which is said to them. (p. 278)

This sophism of acceptance without examination, at least of refusal to take a position, corresponds exactly to the *ad verecundiam* fallacy of Locke, **S. Modesty**. This is different but nonetheless related with the blamed character alluded to in (7), who "in the midst of [the discussion] become obstinate and are silent, affecting a proud contempt, or a stupid modesty of avoiding contention" (p. 277). **S. Modesty; Contempt**.

(9) "The determination to defend our opinion" leads us to
> no longer to consider whether the reasons we employ are true or false, but whether they will avail to defend that which we maintain. We employ all sorts of reasons, good or bad, in order that there may be some to suit everyone. (p. 279).

The whole section closes with a kind of *final recommendation*:
> To have no end but truth, and to examine reasonings with so much care, that even prejudice shall not be able to mislead us. (p. 276)

As observed in (5), each discussant will say that is precisely what he or she does. The attempt to expose the sophism seems to be doomed from the start, as if, in a conflictual dialogue, we were condemned never to know who speaks the truth.

2.2 "Of the false reasoning which arise from objects themselves"
This section insists on the following points:

— There is only a small margin between truth and error; cf. supra (5):
> In the majority of cases, there is a mixture of truth and error, of virtue and vice, of perfection and imperfection (p. 277)

— Rash induction also applies to human affairs; cf. supra §1, "incomplete enumeration", and "defective induction":
> [Men] judge rashly of the truth of things from some authority insufficient to assure them of it, or by deciding the inward essence by the outward manner. (p. 284)

Decisions are made on the basis of "exterior and foreign marks." (*ibid.*), that is peripheral arguments.

— "We rarely avoid judging purposes by the event", a very relevant point:
> If somebody succeeds, he had carefully planned his deeds; if he fails, he miscalculated. (p. 283)

No distinction is made between "the fortunate and the wise." (*Ibid*)

— About "pompous eloquence", **S. Verbiage**.

Fallacies as Sins of the Tongue

When taking sides truth and rationality, fallacy@ theory calls for a criticism@ of language and speech as vectors of error and deceit, **S. Evaluation; Norms**. Other cultures gave other foundations to the criticism of speech. Reconstructing the history of the "sins of language" in the Middle Ages, Casagrande & Vecchio (1991) have demonstrated the link between *speech* and *sin*. The issue then was not to build a *rational* discourse, but a *sinless*, "impeccable" discourse, if not a holy one. The nature of the misconduct has shifted: what was declared sinful in the name of religion is now considered to be fallacious or sophistical in the name of rationality. Whether sin or fallacies, salvation of the soul or rational guidance of the mind, it is always a matter of regimenting verbal behavior, disciplining one's speech and pen.

Casagrande and Vecchio synthesize data from various medieval treatises into a list of fourteen sins. This list can be widely interpreted in terms of misleading interactional argumentative behaviors. These sins-fallacies intend to rule the interaction in a religious context where hierarchy and valorization of authority occupy a central position, **S. Politeness**

Making a connection between fallacy theory and "sins of the tongue" is not indulging in any kind of *derisio*, neither to one nor to the other party. This connection, on the contrary, is intended to show how deep the anthropological roots of discourse criticism are.

1. Sins against truth

1.1 Lying
Telling the truth, all the truth and nothing but the truth is certainly a basic commitment for a non-fallacious debate. Lying, as saying something false to someone who has no access to truth, is a sin in the system of theological norms, and a fundamental violation of Grice's cooperative@ principle S. **Manipulation**.

1.2 Aggravated lying: perjury and false testimony
In judicial rhetoric, oath and testimony, two major instruments to establish the truth, are considered to be non-technical proofs, S. **Technical and Non-technical proofs**. Their violation corresponds to two aggravated lies, the sins of perjury, *perjurium*, and false testimony, *falsum testimonium*.

2. Six sins of interaction

2.1 Against disputes
Rivalry, conflict, fight (*contentio*), and discussion (*disputatio*) are names denoting the very activity of *disputing*. It can thus be said that arguing is potentially considered sinful at its very core. It is the sin of the intellectual monks, and no doubt, that of Abelard. The passage from the peccaminous to the fallacious is explicit in the Port-Royal *Logic*, in which the excessive love of dispute, the spirit of contradiction, is condemned as a sophism of self-esteem (n°6 and 7), a fundamental feature of the character of "those who contradict" (Arnauld and Nicole [1662], p. 272); S. **Fallacies (IV)**. The debate is subject to a moral imperative: the contradiction must be genuine, not "malignant and envious" (*ibid.*) - or, to move on to judicial pathology, querulous. Such a debate might be legitimately declined.

We then discern two families of sins of interactional positioning, on the one hand, the sins "towards the other", the partner with whom we argue (§ 2.2), and, on the other hand, the sins committed "towards oneself" as a speaker (§2.3). In both cases, it is a question of banishing illegitimate treatments of the partners of the interaction, S. **Politeness**.

2.2 Three kinds of sins towards the partner
Undue negative treatment: offensive remarks (*contumelia*) or slander (*detractio*). These two sins are a form of personal attacks, or *ad personam* fallacies. The *derisio*, as a contemptuous mockery, could be associated with to this fallacy, S. ***Ad personam***; **Contempt**.

Negative treatment under the cover of the positive: this is the mechanism of refutation by self-evidence as implemented through irony@, *ironia*. This intention to hurt the other is dealt with only laterally in contemporary theories of irony, S. **Irony**.

Undue positive treatment: flattery (*adulatio*), and even simple praise (*laudatio*). These two sins involve the same interactional mechanisms as found in the fallacy of modesty@, *ad verecundiam*, where the speaker humiliates himself unduly before his partner. *Adulatio* and *laudatio* encourage pride, and pride is a sin. Logic, religion, and politeness speak with one voice, S. **Modesty**; **Politeness**.

2.3 Two kinds of sins against oneself
Undue positive treatment, in other words, boasting, *iactantia*. This ethotic sin stigmatizes the projection in the discussion of an overly positive self-image, S. **Ethos**. According to politeness@ theory, the *iactantia* sins against modesty.

Undue negative treatment is the symmetrical sin of the sin of undue positive treatment of the partner, S. **Modesty**. The *taciturnitas*, sin of the person who keeps silent when he should speak, can be related to the *ad verecundiam* fallacy in which "human respect" inhibits criticism.

4. Murmuring: a sophism of insubordination
A person who complains against authority commits the sin of murmur (*murmur*), S. *A fortiori*. A person who refuses to yield to the force of the best argument having little to oppose to it, save an intimate conviction or sense of justice, is guilty of the same kind of fallaciousness. S. **Dissensus**; **Rules**. Insubordination is irrational, illegal, peccaminous.

5. The sin of eloquence
Eloquence, seen as an abundance of words, amplification, repetition, magnification, is the source of all fallacies, S. **Verbiage**. The same evaluation should apply to *idle speech* (*vaniloquium*), as well as to *chatter* (*multiloquium*).

6. Flaring into a passion: *ad passiones*
Some remaining sins are difficult to connect to the problematic of fallacies, perhaps because they directly involve the relation to the sacred: the prohibition of *obscene words* (*turpiloquium*), *blasphemy* (*blasphemia*) and the *curse* (*maledictum*). Nonetheless, these sins can have an *ad personam* function. Above all, they have an emotional import, so they certainly relate to the *ad passiones* group. Blasphemy is anger against god, and cursing, anger against the other; obscene words can be used to support many passions, including insulting.

To sum up, the theory of the sins of language is a critical theory of discourse taking into account:
— The "non-technical" problems of lying or attesting the truth.
— The spirit of the discussion.
— The relative interactional positions of the participants.

7. The "rules of the devil"

This list of fallacies-sins does not mention violations of logical rules, such as the assertion of the consequent (confusion of necessary and sufficient conditions, S. **Deduction**). One would think that it is because the logical domain, by nature escapes the religious norm. In the Muslim tradition, however, one can find the vocabulary of sin applied to paralogisms, which Al-Ghazali considers as "rules of the devil" (*Bal.*, p. 171; *Deg.*). A medieval *exemplum* also puts the logician into hell, assimilated to the sophist, S. *Exemplum*.

False Cause ▶ Cause-Effect

Figure

The term *figure* is used in syllogistic, in fallacy theory and in rhetoric with different meaning

1. Figures of the syllogism

The figures of the syllogism@ correspond to the different forms of the syllogism, according to the position of the middle term in the premises.

2. Fallacy of "figure of speech"

The fallacy of misleading expression@ is sometimes referred to as the fallacy of figure of speech.

3. Figures of Rhetoric

The figures of rhetoric are variations in the manner of signifying "which give to the discourse more grace and vivacity, luster and energy" (Littré, *Figure*). Dictionaries of rhetoric include entries in the field of argumentation, even though they are primarily concerned with literary rhetoric. For example, the dictionary "*Gradus. The literary processes — Dictionary*" (Dupriez, 1984), includes the entries *argument, argumentation, argument, deduction, enthymeme, epicheirema, example, induction, refutation, paralogism, premise, reasoning, sophism*… These basic concepts within the field of argumentation do not belong specifically to the literary domain.

The word *figure* is used to cover tropes and figures of speech. *Metaphor@, irony@ metonymy@* and *synecdoche* are considered to be the "four master tropes". The *metaphor* as a *model* has a clear argumentative function. There is correspondence between the mechanisms of *metonymy* and *synecdoche* and those that legitimize the passage from an argument to a conclusion. Moreover, *irony* argues from a self-evident situation.

The expression *figure of speech* can actually refer to any salient and recurrent form of discursive organization. This is why the *enthymeme@* can be considered as a figure, the *enthymemism*, along with *refutation*@ or *prolepsis*@. Other figures of rhetoric, from *antanaclasis* (S. **Orientation**) to *analogy*@ and *interpretation*@ correspond to well-identified argument schemes.

Other figures play a role in the construction of argumentative structures. For example, a *figure of syntactic disposition*, such as *parallelism*, can act as an analogy or antithesis indicator, S. **Analogy; Antithesis**.

The *figures of opposition*@ are all directly interpretable as argumentative, insofar as they correspond to various modes of presentation of the discourse *vs.* counter-discourse confrontation.

Without reducing each and every figure to a feature of the argumentative situation, it can be observed that the classical definition of argumentation is based on the idea that arguing constitutes an attempt to gain acceptance for a discourse (conclusion) on the basis of good reasons (argument). A clear index of such acceptance is the *resumption, repetition, and development* of the convincing discourse, particularly as fragments or slogans. Since to have things repeated, it is necessary to facilitate their memorization, *figures of sounds* and every kind of *rhetorical pun* can be used to that effect, and must be viewed as a feature of argumentation.

Follow-the-Leader ▶ *Ad Populum*; Consensus

Force

The word *force* is used with three distinct meanings:
    1. Argument from *or* by force, S. **Threat**.
    2. Force of things, S. **Weight of Circumstances**
    3. Force of an argument.

The graduated concept of *force of an argument* exists in opposition to the binary notion of *valid or invalid argumentation*. An argument is strong (or weak) either in itself or relative to another argument. This force is evaluated according to different criteria.

1. Inherent strength of an argument scheme
In scientific fields, to be *strong* an argument must first of all be *valid*. That is to say that it must develop according to a method which is accepted in the given scientific field. Yet, an argument can be valid and not so *strong*, that is to say, really relevant and interesting for the discussion of such and such hypothesis.
From a philosophical point of view, one might consider that some argument schemes are by nature stronger than others. The strength of an argument is thus determined on the basis of ontology. An adept of moral realism will con-

sider that an argument based *on the nature of things and their definition* is stronger than a *pragmatic* argument; a practical mind will think the opposite; **S. Definition; Pragmatic.**

2. Strength and effectiveness

In relation to a goal such as persuasion, the strongest argument will be the most efficient, the argument that most quickly achieves the arguer's goal, whether it be selling a product or electing a president. A degree of strength can be attributed to the argument on the basis of an impact study carried out on the relevant target population, **S. Persuasion.**

3. Strength of an argument and acceptability by an audience

The New Rhetoric defines the strength of the argument according to the extent and quality of the audiences that accept it, **S. Speaker and Audience.**

4. Strength and linguistic reinforcement of arguments

Two arguments oriented towards the same conclusion belong to the same argumentative class, **S. Orientation**. Both bring some support to this conclusion. Within the same argumentative class, the strength of an argument can be determined by reference to an objective gradation, such as the scale@ of temperature, or it may simply be allocated to the argument by the speaker, who value such argument over another. The hierarchization is marked by the means of argumentative morphemes (for example, *even*) and realizing or de-realizing modifiers. The arrangements of the arguments on argument scales are governed by the laws of discourse.

Forum

Some argumentative questions can be quickly and privately solved ("*who is going to take out the trash?*"); others cannot be solved so easily and are brought before specialist, established social institutions. An *argumentative forum* is a more or less institutionalized physical social space dedicated to the treatment of argued issues. Such a space may or may not have a decision-making capacity. Interventions are ruled by the norms and customs that characterize the forum, in the first place the specific codification of the turns at speech as defined by the rights to the floor, **S. Rules.** Such rules give meaning and consistency to the expression "local rationality".

The concept of a forum, with its institutional accompaniment and its concrete regulations, must be taken into account for the analysis of the social exercise of argument. This approach enables us to go beyond an idealized view of argumentation as an exercise subject only to the law of dialectical reason, regulating verbal exchanges between two artificially de-socialized actors, **S. Roles.**

The crucial question of the *burden of proof* relates not only to the state of opinion (doxa) at the time of the discussion, but also to the forum where the discussion takes place. **S. Burden of proof.**

Tribunals and political assemblies can be seen as typical forums. There are many others "argument marketplaces", where viewpoints are calculated, expressed and traded to inform practical decisions, are part of the fabric of democratic societies. Consider the dispute over the legalization of drugs in Syldavia, a true participatory democracy. The issue will be discussed in a huge range of forum, from the subway carriage, to the family table, at the pub on the corner, in the city conference room, by the commissions drawing up the political parties' official positions, by the National Congress, the Law Commission, etc. Some of these forums are intended for the expression of disputes and have the power to voice a decision or opinion on the matter, others serve simply to amplify and popularize the debate rather than close it.

The following passage is taken from a 2002 speech given by Alfredo Cristiani, President of El Salvador from 1989 to 1994. In 1992, under his presidency, the Chapultepec Peace Agreements were signed, ending a twelve-year civil war between the extreme right and Marxist guerrillas. His 2002 speech was delivered on the occasion of the tenth anniversary of these agreements[1].

> We cannot understand the importance of what happened in El Salvador if we limit ourselves to the recent past. The crisis that swept the Salvadorian nation over the last decade did not come out of nothing, nor has it been the fruit of isolated wills. This crisis, so painful and tragic, has ancient and profound social, political, economic and cultural roots. In the past, one of the pernicious flaws in our national form of life was the lack or insufficiency of the spaces and mechanisms [*de los espacios y mecanismos*] necessary to allow the free play of ideas, the natural development of the various political projects which stem from freedom of thought and to act; in short, the lack of a real democratic living environment.

According to Plato, sophistic discourse reigns over *public forums* and institutional places, in particular, over the court and the assembly, dominated by professional sophists. That is why Socratic dialectic interaction, oriented solely by the search for truth, takes place in a very special, *de-socialized* argumentative place, in the *natural* setting of a *locus amœnus*: a hot day, a stream, a tree, a light breeze and grass to lie down on:

> Phaedrus: — [...] All right, where do you want to sit while we read?[2]
> Socrates: — Let's leave the path here and walk along the Ilisus; then we can sit quietly wherever we find the right spot.
> Phaedrus: — How lucky, then, that I am barefoot today—you, of course, are always so. The easiest thing to do is to walk right in the stream; this

---

[1] archivo.elsalvador.com/noticias/especiales/acuerdosdepaz2002/nota18.html (09-20-2013)
[2] The speech of Lysias, that Phaedrus "[holds] in [his] left hand under [his] cloak".

	way, we'll also get our feet wet, which is very pleasant, especially at this hour and season.
Socrates:	— Lead the way, then, and find us a place to sit.
Phaedrus:	— Do you see that very tall plane tree?
Socrates:	— Of course.
Phaedrus:	— It's shady, with a light breeze; we can sit or, if we prefer, lie down on the grass there.
Socrates:	— Lead on, then.
Phaedrus:	— Tell me, Socrates, isn't it from somewhere near this stretch of the Ilisus that people say Boreas carried Orithuia away?
Socrates:	— So they say.
Phaedrus:	— Couldn't this be the very spot? The stream is lovely, pure and clear: just right for girls to be playing nearby.

Plato, *Phaedrus*, 1229a-c. *CW*, p. 509.

G

Generality of the Law

Lat. *a generali sensu*; Lat. *generalis* "general", *sensus* "thought, idea".

In law, the argument of the *generality of the law* posits that the law must be applied in all its extension, "we must not introduce distinctions where the law does not". General terms should not be given a particular meaning. In other words, law is non-negotiable. Possible exceptions must be explicitly laid down in the relevant regulation, for example, while generally prohibited, the consumption of cannabis may be *tolerated* in some specific places complying with the existing regulation.

In public places, people's behavior must comply with *law* plus specific *rules* of the place. Rules are by nature more flexible than laws, but, when strictly enforced, these rules also obey the principle of generality. If the rule of the school states in general terms that *"the use of mobile phones is prohibited during the course"*, then its application is general, and admits no exception or distinction. One cannot argue that the regulation is especially valid for *"the lower grades"*, or that an exception must be made for students *"urgently managing their bank account"*, or for *"students who have a good academic standing"*.
S. Strict meaning

Genetic Argument ▶ Intention of the Legislator; Fallacy (I)

Genus

Lat. *ejusdem generis* argument. *Lat. idem,* "identical"; *genus*, "genus".

1. Argument from genus

The argument from the *genus* is based on essential definition. It transfers to the species, and ultimately to the individuals, the properties, duties, representations, any and all characteristic attached to the genus they belong to, S. **Taxonomy; Categorization; Definition.**

2. Extending to the genus: the generic clause "... *and the like*"

Generic clauses are phrases such as "... *and the things of the same kind*", "... *and the like*". The text has the form:

> This provision concerns **a, b, c**, and *things of the same kind*.

> *Universal Declaration of Human Rights, Article 2*[1].
> Everyone is entitled to all the rights and freedoms set forth in this Declaration, without distinction of any kind, such as race, color, sex, language, religion, political or other opinion, national or social origin, property, birth *or other status*. (My italics)

If an object **x** is not included in the enumeration "**a, b, c...**" but if it is possible to consider that it belongs to the category defined by the enumeration, then the generalizing clause "*and all beings of the same kind*" applies the provision concerning **a, b** and **c** to **x**. This illustrates that the individuals enumerated are mentioned not only for their own sake, but also as prototypes from which a new category must be derived, S. **Analogy (II).**

> This provision concerns cars, motorcycles, and all private means of transportation.

Cars and motorcycles are considered to be prototypical members of the category "private means of transportation" to which the provision applies. Note that the particle *etc.* would also open the list to new categories of individuals, but would not give any indication about the relevant common feature constituting them into a specific genus, as the provision "*all private means of transportation*" does quite clearly. The generic provision may either create a new category out of the enumeration of specific individuals, or explicitly mention an existing genus:

> One must pay the tax on chickens, geese, *and other poultry*.
> *Conclusion*: therefore on ducks and turkeys.

Chickens and geese are mentioned only as prototypical examples of the category "poultry". One can discuss borderline cases, for example whether a peacock is really a backyard animal or a pet. In any case, there is no levy on rabbits, which don't qualify as poultry.

---

[1] Quoted after www.un.org/en/universal-declaration-human-rights/ (01-07-2017)

On the other hand, the absence of a generic provision limits the application of the measure to the categories that are explicitly mentioned:
> You have to pay the tax on chickens and geese.
> *Conclusion*: So not on ducks.

Unless the legislator's intention@ is invoked.

The use of the extensive clause is not limited to the legal field:
> *Fixed concrete barbecue*
> Warning! Do not use alcohol, gasoline or similar liquids to light or reactivate the fire.

Gradualism and Direction

The *argument of direction*, or *slippery@ slope argument*, is based on the *device of stages* and is used to counter the *gradualist strategy*. It is classified as an argument "based on the structure of reality" by Perelman Olbrechts-Tyteca.

1. The device of stages as a general action strategy

Generally speaking, the process of stages is implemented when the overall goal is judged as being directly unattainable; it is then divided into smaller, more easily achievable goals. This process of division corresponds to a common action strategy, which is not necessarily manipulative. Experienced explorers explain that when lost in the desert, dying of thirst, and trying to reach a desperately distant town (final goal) one must set oneself a manageable goal, say the next dune, and then the next cactus, and so, step-by-step finally reach the distant town. More relevant to everyday life perhaps is the solution to trying to carry a heavy weight. If I cannot carry this one hundred pound object, I dismantle it and carry each of its parts separately. Such small but achievable goals might be ordered, as is the case in every learning process: one first learns to drive on a normal road for example, before learning to drive on an icy road. In these different cases, the actor keeps the ultimate goal in mind, in relation to which the partial goals are determined and organized.

2. The gradualist strategy

To get something from another person, an actor can apply the process of stages. In that case, the *gradualist* process should not be considered to be an argument but an intentionally opaque, manipulative *strategy*, **S. Manipulation**.
> It is often found to be better not to confront the interlocutor with the whole interval separating the existing situation from the ultimate end, but to divide this interval into sections, with stopping points along the way indicating partial ends, whose realization does not provoke such a strong opposition. (Perelman & Olbrechts-Tyteca [1958], p. 282).

Step-by-step strategy, in this second sense, is commonly referred to in sales as a *priming strategy*:

The newlywed Joneses want to buy a flat; the real estate agent proposes a modest, fully sufficient two room flat, and they agree to buy it. Now the agent has got a foot in the door, and observes that very soon a baby will come; so they really need a three-room flat, and they change their mind and agree to buy one. But the agent observes that Mrs. Jones is developing a promising start-up, she needs an individual office; so they need a four-room flat, etc.

Arguing with the Lord to convince him to hold his wrath toward Sodom, Abraham uses such a *priming strategy* and step-by-step process — somewhat manipulative, but nonetheless laudable. The argument goes not from *the few* to *the many* but from *just some* to *a very little few*.

> [...] Abraham remained standing before the Lord. Then Abraham approached him and said: *"Will you sweep away the righteous with the wicked? What if there are fifty righteous people in the city? Will you really sweep it away and not spare the place for the sake of the fifty righteous people in it? Far be it from you to do such a thing—to kill the righteous with the wicked, treating the righteous and the wicked alike. Far be it from you! Will not the Judge of all the earth do right?"*
>
> The Lord said, *"If I find fifty righteous people in the city of Sodom, I will spare the whole place for their sake."*
>
> Then Abraham spoke up again: *"Now that I have been so bold as to speak to the Lord, though I am nothing but dust and ashes, what if the number of the righteous is five less than fifty? Will you destroy the whole city for lack of five people?"* *"If I find forty-five there,"* he said, *"I will not destroy it."*
>
> Once again he spoke to him, *"What if only forty are found there?"* He said, *"For the sake of forty, I will not do it."*
>
> Then he said, *"May the Lord not be angry, but let me speak. What if only thirty can be found there?"* He answered, *"I will not do it if I find thirty there."*
>
> Abraham said, *"Now that I have been so bold as to speak to the Lord, what if only twenty can be found there?"* He said, *"For the sake of twenty, I will not destroy it."*
>
> Then he said, *"May the Lord not be angry, but let me speak just once more. What if only ten can be found there?"* He answered, *"For the sake of ten, I will not destroy it."*
>
> When the Lord had finished speaking with Abraham, he left, and Abraham returned home.
>
> Genesis 18:22-33 New International Version.[1]

Unfortunately, the Lord will not find ten righteous people in Sodom.

3. Argument of direction, or slippery slope argument

The term *argument of direction* is an alternative name for the *slippery@ slope* argument. It is used to prevent the application of a *gradualist strategy*: "[it] consists, essentially, in guarding against the use of the device of stages. If you give in this time, you will have to give in a little more next time, and heaven knows where you will stop" (Perelman & Olbrechts-Tyteca [1958], p. 282).

---

[1] Quoted after www.biblegateway.com/passage/?search=Genesis%2018:16-33

H

Hasty Generalization ▶ Induction

Historic Argument (Law) ▶ Legislator Intent

I

Ignorance

Ad ignorantiam argument, Lat. *ignorantia*, "ignorance"

1. Argumentation from ignorance and legitimacy of doubt

Argumentation from ignorance is defined by Locke as one of the four fundamental forms of argumentation, **S. Collections (II)**:

> Secondly, another way that men ordinarily use to drive others, and force them to submit to their judgment, and receive the opinion in debate, is to require the adversary to admit what they allege as a proof or to assign a better. And this I call *argumentum ad ignorantiam*. ([1690]; Vol. II, p. 410-411)

This argument is considered to be fallacious:

> It proves not another man to be in the right way, nor that I ought to take the same way, because I know not a better. (*Id.*, p. 411)

The following dialogue schematizes the situation where **S1**'s conclusion relies on the ignorance of **S2**:

S1_1: — **C**, *since* **A**.
S2_1: — *This is a bad argument. I do not admit that* **A** *proves* **C**.
S1_2: — *Do you have any reason to conclude anything different from* **C**? *Do you know a better argument for* **C**?
S2_2: — *Well, no*
S1_3: — *Then you have to accept my own proof and my conclusion.*

(i) First turn: **S1_1** proposes a justified claim **C**.

(ii) Second turn: **S2_1** refuses to ratify the claim **C**.

(iii) Third turn: **S1_2** asks **S2** to explain the reasons for his or her doubt. According to the conversational principle which requires justification for non-preferred second turns, **S1** is perfectly justified in doing this. **S2** could answer:
(a) by presenting objections against the alleged argument, **A**, or by utterly refuting **A**;
(b) by constructing a counter-discourse by providing what Locke terms "a better proof". The text does not tell for what conclusion; so we can therefore assume the following two cases:
  (b1) Concluding something different from **C**;
  (b2) Providing "better evidence" for **C**.

(iv) Fourth turn: **S2_2** admits that he or she cannot elaborate anything along the (a), (b1) or (b2) lines.

(v) Fifth round: **S1** may accordingly:

(a) Admit **S2**'s reluctances, while maintaining his argumentation: "*Okay, this is not a very good argument, but it is still interesting, it is even the only one we have*";
(b) Summon **S2** to accept his (**A**, **C**) argumentation, considering that his partner's incapacity is a kind of second order proof to add to his former substantial one, **A**, and so committing an *ad ignorantiam* fallacy (even if his former argument is, after all, not so bad).

A pure *ad ignorantiam* fallacy would be based only on the partner's failure "to assign a better [proof]". Under conversational circumstances, **S2_1** does not ratify **S1_1**'s turn; normally, this should urge **S1** to clarify and elaborate upon his proposal. The crude reaction seems rare: "as you cannot articulate anything against my argumentation, *you have to accept it wholesale*".

Seen from **S2**'s perspective, this situation also seems a little bizarre, a kind of borderline case where **S2** has only his or her inner conviction to oppose to an argumentation. Under standard conditions, a conversationalist and a fortiori a dialectician, knows how to elaborate upon a strong inner conviction. In essence, Locke seems to attribute to **S2** a kind of radical clause of conscience.

Leibniz mitigates this radical stand: "The argument *ad ignorantiam* is valid in cases of presumption where it is reasonable to hold to an opinion till the contrary is proved" ([1765], p. 576).

Presumption here has the meaning of "burden of proof". The pretension of **L1** may be excessive and misleading, but his argument nevertheless creates a preference in the field concerned, and in practice we can stick to it until something else has been proven.

This "for lack of anything better" reasoning seems to be the standard case in practical argumentation when a decision has to be made and a possibly urgent action has to be taken:

S1_1: — Upon such and such basis, I propose
1) that we take such and such a disposition;
2) that we explore such and such a hypothesis;
Now, the floor is yours
S2: [Long silence]
S1_2: — Nothing to say? Silence meaning consent,
1) In the absence of contradiction, my proposal is adopted.
2) In the absence of any other hypothesis, mine will serve as a working hypothesis.

It is difficult to object to **S1_2**'s conclusions. He or she does not claim that his proposition is the only viable one, nor that his hypothesis should be held to be true.

2. Ignorance and principle of the excluded middle

The argument from ignorance is also defined, without consideration of the quality of the argumentation, as an illegitimate application of the principle of the excluded middle:

P is true since you are unable to prove that it is false.

The argument is not conclusive. If we consider that "**not-P** is not proven" is equivalent to "**not-(not-P)**" we conclude that **P**, by application of the principle of the excluded middle. But the two *nots* are not of the same nature: "**not-P** is not proved" does not mean "**not-P** is false", which would be a confusion between what is true (alethic) and what is knowable (epistemic), **S. Absurd.**

3. Ignorance, burden of proof, precautionary principle

I am innocent since you are incapable of proving that I am guilty.
You are guilty because you are incapable of proving your innocence.

Admitting that **P** is true, or acting "as if" it is true in the absence of proof of **not-P** is a decision that falls to the institution empowered to discuss and rule on such matter in the field concerned. In the judicial field, presumption of innocence places the burden of proof on the prosecution and gives the benefit of doubt to the defendant.

In the debate on the safety or toxicity of new products, a decision has to be made in a situation of insufficient knowledge. The *presumption of safety* would be:

Possibly the product has toxic effects, but this is not proven. So it has no toxic effects.

The precautionary principle is easiest to rebut when maximized:

Every new product is assumed toxic and will remain forbidden until its safety has been proved.

Under its common form, it simply reverses the burden of proof:

The *precautionary principle* (or *precautionary approach*) to risk management states that if an action or policy has a suspected risk of causing harm to the public, or to the environment, in the absence of scientific consensus (that the action or policy is not harmful), the burden of proof that it is *not* harmful falls on those taking an action that may or may not be a risk.

Wikipedia, *Precautionary Principle*

4. Argument from ignorance and argument from silence@.

Ignorance of Refutation, *Ignoratio Elenchi* ▶ Relevance

Imitation - Paragon – Model

1. Paragons

When it comes to political thinking, some events act as *paragons*: Munich and the diplomatic defeat of democracies facing Nazi expansionism, the genocide of Jews, Gypsies and homosexuals, are all great analogues that function as an anti-model for all current conflicts. For the United States before the Iraq war, Vietnam was the great analogue called to the rescue when it came to opposing military intervention abroad. Paragons serve as "models" for understanding the new events; they work on the principle of precedent, **S. Analogy (IV); Precedent; Example**.

A "great analogue" can stage characters that are a source of *antonomasia*. The antonomasia is the figure of speech by which a member of a category is designated by the name of the paragon of this category: *a Daladier* or *a Chamberlain* is a politician who capitulates to a dictator instead of fighting him. This references the behavior of the European politicians Edouard Daladier and Neville Chamberlain in Munich in 1938, as they dealt with Hitler.

The model, person or event, creates a class by analogy, **S. Categorization; Analogy.**

2. Model

The *model* is the single most valued member of a hierarchical category.

— It functions as the root of the class, generating the other members of the class.

— It is the most representative element of the category.

— As such, it is the criterion for the evaluation of the other class members and for integrating new individuals into the category.

— It is considered to be the ideal form, towards which all members of the class tend.

The argument by the model supports the conclusions of the type "this is (not) a good (real, true) X" by comparing the item to evaluate and reference.

In classical culture, the doctrine of imitation is based on the authority of a model. Literary genres are defined by the relationship of their members to a founding model, a founding "father": Thucydides for history; Aesop and La Fontaine for fables; Aristotle and Cicero for argumentation, etc.

3. Setting an example@
When chosen as a model by an individual, the model is not necessarily conscious to be a model, and the situation is not clearly argumentative.

To get an individual to do something, one can proceed argumentatively, that is to say, expose discursively, every reason to do so, and particularly *argue by the model*, giving as an example important people, either real or fictional, who have committed the same deed. This "argument from exemplarity" can be seen as variant of the verbal argument of authority, a metonymic *exemplum@*.

In addition, one might set an example in order to demonstrate to the other what is wanted. One might stop smoking for example, to encourage a friend to stop smoking. Metaphorically speaking, this is an "argument by example", as one speaks of an "argument by strength" (appeal to force) when one tries to open a recalcitrant can with a screwdriver.

The example strategy can be applied to all forms of behavior we wish to change; how to eat properly, talk properly, lead a dignified life worthy of reward in the afterlife. During this process, there may be some kind of persuasion, that is transformation of belief correlated with the transformation of behavior, but not all persuasion comes from argument, S. 'You too'.

Setting an example, the person hopes to set in motion *alignment* mechanisms. The *argument by the example given*, plays on non-verbal mechanisms of social imitation, ripple effect, identification, empathy, charisma. Seduction and repulsion are forces distinct from argumentation that push individuals to align or to distance themselves from another person.

The ethotic argument combines with the argumentation by example, thereby pushing the audience to fully identify with the orator as a model, committing themselves to full belief in what he or she says and doing what he or she does.
S. Ethos; Consensus; *Ad populum*.

The *anti-model* represents everything one should not do (Perelman, Olbrechts-Tyteca [1958], p. 362).

Implication ▶ Inference; Deduction; Connective

Index ▶ Natural Sign

Indicator

Ancient rhetorical theory is not particularly concerned with the connecting words structuring the argumentative passages. In contemporary times, Perelman & Olbrechts-Tyteca ([1958]) do not mention connectives, nor does Lausberg (1960) in his monumental re-creation of the classical system.

Toulmin's "layout@ of argument" emphasizes the role of linguistic connectives in the articulation of the element of the argumentative cell (1958) whereby the Warrant is introduced by *since*; the backing by *on account of*; the claim (conclusion) by *so*; the rebuttal (counter-discourse) by *unless*. Toulmin does not however discuss the connectives in any further detail.

Connecting words are a central issue for the linguistic theory of argumentation (Ducrot *& al.* 1980).

1. Indicators

Indicators are relevant to argumentative analysis on three levels
(1) *Boundary indicators*, helping to delineate the argumentative sequence.
(2) *Internal indicators*, helping to identify and articulate the argument and the conclusion within the argumentative sequence.
(3) *Argument scheme indicators*, helping to identify the argument scheme embodied in a specific argumentation.

All linguistic phenomena that can be exploited for any of these operations can be considered to be argumentative indicators, not only discourse particles and full semantic words. The label most often refers to the intermediate level, that of the argument-conclusion structure, where connectives play a prominent role.

1.1 Multifunctionality of connective particles

The terminology used for connectives and markers of discursive or argumentative structure is overabundant. Schematically, the framework for the discussion is as detailed below.

— *Logical connectives@* build complex propositions from simple or complex propositions.

— *Connective words* belong to the category of *discursive particles*. From a grammatical point of view, *discursive particles* are conjunctions, prepositions, adverbs, interjections… Some discourse particles are particularly attached to conversational speech: *well, hm, right*…

— *Natural language connectives* are multi-functional. Some connectives have essentially *non-argumentative* functions, even in argumentative contexts. For example, enumerative and ordering connectives, "*firstly, secondly, and finally*" can be used to list a series of agenda items as well as a successions of arguments. In an argumentative context, the "list effect" can itself be argumentative.

Other connectives such as *since, because, so, therefore*… are particularly helpful for tagging a segment of discourse as an argument or as a conclusion. However, it must be born in mind that their argumentative function although *prevalent*, is not *exclusive*.

To sum up, connectives are multi-functional particles that can signal an argument-conclusion relation.

1.2 Connective verbs
The argument-conclusion structure, "**A** *so* **B**" can also be articulated by a full verbal construction:

[**A**]; which leads me to conclude that [**B**]

1.3 Connectives articulating the semantic contents of whole discourses.
Logical connectives articulate precise sets of well-defined logical *propositions*, whereas natural language connectives articulate not only propositions but also discourse of undetermined length:

[**A**]. From this, we can conclude that [**B**].

In reality, connectives articulate *meanings* inferred from such indeterminate spans of discourse. A statement like "*and* so [Fr. *ainsi*] *Commissioner Valentin jailed the whole gang*" may close a novel. The left scope of *so* sums up all the events since the beginning of the investigation of Commissioner Valentin. The same is true for the connector *but*, which does not articulate propositions but semantic-pragmatic contents (example infra, §3.1); **S. Orientation**.

1.4 Multifunctionality of argument indicators
Argument indicators are not unifunctional words; not all their occurrences are argumentative. The discourse following *so* or *thus* is not necessarily a conclusion, and the discourse following *because* is not necessarily an argument pointing to a conclusion. There are non-argumentative cases of *thus* and *because*, and there are excellent arguments which feature neither *therefore* nor *because*. This means, on the one hand, that peppering a speech with *because* and *therefore* will not necessarily turn it into an argumentation. Aristotle had already spotted this strategy and rightly considered it as vain, **S. Expression**. On the other hand, if the interpreter waits for a *so* or a *because* to realize that he or she is involved in an argumentative situation, he or she can be said to be seriously lacking in argumentative, interpretative and interactional competence. The connective particles restrict the possibilities of interpretation by evoking a possible argumentative structure, but they are not summons addressed to a sleepy recipient to awake him from his or her interpretive torpor.

The discussion of the argumentative value of a particle must be related to the argumentative sequence itself. It must be independently defined, that is, insofar as it is organized by an argumentative question articulating discourse and counter-discourse. The argumentative character of a particle is context dependent. The fact of occurring in argumentative contexts activates its argumentative function. This general condition does not preclude the practice of the *ars subtilior* of reconstructing implicit arguments and conclusions.

In practice, the analysis of the connecting phenomenon should first give full consideration to the complexity of the grammar of connecting words and connected discourses:
— Their grammatical category, full words as well as discursive particles.

Indicator

— Their syntactic characteristics.
— Their idiosyncratic semantic and syntactic properties.
— Their multifunctionality as argumentative particles: a particle like *but* can mark an argument, a conclusion, a contradiction or an argumentative dissociation.

Therefore, but, because are prime examples of particles with an argumentative function.

2. *Thus, therefore, so... since, because...*

So can be a conclusion marker, and many other things. It may for example, mark the resumption of a topic already introduced and forming the ratified topic of the text or of the interaction, but momentarily left aside. To make matters worse, this *non-argumentative* resumption can be found everywhere, and in particular in *argumentative* contexts. The following example is taken from a lively debate about the attribution of French nationality to immigrants living in France[1]:

> I think that:: all these people— and then also the people who came thus, so [Fr. *donc* "therefore"] during the post-war boom years, we still owe them a certain form of respect.

No participant ever doubted that "these people" came "during the glorious thirties". The reasoning here is that *since* they came during the "during the post-war bloom years", as *workers*, they are *therefore* entitled to respect. Actually, *so* [Fr. *donc*, "therefore"], resumes a statement that is, functionally, not a *conclusion* but an *argument*. The structure is {[*we owe respect to all these people*, Conclusion] [*they came to work (during the post-war bloom years)*, Argument]}, and certainly not:

> * we owe respect to all these people, so [Fr. *donc*] they came during the post-war boom years.

The following intervention is made by a property manager, **M**, during a conciliation session with his tenant, **T**. The manager recapitulates his position: he requests a 80F (14 $) monthly increase of the rent[2].

> [I asked/ Mrs. **T** certainly remembers\ I asked if you want uh, so uh: eighty francs if you want to get to a thousand thirty a month=]claim [that seemed very reasonable, VERY REASONABLE]modal considering the apartment/ and considering its location/ (..) you know a three room apartment let's say all the same on the second floor' (..) relatively comfortable\]argument
> Corpus *Negotiation on rents (conciliation commission)*, Clapi Data Base of Spoken French. Our parenthesis, italics and tagging.

---

[1] Corpus *Debate on Immigration, Clapi* Data Base of Spoken French
http://clapi.univ-lyon2.fr/V3_Feuilleter.php? Num_corpus = 35]. (09-30-2013)
[2] Corpus *Negotiation on Rents - Conciliation Commission)*, Clapi Data Base of Spoken French.
http://clapi.univ-lyon2.fr/V3_Feuilleter.php?num_corpus=13]. (09-30-2013)
Our parenthesis, italics and tagging.

T.'s claim is articulated to the context by *so* [Fr. *donc*, "so, therefore"], which sounds quite standard. But this claim is not inferred from what comes before, which has already been expressed and repeated. The *so* [*donc*] is in its classical *recall*, *resumptive* function; it just happens that the repeated segment is a claim. Thus, this is the case of a *non-argumentative so*, in a strongly argumentative context.

So, then... because... can be used to extract and thematize the implicit content of a sentence:

— An encyclopedic content:
> *All this happened in Greenland, so far in the North*

— A semantically presupposed content:
> S1 — *Peter stopped smoking*
> S2 — *then you know he used to smoke (in the past)?*

— An implication of the act of saying such and such thing:
> S1 — *this dress suits you very well!*
> S2 — *because the others don't?*

3. But

But reverses the argumentative orientation@ of the propositions it connects. Nonetheless, no more than *so*, *but* is not an inherently argumentative particle, and the argumentative framework and vocabulary cannot account for all its occurrences. In particular *but* reverses not only argumentative orientations but also narrative and descriptive orientations.

3.1 *But*, reverser of narrative and descriptive orientations

Generally speaking, *but* serves to reverse the *orientation*, regardless of the kind of orientation: narrative, argumentative, or descriptive.

But is used to introduce a new narrative development:
> August 27: On Friday, I remembered that the annual tax on my car was due to expire. Since I am not one of those who wait until the last minute to renew it, I went to the tax office. An employee was there, waiting for me, or almost. In just a few minutes, via the Internet, everything was done. I'm set until next year. <u>But</u> in the meantime...
> He walked, and while he walked, tirelessly, with his head held high, rocked by his regular rhythm, he dreamed of next year [...] (¹)

Such non-argumentative occurrences of *but* are quite common. The following passage contains perhaps the most famous *but* in all of French literature. Emma is the heroine of Gustave Flaubert's novel, *Madame Bovary*. The whole passage is narrative-descriptive. First, it develops a semantic isotopy, "travel, love, beauty, exotic life, hammocks and gondolas". *But* articulates this first isotopy to a se-

---

[1] http://impassesud.joueb.com/news/mali-pendant-ce-temps-la-il-il-marchait]. 07-28-2010. Our emphasis.

cond one, "husband snoring, children coughing, irritating screeching noises and provincial life". It would not make sense to impose an argumentative analysis upon such a *but*.

> Emma was not asleep; she pretended to be; and while he dozed off by her side she awakened to other dreams.
> To the gallop of four horses she was carried away for a week towards a new land, whence they would return no more. They went on and on, their arms entwined, without a word. Often from the top of a mountain they suddenly glimpsed some splendid city with domes, and bridges, and ships, forests of citron trees, and cathedrals of white marble, on whose pointed steeples were storks' nests. They went at a walking-pace because of the great flag-stones, and on the ground there were bouquets of flowers, offered you by women dressed in red bodices. They heard the chiming of bells, the neighing of mules, together with the murmur of guitars and the noise of fountains, whose rising spray refreshed heaps of fruit arranged like a pyramid at the foot of pale statues that smiled beneath playing waters. And then, one night they came to a fishing village, where brown nets were drying in the wind along the cliffs and in front of the huts. It was there that they would stay; they would live in a low, flat-roofed house, shaded by a palm-tree, in the heart of a gulf, by the sea. They would row in gondolas, swing in hammocks, and their existence would be easy and large as their silk gowns, warm and star-spangled as the nights they would contemplate. However, in the immensity of this future that she conjured up, nothing special stood forth; the days, all magnificent, resembled each other like waves; and it swayed in the horizon, infinite, harmonized, azure, and bathed in sunshine. <u>But</u> the child began to cough in her cot or Bovary snored more loudly, and Emma did not fall asleep till morning, when the dawn whitened the windows, and when little Justin was already in the square taking down the shutters of the chemist's shop.
>
> Gustave Flaubert, *Madame Bovary*, [1856][1]

In these two examples, *but* is not argumentative, it marks an isotopic shift.

3.3 *But*, indicator of an unresolved contradiction

While in the standard case of an argumentative *but*, the inferred contradiction **E1** *but* **E2** is resolved, the coordinated construction being cooriented with **E2**, in other cases *but* articulates two anti-oriented arguments without argumentative resolution:

   S1 — *What shall they do today?*
   S2 — *Some want to go to the woods, but others to the beach.*

Discourse (a) sounds strange, (b) more standard:
   *(a) *so we'll go to the beach.*
   (c) *so we do not know what to do, we'll have to talk about that*

---

[1] Quoted after Gustave Flaubert, Madame *Bovary*. Trans. by Eleanor Marx-Aveling. Ebook, 2006. http://www.gutenberg.org/files/2413/2413-0.txt

3.4 *But*, indicator of argumentative dissociation
    S1   — *I thought you wanted reform?*
    S2   — *We do want reform, but real reform.*

The concept of argumentative dissociation was introduced by Perelman & Olbrechts-Tyteca, who define it as the splitting of an elementary notion, operated by the arguer to escape a contradiction ([1958], 550-609), **S. Dissociation**

3.5 Other functions
— Rectification: with reference to *"Beautiful blue Danube"*
    In Vienna, the Danube is not *blue* but dirty gray

— Preface to a second turn at speech, aligned with the first turn:
    S1   — *Once again, Peter failed to get his degree*
    S2   — *But that's exactly like me!*

4. Other constructions articulating an argument to a conclusion
An argumentative *thus* can be paraphrased by a set of verbal constructions connecting an argument to a conclusion:

    [Left Context]   *therefore, from where, hence, that is why,*   **[Conclusion]**
                               this means, proves, shows clearly that,
                                one can (then) conclude that

The conclusion appears as the completion of a "connective predicate". Markers of argumentative structure would thus be unduly restricted to "small connectives words"; other constructions, combining anaphoric terms, verbs, or substantives can play this role.

3.1 Connective predicates
Some verbs predicate a conclusion upon an argument or an argument upon a conclusion. In reality, these connective predicates are the only indisputable and univocal *argumentative indicators*. We must distinguish between two cases (*argument* is taken in the sense that it has in theory of argumentation, not as "argument of a mathematical function", **S. Argument**)
(1) *Conclusion Predicate*: the conclusion is predicated upon the argument.
    Subject (Argument) + Pred (Conclusion)

— *from* [Argument] *I conclude (that)* [Conclusion]:
    V = *to conclude, infer, deduce…*

— [Argument] *allows to deduce (that)* [Conclusion]:
    V = *to induce, show, demonstrate…*

— [Argument] *proves* [Conclusion]
    V = *to prove, demonstrate, support, corroborate, suggest, go in the direction of, motivate, legitimate, justify, entitle to believe (say, think…)*

(2) *Argument predicate*: the argument is predicated upon the conclusion.
    Subject (Conclusion) + Pred (Argument)

[Conclusion] ensues from [Argument]:
V = to ensuing, result, follow, derive…

To argue is not a conclusion predicate, but a simple verb of speech activity. In "**X** argues for such a conclusion", the subject **X** must be [+ Human]; it cannot be an argument, a description of a state of affairs. This construction contrasts with the construction "**X** suggests such a conclusion" where **X** can be a discourse or a human, S. *(To) argue*.

Overlooking this set of constructions is particularly damaging in the teaching of argument.

3.2 Constructions framing an argumentation

All the words used to talk about arguments and argumentation can serve as markers of argumentative structuration and argumentative function. This class of nominal indicators includes all the ordinary lexicon of argumentation: *(counter-)argument, (counter-)conclusion, point of view…, premise, objection, refutation…*
>this is my conclusion, a consequence, a serious objection, an argument to be taken into consideration…
>
>[**D1, argument**] *is given as a good reason to admit, to do… is stated, said for, with a view to, to make acceptable, to make, to say, to feel…* [**D2, conclusion**]
>the conclusion, the premise, the objection that…; against this point of view…

We can be certain that "*building the school here, the land is cheaper*" is an argumentation, because it can be satisfactorily paraphrased as follows:
>A good reason to build the school here is that the land is less expensive.
>The fact that the land is cheaper legitimizes the decision to build the school there.

Induction

Induction is one of the three classical modes of inference@. Induction goes from the particular to the general; it generalizes to all cases findings and information gathered from a limited number of cases.
>I draw *a marble* from the bag; then *a second marble*, … then still *another marble*…
>The bag is still not empty; nonetheless, I conclude that this is *a bag of marbles*.

To conclude with certainty, all the remaining items would have to be examined, but it would take a long time. A trade-off must be found between 1) the margins of uncertainty I can tolerate and 2) the amount of time that would be needed to check the entire bag. I decide to save time I check some items and conclude, "*this is a bag of marbles*".
Induction rests on similarity between the individuals, possibly based just on one feature, deemed relevant by the arguer.
An induction based on just one case is an example, **S. Example**.

1. Forms of induction

Complete induction — Induction is said to be complete and its conclusion positive (valid, certain), if one proceeds by an exhaustive inspection of each individual. Such a process is possible only if one has access to all the members of the set.

Induction from a representative subset to the set — A proposition found true in a carefully selected sample can be extended to the whole:

> 40 per cent of a representative sample of voters polled declared their intention to vote for candidate Joni. So Joni will get 40 per cent of the vote on Election Day.

Depending on whether or not the sample is truly representative, whether or not people have given fanciful answers, the conclusion varies from almost certain to vaguely probable.

Induction from an essential characteristic — The generalization from an accidental property of a specimen to all other specimen is hazardous, but when based on an essential property, the conclusion is positive, S. Example:

> This is a normal Syldavian passport.
> This passport mentions the religious affiliation of the holder.
> So all Syldavian passports mention religious affiliation.

2. Refutation of induction

A conclusion obtained by induction is refuted by showing that it proceeds from a *hasty generalization*, based on the examination of an insufficient number of cases. To that end, one exhibits members of the collection that do not possess the property.

3. Induction in mathematics: recursive reasoning

In mathematics, *recursive reasoning* is a form of induction which leads to positive conclusions (Vax 1982, [*Mathematical induction or recursive reasoning*]). It is operative in domains such as arithmetic, where a relation of succession can be defined. First, it must be shown that the property holds for **1**; then, that if it holds for an individual "**i**", it also holds for its successor "**i + 1**". The conclusion is that all the members of the set possess the tested property.

4. Induction as a positive method in literary history

An inductive argument consists of establishing a general law or tendency and applying this to a large number of examples. This process is typical of the positivist science of literature and ideas.

> § 2 Diffusion of Irreligion in the Nobility and the Clergy
> Diffusion of irreligion is considerable in the high nobility. General testimonies abound, 'Atheism', says Lamothe-Langon, 'was universally spread in what was called high society; to believe in God was becoming ridiculous, and we were careful to guard ourselves'. The *Memoirs* of Ségur, those of Vaublanc, those of

> the Marquise de la Tour du Pin confirm what Lamothe-Langon writes. At Madame d'Hénin's, the Princess de Poix, the Duchesse de Biron, the Princess de Bouillon, and in the ranks of officers, people are, if not atheist, at least deist. Most members of the salons were "philosophers", and adopted the spirit of the philosophers, and the great philosophers are their most beautiful ornament. This may be seen not only in the salon of the philosophers themselves, at d'Holbach's, Mme Helvetius's, Mme Necker's, Fanny de Beauharnais's (where we see Mably, Mercier, Cloots, Boissy d'Anglas), but also among the great nobility. At the Duchesse d'Enville's, one meets Turgot, Adam Smith, Arthur Young, Diderot, Condorcet; at the Count de Castellane's, D'Alembert, Condorcet, and Raynal. In the salons of the Duchesse de Choiseul, the Maréchale de Luxembourg, the Duchesse de Grammont, Madame de Montesson, the Comtesse de Tessé, the Comtesse de Ségur (her mother), Ségur meets or hears Rousseau, Helvétius, Duclos, Voltaire, Diderot, Marmontel, Raynal, Mably. The Hôtel de la Rochefoucauld is the meeting place of the more or less skeptical and liberal great lords, Choiseul, Rohan, Maurepas, Beauvau, Castries, Chauvelin, Chabot, who meet with Turgot, d'Alembert, Barthélémy, Condorcet, Caraccioli, Guibert. There are many others who might be mentioned here: the salons of the Duchesse d'Aiguillon, who was 'very infatuated with modern philosophy, that is to say, with materialism and atheism', Madame de Beauvau, the Duke of Levis, Madame de Vernage, the Comte de Choiseul-Gouffier, the Vicomte de Noailles, the Duke de Nivernais, the Prince de Conti, etc.
>
> Daniel Mornet, [*The Intellectual Origins of the French Revolution*], 1933[1]

The affirmation to be justified asserts that, *"the diffusion of irreligion is considerable in the high nobility"*. It is supported by an explicit testimony, accompanied by three others, which are merely evoked. This is followed by an affirmation of the same order, *"most members of the salons are "philosophers" and philosophers are their most beautiful ornament"*, supported by twenty-eight names of philosophers. The reasoning is irresistible, but the reading can be boring.

The strength of the asserted principle depends on the number of cases considered. Their small number gives some reasons for skepticism:

> It hasn't been sufficiently appreciated how insignificant is the number of these historical examples upon which are asserted the "laws" claiming to be valid for all the past and future evolution of the humanity. [Vico] claims that history is a succession of alternations between a period of progress and a period of regression; he gives two examples. [Saint-Simon] that it is a succession of oscillations between an organic epoch and a critical epoch; he gives two examples. A third, [Marx], that it is a series of economic regimes, each of which eliminates its predecessor by violence; he gives one single example!
>
> Julien Benda, *The Treachery of the Clerks*, [1927].[2] Our emphasis.

---

[1] Daniel Mornet (1933). *Les origines intellectuelles de la Révolution Française, 1715-1787*. Paris: Armand Colin, p. 270-271.
[2] Quoted after Julien Benda, *La Trahison des clercs*. Paris: Grasset, 1975, p. 224-225.

It should be noted that Benda's own claim that, "the number of these historical examples upon which is asserted a "law" claiming to be valid for all the past and future evolution of humanity is insignificant", is itself backed by *three* examples.

Inference

The concept of inference is a primitive, that is, it can be defined on the basis of concepts of an equal complexity, or by an example of inference taken from a special field, logic: an inference is "the derivation of a proposition (conclusion) from a set of other propositions (the premises)" (Brody 1967, p. 66-67). Inference is used to establish a new truth on the basis of truths already known or accepted.

There are two kinds of inference, *inference* strictly speaking and *immediate* inference.

— In *immediate* inference, the conclusion is derived from a single proposition, S. **Proposition**.

— Strictly speaking, *inference* is based upon several propositions, its premises. Traditional logic distinguishes between *deductive* inference (deduction) and *inductive* inference (induction). In Aristotle's vision of rhetoric, the *enthymeme* is the argumentative counterpart of *deductive* inference and the *example* is the counterpart of *inductive* inference, S. **Enthymeme; Example**.

1. Analogy, deduction, induction and conduction

Analogical inference is accepted only as a heuristic instrument, it has no probative value, S. **Analogy**.

Deduction and induction are traditionally opposed on two bases.

— The *particular / general* orientation. Deduction and induction are considered to be two complementary processes, induction going from the general to the most general:

> This Syldavian is red-haired, so all Syldavians are red-haired.

Whereas, the deduction would go from the most general to the least general:

> Men are mortal, so Socrates is mortal.

But syllogistic deduction can be generalizing:

> All horses are mammals, all mammals are vertebrates, therefore all horses are vertebrates.

— The *degree of certainty*. The valid conclusion of a syllogistic deduction from true premises is apodictic, i.e., necessarily true, while induction only concludes in a probable way.

Conduction@ is considered by Wellman (1971) as a kind of inference on a par with deduction and induction.

2. Immediate inference and analytical statements

An analytic statement is a statement deemed true "by definition", i.e., in virtue of its meaning. Good definitions are analytically true *"a single person is an adult unmarried person"*. While *logical* immediate inference proceeds from quantifiers or "empty words", immediate *analytical* inference operates upon the meaning of the "full words" of the basic utterance:

> He is single, so he is not married.

In arguments such as, *"this is our duty, so we must do it"*, the proposition introduced by *so*, *"we must do it"* is contained in the argument *"it is our duty"*; by definition a duty is something people must do. The conclusion, if a conclusion at all, is *immediate*.

More broadly, an analytic inference is an inference where the conclusion is some way embedded in the argument; the conclusion only develops the semantic contents of the argument. If I'm advised that my colleague recently *"quitted smoking"* I can analytically infer that he or she smoked in the past, **S. Presupposition**.

Consider the example:

> You talk about the birth of the Gods; this implies that at one time the Gods did not exist. This is just as impious as talking about the death of Gods, for which your colleague was recently sentenced to death.

Birth is defined as the "beginning of life". The conclusion does not directly follow from the definition of the word; an additional step is needed to make explicit the meaning of "beginning", chosen so as to imply equivalence between the *times after death* with the *times before birth*. For this reason, the conclusion does not seem so obvious as in the preceding cases.

3. Pragmatic inference

The concept of pragmatic inference is used to account for the interpretation of utterances in discourse. In the dialogue:

> S1 — *Whom did you meet at the party?*
> S2 — *Paul, Peter and Mary*

From **S2**'s answer, **S1** will infer that **S2** encountered *no other person they both know*. This inference is based on a transition law, the maxim of quantity, or completeness: *"when you are asked something, give the best information you have, both quantitatively and qualitatively"*. If **S2** met Bruno at the party, a person well-known to **S1**, then **S2** can be said to have lied to **S1** by omission, **S. Cooperation**.

Intention of the Legislator

In law, the argument of the legislator's intention (or teleological argument) is based not on the strict literal meaning of the law as actually expressed, but on the *intention of the legislator*, that is the social and historical context of the legislative act, the kind of problems the legislator wanted to address, and the solution he or she wanted to achieve. This form of argumentation is recognized as relevant, S. **Juridical Logic; Strict Sense.**

1. Historical argument, genetic argument, psychological argument

The intention of the legislator can be established by *an historical, or genetic argument*, using the data provided by the *history of the law*. This history is known by the *preparatory works*, the *"whereas"* section of the law, the *parliamentary debates* having led to the drafting of this law, and so on. When relying on the previous state of legislation, the historical argument assumes that the legislator is *conservative* and that the new texts must be interpreted in the context of the legal tradition (*presumption of continuity* of law).

The intention of the legislator can also be sought in reference to *the spirit* of the law: one will then speak of a *psychological argument* (Tarello, quoted in Perelman 1979, 58).

2. General principles of interpretation

The scope of this class of arguments extends beyond the legal field. They can be used in relation with any written standards, when the institution recognizes the validity of an argument based not on the letter of the text but on the intention of the author. For example, in the philosophical or literary field, the interpretation of a text can appeal to the author's intention, which is itself based on *preparatory work and historical data* (notes, manuscripts, declarations of the author), or on *psychological* data (the spirit of the work and the author's mind at the time, as understood by the interpreter).

Such arguments are considered fallacious in structuralist literary analysis, which advocates an immanent approach of literary texts, S. **Fallacy (II)**,.

Interaction, Dialogue, Polyphony

Rhetorical approaches to argumentation focus on monological data; *dialectical* approaches, focus on conventionalized dialogues; *interactional* approaches to apply to everyday argumentation, when needed, the concepts and methods of verbal interaction analysis. Argumentation is necessarily two-sided, developing both as a monological and as an interactional activity, and it would be pointless to oppose these two kinds of argumentative activities. Argumentative questions can be relevantly discussed under a variety of speech formats, from the philosophical treatise to the internet forum and dinner conversation, S. **Argumentation (I)**.

1. Interaction, dialogue, argumentative dialogue

Ordinary exchanges, *dialogues* and *conversations* are two special kinds of *verbal interactions*. Verbal interactions are characterized by the use of oral language, the physical presence of face-to-face interlocutors, and a key feature, the organized, continuous chain of alternate turns of speaking.

Dialogue is practiced first among humans, and, by extension, between humans and machines. This is not necessarily the case for *interaction*: particles interact, they do not engage in dialogue. Human interactions are both verbal and nonverbal. One can reject a dialogue, but one cannot reject interaction. Social organizations necessarily interact; they can open dialogues in view of promoting their respective interests or solving their disputes.

Dialogue is chiefly verbal with some nonverbal aspects, and this implies a kind of egalitarian situation. The concept of *interaction* takes the inequalities of the participants' social statuses and their specific participations in the ongoing common task into account. It focuses on the coordination between language and other forms of action (collaborative or competitive) carried out by the participants, in complex material environments, including manipulation of objects. *Language at work* is interactional, not dialogical or conversational; *conversations at work* exclude work.

The interactive perspective paved the way for the study of argumentation in the work place or its role in the acquisition and development of scientific knowledge in labwork activities, where argumentative sequences are produced as regulatory episodes, in coordination with the manipulation of objects.

Dialogue has an "about-ness", which makes it quite distinct from ordinary *conversation*, which tends to jump from topic to topic. In ordinary usage, the word *dialogue* has a quasi-prescriptive positive orientation: dialogue is good, we need dialogue. The philosophies of dialogue have a marked humanistic color. Personalities *open* to dialogue are opposed to the fundamentalists, *closed* to dialogue. When two parties enter into dialogue they commit to negotiating; breaking the dialogue may give way to violence. In this sense, as can be seen from the title of Tannen's book, *The argument culture: Moving from debate to dialogue* (1998), *debate*, as a potentially acrimonious and vindictive *argument_2* quasi-deprived of *argumentation*, can be contrasted with *reasoned dialogue*. We see a progress in the transition from the first to the second.

The formal approaches of argumentation as a dialogue game first appeared in the second half of the twentieth century, as a development of the Aristotelian dialectical rules. **S. Dialectic; Logics of dialogue**.

2. Dialogue and polyphony

The concepts of *dialogism*, *polyphony* and *intertextuality* make it possible to apply the interaction-based vision of argumentation to monological argumentative discourses and written texts more generally. *Monological* discourse is defined as a possibly long and complex, spoken or written, one-speaker discourse.

2.1 Dialogism

In rhetoric, *dialogism* is a figure of speech featuring the direct reproduction of a dialogue as a passage in a literary or a philosophical composition.

Mikhail Bakhtin introduced the concept of *dialogism*, or *polyphony*, to refer to a specific fictional arrangement. In a nineteenth century classical perspective, the fictional characters are in some way, if not the puppets of the narrator at least supervised by him or her; all of their acts and speeches are framed according to their contribution to the intrigue. In a dialogic disposition, the narrator is less dominant; the characters tend to develop autonomous discourses and are relatively free of the duty to contribute to the intrigue.

2.2 Polyphony

In music, a polyphony "consists of two or more simultaneous lines of independent melody, as opposed to a musical texture with just one voice, monophony" (Wikipedia, *Polyphony*)

In relation with the Bakhtinian concepts of dialogism and polyphony, the word *polyphony* can be used metaphorically to designate a set of phenomena corresponding globally to the *monological staging* of a dialogue situation, in the mouth of a single physical *speaker* (Ducrot, 1988), called the *animator* of speech, in Goffman's vocabulary.

The theory of polyphony conceptualizes monological discourse as a *polyphonic space*, articulating a series of clearly identified *voices*, each one singing its tune, that is voicing a specific *viewpoint*. These voices are not attributed to identified persons, as they are in direct quotations.

The polyphonic approach to connectives and negation have proved particularly fruitful; for example, "Peter will not attend the meeting" stages two voices, the first voicing the positive *"Peter will attend the meeting"*, and a second one rejecting the first: *"No!"*; and the speaker identifies with the second voice, which is, in Goffman's words, the Principal, assuming responsibility for the talk. **S. Connectives; Denying.**

It is particularly worth noting that one specific *Animator* can develop a two-sided discourse, staging two voices, articulating arguments and counter-arguments, as in a regular argumentative two-person interaction. The argumentative dialogue is then *internalized*, in an inner confrontation free from the constraints associated with face-to face interaction. This is the case when, as in the theater, a character engages in a monologal deliberation. The polyphonic speaker speaks in a voice, then in another, opposed to the first, to finally reject one side of the argument and accept the other, therefore identifying with that voice. According to Ducrot, the polyphonic speaker acts as a theater director, staging the voices, and choosing to identify with one of them, **S. Roles; Persuasion.** This concept of identification is central to the theory of Argumentation within Language. First, the speaker sets out a range of *enunciators*, the sources of the points of view evoked in the utterance. In a second stage, he identifies himself or herself with one of these enunciators, this identification being marked in the grammatical structure. For example, *denying* implies the staging of two voices, and identification of the speaker to the denying voice (cf. supra); the same for the "**P, but Q**" coordination. It must be emphasized that this concept of identification is totally foreign to the psychological concept of identification that is discussed in connection with the issue of persuasion.

Polyphony is not restricted to developed monologues. A conversational turn, necessarily dialogical, can also be polyphonic, as shown by the use of negation. The possible discrepancies between the *interlocutor as such* (as a real person) and the *interlocutor as framed by the speaker* can be seen in a polyphonic perspective, S. **Straw Man**.

The two adjectives, *dialogic* and *dialogical*, both refer to dialogue. It could prove interesting to use one of these words, perhaps *dialogic* to cover the polyphonic and intertextual aspects of discourse on the one hand, and *dialogical* to cover the interaction related phenomena (including their *dialogic* aspects) on the other. Either way, full-blown argumentation articulates two disputing voices, it is a *dialogical* activity.

2.3 Intertextuality

In line with the classical monolithic vision of the speaker, rhetoric considers that the arguer is the source of the speech that he or she masters and pilots at will. According to the concept of intertextuality, speech and discourse have their own permanent reality and dynamics, preexisting to their voicing by some individual. Speakers are, as it were, second to their speech. Intertextuality decreases the role of the speaker, who is considered only as an instance of coordination and reformulation of discourses already elaborated and concretized elsewhere. The speaker is not the intellectual source of what is said, but merely the conscious or unconscious vocalizer of pre-existing contents. The discourse is not produced by the speaker, but *the speaker by the discourse*. This vision of the arguer as a machine to repeat and reformulate inherited arguments and points of views is particularly humbling when compared with the classical image of a creative, "inventive" orator.

In the case of argumentation, these relations of intertextuality are specifically taken into account through the notion of argumentative script, S. **Script**.

Interpretation

The concept of interpretation refers to:
— The general process of understanding complex text, S. **Interpretation, Exegesis, Hermeneutics**.
— In rhetorical argumentation, the word *interpretation* can refer to:
    1. A special kind of stasis.
    2. A figure of repetition.
    3. An argument scheme, S. **Motives**.

1. Stasis of interpretation

In stasis theory, the *stasis of interpretation* corresponds to a special case of contradiction between the parties, the "legal question", S. **Stasis**. In court, or, more broadly, whenever a debate is based on a written text, and especially a normative rule, a "question of interpretation" arises when the two parties base their conclusions on different readings of the text. One party, for example, may base his or her argument on the *letter* of the law, whilst the other will argue from its *spirit*.

2. A figure of repetition

As a figure of discourse, interpretation consists in duplicating a first term in the form of an immediately following second term, quasi-synonymous and more easily understood than the first. In the sequence "**Term$_1$, Term$_2$**", **T$_2$** *interprets* **T$_1$**, i.e., *explains, clarifies* the meaning of **T$_1$**. **T$_2$** can be a common language equivalent of a technical term **T$_1$**:

> We found *marasmius oreades*, I mean, *Scotch Bonnets*.

The interpretation applied to a word or an entire expression may maintain its argumentative orientation:

> The President announced an expenditure control policy, a "sober state" policy.

Phrased by an opponent, the interpretation can reverse the argumentative orientation of **T$_1$**:

> The President announced an expenditure control policy, that is, a policy of austerity.

This change is marked by the introduction of a *reformulation* connective (one might say an *interpretive* connective): *in other words, i.e., that is to say, which means that…*

3. Refutation by interpretation

The *Treatise on Argumentation* classifies the *interpretatio* as a "figure of choice", and offers an example borrowed from Marcus Annaeus Seneca, (known as the Seneca the *Elder*, or the *Rhetorician*). Seneca the Elder is the author of the *Controversies*, a collection of more or less imaginary judicial cases, treated by different rhetors of his time (1st century), in a kind of speech contest. Perelman & Olbrechts-Tyteca's example is taken from the first case of the collection ([1958] p. 233), where the question proposed to a score of expert orators is an ingenious story of a son *who fed his uncle despite his father's ban*. The wheel of fortune having turned, it is now the father who is in difficulty, and the son has now *fed his father in spite of his uncle's prohibition*. The unhappy son is thus driven out for the same reason, first by the father, then by the uncle. In the following passage, the author reports the words of the lawyers addressing the father on the son's behalf, first Fuscus Arellius, then Cestius:

> Arellius Fuscus, in concluding, suggested, as a question: '*I thought that, in spite of your prohibition, you wanted your brother to be fed: you had this air in pronouncing your defense, or at least I believed.*' Cestius was bolder: he did not just say '*I thought you wanted it*', he said, '*You wanted it and you still want it today*', and by means of this figure, he pointed to all the motives which forced [the father to want it so] [*and concluded:*] '*Why do you drive me away? Doubtless you are indignant at the fact that I took your part*'.
>
> Seneca the Elder, or the Rhetorician (54 BC - 39 AD), [Controversies and suasories], (written at the end of his life).[1]

The interventions of the two lawyers are co-oriented. Fuscus Arellius argues that the father may have given his order reluctantly. Cestius then goes farther, and attributes to the father *an intention contrary to his words*, "*you wanted it and you still want it today*". Perelman & Olbrechts-Tyteca see here an "argumentative figure or a stylistic figure depending on the effect it has on the audience" ([1958], p. 172), **S. Figure**. Actually, the counsel's words are clearly argumentative. Firstly, they introduce a typical stasic situation, a question about the *qualification* of the act under examination, "*you wanted me to disobey you. So, don't punish me, rather congratulate me in having accomplished your secret wish!*" **S. Stasis**. Second, it implements the private *vs.* public scheme, substituting the private, sincere, will, to the publicly affirmed will, made under social pressure, **S. Motives**.

4. Refutation by interpretation *vs.* performative analysis

The discussion of this example involves the analysis of the order as a performative act. Interpretation is an instrument of refutation and defense that, interestingly, opposes a charge based on a performative analysis of such a speech act. Austin illustrates his discovery of performativity with an example borrowed from the *Hippolytus* of Euripides (I, 612). According to Austin, an order, is valid as soon as it is uttered, regardless of what the speaker is actually thinking:

> Surely the words must be spoken 'seriously' and so as to be taken 'seriously'? [...]. But we are apt to have a feeling that their being serious consists in their being uttered as (merely) the outward and visible sign, for convenience or other record or for information, of an inward and spiritual act: from which it is but a short step to go on to believe or to assume without realizing that for many purposes the outward utterance is a description, true or false, of the occurrence of the inward performance. The classic expression of this idea is to be found in the Hippolytus (1. 612), where Hippolytus says, "*my tongue swore to, but my heart* (or mind or other backstage artiste) *did not*". Thus "I promise to..." obliges me — puts on record my spiritual assumption of a spiritual shackle.
> It is gratifying to observe in this very example how excess of profundity, or rather solemnity, at once paves the way for immorality. (Austin, 1962, p. 9-10)

---

[1] Translated after the French edition used by Perelman, Sénèque le Rhéteur, *Controverses et Suasoires*. Trans. by H. Bornecque. T. 1. Paris: Garnier Frères, 1932, p. 23-24.
https://archive.org/details/ControversesEtSuasoiresTraditionNouvelleTexteRevueParM.Henri

As the son and the father in Seneca's example, Hippolytus and the nurse, are engaged in highly argumentative interactions. In such situations, semantics, pragmatics, and morality can all be discussed and argued. The son acknowledges the facts (he fed his uncle) and pleads not guilty to the charge of disobedience, maintaining that the *verbal* order, what the father *said*, did not expressed the true *will* of the father. This is exemplary of the case of opposition described by Austin, which exists between what *language actually does* and what *goes on in the mind* of the speaker. It should be noted first that Austin's binary distinction knows only the verbal aspects of language, and excludes all paraverbal (mimic) modalization of the order.

There remains the question of the validity of the father's prohibition. For the father and for Austin, the prohibition is valid because the father uttered the relevant formula, and the son is guilty of the double Austinian sin, analytical fallacy and moral perversity. Yet the analysis offered by the Austinian father is rather questionable; what the father really said is problematic and must be subject to an *interpretation* which will takes into account the pragmatic environment of the speech act utterance. The situation is analogous to that of ironic utterances, **S. Irony**. The addressee hears something contextually incongruous, said by someone who usually talks seriously, and this forces him to engage in *interpretation* of this puzzling utterance. Similarly, the father uttered a prohibition which contradicts the natural (doxic) law of brotherly love, which the son esteems inconsistent with his father's true character; the son is obliged to interpret the incongruity — maybe the verbal utterance of the father was accompanied by a paralinguistic sign, which pointed to another intention? He thus infers that the order was not given in the *true*, *natural* voice of the father but in his *social* voice, as argued by Fuscus Arelius. As a consequence, the son feeds his uncle. To decide that this latter interpretation is "the right one" is to side with the son and oppose the father; to decide that the Austinian interpretation is the correct one is to take the side of the father and oppose the son. In either case, to take a stand for an analysis is to side with a party or another.

Interpretation, Exegesis, Hermeneutics

1. The arts of understanding

Hermeneutics, exegesis and interpretation are the arts involved in the understanding of complex texts such as the *Bible*, the *Criminal Code*, the *Koran*, the *Iliad*, the *Communist Manifesto*, the *Talmud*, the *Upanishads*, etc. (Boeckh [1886], p. 133 ; Gadamer [1967], p. 277 ; p. 280). Texts require an exegesis because they are written in forgotten languages, or are historically distant, or hermetic. The community considers that vital things depend on what such texts precisely say and mean. This meaning is not immediately accessible to the contemporary reader. It must be established and preserved to be transmitted as well as possible.

Hermeneutics is a philosophical approach to interpretation, defined as an effort to share a form of life, a search for empathy with the text, its author, the language and culture in which it was produced. The hermeneutical *understanding* is thus opposed to the physical *explanation* sought in the natural sciences, where "to explain" has the meaning of "subsuming under a physical law".
Psychoanalysis and linguistics have shown that ordinary acts and words also may require interpretation.

The theoretical language of interpretation is complicated by the morphology of the lexicon, as is always the case when a theory develops within ordinary language. What difference should be made between *hermeneutics, exegesis* and *interpretation*? Their three respective lexical series include a term designating the agent *exegete, hermeneutist, interpreter*; two of them include a noun referring to the process and result, *interpretation, exegesis*, which, as *hermeneutics*, can also refer to the field of investigation. Only one series includes a verb, namely *to interpret*. This verb will therefore be used for the three series, imposing its meaning upon the whole lexical field.

In the philological and historical sense, exegesis is a critical activity whose object is typically a text belonging to a cultural or religious tradition taken in its material conditions of production and original practices, linguistic conditions (grammar, lexicon), rhetorical conditions (genre), historical and institutional context, genesis of the work in its links with the life and milieu of the author. *Philological exegesis* establishes the text, reveals its meaning(s), contributing thus to resolving conflicting interpretations or articulating different levels of interpretation. It stabilizes the "literal meaning", that is the core meaning of the text, and thus lays down the material to be interpreted. In a broad sense, exegesis encompasses interpretation; both endeavor to overcome the distance carved by history, between the text and its readers.
The purpose of *philological exegesis* is to express the meaning of the text; it tries to create the conditions for a certain projection of the reader into the past. *Interpretative exegesis* (or *interpretation, hermeneutics*) seeks to reformulate this meaning to make it accessible to a contemporary reader; it actualizes the meaning of the text. This is where the link between hermeneutics and the rhetoric of religious preaching lies.
Exegesis aims at understanding the meaning as expressed by the text; interpretation and commentary push the meaning of the text beyond the text itself. Contrary to philological exegesis, interpretation can be allegorical. The philological interpretation is exoteric, whilst hermeneutics can be esoteric.

2. Rhetoric and hermeneutics
The hermeneutic task is to make intelligible to one person the thought of another via its discursive expression. In this sense, rhetoric as the "art of persuading" is the counterpart of hermeneutics as the "art of understanding"; their

directions of fit are complementary: rhetoric adopts the perspective of a speaker/writer striving *to persuade an addressee*, the listener/reader. In contrast, hermeneutics adopts the perspective of a reader/listener striving *to understand a speaker/writer* addressing him or her through a text. Rhetoric is related to *live speech*, taking into account the listener's beliefs, trying to minimize his or her efforts; hermeneutics is linked to distant speech, to *reading*; the reader having to adapt to the meaning of the text. Taken together, hermeneutics and rhetoric establish a dual cultural communicative competence, *to understand* and *to be understood*. The rejection of rhetoric in the name of pure intellectual demand results in the transfer of the burden of understanding to the reader, and so requires hermeneutics.

3. Interpretation and argumentation

The interpretative process applies to any discourse component, from words to whole texts, in order to derive their *meaning*, and this meaning is necessarily (expressed in) another discourse. The interpretive relationship thus binds two discourses, the link between the *interpreting* and *interpreted* utterance being made according to *transition rules* that are not different from the general *argumentation schemes*, **S. Schemes**.

In the case of argumentation, the argument might be any statement expressing a true or accepted vision of reality. In the case of interpretation, the data, the argument statement, is the *utterance* to be interpreted, in view of the precise form it has in the text. Once this statement is available, the linguistic mechanisms are the same. If we consider the argument-conclusion relation in its greatest generality, we shall say that the conclusion is what the speaker *has in view* when the argument is stated, the conclusion being the *meaning* of the argument. The argumentative relation is therefore no different from the interpretative relation. When the listener/reader has grasped the conclusion of the text, he or she has achieved an authentic understanding of this text. This amounts to considering that meaning is always lacking within the statement, and the statement will be allocated a meaning only in relation to a later statement. Meaning is thus construed within an endless process, **S. Orientation**.

Just as with argumentation, interpretation is valid insofar as it is based on principles that correspond to a transition law accepted by the interpretative community concerned, the community of jurists or theologians for example:

> The rabbis saw the Pentateuch as a unified, divinely communicated text, consistent in all its parts. It was consequently possible to uncover deeper meanings and to provide for a fuller application of its laws by adopting certain principles of interpretation (*middot*; "measures," "norms").

Jacobs & Derovan, 2007, p. 25

The same principles apply to the Muslim legal-religious interpretation (Khallaf [1942]), or to legal interpretation. The argumentative forms used in law are the same as those which govern the interpretation of all texts to which, for whatever reason, a systematic character is attributed. This is because they are considered as the best expression of the legal-rational views of the time, because they flow from a divine source or from an individual genius, S. **Juridical Arguments**.

This postulate of strong, even perfect coherence is fundamental for the structuralist interpretations of texts, as for the interpretation of legal texts or religious texts, as mentioned in the preceding quotation. It may conflict with the *genetic argument* constructing the meaning of a text by derivations justified by "preparatory works", such as the manuscripts, or the intentions@ of the writer, as they can be grasped through his or her correspondence, for example. Arguments from genetic evidence are one aspect of the philological interpretive work on the text. They may be regarded with suspicion by true believers, for genetic arguments suppose that a non-divine origin, at least partly human, can be attributed to the text.

Irony

Irony can be considered a pivotal strategy, positioned somewhere between discourse destruction and refutation. Irony ridicules a speech that pretends to be dominant or hegemonic, by implicitly referring to some contextually available irrefutable rebutting evidence.

1. Irony as refutation

Ironic development originates from a hegemonic D_0 discourse. A hegemonic discourse is a discourse which prevails within a group, which has the power to direct or legitimize the actions of the group, and which opposes the discourse of a minority.

In a situation **Sit_1**, the participant **S1**, the future target of the irony, claims that D_0, with which **S2**, the future ironist, disagrees. **S2** submits to D_0, although he or she is not convinced of the validity of the argument.

S1_1 (future Target) — *What about taking a shortcut to reach the summit?*
S2_1 (future ironist) — *Hmm ... It seems that there might be icy zones...*
S1_2 — *No problem, I know the place, it's easy going!* (= D_0)
S2_2 — *Oh well then...*

Later, in situation **Sit_2**, when the group finds itself on a rather slippery slope, the ironist takes up **S1**'s discourse, as the circumstances make this discourse indefensible:

S2_Ironic — *No problem, I know the place, it's easy going!*

This last statement sounds strange:
— In the present circumstances, the statement is absurd.
— If the original discussion has been forgotten, it is interpreted as a humorous euphemism or antiphrasis.

— If it is still present in the memory of the participants, then the statement is entirely ironic: **S2_Ironic** repeats **S1_2**, whereas the circumstances show that the statement is obviously, and tragically, false. The mechanism is rather similar to an *ad hominem*@ argument, what the adversary says is opposed to what he or she does, and this is clear to all parties involved. The facts being self-evident, **S1** is now shown to be wrong and is seen to have misled the company. Irony combines malice and humor, S. **Dismissal**.

Ironic destruction and scientific refutation can be opposed as follows:

Scientific Refutation	Ironic Discourse Destruction
S1 says 'D_0'	S1 says 'D_0' in situation **Sit_1**
The opponent **S2** quotes D_0, and explicitly attributes D_0 to **S1**	The ironist, **S2**, says '**D**' in situation **Sit_2**: — **D** resumes, echoes D_0 — **D** = D_0 is not explicitly referred to its occurrence in **Sit_1**, but the link is easy to make; either everybody recalls, or **S2** gives a cue to recall (for ex. **S2** mimics **S1**'s voice)
The opponent refutes D_0 with explicit and concluding arguments	Contextual evidence drawn from **Sit_2**, destroys **D** = D_0. This evidence is so obvious that (**S2** thinks that) it does not need explanation.

2. Countering the ironic move

Ducrot uses the following example, consisting of a statement and a description of its context:

> I told you yesterday that Peter would come to see me today, and you didn't believe me. Peter being physically present today, I can tell in an ironic way '*You see, Peter did not come to see me*'. (Ducrot 1984, p. 211).

Some times ago, in $S°$, the speaker and his or her partner "*You*" had a debate about whether or not Peter will be coming. The speaker, the (future) ironist lost this debate. Now, the evidence of Peter's presence "*you see*" is given as a conclusive argument, as concrete proof, supposed to silence *You*, proving *You* wrong. But the game is not necessarily over. Irony is mainly studied on the basis of the isolated ironic statements, whereas it is a sequential phenomenon with two kinds of developments, depending on the target reaction. If he or she stays mute and embarrassed, the ironist wins the game; if he or she retorts, then the game continues. Here, *You* can reply that he or she can certainly see that Peter is actually there, but that does not prove that Peter came to see the interlocutor:

— No, Peter did not come to see you. He actually came to see your sister.

This refutation or reversal of irony applies the scheme of substitution of motives@.

3. Irony can dispense with markers

In Zürich, in the years 1979-1980, a youth protest movement made quite an impression on the city's people.

> There are two television shows which caused extreme shock in German-speaking Switzerland. The first, a popular show, was disturbed by members of the "Movement", who put a stop to it. The second, later referred to as "Müller's Show"[1], showed two militants dressed as members of the bourgeoisie from Zurich, and seriously voicing the opinion that the "Movement" should be repressed with the utmost severity, the autonomous center should be closed etc. The sensationalist media and some individuals orchestrated a campaign of defamation after the shock of this second show. Let us note in passing that the term *müllern* entered the vocabulary of the movement [...]. The creation of paradoxical situations was one specialty of the "movement".
> Gérald Béroud, [*Work Values and Youth Movement*], 1982[2]

The ironic discourse **D** consists in the strict repetition, with a straight face, of the primary discourse **D₀**, as held by the opponents; **D** and **D₀** coincide perfectly. The ironized discourse **D₀** is the typical bourgeois argumentative discourse, taken with its contents, its modes of expression, its dress codes, gestures, body postures, modes of arguing following the bourgeois norms of maintaining a calm and courteous atmosphere, ritually invoking some counter-discourse in the role of the "honorable opponent" while ignoring the real existing strong, deep disagreement as well as power and strength relations. The entire practice of the argued, contradictory, quasi-Popperian mode of discussion is ironized and negated by Müller's sarcastic behavior.

Irony is a borderline case of an argument based on self-evidence. It becomes dramatically prominent in situations where argumentation is vain or impossible. The following remarks were written in Czechoslovakia, a country which at that time, was under the dictatorial rule of a communist regime:

> In intellectual circles, the attitude towards official propaganda often results in the same contempt that one feels for the drunkard's drunkenness or the graphomaniac's lucubration. As intellectuals particularly appreciate the subtleties of a certain absurd humor, they may read the *Rude Pravo* editorial or the political discourse printed there for pleasure. But it is very rare to meet someone who takes this seriously.
> Petr Fidelius, [*Lies Must be Taken Seriously*], 1984[3]

---

[1] Name of the two delegates of the movement, Hans and Anna Müller.
[2] Gérald Béroud, "Valeur travail et mouvement de jeunes", *Revue Internationale d'Action Communautaire* 8/48, 1982, note 62, p. 28. Television program (in German) available at: [http://www.srf.ch/player/video?id=05f18417-ec5b-4b94-a4bf-293312e56afe] (09-20-2013).
[3] Petr Fidelius, *Prendre le mensonge au sérieux*. *Esprit*, 91-92, 1984, p. 16. The *Rude Pravo* was the newspaper of the Communist Party of Czechoslovakia, during the Communist period.

J

Juridical Arguments: Three Collections

Juridical arguments are argument schemes considered by law professionals as the most important and typical in their field, and presented as the basis of *"juridical logic"* (Perelman, 1979). Such arguments are important for the general theory of argumentation insofar as they illustrate the explicit and controlled implementation, in the field of law, of general principles currently met in ordinary argumentation. They are presented here from this perspective. Cicero' *Topica* is perhaps the first essay to bring together a list of legal inferential principles, which are historically significant in all the classical fields of argumentation study. S. **Interpretation, Exegesis, Hermeneutics; Typologies (I)**.

These juridical arguments rule the interpretation of legal texts and their application to concrete cases. They allow the application of a text to a case, possibly by extending its meaning and legal force. Given a fact "**f**" submitted to legal evaluation on the basis of a code (legal, religious…), it most often happens that the judge can attach "**f**" to a category **M** mentioned in the code in order to apply to "**f**" the legal provisions concerning **m.s**, the members of the category **M**. It may also be the case, however, that the code does not contain a category which is immediately relevant to the case at hand. This may occur, for example, if there are equally good reasons to categorize "**f**" as an **M** or as an **X**. This situation corresponds to a *stasis@ of categorization*. Such stasis might evolve into a *stasis of definition@*, where the code must be interpreted in order that it also applies to

"**f**'. In such cases, the judge does not simply *apply* the law, but *produces* the law.
S. Categorization, Definition, Stasis.
The process of interpretation is not limited to the juridical domain. Generally speaking, it starts from a proposition **P**, which is to be interpreted. In the interpretation process, **P** takes the status of argument, accepted because it belongs to a stock of statements, a Code, a Regulation, a Sacred Text... itself accepted by the community of interpreters or believers. A proposition **Q** is then derived from **P**. **Q** has the status of a conclusion, which corresponds to an *interpretation* of **P**. The juridical argument schemes are the basic tools that rule such derivations in the domain of law. The limit of interpretation is fixed by the principle "what is clear must not be interpreted". This principle enshrines the existence of a literal meaning, based on grammatical data. If, in order to vote in a Syldavian presidential election, a citizen must be 18 and a Syldavian national, nobody meeting only one of the two criteria will be admitted to vote. There is nothing to interpret.

1. Three collections of juridical argument schemes

Specialists in legal arguments offer lists of argument schemes that are particularly important in law. The lists provided by Kalinowski and Tarello are frequently included in the general framework of argumentation studies (Perelman 1979, Feteris 1999, Vannier 2001). We have added the list provided by *lawoutlines.com*[1] (no author's name). These three lists make extensive use of Latin terminology.

— **Kalinowski** (1965) lists 11 argument schemes:
- *A pari*
- *A contrario sensu*, or *a contrario*
- *A fortiori ratione*, or *a fortiori*
- *A maiori ad minus*, or from the biggest to the smallest
- *A generali sensu*, or argument of the generality of the law
- *A ratione legi stricta*
- *Pro subjecta materia*, or consistency argument
- From preparatory work
- *A simili*, or argument by analogy
- *Ab auctoritate*, or argument from authority
- *A rubrica*, or argument from the title

— **Tarello** (1974 ; quoted in Perelman 1979, p. 55) lists 13 argument schemes:
- *A contrario*
- *A simili*, or argument by analogy
- *A fortiori*
- *A completudine*
- *A coherentia*

---

[1] Legal tradition-Trahan.doc. P. 21-22.
www.lsulawlist.com/lsulawoutlines/index. php?folder=/TRADITIONS (09-20-2013)

- Psychological a.
- Historical a.
- Apagogical a.
- Teleological a.
- Economical a.
- *Ab exemplo* a.
- Systemic a.
- Naturalist a.

— ***Lawoutlines*** considers 10 argument schemes:
- By analogy or *argument a pari*
- Of greater justification or *argument a fortiori*
- By contrast or *argument a contrario*
- Of absurdity or *ab absurdum*
- From generality or *a generali sensu*
- From superfluity or *ab inutilitate*
- From context or *in pari materia*
- From subject matter or *pro subjecta materia*
- From title or *a rubrica*
- From genre or *ejusdem generis*

2. How many argument schemes?

34 argument schemes are specified.

— Four argument schemes are included in the three lists:
- *A contrario*; *a contrario sensu*; by contrast or *a contrario*, S. ***A contrario***.
- *A fortiori ratione, a fortiori*; of greater justification or *a fortiori*, S. ***A fortiori***.
- *A pari* argument is considered separately, or as equivalent to the argument by analogy ("by analogy or *a pari*").
- *A simili* argument is assimilated to analogical argument, S. **Analogy; *A pari***.

— Three argument schemes are common to two lists.
- *A generali sensu*, generalizing argument ; or argument from *generality*@.
- *Pro subjecta materia* ; from *subject*@ *matter* or *pro subjecta materia*.
- *A rubrica* ; from title@.
- Apagogical ; from the absurd@, *ad absurdum*.

— Fifteen (or twelve) are specific to one of the three lists. Arguments:
- From preparatory work, historical, psychological, teleological, **S. Legislator's Intention**.
- From context or *in pari materia*, **S. Consistency**.
- *Ratione legi stricta*, **S. Strict Meaning**
- *Ab auctoritate*, **S. Authority ; Precedent**
- *A completudine*, **S. Completeness**
- *A coherentia*, **S. Non-contradiction ; Consistency**
- Economical, **S. Superfluity**
- *Ab exemplo*, **S. Precedent ; Example**
- Systemic, **S. Systemic**
- Naturalist, **S. Weight of things**

- From superfluity or *ab inutilitate*, **S. Superfluity**
- From genus, or *ejusdem generis*.

We thus obtain 22 different *legal topics*, which may be reduced to 19 if we admit that, under various labels, the argument *from preparatory work*, the *historical, psychological* and *teleological* arguments refer to what Perelman globally terms the "legislator's intent" (1979, p. 55).

3. Groupings

These 22 legal topics can be divided into sub-groups as follows.

(i) General arguments, not specific to law, operative in any controlled argumentative situation:
- From consistency (*a coherentia*)
- *A pari, a simili*, analogy
- From genus
- *A contrario*
- *A fortiori*
- Fom the absurd
- From precedent
- From authority.

In law, these last two forms of argument are based on, and reinforce, the historical continuity of legal practice.

(ii) Arguments legitimizing interpretations based on the conditions of production of the law. Arguments based on:
- Preparatory work
- History (of the law)
- The legislator's intention, teleological argument
- Psychological argument.

(iii) Arguments appealing to the systemic character of the code of laws to legitimate an interpretation. Arguments based on
- Systemic considerations
- Coherence (*a coherentia, in pari materia*)
- Comprehensiveness
- Necessity (all the articles of the code are necessary)
- The title of a section of the code, *a rubrica*.

These argument schemes are based on the assumption that the text to be interpreted is "perfect", in the sense that it contains neither contradiction nor redundancy. All content is necessary; the text contains nothing superfluous, or redundant. All elements hang together; they have meaning only by their relation within the structure. This insistence on the systemic character of the legal code could lead to a mechanical view of the law and its application. Ultimately, all the properties of a formal system are attributed to the code.

The establishment of precise definitions of these forms of argumentation within the field of law, their illustration with examples, the determination of the conditions for their application, and the problems connected with their construction and use, fall within the jurisdiction of lawyers.

4. Prescriptive scope of the topics

This set of arguments legitimizes the interpretation of the law in view of application to specific cases. When used in the imperative form, this set of arguments becomes a guide for the drafting of laws. For example, as the argument from superfluity@ (*economic* argument, or argument *from uselessness*) assumes that the laws are not redundant; the legislator will endeavor to avoid any redundancy in the drafting of the law, and the same for the other interpretative principles.

5. Generalization to other fields, S. Interpretation.

Justice: *Rule of Justice*

Perelman & Olbrechts-Tyteca introduce the rule of justice as a fundamental argumentative principle, *"all beings of the same category must be treated in the same way"*. The rule is illustrated by some categories that have historically regulated the distribution of benefits, *"to each according to his merit; to each according to his birth; to each according to his needs"* (Perelman [1963], p. 26).

The rule founds claims such as *"equal pay for equal work"*. It involves distinct operations.

(i) A categorization — First, individuals are categorized as members of a general category, *"to be born"*; *"to have needs"*; *"to have merit"* (admitting that one can deserve a punishment and that to demerit is to have a negative merit); *"to be an employee, having worked such and such hours and produced such and such assessable products"*.

General rights and duties can be defined with recourse this first level, *"all born human beings have the same right to life"*. The following practice refers to a strict *a pari* argument, referring to thieves as a non-hierarchized category, *"a thief is a thief"*.

> General Baclay was also quite a character, but a funny woman, very just in her own way. She shot in the same way women and men, all thieves, whether they had stolen a needle or an ox. A thief is a thief and all were shot. It was fair.
>
> Ahmadou Kourouma, *Allah is not obliged*. 2000.[1]

(ii) An equality relation — Secondly, there is an *equality relation* defined as *"equality of birth; of needs; of merits; of work"*. This relation determines a *hierarchy* between workers, "**P** has worked as much as **Q** or **R**...; *more than* **A** or **B**...; *less than* **X** or **Y**...".

---

[1] Ahmadou Kourouma, *Allah n'est pas obligé*. Paris: Le Seuil, p. 111

Such equipped categories can be represented on oriented scales@. The position of an individual upon this scale can be debated, *"has* **X** *more/less merit than* **Y**?*"*. The metric is easy to define in cases of work, when determined by the weight of the fruit picked from the trees for example. Things become more complicated when it comes to scientific production, or when it comes to needs and merits. In any cases, the criteria for prioritizing one individual over another one must be set.

(iii) An allocation scale — Another quite different scoring method must be established in order to define the parallel scales of punishments and rewards (*what wages for that level of work?*), and the two scales must be coupled.

These two independent rankings ((ii) and (iii)) make the rule of justice more complex than an *a pari*@ argument. Gross *a pari* holds that *"work must pay"*:
> if **P** works, **P** has a right to be paid for this work (except if **P** is a voluntary worker serving a non profit organization),

while the rule of justice connects two graduated scales.

In addition, it is supposed that the rule of justice is to be applied to all members of the group in a linear order, but actual rules include thresholds. Regarding a tax level, the rule *"to each according to his or her income"* applies only beyond a certain threshold, and contains tax brackets and smoothing principles.

Other categories may be considered, showing that the rule of justice can also serve in support of injustices:
> To each according to his or her gender
> To each according to his or her color of skin

The rule of justice excludes arbitrariness, but not injustice. According to the principle *"who favors disfavors"*, the rule of justice, necessarily creates innumerable injustices. If the benefits are distributed according to *merit*, they are not distributed according to *birth* or according to *need*.

The rule of justice is said to be "just" because it excludes the arbitrariness of the principle *"to everyone according to my convenience"*; and because the category and the hierarchy have been defined by disregarding the cases to be judged, *"the decision is just because the rule existed before your case."* This "justice" is formally just because it allows the application of a legal syllogism.

Justification and Deliberation

People *justify* an answer already given to an argumentative question, while they *deliberate* on an open argumentative question, when they do not know its answer, either individually (Third parties, **S. Role**), or when the group has not yet reached a decision.

Deliberation takes place in a situation of doubt about what to do, while *justification* bears on a decision which has already been taken. The starting point determines the difference between justification and deliberation.

— Deliberation intervenes in contexts of *discovery*. It develops from argument to conclusion. A decision is to be taken, and I *deliberate* to construct it through an inner or collaborative deliberation; the arguments condition the conclusion. The argumentation:

> Question: *Should I resign?*
> [Deliberation: I weigh up the pros and cons]
> The answer states the conclusion: *I resign.*

— In contexts of *justification*, the discourse proceeds from conclusion to argument. I resigned, this is a practical reality:

> Question: *Why did you resign? Justify your decision!*
> Justification: *I was sick and didn't get along with my boss.*

A decision has been taken, and, when required to account for it, I *explain* why I took this decision or made this choice, I recall all the good reasons I had to do so, and, if necessary, I invent new ones. Now, the conclusion determines the arguments.

Deliberation leads to a conclusion introduced by *so, therefore*; justification enumerate good reasons introduced by *since*.

The mechanisms of argumentation are valid for justification and deliberation. I deliberate, I reach a conclusion and make my decision. When I am asked to justify this decision, the same arguments, which were *deliberative*, become *justificatory*, and *explain* the decision taken, **S. Explanation**.

Argumentation	Deliberative
	Justificatory

In the case of deliberation, there is real uncertainty about the conclusion, which is constructed in the course of a cognitive and interactional argumentative process. In the case of justification, the conclusion is already there. *Justification* tends to erase doubt and counter-discourse, whilst stimulating *deliberation*.

Private arguments put forward during an inner deliberation may have nothing to do with the arguments put forward publicly as a justification for the same conclusion, **S. Motives**.

Situations of pure deliberation and pure justification are border cases in which I do not know what I will conclude or do (full deliberation), and I'm sure I did well (full justification). The same arguer may oscillate between justification and deliberation, for example if, during the justification, he or she questions the decision already taken, or is about to change his or her mind.

If we postulate that any argument that presents itself as deliberative is in fact oriented by a decision which has been taken unconsciously, then anything and everything is in fact justification. Yet the institutional organization of debates reintroduces deliberation. A debate may well be *deliberative* when each of the parties comes with *firmly entrenched and duly justified* positions and conclusions. The shock of justifications produces deliberation.

K

Kettle Argumentation

A co-orientation condition does not suffice to characterize a well articulated *convergent@* argumentation; co-oriented arguments must be *consistent*. This is the thrust of Freud's point in *The Interpretation of Dreams* [1900], in which he uses *kettle argumentation* as an analogue in order to interpret the content of his dream about "the injection made to Irma". Both his dream and the following argument are *incoherent* defense systems putting forward *good but incompatible* justifications:

> I noticed, it is true, that these explanations of Irma's pains (which agreed in exculpating me) were not entirely consistent with one another, and indeed that they were mutually exclusive. The whole plea — for the dream was nothing else — reminded one vividly of the defense put forward by the man who was charged by one of his neighbors with having given him back a borrowed kettle in a damaged condition. The defendant asserted first, that he had given it back undamaged; secondly, that the kettle had a hole in it when he borrowed it; and thirdly, that he had never borrowed a kettle from his neighbor at all. So much the better: if only a single one of these three lines of defense were to be accepted as valid, the man would have to be acquitted. (Freud [1900], p. 143-144)

The neighbor collates all the possible defensive replicas, as laid down by stasis@ theory. More justifications could be added, "*I am not the one who holed the kettle*"; "*it's really a tiny hole*", "*very easy to fix*" etc.

L

Laughter and Seriousness

Laughter and seriousness are the manifestations of two antagonistic psychic states. *Laughter* is a manifestation of a positive emotion@, such as *joy*. Laughter is the opposite of *tears* and *grief*, which are manifestations of negative emotions, and also the opposite of *seriousness*, denoting *calm*, S. **Pathos**.

Laughter is a major instrument of discourse disorientation and destruction@, S. **Orientation; Irony**. Laughter and entertainment are classed along with *rhetoric*, whilst seriousness and austerity are associated with *argumentation*. In a debate, laughter and seriousness correspond to two antagonistic positioning strategies: if the opponent jokes and laughs, let your answer be stern and to the matter; to an austere technical discourse, answer with a smile and make a pun everybody can understand.

Hamblin mentions three standard *ad* fallacies of entertainment, which occur in two different discursive and interactional organizations (Hamblin 1970, p 41).

1. The arguer as a public entertainer

Ad ludicrum, Lat. *ludicrum*, "game; show", which Hamblin translates as "dramatics".

Ad captandum vulgus, Lat. *vulgus*, "the populace"; *captare*, "to seek to seize".

Rational criticism rejects discursive histrionics, which spare no form of public speech, even conference communications. An address is transformed into a performance. Such shows were put on first by the ancient sophists as staged in Plato's *Euthydemus*, S. **Sophism**. The arguer becomes an actor, "playing to the gallery" or "to the crowd", referring to an actor whose demagogic play appeals to easy popular tastes, S. *Ad populum*.

2. The arguer makes fun of the opponent
Ad ridiculum, Lat. *ridiculum* "ridiculous"

This latter kind of talk is quite distinct from the former. Hamblin uses the labels "appeal to ridicule" and "appeal to mockery" (*ibid.*). Strictly speaking, this is a kind of refutation by the absurd, whereby the advanced proposition is rejected by indicating that it has unacceptable, counter-intuitive, amoral and laughable consequences, **S. Absurd**. The *ridiculous* is not necessarily *comic*, and laughter may be sarcastic rather than joyful.

Hedge's seventh rule explicitly excludes laughing about the opponent, "any attempt to [...] lessen the force of his reasoning, by wit, caviling, or ridicule, is a violation of the rules of honorable controversy" (1838, p. 162); **S. Rules**. This is a special case of the prohibition to substitute *discourse destruction* to *argument refutation*, **S. Destruction**.

Lucie Olbrechts-Tyteca's book, *The Comic of Discourse* (1974), is devoted to the comic exploitation of argumentative mechanisms as jokes.

Laws of Discourse ▶ Scale

Layout of Argument (Toulmin)

In *The Uses of Argument*, Stephen Toulmin presents a general description of the structure of argumentative passages, "the layout of argument" (1958, Chap. III, p. 94-145). This very influential representation is also known as "Toulmin Schema", "Toulmin Model of Argument" or "Toulmin Argument Pattern" (TAP).

1. The structure of the prototypical argumentative dialogue and monologue

1.1 Argumentation as a polyphonic monologue
The following passage is an elementary argumentative cell, putting together the basic components of argumentative discourse according to Toulmin:
— Harry was born in Bermuda, so, presumably, Harry is a British subject
— Since a man born in Bermuda will generally be a British subject, on account of the following statutes and other legal provisions '...' —
— Unless both his parents were aliens / he has become a naturalized American/...(*id.*, p. 103)

The layout of argument combines two major components:

Layout of Argument (Toulmin)

— A central, affirmative component.
— A negative component, staging a challenging voice, that details the "circumstances in which the general authority of the warrant would have to be set aside." (*Id.*, p. 101)

1.2 Argumentation as dialogue

This discourse can be re-played as a prototypical argumentative dialogue, starting from a question, asked by some investigating third party, and developing under the pressure exerted by a challenger.

(i) An Issue
   Question: — *What is the nationality of Harry?*

(ii) A Claim — The arguer answers that:
   Arguer: — "*Harry is a British subject*" (ibid., p. 99).

Making this assertion, the arguer "[is thereby committed] to the claim which any assertion necessarily involves". As a Claim (**C**), it can be "challenged":
   Challenger: — *"What have you got to go on?"* (ibid. p. 98)

(iii) Data — In defense, the arguer "must be able to establish [the Claim] — that is, make it good and show that it was justifiable. How is this to be done?" (*Id.*, p. 97): "we shall normally have some facts to which we can point in its support" (*ibid.*). Here, the arguer gives a fact, or *Data* (**D**) to justify the answer:
   Arguer: — *Harry was born in Bermuda.*

Toulmin's layout is clearly built on a dissensus background. A *Claim* is "a demand for something rightfully or allegedly due" (WCD, *Claim*): a claim is put forward in the context of a contestation "to lay claim to, to assert one's right or title to" (*Ibid.*).
Data are "things known or assumed; facts or figures from which conclusions can be inferred" (WCD, *Data*). The quest for *data* is led with some *claim* in mind, **S. Justification**.
As words, *Data* and *Claim* are correlative words: *Claims* require *Data*, and *Data* is sought for and selected in function of *Claims*; they are explicitly connected through a *Warrant*.

(iv) Warrant — The challenger can still consider that the answer is not fully satisfactory, and "[require]" the speaker to indicate "the bearing on [his/her] conclusion of the data already produced" (*id.*, p. 98):
   Challenger: — *"How do you get there?"* (*Ibid.*)

The arguer is now required to give a *Warrant* (**W**), that is "some rule, principle or inference license" (*Ibid.*):
   Arguer: — "*A man born in Bermuda will be a British subject*" (*id.*, p. 99).

Now the inquisitive challenger may be "dubious" "whether the warrant is acceptable at all" (*id.*, p. 103):

> Challenger: — *"You presume that a man born in Bermuda can be taken to be a British subject; [...] why do you think that?"* (Ibid.).

A warrant is an "authorization or sanction, as by a superior or the law" (WCD, *Warrant*): the "argument — conclusion" gap is sutured by some authority. It can also be "a justification or *reasonable grounds* for some act, course, statement or belief" (*ibid.*). In that case, the warrant would correspond itself to a good reason added to the data; it is generally a law orienting the fact as *a data for this claim*.

Another warrant would give a different orientation to the same data. For example, the warrant "*In Bermuda from late May to October, the climate can be uncomfortably hot and with especially high humidity*" would orient the same fact toward the claim "*Harry certainly knows how to behave under a humid subtropical climate*".

(v) Backing — The arguer is now required to give a *Backing* (**B**), making the Warrant acceptable

> Arguer: — I say that "*on account of the following statutes and other legal provisions: …*" (*id.*, p. 105).

(vi) Rebuttal — For the preceding moves, the challenger asked for formal clarifications; now, he or she turns to substantial objections, such as:

> Challenger: — But "*special facts may make this case an exception to the rule, or one in which the law can be applied only subject to certain qualifications*" (*id.*, p. 101).

Finally the arguer acknowledges these reservations. His or her Claim is a "presumption", only "presumably" true, not "necessarily" so. This must be clearly expressed by a *Qualifier* (**Q**), "indicating the exceptional conditions which might be capable of defeating or rebutting the warranted conclusion (**R**)":

> Arguer: — My claim (C) is probably true, insofar we don't know if "*both his parents were aliens [or] he has become a naturalized American*" (*id.* p. 102-103).

The *Rebuttal* articulates the conditions that, if met, would cancel the reasoning. In integrating the challenger's contributions into his or her reasoning, the speaker introduces co-operation in a situation of inquiry.

The *Qualifier* should not be considered as the expression of a vague mental restriction, just in case things do not turn out as expected. It is the trace of substantial *Rebuttals*, not just any face-saving *softener* or *mitigator*; these terms would not express the link with the substantial rebutting counter-discourse.

2. Representation

Toulmin articulates these six basic elements in the following diagram

Layout of Argument (Toulmin)

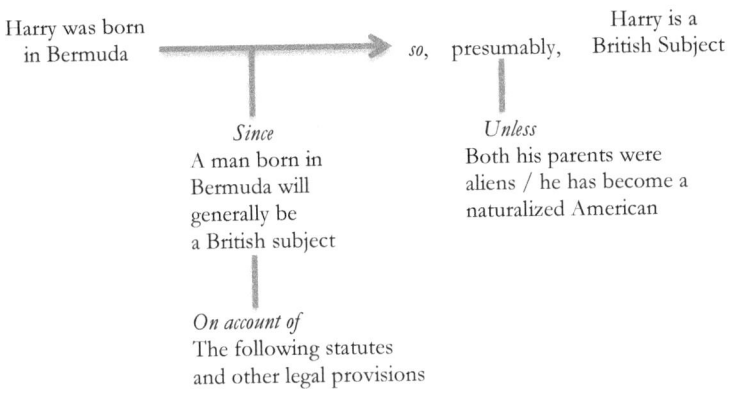

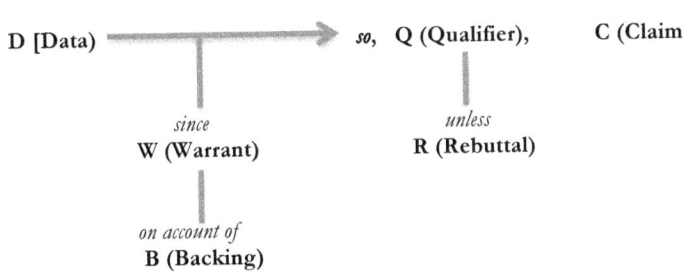

The chain "Data — Warrant — Backing — Claim" represents the *positive* component of the model.
The combination "Modal + Rebuttal" represents the *default@* component of the model.

3. Corollaries

3.1. A legal syllogism

Toulmin speaks of his approach to argument as "generalized jurisprudence" ([1958], p. 7). The instance of reasoning illustrating the layout of argumentation corresponds to a *legal syllogism*, where a law is applied to a fact.

Positive component:
> Law: *Any motorist crossing the yellow line is an infraction and will be fined*
> Recorded fact: *X has crossed the yellow line*
> Conclusion: *This is a violation of the Law and will be accordingly fined*

Default Component:
> *Unless X was driving a fireman's car, an ambulance… on a mission; was participating in a formal parade…; road works were in progress…*

The positive component articulates a premise with a general subject (a law), a premise with a concrete subject (or singular proposition, the argument) in order to deduce a proposition with a concrete subject (the conclusion). It corresponds to a categorization@ process, including an individual into a class, and therefore authorizing the attribution to the individual of the properties and stereotypes characterizing the class. Toulmin's basic example draws attention to the importance of categorization and intracategorial deduction in ordinary argumentative activity. Nonetheless, the warrants are not restricted to categorizing principled. Actually, a Warrant is an instantiation of an *argument scheme@*.

3.2 The "rediscovery of the topoi"

The Warrant corresponds to the traditional argumentative notion of *topos* (Bird 1961), or *argument scheme@*. A topos is a general statement "warranting" the acceptability of the argument and capable of generating an infinity of particular arguments or enthymemes having the same form.

Ehninger and Brockriede have shown how the concept of warrant could cover the main forms of argument schemes, for example "authoritative arguments" ([1960], p. 293):

> — (D) Klaus Knorr states "Soviet leaders calculate that a minor build-up of nuclear power in the NATO countries of Western Europe will add only marginally to the danger of American striking power.
> — *therefore* (C) Soviet leaders calculate that a minor build-up of nuclear power in the NATO countries of Western Europe will add only marginally (to the danger of American striking power).
>
> — *Since* (W) what Knorr says about the power of nuclear weapons is reliable
> — *Because* (B) Knorr is a professor at Princeton's Center of International Studies / is unbiased / has made reliable statements on similar matters in the past / etc.
>
> *Unless* (R) Other authorities more qualified than Knorr say otherwise / special circumstances negate or reduce Knorr's usual reliability as a witness.

Accordingly, the specific objections and counter-discourses attached to a given argument scheme will come under the Qualifier - Rebuttal subsystem.

3.3 Open foundations

Let us suppose that Harris was born not in Bermuda but in the *Falkland Islands* (English name) also called *Islas Malvinas* (Argentine name). Then, the Backing mentioning the statutes on British nationality, would possibly be supplemented by an evocation of the right of occupation, conquest and the right of the strongest", considering the complex history of the islands.

Basing the Warrant on a Backing opens a potential regression to infinity, the guarantee needing itself to be guaranteed. The same regression could be observed on the argument, which may also be challenged.

3.4 Scientific calculation and the erasing of the *rebuttal* component

Toulmin's layout is a favorite among scientists interested in argument. The following example, which is less often quoted than the preceding one, corresponds to the expression of a scientific prediction based on a calculation involving laws derived from experience and observation (1958, p. 184):

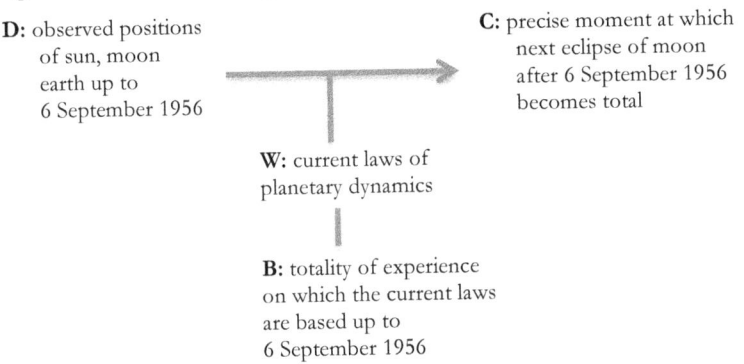

The general premise is replaced by a calculus based on physical laws. The disappearance of counter-discourse (Modal + Rebuttal) characterizes the transition to mathematical calculus based upon stabilized scientific content

Legal Syllogism ▶ Layout; Categorization; Definition

Likely ▶ Probable

Linked Argumentation

Linked (or *coordinate*) argumentation is defined in relation with two different issues, as:
(i) An argumentation whose conclusion is based on several statements *combining* to produce an argument (whose conclusion is supported by a set of *interrelated* premises). The issue is about the link between *statements*, the sum of which constitutes a single argument; the notion of link being then constitutive of that of *argument*.
(ii) An argumentation whose arguments are *sufficient* for the conclusion only if they are *taken jointly*. The issue is about the mode of combining *arguments* so as to produce a conclusive conclusion. The notion of link is then constitutive of that of *conclusive argumentation*.
S. Convergence, Linked, Serial

1. Statements combined so as to build an argument
A *linked argumentation* is defined as an *argumentation based on linked premises*. A premise (major, minor, S. Syllogism) is defined in relation to a conclusion:
> Logic. a proposition supporting or helping to support a conclusion (Dic., *Premise*)

Linked Argumentation

The expression "linked premises" can therefore sound pleonastic. In reality, *propositions* or statements are linked so as to function as *premises* supporting a conclusion.

Syllogistic reasoning has a linked structure: "*all members of this Society are more than 30 years old*", is an argument in favor of "*Peter is more than 30 years old*" only when combined with the proposition "*Peter is a member of this Society*".

Representation:

Similarly, according to Toulmin's representation the assertive component has a linked structure. The "data" statement becomes an argument only insofar as it combines with "warranting" and "backing" statements. S. Layout.

Representation:

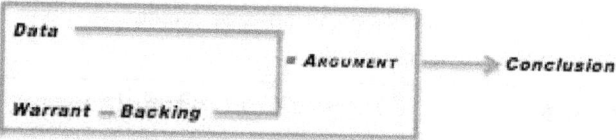

2. Convergent and linked argumentation

The concepts of link and convergence do not describe same-level phenomena: several *arguments* converge to (point to) the same conclusion, whilst several *statements* are linked in order to build an argument for a given conclusion.

Convergent arguments are made of two or more co-oriented arguments, each of them having, by definition a *linked* structure, as shown in the preceding paragraph. The complete schema of convergent argumentation therefore looks as follows:

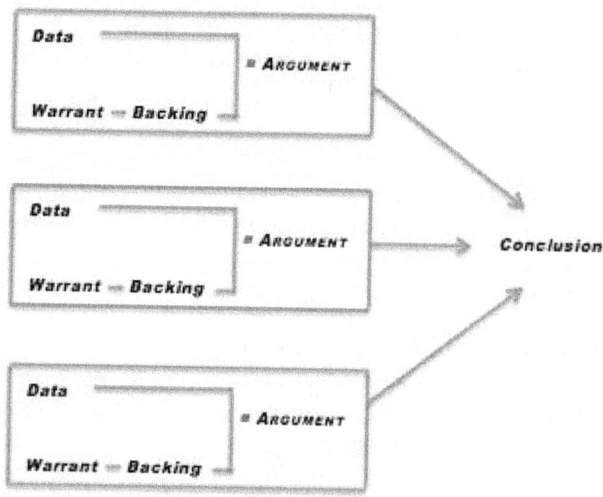

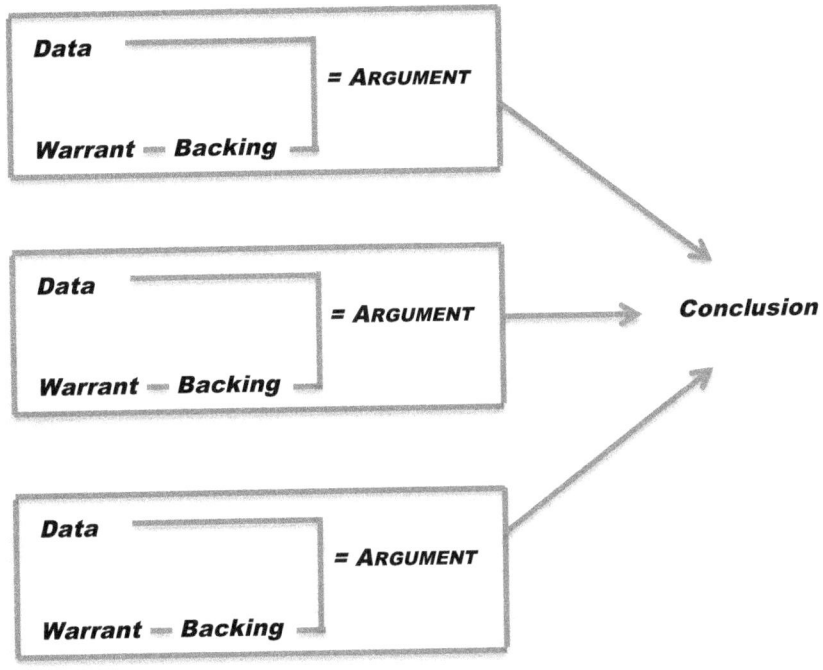

2.1 Arguments linked to produce a conclusive conclusion

The linking effect also affects convergent argumentation, the strength of which is not just in the addition of the individual strength of the added arguments. For example, an argument *from necessary signs* can combine necessary indices into a necessary and sufficient bundle. Likewise, *case-by-case* arguments, when exhaustive, benefit from a binding effect, giving to the whole greater strength than would be achieved by the mere addition of each of the parts. **S. Signs; Case-by-case**.

2.2 Convergent or linked argumentation?

The technique used to answer this question is a) consider a conclusion supported by a set of statements, b) consider a particular statement, c) look what happens if it is false or suppressed (Bassham 2003):
— If what remains is still an argumentation, we are dealing with a *convergent* argumentation:
        Peter is clever and personable, he will be a great negotiator
        Peter is clever, he will be a great negotiator
        Peter is personable, he will be a great negotiator

All these argumentations are admissible; *"Peter is clever"* and *"Peter is personable"* are two convergent, co-oriented arguments giving rise to the same conclusion " *Peter will be a great negotiator"*.

— If what remains is not an argumentation, we are dealing with a linked argumentation:

>(1) It rained and the temperature is below 0°C, there should be black ice on the road.
>(2) It rained, there should be black ice on the road (wrong)
>(3) The temperature is below 0°C, there should be black ice on the road (wrong, unless one adds the premise *"low temperatures generally goes with wet roads"*).

Discourse (1) is an explicit, valid and sound argumentation. Discourses (2) and (3) are still argumentations, but they are not valid and sound as they are. To make them sound, missing premises, corresponding precisely to the suppressed statements, must be added.

The usefulness and practicability of the convergent / linked distinction is challenged (Goddu, 2007). Walton considers that its merit is in its ability to capture the different conditions of the refutation for the two constructions. To refute a linked argumentation, one must simply show that one of the premises is false or inadmissible; to refute the conclusion of a convergent argumentation, each converging argument must be tested separately (Walton on 1996, p. 175). The arguer can grant one of the arguments in the case of convergent argumentation, but cannot give up a premise in the case of linked argumentation.

Essentially, one must decide whether one or more good reasons are involved in the argumentation, that is to say, one must structure the verbal flow by proposing coherent semantic blocks supporting the conclusion.

Logic: A Branch of Mathematics, an Art of Thinking

1. Traditional logic

1.1 The Aristotelian framework

Aristotle does not use the word "logic" in his logical and ontological writings gathered in the *Prior* and *Posterior Analytics*. In his own words, he deals with "demonstrative analytical behavior (reasoning, discourse)", which corresponds "to the current meaning of the term *logic*." (Kotarbinski [1964], p. 5; Woods 2014). The *Posterior Analytics* defines scientific knowledge:

> We attain knowledge through demonstration [...] I call demonstration a scientific syllogism. (*Post. An.*, I, 2; Owen, p. 247)

It follows that "it is necessary that demonstrative science should be from things true, first, immediate, more known than, prior to, and the causes of the conclusion" (*ibid*).

In a note added to this passage, Tricot points out that "syllogism is the *genre* ("producer of science") common to demonstrative, dialectical and rhetorical syllogisms; scientific is the specific *difference* separating demonstration from dialectical and rhetorical syllogisms" (In Aristotle, *SA*, I, 2, 15-25; Note 3 p. 8). The scientific syllogism produces *categorical* knowledge, the dialectical syllogism produces *probable*, that is criticized, knowledge where no categorical knowledge is available, and the rhetorical syllogism produces *persuasive* representations. The position of persuasion@ in the rhetoric of Aristotle should be understood within this framework.

Aristotle's logical theory is based on an analysis of propositions@ as subject-predicate constructions, on a definition of the relations between the four forms of a general proposition and of a theory of syllogism@.

1.2 Neo-Thomist logic

In the Middle Ages, Thomas Aquinas took up the Aristotelian definition of logic and defined it in relation to the reflexivity of the act of reasoning, that is "its ability to reflect upon itself":

> an art is needed to direct the act of reasoning, so that by it a man when performing the act of reasoning might proceed in an orderly and easy manner and without error. And this art is logic, i.e. the science of reason. (*Com.* Post. *An.*, "Foreword")

This definition is taken up by the Neo-Thomist tradition, especially by Maritain, who defines logic as:

> "the art WHICH DIRECTS THE VERY ACT OF REASON" (Maritain 1923, p. 1; capitals in the text)

This definition is taken up by Chenique in his *Elements of Classical Logic* (1975).

The following definition stresses the normative value of logic; "formal logic" is then defined as "a science that determines the correct (or valid) forms of reasoning." (Dopp 1967, p. 11, italics in the original).

1.3 Logic and inference@

Logician-mathematicians define logic as:

> The discipline that deals with correct inference. (Vax 1982, *Logic*)
> Logic is concerned with the principles of valid inference. (Kneale and Kneale, [1962], p. 1)

Or, in a very general way, as the study of the valid forms of deduction@:

> Logic has the important function of saying what follows from what. (Kleene, 1967, Chap. 1, §1)

1.4 Logic is a science

> Logic, like any science has as its business the pursuit of truth. (Quine, 1959, p. xi)

The Stoics first defined logic not in the manner of Aristotle as an organon, an *instrument* (in the service of the sciences), but as a *science*.

1.5 Classical logic

Classical logic (or *traditional logic*, according to Prior 1967) is by nature a *formal* logic: it is one of the revolutionary merits of Aristotle to have introduced a systematic use of variables. Classical logic covers a set of theses and techniques synthesizing proposals of Aristotelian, Stoic or Medieval origin. It consists in two parts:
— The logic of *analyzed* propositions@ or *predicate calculus*, and the *theory of the syllogism*@.
—The logic of *unanalyzed* propositions or *propositional calculus*, which deals with the construction, using logical connectives@, of complex propositions on the basis of simple or complex propositions, and with the determination of *valid formulas* (*logical laws, tautologies*).

Classical logic is based on a set of principles, considered to be laws of thought and rational discourse:
— *non-contradiction*@, "**non-**(**P** and **non-P**)"; a proposition cannot be true and false.
— *excluded middle* (*excluded third*), "either (**P** or **non-P**)"; a proposition must be true or false.
— *identity* "**a = a**", and its practical consequences, such as the principle of *indiscernibility* and *intersubstitutability* of the identicals, and the *unicity and stability* of meaning of the logical symbols in the same universe of discourse (same reasoning).
Contemporary logics no longer regard these principles as laws of thought, but as possible axioms, among others.

The contemporary era saw the multiplication of "unconventional" logical formalisms, sometimes inspired by certain phenomena of ordinary language not taken into account by classical logic, such as time or modality.

2. Logic: An art of thinking, a branch of mathematics

2.1 The three operations of the mind
From Aristotle to the end of the nineteenth century, classical logic was considered the art of thinking correctly, that is, of combining propositions in such a way as to convey the truth of the premises to the conclusion, in a universe of shared and stable symbols and meanings. Logic provides the theory of rational discourse and of scientific argumentation by defining and determining the valid reasoning schemes.

The theory of the three operations of the mind comes from Maritain (1937, §2-3). For a long time, such an approach was abandoned by logicians, who were legitimately motivated by the fantastic potential of expansion and discoveries offered by mathematical models. Nonetheless, it certainly has its place in relation to ordinary thinking, anchored in ordinary language. It indeed illuminates the necessity to take into account the progressive and multi-dimensional construction of an argument, articulating words and concepts into judgments, and propositions into arguing and reasoning. Such a model is quite compatible with the idea of schematization@ as defined in Grize's Natural Logic.

(i) Argumentation as a mental process
As a mental process, *argumentation* is defined as the third "operation of the mind", *apprehension, judgment* and *reasoning*.
— *Apprehension*: the mind grasps a concept, "man", then delimits it scope: "some men", "all the men".
— *Judgment*: the mind constructs a proposition, affirming or denying something about this delimited concept: *"some men are wise"*. This judgment is *categorical*, it is true or false and nothing else.
— *Reasoning*: the mind concatenates the judgments without any loss of truth, so as to develop new truths on the basis of known truths.

(ii) Argumentation as a discursive process
In the discursive process, *argumentation* is defined as the third of the three basic linguistic operations: *naming* the concept; *predicating* something of this concept in a statement; and *arguing*.
— *Naming*: Speaking of something clearly delimited. The concept is anchored in language by a term according its *quantity*, **S. Proposition**.
— *Predicating*: Saying something about this delimited concept, that is constructing a proposition (a linguistic *statement*) by imposing a *predicate* on this term.
— *Arguing*: Composing the statements orderly into the premises of a discourse so as to produce a new proposition, the conclusion, developed exclusively from the premises which are already known. *Argumentation* on the discursive level thus corresponds to *reasoning* on the cognitive level.
In Aristotelian logic, the rules of correct reasoning are given by the theory of syllogism@, which distinguishes between *valid* syllogisms and *paralogisms@* (vicious reasoning, fallacies, sophisms).

2.2 Logic as *the* art of reasoning and the emergence of scientific method
In modern times, this view of logic as a theory of discursive reasoning and the assimilation of discursive reasoning with scientific reasoning has been destabilized by the emergence of natural sciences and experimental reasoning, based on observation, measurement, prediction and experimentation, all regulated by mathematical calculation. In contemporary times, this evolution has been complemented by the integration of logic into mathematics. The rules of scientific method include and exceed logic.
From the point of view of argumentation, this evolution began in the Renaissance, and can be traced back to Ramus (Ong 1958), for whom judgment, logic and method must be considered as stand-alone operations we would call epistemic or cognitive, independent from rhetoric and language. The mutation

appears clearly if one compares the Port-Royal Logic, in its full title: *Logic, or, the art of Thinking: Containing, Besides Common Rules, Several New Observations Appropriate for Forming Judgment* of Arnauld and Nicole ([1662]) to Condillac's *Treatise on the Art of Reasoning* ([1796]). In the latter work, the language of the "art of reasoning" is not syllogistically organized natural language, but *geometry*. Rhetorical argument is never considered, as shown by the case of analogy, which is reduced to mathematical *proportion*.

2.3 Mathematization of logic

Logic is by its nature *formal*, it is interested not in the content (in substance, in the particular objects) of reasoning, but in the form. In contemporary times it has been *axiomatized* and *mathematized*. The publication of Frege's *Begriffschrift*, "Concept Writing" in 1879 set the point from which logic cannot be seen as an "art of thinking", but as an "art of calculating", that is, as a branch of mathematics. At the beginning of the twentieth century, classical logic was overwhelmed by the "twilight of self-evidences" (Blanché 1970, p. 70):

> We move from *Logic* to *logics* that can be built at will. And this plurality of logics withdraws its privileges to classical logic, which is now merely one system among others, like them a simple formal architecture whose validity depends only on its internal coherence. (*Id.*, p. 71-72)

To become an axiomatic exercise, logic had to renounce its reflexive and critical function over common thought and discourse. It could no longer provide the model of rationally argued discourse or dialectical exchange. Logic is now the mathematical discipline, which was questioned, in the 1950s and 1970s, by the *Natural*, *Non-formal* and *Substantial* logics. *Classical* logic can indeed also be appended to this list.

2.4 Neo-Thomism: Resistance to the formalization trend

In 1879, the year when Frege published the *Begriffschrift*, Pope Leo XIII established Thomas Aquinas and his interpretation of Aristotelianism as a quasi-official philosophy of the Catholic Church in the *Aeterni Patris* Encyclical. This decision was certainly unfortunate, insofar as it promoted an outdated vision of logic. Nonetheless, it has brought about a powerful trend of research and teaching on classical logic as a method of thought and as an analytic frame for natural language cognition. Substantial developments relating to classical logic constructions and interesting considerations on arguments schemes and sophisms can be found in textbooks for the Neo-Thomist philosophical curriculum at a higher level.

Under various agendas, Maritain's *Logic* (1923), Tricot (1928), Chenique (1975) reflect this continuing interest in classical logic. This trend may be compared and contrasted with the so-called revivals of rhetoric that developed from the fifties onward.

3. Pragmatic logic and argumentative calculations

In a quite different tradition, that of the philosophy or ordinary language, Toulmin was the first to suggest that the formalization movement in logic required an accompaniment and counterpart able to address "logical practices", ([1958], p. 6), mobilizing "substantial" and "field-dependent" argument (*id.,* p. 125; p. 15). He sought a logic which would be a "generalized jurisprudence" (*id.,* p. 7), whose primary purpose would be "justificatory" (*id.,* p. 6).

The logico-pragmatic movement including *non-formal, substantial, natural,* and generally *dialogue* logics, distances itself from axiomatized formalisms to take into account the ecological conditions of argumentation. People argue in natural language, and in a given context; *classical* logic does not meet the second condition, but does meet the first, at least for the restricted aspects of language it can deal with.

Unlike other theories of argumentation, and perhaps in opposition to the utter rejection of logic by the New Rhetoric, *Informal Logic* and *Natural Logic* have retained the word *logic* in their name, perhaps to stress the fact that, beyond their specific difference they do belong to a common genre, S. **Argumentation Studies; Demonstration; Proof.**

These pragmatic logics must combine with ordinary language and subjectivity@. Classical logic has its roots in a severely regimented ordinary language, whilst the speaker of natural language is a virtuoso of contextualization, implicitness and polysemy. These characteristics are constitutive of the efficiency, dynamism and adaptability of natural language in ordinary life circumstances and the possibilities of strategic management of the worlds of action and interaction. Nevertheless, these observations do not imply any rejection of logic: the practice of ordinary discourse necessitates logical competences, just as it necessitates some arithmetical capacities: "*It takes about two hours to reach the refuge, night will falls in about one hour, we will arrive at the refuge in the dark; that is risky*"; "*some mushrooms are edible, not all: you can't cook any mushroom like that, that is risky*".

4. Entries concerning classical logic
— Predicate Logic: S. **Proposition; Syllogism**
— Propositional Logic: S. **Connectives**

Logics for Dialogues

In the second half of the twentieth century, different systems of logic were constructed to give a formal representation of argumentative dialogue.
— In addition to his historical presentation, discussion and critique of the "standard treatment of fallacies" Charles L. Hamblin proposed a "formal dialectic" (1970).
— Paul Lorenzen and Kuno Lorenz developed a *dialogical logic* (Lorenzen, Lorenz, 1978).

— Else Barth and Jan L. Martens constructed a *formal dialectic* for the analysis of argument (Barth, Martens, 1977).

— Jaakko Hinttika studied the *semantic of questions*, and the logic of *information-seeking dialogs* (1981).

— Taking Hamblin's work as a starting point, Douglas Walton and John Woods developed a logical approach to fallacies (Woods, Walton 1989) and to argumentative dialogues (Walton 1989).

The dialogical logic (*Dialogische Logik*) of Lorenzen and the school of Erlangen was developed as a contribution to formal logic. This model extended to apply to the definition of rational dialogue, is a precursor of the pragma-dialectic approach to argument.

1. Logical dialogue game

The logical contribution consists in a method of no longer defining logical connectives@ by the traditional method of truth tables, but by means of permissible or prohibited moves in a "dialogical game". Consider, for example, the connector "&", "and". It can be defined by the truth table method, S. **Connectives**. In dialogical games, "&" is defined by the following moves:

(a) First round:
>   Proponent: **P & Q**
>   Opponent: Attacks **P**
>   Proponent: Defends **P**

If the proponent defends **P** successfully, he wins round (a). If his or her defense fails, the game is over, and the proponent has lost the game. In the language of truth tables, this corresponds to the truth-table line "if **P** is false, then the conjunction '**P & Q**' is false". In other words, the line "if **P** is false, then the conjunction '**P & Q**' is false" is excluded.

If the proponent won round (a), in relation to **P**, the game continues.

(b) Second round, the opponent attacks **Q**.
>   Proposing: **P & Q**
>   Opponent: Attacks **Q**
>   Proposing: Defends **Q**

If the proponent defends **Q** successfully, he wins round (b), and, as round (a) has already been won, the game is won for the proponent. If his or her defense fails, the game is over, the proponent lost the game, and the opponent won it.

In the language of truth-tables, this translates as "**P & Q**" is true: the proponent won; and "**P & Q**" is false: the opponent won.

2. Dialogue logic rules and Pragma-Dialectical rules

Dialogical logic uses three kinds of rules (van Eemeren *& al.* 1996, p. 258)

— *Starting rule*: the proponent starts by asserting a thesis.

— *General rules* on legal and illegal moves in dialogue (see above).

— *Closing rule*, or winning rule, determining who has won the game.

Similar rules apply in Pragma-Dialectic:
— The starting rule corresponds to "Rule 1. *Freedom* — "The parties must not interfere with the free expression or questioning of points of view" (van Eemeren, Grootendorst, Snoeck Henkemans 2002, 182-183).
— The closing rule, or the winning rule corresponds to "Rule 9. *Closing* — If a point of view has not been conclusively defended, the advancing party must withdraw it. If a point of view has been conclusively defended, the other party must withdraw the doubts it has expressed with respect to that point of view" (*ibid.*).
The other rules are intended to ensure the smooth running of an argumentative dialogue in ordinary language aimed at eliminating differences of opinion.

3. A contribution to the theory of rationality

In a work entitled *Logical Propaedeutic: Pre-School of Reasonable Discourse* ([1967] / 1984), Kamlah and Lorenzen aim to provide "the building blocks and rules for all rational discourse" (quoted in van Eemeren *& al* 1996, p. 248). Their basic assumption is that, "in order to prevent them from speaking at cross purposes in interminable monologues, the interlocutors' linguistic usage in a discussion or conversation must comply with certain norms and rules. Only when they share a number of fixed postulates with respect to linguistic usage can they conduct a meaningful discussion" (van Eemeren *& al.* 1996, p. 253). The goal of the enterprise is therefore the construction of an "ortholanguage" (Lorenzen & Schwemmer, 1975, p. 24; quoted in *id.*, p. 253), defining the rational dialogical behavior capable of resolving inter-individual contradictions.

There is obviously a great difference between this approach and the interactional approaches to speech in interaction that began to develop at the same time.

Logos - Ethos – Pathos

In order to build a correct representation of the world, knowledge oriented theories of argumentation focus on phenomenon concerning the *objects* of debate (categorizations; physical surroundings of the facts; probable and necessary signs; causal and analogical networks, etc.), and the representational function of language (well-built definitions, univocity, etc.). The construction and strategic management of *people and their emotions* is essential in the overall orientation of rhetorical discourse towards persuasion and action: its goals are to make people think, feel and act. The accomplished action is the only criterion of successful persuasion, which would be unduly reduced to creating or strengthening the mind's adherence to a thesis, **S. Persuasion**. The rhetorical judge is not persuaded if he does not pronounce in favor of the party who convinced him.
The connections between convictions and actions are far from clear, **S. Motives**. It is said that a MP once replied to someone who tried to convince him to alter

his opinion, *"you can certainly change my opinion, but you will not change my vote"*; this quip highlights the crucial difference between the determiners of representation and those of action. The rhetorical technique provides three instruments of persuasion (*pistis*) respectively drawn from the logos, the ethos and the pathos. These instruments, sometimes called "proofs", are used by the speaker not only to make believe, but also to guide the will and determine the action.

> Of the modes of persuasion offered by the spoken word there are three kinds: the first kind depends on the personal character of the speaker; the second on putting the audience in a certain frame of mind; the third on the proof, or apparent proof, provided by the words of the speech itself. (*Rhet.*, I, 2, 1356a1; RR, p. 105).

All three forms are discourse dependent; *logo-ic* evidence is purely discursive, while *ethotic* and *pathemic* evidence is discursive and para-discursive. The parallel, "ethos, pathos, logos" tends to assimilate these three kinds of evidence, which leads to define rhetorical evidence as any sign, verbal or non-verbal, capable of inducing a belief, **S. Persuasion.**

Cicero and the later rhetorical catechisms assign three goals to the speaker engaged in a persuasion process, respectively achieved through the logos, the ethos and the pathos: speech must *prove (probare)*, *please (conciliare)*, and *move (movere)* (*De Or.*, II, XXVII, p. 114). The speech must first *teach*, via the *logos*. That is to say that the speech must inform, narrate and argue. This teaching thus takes an *intellectual* approach in achieving persuasion, that of *evidence* and *deduction*. Yet information and argumentation may be weakened by the boredom and incomprehension of the audience. The listener must therefore be given peripheral indications, and this is the function of ethos (*"maybe you understand nothing, anyway you can trust me"*). But logos and ethos do not have the power to trigger the "acting out", hence the recourse to pathos. It is not enough to see the good, it is still necessary to want it; the almost physical emotional stimuli produced by the orator, that is the pathos, are supposedly the final determinants of the will and action, **S. Emotion; Pathos; Persuasion.**

Evidence based on logos is considered to be objective, at least the only one of the three to serve as proof in the proper sense of the term. Firstly, it meets, at least partially, the propositional condition for reasoning (to be expressed in an identifiable statement, evaluable independently from the conclusion is supports), so it is open to refutation. In contrast, pathemic and ethotic evidences, by nature subjective, are expressed indirectly, through the subtlest of channels, and are therefore hardly accessible to verbal refutation.

Classical texts insist on the practical superiority of the subjective proofs, ethos and pathos, over objective ones. Aristotle poses the primacy of the ethos@, S. ethos: "[the speaker's] character may almost be called the most effective means of persuasion" (*Rhet.*, I, 2, 1356a10; RR, p. 106), and warns against the overly effective use of the pathos@. Cicero and Quintilian quasi assimilate ethos to pathos@, in order to affirm the practical supremacy of emotions.

M

Manipulation

1. Word and Domains

The transitive verb to *manipulate*, "**No** *manipulates* **N1**" functions within two structures:

Manipulate$_1$: **N1** refers to an object (non-human, inanimate) (*container manipulation*) or body parts (*spinal manipulation*).

Manipulate$_2$: **N1** designates a person as a synthesis of representations and capable of self-determination. *Manipulating$_2$* is exploitative; manipulating people is using them as objects or instruments.

To manipulate is the head of a rich and homogeneous derivational family: *manipulation, manipulator, (non-)manipulatory, (non-)manipulative, outmanipulate*, "to outdo or surpass in manipulating", (MW, *Outmanipulate*).

Manipulation_2 can influence all domains of human activity.
— Political, ideological and religious fields.
— Everyday psychology: *a manipulator, manipulative behavior*.
— Military domain: *White* propaganda comes from domestic source and targets domestic public opinion; it may be misleading. *Black* propaganda has a concealed origin and purpose. It appears to come from a well-meaning and harmless source, although it comes from an evil or enemy source.

Manipulation

— Commercial action and marketing techniques are used to encourage or manipulate people to buy this rather than that or nothing, using different techniques to "bait and hook" the customer, S. **Priming**.

In these different fields, manipulative influence may cross, combine or contradict argumentative persuasion.

2. *Doing together*: from collaboration to manipulation

Manipulation is a resource that may be activated in any situation where a person **M** pursues a goal ϕ. To achieve this goal, **M** requires a contribution to be made by another person, **N**.

2.1 Overt purpose negotiation

(i) M considers that ϕ is in the interest of N, and N agrees

N has a positive representation of ϕ; ϕ is considered important, pleasant, in the individual's interest; **N** pursues ϕ spontaneously, for independent reasons. So, **M** needs **N** and **N** needs **M**; **M** and **N** co-operate to achieve ϕ.

If **N**'s commitment is less immediate, **M** will take a more open approach and will seek to persuade **N** to associate with him or her in order to realize ϕ. **N** knows that **M** intends to make him or her do ϕ, and they will discuss this with one another.

(ii) Doing ϕ is not really in the best interest of N

N doesn't care about ϕ. He or she will not spontaneously collaborate with **M** in order to achieve ϕ. **M** may then act on the will or on the mental representations of **N**.

(a) *Action on the will to do*

In this situation, **M** may undertake to persuade **N** to do ϕ. **M** threatens **N** (*ad baculum*), tries to blackmail or bribe **N** (*ad crumenam*), to move N to pity (*ad misericordiam*), to charm or seduce **N** (*ad amicitiam*), S. **Threat; Emotion**.

N still has a rather negative view of ϕ. But **M**'s arguments, if they are arguments at all, have transformed **N**'s willingness to act, and he or she will ultimately agree to act in favor of ϕ even if he or she does not like it. **N** does ϕ reluctantly, *as a favor to M*. The question arises as to whether **N** has been manipulated.

(b) *Action on representations of the action to be taken*

M may reframe ϕ so that it seems to be pleasant or favorable, in **N**'s in best interests. As in case (i), **N** agrees to do ϕ because it seems beneficial.

In case (a), **N** will do a job that he or she knows to be dangerous, because it is well paid. In case (b), **N** will do a job, hazardous or not, which he or she does not consider to be dangerous. **M** can combine the two strategies: *"you can do this for me, it's not so dangerous"*. These two situations are not necessarily manipulative. **M** has openly presented the goal ϕ to **N**; **N** was persuaded to do ϕ for arguably good reasons; the work may not actually be all that dangerous, and it is well paid.

M behaves manipulatively only if he or she knows that the work is dangerous, but knowingly misrepresents it, concealing the danger to **N**. Lying is the basis of manipulation.

(iii) Doing ϕ is against the interests and values of N.

Now, ϕ is clearly contrary to the interests of **N**. In normal circumstances, **N** would automatically oppose **M** in his or her attitude to ϕ. Nevertheless, it is still possible for **M**:

— To persuade **N** to willfully do something contrary to his interests or values. In an extreme case, for example, **N** might be persuaded to commit suicide or sacrifice him or herself, even if he or she does not wish to die, in the name of a higher interest or value, *"God, the Party, the Nation, asks you to..."*; *"You must sacrifice your children to make our cause prevail"*.

— To persuade **N** that the action to which he or she is urged is good, and in his or her best interest. **M** urges **N** to sacrifice him or herself for example, even if **N** is not eager to die, *"you will go to le se"*. The discourse and arguments through which **M** persuades **N** to consent to ϕ are manipulative because they do not respect a hierarchy of values that is considered natural. On the basis of highly questionable arguments, **N** was induced to do something to which no person would reasonably commit. This is a case of brainwashing.

2.2 Covert purpose negotiation

In the cases described above, **N** is more or less aware of what he or she is committing to doing. *Deep manipulation*, however, is characterized by **M**'s hiding his or her actual intentions or the true nature of the goal ϕ, which in reality is unacceptable to **N**. **M** will use a secondary goal, as a decoy (ϕ_d):

(i) ϕ_d is positive for **N**: **N** is led to believe that it is in his or her interests to do ϕ_d.
(ii) ϕ_d leads fatally to ϕ.
(iii) **N** ignores (2).
(iv) **N** achieves the decoy goal; **M** pockets the bet.

There is not necessarily a verbal exchange, or even contact between **M** and **N** during this process. **N** suffers any damage, and may or may not understand that he or she has been manipulated. **N** may lose the game without even knowing he or she was playing a game. One example might be that of a salesman. A large encyclopedia, for example, is sold to consumers who, although delighted by its purchase, hardly know how to read, have no use for this type of book, and, in any case, cannot afford to pay the bill. The salesman has achieved the feat of framing the *sales interaction*, ϕ, as an ordinary, *friendly conversation*, ϕ_{decoy}.

3. "Pious lies"

Manipulation achieved via a *pious lie* is what we see in action when, for example, we put sweeteners in cod-liver oil administered to children; or what Calvin attributes to monks who wish to bring people to their salvation by any means, because *the end justifies the means*. The following excerpt is about the multiplication of the relics of the true cross:

> Now, what other conclusion can be drawn from these considerations but that all these were inventions for deceiving silly folks? Some monks and priests, who call them pious frauds, i.e., honest deceits for exciting the devotion of the people, have even confessed this.
>
> John Calvin, *A Treatise on Relics*, [1543][1]

The concept and practice of "patriotic fraud" in elections might be seen as a modern day version of the practices that Calvin attributes to medieval monks.

4. Manipulation and power practices

The status accorded to manipulation is based on ideas of power and action. Should power be exercised by reason and valid argument, or, in a Machiavellian perspective, does it necessarily require the use of force and lies?

> I must confess that what is called the cultured circles of Western Europe and America are incapable of understanding the actual balance of power. These people must be considered deaf-mutes.
>
> To tell the truth is petty bourgeois prejudice, while lying is often justified by the objectives. (Lenin, quoted in V. Volkoff, [*Disinformation, A Weapon of War*], 1986[2]

Discussing the vital necessity of keeping the place and time of the Normandy landing a secret, Churchill said:

> *"In war-time"*, I said, *"truth is so precious it should always be attended by a bodyguard of lies"*. (Discussion of Operation Overlord with Stalin at the Teheran Conference, Nov. 30, 1943[3])

The answer to the previous question may be that:

> [The] truth is incontrovertible. Panic may resent it, ignorance may deride it, malice may distort it, but there it is.
>
> Winston Churchill, Speech in the House of Commons, May 17, 1916[4]

---

[1] John Calvin, *A Treatise on Relics*. Trans. and introd. by Valerian Krasinski. 2nd ed. Edimburg: Johnstone, Hunter & Co, 1870. Quoted after http://www.gutenberg.org/files/32136/32136-pdf.pdf (08-17-2017)
[2] Vladimir Volkoff, *La désinformation, arme de guerre*. Lausanne: L'Âge d'Homme, 1986, p. 35.
[3] In *The Second World War*, Volume V: *Closing the Ring* (1952), Chapter 21 (Teheran: The Crux), p. 338.
[4] Quoted after https://en.wikiquote.org/wiki/Winston_Churchill

5. Argumentation and manipulation

5.1 Argumentation and propaganda

The study of discursive schematizations is the study of the processes through which the speaker arranges a synthetic, coherent, stable meaning. This constructed meaning is neither a manipulation$_2$, nor reality itself, nor an illusion of reality, but simply a significant view taken of reality, S. **Schematization**. To communicate, the speaker must necessarily *manipulates$_1$* the discursive material, but this process is not necessarily intended to *manipulate$_2$* the interlocutor. Manipulation$_2$ presupposes deliberate falsehood. Considering that all speech is necessarily manipulative would amount to an undue dramatization of the process of signification.

A very tenuous thread separates the study of argumentation as defined by the *Treatise on argumentation* and that of political propaganda, as defined by Domenach. For Perelman & Olbrechts-Tyteca, "the object of the study of argumentation is the study of the discursive techniques allowing us *to induce or to increase the mind's adherence to the theses presented for its assent.*" ([1958]/1969, p. 4; italics in the original). Domenach defines the object of propaganda as "to create, transform or confirm opinions" by means of multi-semiotic processes (image, music, demonstration and crowds) (Domenach 1950, p. 8). This difference may be that between ratio-propaganda and senso-propaganda as defined by Tchakhotine (1939, p. 152). The former is effective "by persuasion, by reasoning", and the second by "suggestion" (*ibid.*), that is, by manipulation$_2$.

5.2 Manipulation and lying

Lies and concealed intentions crucially oppose argumentation to manipulation; a lie being understood as an active lie, asserting a known falsehood, and a passive lie, as failing to tell the whole truth, or relevant parts of it. Manipulative discourse is based on *lies*, which may be presented as "alternative facts". Disorienting hints, false cues and misleading prospects are put forward as truths. Even some *true* information may be mixed with *false* information to make it believable.

The denunciation of manipulative discourse is a denunciation of lies; but there is no formal mark of errors and lies; exposing lies necessitates a substantial knowledge of the issue. For this reason, as Hamblin says, "[the logician] is not a judge or court of appeal: and there is no such judge or court" (1970, p. 244); but, as a responsible citizen, he or she must denounce manipulation in favor of a better-informed picture of reality, S. **Evaluation**.

Many Questions

1. Many questions as a dialectical fallacy

Dialectical games use an ortho-language (**S. Logics for Dialogue**), that is to say a language game derived from ordinary language and interaction supplemented by a system of conventional rules. The problem about the so-called fallacy of "many questions" originates first in two specific rule of the dialectical game. Firstly, in ordinary language, one single interrogative sentence may contain many questions and many answers, a property derived from the fact that sentences have several layers of meaning. Logical dialectical game prohibits the exploitation of this linguistic resource, and requires the use not of ordinarily phrased sentences, but of *propositions*, a proposition being defined as "a single statement about a single thing" (*id.*, §6). Secondly, logical dialectical game authorizes only yes/no answers.

The linguistic phenomenon of *loaded questions* (also known as *many* or *multiple* questions) is examined by Aristotle in the context of a dialectical exchange, where they are considered a fallacious discursive maneuver, **S. Fallacies**. It consists in "the making of two questions into one" (Aristotle, *SR*, part 5).

Consider a set composed of bad things and good things (*id.*, §5). The misleading question is: "*is the set good or bad?*". The answer "good" will be rejected by alleging a bad thing, and the answer "bad" by alleging a good thing. (*ibid.*). The correct answers are *yes* for the first component and *no* for the second one; but the smart sophist will refute the *yes* by alleging the second component, and vice-versa.

The case of the half white and half black picture might be more convincing. The sophistical dialectical question is: "*is it (=the picture) black* (resp. *white*)?". As there are only two authorized answers "*yes*" or "*no*", they will be refuted respectively by focusing on the white (resp. black) part of the picture (*id.*, §5). Ordinary language would simply give the sensible non-dialectal answer, "*this part is white and the other black*".

One can imagine that the question "*is anger a good thing?*" exhibits that kind of problem. The answer *yes* is refuted by any negative aspects of anger such as violence or lack of self-control, whilst the *no* is undermined by any case of "righteous anger".

The fallacy of many questions is thus a clear example of fallacy defined as a breach of dialectical rules, **S. Fallacies (II)**. The issue of many questions arises as a by-product of the rules of the dialectical game, and there is no need to import it as such in the analysis of ordinary argumentation. Rhetorical argumentation has no problem with confusing questions; they are answered with a conceptual *dissociation*@ or a *distinguo*@.

2. Presupposition

Natural language questions might concern statements containing presuppositions that are, or are not, considered acceptable by their recipient:

    S1: — *You should think about the reasons for the failure of your policy.*
    S2: — *But my policy has not failed!*

S2 rejects the presupposition of S1 "your policy has failed".

The imposition of a presupposed judgment is contrary to the logical principle that a statement expresses a single judgment (if it contains several judgments, each must be asserted separately). Consequently it contradicts the dialectical rule requiring that each proposition be explicitly accepted or rejected by the respondent. S1 could therefore ask S2 the question "why P?" only if S1 and S2 previously agree on the existence of P. From a Perelmanian perspective, the question of presuppositions should be settled within the framework of prior agreements, S. **Conditions for discussion.**

The problem is that, in ordinary language, all statements are more or less "loaded" not only by their orientation@, but also by their implicit contents of various kinds, some of them inferred from oriented words. In reality, it is always possible to extract litigious presupposed or infer propositional contents from a statement and to subsequently hold the interlocutor liable for it. Let us consider a discussion between a banker and a recriminating customer trying to get a better interest rate:

    S1_1:   — *I went to the bank just across the street from my house, and they immediately offered me a loan at a lower rate than the one you proposed to me!*
    S2:   — *It's because they wanted to have you as a customer.*
    S1_2:   — *Because you do not want to* keep me *as a customer?*

S1_2 extracts from or infers from S2's intervention an implicit content that S2 certainly rejects, but nevertheless shows the banker that a different explanation is needed. This move can be considered to be a special straw man@ maneuver (de Saussure 2015).

Map ▶ Script

Matter

    Ad judicium argument; Lat. *iudicium*, "judgment".
    Ad rem argument; Lat. *res*, "reality, thing; point of discussion, question".
    Ad orationem argument; Lat. *oratio*, "language, speech".

Three *ad* – arguments try to capture the general idea of *relevant* argument, S. **Relevance**: argument appealing to *the judgment* (*ad judicium*); to *the matter* (*ad rem*); to *the discourse itself* (*ad orationem*). These labels convey a positive appreciation of the arguments they refer to, as opposed to other kinds of arguments deemed irrelevant.

1. Appeal to judgment (*ad judicium*)

Locke [1690] opposes the *ad judicium* argument, declared valid, to three kinds of argument he considers fallacious, the arguments *ad ignorantiam*, *ad hominem*@ and *ad verecundiam*, S. **Collections (II)** ; **Ignorance; Humility**. The argument *ad judicium* is defined as:

> The using of proofs drawn from any of the foundations of knowledge or probability. This I call *argumentum ad judicium*. This alone of all the four, brings true instruction with it, and advances us in our way to knowledge. (Locke [1690], Vol. 2, p. 411)

The following declaration shows that this validity is derived not only from judgment but also from "the things themselves", which can correspond to *ad rem* arguments:

> [truth] must come from proofs and arguments, and light arising from the nature of things themselves. (*Id.*, p. 411-412).

This is the reference definition for the *ad judicium* argument(s), as based on scientific procedures and criteria, and developing object-based knowledge. This is why *ad judicium* is not strictly speaking an argument scheme in itself, but instead covers the whole scientific methodology.

Fallacious *ad hominem* and *ad verecundiam* arguments also appeal to judgment, at least to a calculus: *ad hominem* appeals to consistency; *ad verecundiam* is based on a sense of humility or personal insufficiency that can be well grounded. Actually, these arguments are said to be fallacious because they are *subjective*. Subjectivity does not here mean "arbitrary", but rather nonuniversal, context-bound, taking the circumstances of the speech situation and the speaker's transitional state of knowledge into account, what he or she knows, believes or dares say or not.

According to Locke's definitions, the correct argumentative method is the name of scientific method when applied to social questions and human projects. Argument thus conceived rejects the speaker and his system of knowledge as consistently relative. It is the antithesis of what Grize calls a logic of the subject.

Ad judicium, **a homonymic label** — Various non-equivalent, definitions are attached to the *ad judicium* label. This can prove somewhat confusing.

(i) Perhaps referring to Locke, Whately considers that the *ad judicium* label designates "most likely the same" as the *ad rem* argument ([1832], p. 170), that is, argument *to the matter*. In this case the terminology would just be redundant, which is relatively benign.

(ii) A dictionary of theology defines *ad judicium* as: "an argumentation calling on common sense and general opinion to validate a position"[1] which is something quite different from Locke's perspective.

---

[1] http://carm.org/dictionary-argumentum-ad-judicium (20-09-13).

(iii) And Bentham uses the *ad judicium* label to designate a series of fallacies of confusion (Bentham [1824]), **S. Political Arguments.**

2. Argument addressing the *subject*, or the *matter*, under discussion (*ad rem*)

The labels *ad rem* and *to the matter* could be taken absolutely, in the Lockian style, as referring to arguments producing knowledge based on natural objects by scientific method. They should rather be taken as addressing the *relevant* facts and the *central issue* under discussion, that is the *substance of the controversy*, the heart of the *matter under discussion*, which defines the argumentative situation. Discussion of the matter is avoided, for example, when somebody accused of *corruption and embezzlement of public money* answers to the charge by a counter accusation of *misogyny*, using a classical argument substituting a private and shameful motive@ for a public and honorable one.

It should be stressed that, in that sense, an argument *to the matter* is a quite different thing from an argument *drawn from the subject@ matter of the law*.

*Indirect proof*s correspond to a reasoning from the absurd@, and they can be *to the matter* or not. The same is true for *peripheral arguments*, exploiting indices accidentally associated with action. A peripheral argument on the person, for example, is to the matter if relevant to the discussion: *a witness saw him near the scene of the crime*; or not really so: *a witness says the suspect is a good friend of his*, **S. Index, Circumstances.**

3. Argument addressing the discourse of the opponent (*ad orationem*)

The infrequent label *ad orationem* refers to an argument addressing what *has actually been said* by the opponent, in opposition to addressing, for example, her person:

> He answers *ad orationem*, and not *ad hominem*.

This label usefully refers to the concept of an argumentative question as defined by the replica addressed to an accusation. It stresses the verbal condition of argumentation, and might be rendered as *argument addressing the question as formulated*, as it has been defined, legally or by mutual agreement. Addressing the letter means here being *relevant@*, which is in line with addressing the matter.

However, taking an alternative meaning, addressing the *letter* is in opposition to addressing the *spirit*, the intention of the discourse. For example putting the focus on the discourse to show that the accusation is badly formulated is an address *ad orationem*, it does satisfy the turn-taking obligation, but doesn't bring in any substantial information about what really happened, **S. Destruction of discourse.**

Taken literally, the label *ad orationem* might correspond to the *ad litteram* argument, **S. Strict sense.**

4. Discussion

Argument to the matter and validity — From an evaluative perspective, arguments to the matter are the only ones whose strength and value are worthy of discussion and should be kept in the record of the case, which does not mean that they are automatically validated. A party can invoke *a precedent*, for example, which is clearly a legitimate and substantial move, when dealing with the matter. The precedent can, however, be criticized and rejected: this argument *on the matter* is finally declared *invalid*, irrelevant to the issue under discussion.

Whether an argument addresses the matter or not, what the relevant criteria are in order to reach a decision on the given point, is often disputable. In formal situations, the role of the third party is to enact this arbitration, possibly against the strong convictions of the parties.

Logos-based arguments — Misleading associations could lead one to think that the arguments related to the *logos* are *logical* and therefore *objective*, dealing with *objects*, and, as a consequence, with the *matter* and *substance* of things. As such, logos derived arguments would be opposed to ethotic and pathemic arguments, which would be linked to subjectivity, as in scientific matters.

In everyday argumentation, as well as *logo-ic* arguments, ethotic and pathemic arguments exploit the *logos,* understood as language and discourse. In an argumentative situation, however, it is the question which determines what the object, the *matter* and substance, of the debate is. Arguments referring to persons, their values and emotions are substantial (*ad rem* and *ad judicium*) to the extent that they are relevant to the question. Recalling the previous convictions of a person is not irrelevant in all contexts. The description of the state of emotional shock in which the victim was found, for example, might be relevant in court. The problem is distinguishing between the aspects of a personality which are relevant, and those which are not. This process is particularly complicated when the persons involved are parties in the argument process.

Metaphor, Analogy, Model

From a rhetorical point of view, metaphor is valued as a cryptic analogy, the clarification of which is entrusted to the audience. The key difference between metaphor and analogy is that, while analogy keeps the two domains it relates separate and distinct, metaphor tends to conflate them.

According to Aristotle, metaphor is the most efficient persuasive instrument of ordinary discourse. From the perspective of an anti-rhetorical theory of argumentation, metaphor is abundantly misleading. But metaphor is also a powerful cognitive tool for building representations, and better understanding complex situations. Metaphor applies the language of a model, i.e. the Resource domain (the metaphorical term) to an actual situation, the Problematic domain to which belongs the (sometimes missing) metaphorized term, S. **Analogy (IV)**.

1. Metaphor put on trial

If metaphor is defined as a figure@, and figures are defined as ornaments, then metaphor is misleading in all its dimensions, **S. Fallacy; Rhetorical discourse**. The metaphorical statement is false: "*The voter is a calf*" said Charles de Gaulle; but the *voter* (proper term) is not a *calf* (metaphorical term) the voter is a human being. Metaphor systematically commits a *category mistake*. One can also accuse metaphor of creating *ambiguity*, because it introduces a parasitic level of signification, the figurative meaning, running parallel to the proper, standard meaning.

Metaphor pops up, creating a surprise and introducing an emotion (*ad passiones* fallacy); it entertains the audience (*ad populum* fallacy), thus sacrificing *docere* to *placere*. It turns the reasonable arguer into an actor (*ad ludicrum* fallacy). Metaphor is therefore the discursive *distractor par excellence*, putting the audience on a false trail, and confusing the honest literal individual in his or her pursuit for truth. **Relevance; Red Herring; Straw Man.**

Therefore, metaphor is, and should be, banished from serious argumentative discourse, as it is from scientific language; it can be helpful only when reformulated as a comparison (Ortony 1979, p. 191). Nevertheless, it should be pointed out that metaphor is active and welcome to stimulate creativity and facilitate science transmission and popularization.

2. Metaphor, the ultimate weapon of persuasion?

Persuasion, *pistis*, is produced in three ways "(1) by working on the emotions of the judges themselves, (2) by giving them the right impression of the speaker's character, or (3) by proving the truth of the statements made" (Aristotle, *Rhet.*, 1403b10; RR p. 397), in the latter case, persuasion emerging "from the facts themselves" (*ibid*).

Ideally, the issue should be discussed on the basis of facts and proofs: "we ought in fairness to fight our case with no help beyond the bare facts: nothing, therefore should matter except the proof of those facts" (*ibid* 1404a1; RR p. 399). But normal people are not perfect, and "owing to the defects of our hearers", and of our "political institutions", "the arts of language cannot help having a small but real importance" in public discourse and education — but not in geometry: "nobody uses fine language when teaching geometry" (1404a1-10, RR. p. 399).

So, refined language is the most effective tool of persuasion. Persuasion by emotion (pathos) and image (ethos) is produced, orally, by the "oratorical action"; in writing, by the stylistic arrangement of facts, "because speeches of the written or literary kind owe more to their diction than to their thought" (1404a15; p. 401). Metaphor is the supreme tool of written discourse "both in poetry and in prose"; it "gives style clearness, charm and distinction as nothing else can" (1405a1-10; RR, p. 405). The conclusion is clear: metaphor is the ultimate weapon of persuasion, defined as the art of "[hiding one's] purpose successfully" (1404b20; RR, p. 403), and charming the audience, **S. Logos, Ethos, Pathos.**

Contemporary approaches to metaphor unanimously consider that metaphor derives this power from the intrinsic element of *surprise*, resulting from the perception of an anomaly in the discourse, a rupture, an inconsistency, an incongruity, a contradiction of logic, in short, a discursive coup, to the audience's delight. Pleasure cannot be rebutted, and metaphor is thus considered to be quasi inaccessible to refutation — in reality, it is: cf. infra, §4

3. Metaphor and interpretative cooperation

Using a metaphor, the speaker openly seeks the interpretative cooperation of the audience; creating cooperation, metaphor strengthens the importance of prior agreements. Note that the same functional explanation is given for the derivation of enthymemes from underlying syllogisms. In both cases, the argumentative (i.e. effective, persuasive) function of the enthymematic or metaphoric condensation is the activation of the partner, **S. Enthymeme §5.**
This analysis assumes that the *non*-argumentative metaphorical language, or the *non*-elliptic syllogism would be transparent, or less complex than the metaphorical language, and that their direct interpretation would not require the same degree of cooperation from the audience, which is not self-evident.

4. Metaphor as analogy

Metaphor finds smart solutions to the riddle of metaphor:
> Metaphor is the dreamwork of language and, like all dreamwork, its interpretation reflects as much on the interpreter as on the originator. The interpretation of dreams requires collaboration between a dreamer and a waker, even if they be the same person; and the act of interpretation is itself a work of the imagination. So too, understanding a metaphor is as much a creative endeavor as making a metaphor, and as little guided by rules. (Davidson 1978, p. 29)

In *The Interpretation of Dreams* (1900) Freud defines dream-work as the process by which the *latent* content of a dream is covered by its *manifest* content, by displacement, distortion, condensation and symbolism. The metaphor "metaphor / dream work" is difficult to reject, even if it commits the fallacy of trying to go *ad obscurum per obscurius*, that is, it attempts to illuminate the dark (metaphor) by the darker (the dream work).
The metaphor is a model, (Black 1962), and an imperialist model, urging one to identify the metaphorized reality within the metaphorical world:
> L1 — *we should do something with the economy...*
> L2 — *with the "economy-casino" you mean*
> L1 — *oh yes, all these addicted traders should be banned from the market!*
> Reconstructed analogy *"as addicted players are banned from casinos"*

Saying that *"the voter is a calf"* is to mean that *"the voter is hesitant, weak and can be manipulated like a calf"*, calves being here the stereotypical animal combining these characteristics. The metaphor opens new perspectives, and legitimates a new set of inferences about voters: if they are categorized as calves, one can make them adopt behaviors directly contrary to their interests, e.g. to lead them to a more or less metaphorical slaughterhouse.

Metaphor draws its argumentative strength from an analogy pushed to identification. Structural analogy *explicitly* brings together two domains, respecting their specificities; the domains are confronted, not assimilated. Metaphor renders the comparison implicit, negates the metaphorized domain, assimilated to the metaphorical one. This is why the reconstruction of the analogy underlying the metaphorical expression betrays metaphor: it splits apart what metaphor has joined together. *Peter is a lion*: the language referring to lions is substituted for the language referring to humans; we are not far from the hyper-unitary coherent Renaissance world where everything is mirrored in everything, **S. Analogy (I): Analogical thinking**.

5. Jumping from analogy to identity?

Analogy can be defined as a partial identity. The question of possible profound identity, underlying immediately discernible differences plays an essential role here:

> Snowdrifts are like corrugated iron.
> Snowdrifts are like dunes.

The syntactic structures of these two statements are identical, both propose the image of "waves" to the interlocutor, and a key common semantic feature, /waving/. But the second comparison is deeper; it opens the way to a theory. It introduces an analogy of proportion:

> snow : snowdrift :: sand : dune

It suggests that the analogy can be explained by the action of wind on, respectively, the snow particles and sand grains. It puts the hearer on the way to the construction of a physical-mathematical model covering the two phenomena (with due respect paid to the differences between the two kinds of particles, grains of sand and snowflakes, and their respective laws of agglomeration). From two apparently distinct phenomena (one can know what a dune is without knowing what a snowdrift is, and vice-versa), we end up with the problem of a unifying abstract representation: can the same physical model account for the two phenomena?

Establishing an analogy may be considered to be the first step toward the affirmation of an in-depth identity. Such a shift, from *explanatory analogy* to *identity* is at the center of a class of arguments about analogy, which fit perfectly into the framework of a vision of metaphor, not only as a model but also as the genuine essence of the metaphorized phenomenon.

6. Mole rats "societies", human society: metaphor or identity?

The following texts and information are taken from S. Braude & E. Lacey, "*A revolutionary monarchy: the society of mole rats*"[1]. Mole rats are mammals, precisely hairless rats, living in "groups" or "communities" (the difference is relevant); they exhibit behaviors evoking those observed among social insects, like ants or bees. But this behavior has never been observed in mammals. Hairless mole rats are thus the first mammals with this kind of "social behavior".

But, when speaking of "social behavior" or "community", do we use a simple analogical-metaphorical lexicon, a pedagogical or explanatory metaphor? Or are we engaged in a process of describing these newly identified animal behaviors in terms of the existing structures of human societies? Do we suggests, as in the case of the dunes and snowdrifts, that both phenomena may well have the same foundations, biological in this case? Does the organization of human societies obey the same biological laws as apply to mole rat "societies"? Are we on the way towards a socio-biological theory of human societies? Have we moved surreptitiously from metaphorical language to identification?

This is a strategy of "slippery metaphor". This strategy is so successful, that it reverses the relationship Target / Resource. Being closer to nature, mole rats, formerly *the Target*, now become a *model* for the study of human society, formerly *the Resource*.

In order to reject this assimilation, the opponent lists the terms coming from the field Resource, the human social lexicon:

> The phrase "*division of labor*" is used four times; the word "*task*" also appears four times; the term "*responsible*" also appears four times, and "*they take care of*" once; the terms "*cooperation*" and "*subordinate*" are used once each. The expression "*sexual status*" is used three times to refer to the reproductive state of the animals. (G. Lepape, *[Investigation]*, 1992)

In their reply to this criticism, the authors of the article set limits to the identification of the two areas:

> G. Lepape also contends that our language introduces unfair comparisons which attribute common behavioral traits to mole rats and social insects. This assertion surprises us, especially when he writes, "*the similarities [between hairless mole rats and social insects] are treated as true homologies*". Our article is clear on that point: we believe that the behavior of hairless mole rats and eusocial[2] insects have striking similarities. However, we do not see how the language used to describe these similarities can suggest that a common origin of these animals would constitute the evolutionary basis of these similarities.
>
> Braude & Lacey, *id.*.

---

[1] Braude, S. & Lacey E. (1989). Une monarchie révolutionnaire: la société des rats-taupes. *La Recherche* [Investigation], a journal of general scientific information] July-August 1989. Comments from G. Le Pape, and reply of the authors in the same journal, Oct. 1992.

[2] "Living in a cooperative group in which usually one female and several males are reproductively active and the nonbreeding individuals care for the young or protect and provide for the group *eusocial* termites, ants, and naked mole rats" (MW, *Eusocial*)

The danger here is that we might be tempted to forget that we are dealing with analogy, which is "never more compelling than when it is abolished and ceases to be perceived as an analogy. Becoming invisible, it merges with the order of things." (Gadoffre 1980, p. 6)

7. Against metaphors

> Politicians [are] catering to a public that doesn't understand the rationale for deficit spending, that tends to think of the government budget via analogies with family finances.
> When John Boehner, the Republican leader, opposed US stimulus plans on the grounds that "*American families are tightening their belt, but they don't see the government tightening its belt*", economists cringed at the stupidity. But within a few months the very same line was showing up in Barack Obama's speeches […]. Similarly, the Labour party […] (*The Guardian* 04-29-2015)[1]

The "stupidity" is that of inference "*families are tightening their belts, SO the state must tighten its belt*". We can reconstruct the warranting principle of this argument as a metaphor:

> A state, a nation, a country is a family.

One could also think of a kind of composition:

> The state is made up of families, families are tightening their belts, the state must tighten its belt.

However, the metaphor "*state, family*" has deep roots; it is based on the etymology of the word *economy*, from the Greek *oikonomia*, "home management"; it is found in the praise of the leader as "father of the nation", "founding father", etc.

Krugman considers that politicians are "*catering*" ("providing what the public wants, desires or what amuses them"; after d.c, *cater*) to a public "that doesn't understand"; so politicians must use metaphors, and metaphors, at least this metaphor – is *stupid* — this is indeed exactly what Aristotle said, cf. supra.

Happy metaphors do serve to charm the audience, but the fact that there are also *unhappy* metaphors must be fully acknowledged. Where they are used, the interlocutors are not only *not* "charmed", showing no pleasure, but they also "cringe at the stupidity", that is, "show on their face and bodies their feeling of disgust and embarrassment" (after MW, *Cringe*). This is exactly how metaphor can be rebutted *as metaphors*.

Then, in a second step, the accounts can be settled with the substantial contents, that is the de-metaphorized claim "in times of economic crisis, the state must turn to austerity". Krugman conducts this substantive rebuttal in the semi-technical language of economics, combining a priori refutation (*theoretically ill-grounded*), falsification (*forecasts contradicted by facts*) and pragmatic refutation (*policies inspired by this theory have failed*). But a second metaphor remains; if words

---

[1] www.theguardian.com/business/ng-interactive/2015/apr/29/the-austerity-delusion (15-08-16)

such as *restriction* and *austerity*, have a clearly negative orientation, the expression *"to tighten one's belt"*, associated with successful diet, weight loss, slimness, has strong visual, irrefutable positive connotations, inaccessible to refutation.

Metonymy, Synecdoche

Traditionally, two main domains are distinguished within the field of rhetoric, one deals with *tropes and figures*, and the other deals with *argument schemes*. A *semantic and ornamental* rhetoric is opposed to a *cognitive and functional* rhetoric. This approximate opposition can be misleading.

1. Tropes

A *trope* is defined as an operation "through which a word is given a meaning which is not precisely the proper meaning of that word" (Dumarsais [1730], p. 69). This definition may be paralleled by that of an argument as an operation "through which a statement (the conclusion) is given a *belief value* which is not precisely the proper belief value of that statement".

The linguistic mechanisms involved in the tropic referential shift bear a significant resemblance to those involved in argument. In both cases, this is a transfer problem. In the case of a trope, the *meaning* of a word is transferred to another. In the case of an argument, the *belief value* of a statement is transferred to another, and the rules of transfer are similar.

Metaphor@, irony@, *metonymy* and *synecdoche* considered to be the four "master tropes" (Burke, 1945), are all relevant to the study of argumentation, although in fairly different ways.

2. Metonymy

2.1 Metonymy as a trope.

Consider the classical example of metonymy, *"the pen is mightier than the sword"*. A pen is "an instrument for writing or drawing with ink..." (MW, *Sword*), and a sword is "a weapon with a long metal blade and a hilt with a hand guard..." (OD, *Sword*). In the quoted proverb, *pen* and *sword* are used metonymically to mean respectively "word, thought and discourse, verbal communication..." and "physical violence, military force...". The global meaning being that *"strength will not prevail over reasoned discourse"*.

Generally speaking, the metonymy semantic scheme can be described as follows.

— There is a word {S / C1}, its signifier is S and its content C1: pen/"instrument for writing".
— The signifier S is used metonymically to designate content C0: pen/"discourse".

— This transfer of meaning operates under a condition: it needs a backing, expressed in a transition law such as "**C0** is in some relation of contiguity with **C1**". Here, "the pen is the *instrument* used to produce discourse"

The subtypes of metonymy schemes are classified according to the kind of contiguity connection between the contents of **C0** and **C1**, for example:
— Effect for cause, "*Death is in the Meadow*".
— Instrument for agent, "*She is the pen of the President*".
— Agent (or "cause") for the work produced: "*A new Shakespeare just came out*".
— Instrument for object produced, "*The pen is mightier…*".
— Name of the place where the object is made for the object itself, etc. "I feel like having a Cognac".
— Relevant ongoing planned action for a participant: "*Sir, your rendezvous just left*".

2.2 Metonymic transfer and argumentative transfer

Figures and arguments require the same kind of backing. This can be suggested by the following examples.
The effect for cause metonymy: "*Death is in the Meadow*"[1] meaning that *phytosanitary products* (also called *crop protection products*) (**Ph**) used in agriculture can cause *death* (**D**). The word (signifier) designating the effect (**D**) now designates (refers to) the cause (**Ph**).
— In the effect-to-cause argument, the (truth-)value predicated upon the effect is transferred back to the cause, or to a series of causes:
>Metals expands when heated
>This metal expanded (*is an established fact*) SO it has been heated (*is an established fact*)
>The tire exploded, so [either **C1**, or **C2**, or…] (*id.*); **S. Case-by-Case**

Effect-to-cause argument transfers the predicate "*— is an established fact*" from the effect to the cause.
The word *death* refers to death; in the case of metonymy, its referential domain is extended so as to include the *cause* of death, "*death* refers to phytosanitary products". In our standard vision of reference, a word refers to an object; actually it refers centrally to an object, and to the objects contextually connected to it; that is, the word (signifier) actually refers to any element belonging to the *cluster* of that objects, **S. Object of discourse**. Ordinary language clearly expresses this fact:
>(1) He has a temperature so he has an infection.
>(2) Give him antibiotics, it will reduce the fever.

The antibiotic in fact acts upon the *infection*, and *fever* in (2) should thus be considered to be an effect-for-cause metonymic designation of the infection. On

---

[1] *La Mort est dans le pré*, youtube.com/watch?v=nAMARhJoFaQ

the other hand, fever is a *natural sign* of an infection: "*he has a fever that* means *he has an infection*": this is precisely what the metonymic analysis says.

A metonymy designating a work by the name of its author corresponds to an argumentation transferring to a work a judgment about its author: "*The author of this book supported the former dictator*". The mechanisms of this metonymic transfer from the person to his or her acts and products have been studied from the argumentative point of view in Perelman (1952), **S. Person.**

3. Synecdoche
As shown by example of the rendezvous above (§1), metonymic naming can operate upon any pair of strongly connected objects, this connection being accidental (local), or essential. Synecdoche operates upon constituents of a whole. The word "metonymy" is sometimes used to refer to both metonymy and synecdoche.

3.1 "Part – whole" and "whole – part" relations
A *roof* is a component of a *house*; in "*looking for a roof*", *roof* means "house", houses being considered prototypical lodgings.
Part – whole arguments transfer to a whole the predicate attached to the part. These are backed by the same kind of connection, **S. Composition and division.**
> The roof is in poor condition, so the house must not be well maintained.

3.2 Genus for species and species for genus
A synecdoche of a genus for a species uses the name of the genus to refer to one of its species; the name of the genus replaces that of the species: "the animal" for "the lion". This use is most common in textual co-referring:
> We saw a lion. The poor animal was gaunt and sick.

Backed by the same relation, the argumentation by the genus attaches to the species the predicates of the genus, **S. Taxonomy and category; Categorization:**
> This is a lion, therefore it is an animal, and therefore, it is mortal.

4. The tree and its fruits
The following argument was advanced in defense of Paul Touvier, leader of the pro-Nazi Militia in Lyon, France, during the German Nazi Occupation. Sentenced to death after the war, he escaped and remained in hiding for 25 years. The following excerpt is taken from a letter to the then President of the French Republic by the Rev. Blaise Arminjon, S. J., on December 5, 1970, in support of Paul Touvier's petition for clemency:
> How are we to believe that he [Touvier] is a "criminal", or a "bad Frenchman", when his conduct for twenty-five years, and the education he has given his children, has been so admirable? A tree is known by its fruits.[1]

---
[1] Quoted in René Rémond *& al., Paul Touvier and the Church*, Paris, Fayard, 1992, p. 164.

A Toulmin style analysis can be applied to this passage, the warrant being provided by the biblical topos, *"a tree is known by its fruits"*:
> For a good tree does not bear bad fruit, nor does a bad tree bear good fruit.
> Luke 6:43-45, New King James Version.

But this transition law also authorizes a metonymy-based interpretation. To speak of *"the [admirable] conduct of Touvier for twenty-five years"* is a way of referring to Touvier metonymically. To say that this conduct is *"admirable"* is to say metonymically that Touvier *is* admirable. Similarly, a positive evaluation of the act, *"the education that Touvier gave his children* is *admirable"* also spreads metonymically to the agent, Touvier, who is necessarily equally admirable. The same phenomenon can be equally expressed in the language of tropes or in the language of argument, both of which implement the same kind of rationality.

Moderation and Radicalism

> Lat. argument *ad temperentiam*, Lat. *temperentia*, "moderation, measure, restraint"

1. Argument to moderation and radicalism

In politics, *moderation* is opposed to *radicalism* or extremism, as reformism is to revolution. The argument from *moderation* is developed in discourses which prioritize the necessity of sticking to practicality, to compromise, of holding inclusive positions, changing things little by little, etc. The appeal to *radicalism* is developed in discourses which foreground the urgency of the decision, the necessity of a new start, of avoiding deadlocks in discussions, the will to be true to one's principles framed as antinomies, *"freedom or death"*.

Two contrasted ethos and emotional states are associated respectively with moderation and radicalism: conservative *vs.* progressive; open to dialogue and compromising *vs.* uncompromising; realist *vs.* idealist; calm / exaltation; etc.

2. Middle ground argument

The *middle ground argument* justifies a measure by showing that it does not satisfy any of the opposing parties. The speaker takes the position of the responsible third party, S. **Roles**.

> Both the far right and the far left attack my policy; it clearly shows that it is a good policy.
>
> Keep away from extremes.
>
> Christianity has reestablished in architecture, as in other arts, true proportions. Our temples, bigger than those of Athens, and smaller than those of Memphis, have that proper balance, in which beauty and taste par excellence prevail.
>
> Chateaubriand, [*The Genius of Christianity*], 1802[1]

---

[1] Quoted after François René de Châteaubriand, *Le Génie du christianisme*. Part 3, Book 1, Chap. 6. Tours: Mame, 1877, p. 194-195.

The intermediate position is valued: reason and virtue "stand in the middle" (Lat. *in medio stat virtus*):

> Neither rash, nor coward, just courageous.

The arguer who opts to take the middle-ground will be stigmatized as a person who is indecisive, or who does not want to examine the arguments of the parties in detail, *"let's stop the discussion, meet in the middle; split the difference"*. The case of Solomon's judgment shows that there are stakes that cannot be so easily split up.

Modesty

Latin *"argumentum ad verecundiam"* lat. *verecundia* "modesty, humility"

1. The *ad verecundiam* argument

The argument of modesty is invoked by someone who bows before the speech and the good reasons offered by a person he considers to be superior to him- or herself. It typically refers to an act of submission to ethos. The *ad verecundiam* argument is the *interactional correlative* of an appeal to authority, not an appeal to authority. Note that, in the following key passage, Locke refers to *ad verecundiam* as coming from a fear of breaching "modesty".

> The first [*fallacious argument*] is to allege the opinion of men, whose parts, learning, eminency, power, or some other cause has gained a name, and settled their reputation in the common esteem with some kind of authority. When men are established in any kind of dignity, it is thought a breach of modesty for others to derogate any way from it, and question the authority of men who are in possession of it. This is apt to be censured as carrying with it too much pride, when a man does not readily yield to the determination of approved authors, which is wont to be received with respect and submission by others: and it is looked upon as insolence for a man to set up and adhere to his own opinion against the current stream of antiquity; or to put it in the balance against that of some learned doctor, or otherwise approved writer. Whoever backs his tenets with such authorities thinks he ought thereby to carry the cause, and is ready to style it impudence in any one who shall stand out against them. This I think may be called *argumentum ad verecundiam*. (Locke [1690], p. 410).

This argument is deemed fallacious:

> It argues not another man's opinion to be right because I, out of respect, or any other consideration but that of conviction, will not contradict him. [1690], p. 411).

In a similar way, topic n°11 of the *Rhetoric* argues "from a previous judgment in regard to the same or a similar or contrary matter". Such a precedent-setting judgment must have been produced by an authority, one of "those whose judgment it is not possible to contradict" (Aristotle, *Rhet.*, II, 23, 12; F. 309), that is to say, "it would be *disgraceful* to contradict him" (*ibid*.; my italics), be he a father, a god, an instructor or a wise man. Politeness is argumentatively oriented in favor of the submission to the status quo.

2. Authority or pusillanimity? *Ad verecundiam*, or misplaced modesty

Locke stages an interaction, where one partner "allege[s]" an authoritative opinion. It appears from the description that the characteristics conferring authority to an opinion have either a social ("*parts, learning, eminency, power ... dignity*") or an intellectual source ("*learning, ... approved author... learned doctor... approved writer*"), **S. Ethos (II)**. Such sources do indeed have a legitimizing power, **S. Dialectic**. Note that religious authorities are not mentioned.

It must be emphasized that Locke does not censor the expression of or the reference to authoritative opinions in a first round of speech, but blames the acritical acceptation of such an authority, that is the lack of a second round of speech in which what has previously been said is criticized and alternative views are expressed. The condemnation *ad verecundiam* is a protest against the censure of this second round by an internal impulse of modesty, the feeling of one's own insufficiency (however legitimate it can be). This censorship is a preventive reaction to a threat that could come from a third round claiming to silence the objection addressed to the authority. This third turn itself does not deal with the substance of the objection made to the second (by an argument *ad judicium*, **S. Matter**). It merely substitutes in the discussion of the critical opinion, a negative evaluation of the person who supports it, an *ad personam* attack invoking "*a breach of modesty, too much pride, insolence, impudence*", that is, an intimidation maneuver, **S. Personal Attack; Respect**. The problem is therefore not located in the authoritative first round, but in the inhibitive foreboding of an aggressive third round. As expressed by the label "argument *ad verecundiam*", the fallacy is committed by the interlocutor, the overly humble individual who expresses no objection for fear of creating a scene. This is not primarily a fallacy of authority but of cowardice or spinelessness. The *verecundia* is the (false sense of) shame that prevents one saying what one thinks out loud.

4. Justified modesty

When it comes to authority itself, the problem is twofold. In the first turn, participant **S1_1** has "alleged" an authoritative opinion, which may be a fairly sensible move. Suppose that **S2** can overcome his or her *ad verecundiam* inhibition, and quite freely voices his or her opinion, in a second turn. Then, if in a third turn **S1_2** bars **S2_1**'s remarks in the name of authority, whilst also criticizing his opponent for his or her boldness and pride then **S1** argues from authority, which certainly is a fallacious move. Some situations are nonetheless embarrassing. If **S1** quotes Einstein in his (Einstein) field of competence, **S1** having a good background in physics and **S2** none, then a humble lay speaker **S2** would be wise to ask for more explanation before voicing his or her doubts. If not, **S1_2** would legitimately give in to an authoritative exasperation.

3. A fallacy in dialogue

The problem of authority is thus reframed as that of authoritarian interaction, that is to say a dialogue where an authority is quoted in the first speech turn, and alleged in the third turn to silence the objections, considering that the quoted authority gives the quoter the power to close the discussion. This use of authority is a direct contrast to the use made of it in a dialectical game. The problem does not lie so much in the quoting of authority as in the possibility of contradicting authority. Modesty, respect@, concern not to cause the other to lose face, rules of politeness@, preference for agreement are all intellectual inhibitors. All these constraints define a typically *anti-dialectical* situation, **S. Dialectic**.

Authority is accepted as fact, the problem lies in the possibility of calling this authority into question. Authority is deceptive only if it claims to escape from dialogue, to silence and not to answer its counter-discourse. The conclusion is that what is fallacious or not fallacious is the *dialogue* itself. It is impossible to say whether a statement such as *"The Master said it!"* is misleading or not; it all depends on the statement's position in the dialogue. If it is an opening statement, it is not fallacious. If it is a closing statement, intending to silence the critic, it is.

Motives and Reasons

The individual's will, intentions, desires, motives, reasons... may be interpreted as *causes* for action, considered to be *effects* or *consequences* of such an "inner" causation. Conversely, actions are evaluated and interpreted according to their motives and reasons are seen as their causes. The consistency requirement imposes this causal structure on human motivation, **S. Consistency**

1. Argumentation from the existence of reasons for action

Two basic Aristotelian topics transpose the law of causality in human conduct, with reasons and motives substituted for causes. When the cause exists, then the effect follows. That is to say that when one party has a motive or a reason to do something, as soon as he or she has the opportunity, he or she will do it. In the wording of the topic n° 20 of the *Rhetoric*:

> To consider inducements and deterrents, and the motives people have for doing or avoiding the action in question. (*Rhet.*, II, 23, 20; RR, p. 373)

The basic topos is:
> You wished it, so you strived for it!
> Who wants the end wants the means.

This topos is also implemented by the pathetic@ argument. Here, it supports a charge:

> You had a motive, you talked about it, the opportunity came up, and you did it!

Or a defense:
> L1: — *You did it!*
> L2: — *I had no reason to do it, I even had reasons not to do it.*

Likewise, in topic n°24, *cause* means "reason to do":
> Another topic is derived from the cause. If the cause exists, the effect exists; if the cause does not exist, the effect does not exist. [...] For example, Leodamas [...] (*id.*, II, 23, 24; F. p. 319).

2. Arguments on the "real reasons"

The following argument schemes substitute a covert motivation for a public good reason, as a *true* cause can be substituted for a *false* one, **S. Interpretation**:
— Topic n° 15 substitutes a covert, underhanded, interested motive for a noble, publicly claimed reason. It is used to charge or to refute the opponent.
— Topic n° 23 rejects the malevolent interpretation given for an act by giving an acceptable, respectable reason for the alleged guilty motive. It is used to clear somebody from a charge.
— Topic n° 19 changes the benevolent interpretation given to an act for a malevolent one.

2.1 Publicly displayed good reasons and real private ugly intentions

According to topic n° 15 of Aristotle's *Rhetoric*:
> The things people approve of openly are not those which they approve of secretly: openly, their chief praise is given to justice and nobleness, but in their heart they prefer their own advantage. {...} This is the most effective of the forms of argument that contradict common opinion. (*Rhet.* II, 23, 15; RR, p. 369)

The argument highlights a (possible) private, hidden, poor motive for refuting the public, honorable, good reason given as justification for an action:
> S1: — *Supporting this Charity, I fight for a noble cause!*
> S2: — *You fight especially for your own advertising.*
>
> S1: — *We wage war to restore democracy and human rights in Syldavia*
> S2: — *You wage war to seize their oil.*

In the second dialogue, **S1** justifies war, **S2** does not oppose war, he or she can simply introduce a realpolitik argument, which could be openly put forward in another situation.

2.2 A commendable motive substituted for a guilty one

This argument corresponds to topic n° 23, "useful for men who have been really or seemingly slandered":
> To show why the facts are not as supposed; pointing out that there is a reason for the false impression given. (*Rhet.*, II, 23, 23; RR p. 375)

embodied in the enthymeme:
> *She hugs him because he's her son, not because he is her lover.*

Topic n° 23 is quite the reverse of topic n° 19; it helps to exculpate by substituting an honorable motive for the offending one:

> I struck him to save him from drowning, not to hurt him.

The action is reinterpreted according to a re-evaluating strategy: "you must congratulate me and not blame me." **S. Stasis; Interpretation; Orientation.**

2.3 The poisoned chalice

The wording of topic n° 19, is rather puzzling, "some possible motive for an event or state of things is the real one"; it matches the enthymemes:

> A gift was given in order to cause pain by its withdrawal.
> *Gods give to many great prosperity, / Not out of good will towards them, but to make / Their ruin more conspicuous.* (Rhet., II, 23, 19; XX p. 371)

The topic operates a dramatic negative reinterpretation of an act which was assessed positively.

> She seduced not by love but by hatred / greed / to make him suffer by leaving him later

This is the principle of the "Dinner Game", "*they invited me not as a friend, but to make fun of me*". This technique for narrowing the cognitive and affective discrepancy, particularly effective for destroying a sense of gratitude, **S. Emotion.**

Multiple Argumentation ▶ Convergent

N

Natural Signs

A *natural sign* is a perceptual datum, an *actual* material fact or item, *materially linked*, necessarily or ambiguously, to another fact or item or state of things not perceptually accessible.

A *natural sign* is typically an indisputable fact, "as certainties, we have, in the first place, what is perceived by the senses, as what we see, what we hear, as signs [*signa*] or indications" (Quintilian, V, 10, 12).

A *natural sign* is quite different from a *linguistic sign*, for which the link between *signifier* and *signified* is social and arbitrary. It is not a global *analogon* of the thing it "represents", as in the case of analogical thinking, **S. Analogy (I)**. Nor is it a symbolic representation of the associated phenomenon.

The natural sign is just *a part* of the phenomenon through which the observer can access the whole phenomenon. The link between the *present* natural sign and its *absent* counterpart can be:

    The very first manifestation of a phenomenon: *a red setting sun* / rainy weather tomorrow
    A remnant of something disappeared: *the leftover* / the meal
    A part of a whole: *a hair* / a person
    An effect to its cause: *being tired* / having worked

1. Natural signs, clues and traces

Clue is an accurate synonym of *material sign*, since to look for clues, one must necessarily have to deal with some "intricate procedure or maze of difficulties", or be seeking "to find something, understand something, or solve a mystery or puzzle" (MW, *Clue*). These description fit well with exploratory argumentative situations. Generally speaking, an argument is indeed *a clue* to a conclusion; etymologically, a clue is 'a ball of thread'; hence, one used to guide a person out of a labyrinth" (OD, *Clue*). Clues are typically looked for "in the detection of a crime"; "police officers are still searching for clues" (*ibid.*). Yet *clue* is also used to refer to a "piece of information" given to someone; and this is not a natural sign in the sense discussed in this entry.

Traces, such as fingerprints (necessary sign), or tire marks (probable sign), are a special kind of natural signs, but, insofar traces are leftovers, "a mark [...] left by something that has passed", not all material signs are traces; smoke is not a trace of fire, whereas ashes are.

Index, indication, indicator can also be used with the meaning of "natural sign".

2. Reasoning on probable and necessary signs

The relation of the natural sign to its counterpart is inferential in nature:

> Anything that *when it is*, another thing is, or *when it has come into being* the other has come into being before or after, is a sign of the other's being or having come into being. (Aristotle, *P. A.*, II, 27; my italics for the sign and underlining for the counterpart).

In the Aristotelian system, enthymemes are developed from *natural signs* and *probabilities* (*P. A.*, II, 27); **S. Enthymeme; Probable.**

These inferences are exploited in concrete argumentations such as:

> I can see smoke, the house is on fire
> Peter's face is flushed, he must have a fever

The quality of the argumentation depends on the nature of the link it exploits. If the sign is *necessary*, the argumentation is conclusive; if it is *probable*, the possible claim is slightly more probable that it would be in the absence of the argument; probable signs reduce uncertainty, **S. Abduction.**

Probable signs are distinct from human and social *probabilities*, **S. Probable.**

— A *necessary sign* (*tekmerion*) is associated with a material being or state of affairs. It corresponds to a *material, empirical necessity* (not a logical necessity):

> A scar, an ancient wound
> Callous hands, being a manual worker
> Smoke, fire
> Footprints on the sand, humans on the island

Such signs thus have the force of proof, the associated syllogism is valid, as in the following *propter quid* argument, **S. *A priori***

> Law (major): A woman who has milk has given birth (if M, then B)
> Sign (minor): This woman has milk.

Conclusion: This woman has given birth.

— *Probable (contingent) signs (semeion)* may correspond to several associated realities. Contingent signs are ambiguous, whereas necessary signs are unambiguous
> *tiredness* is a possible sign of *having worked*
> *being flushed* is a possible symptom of *having a fever*

Typically, peripheral indicators are non-necessary signs: "*he has a guilty look so he feels guilty, so he is guilty*", **S. Circumstances**. The associated syllogism is not valid:

Law:	*Women who have given birth are pale.*
Sign:	*This woman is pale.*
Conclusion:	*This woman has given birth.*

A necessary condition is taken for sufficient: one might simply have a naturally pale complexion, or one might be pale because one is ill. The probable sign brings only one piece of evidence (judicial); it can support a suspicion, it is not a proof.

The human body is an inexhaustible source of natural signs; white hair and flexibility of the skin are natural signs indicative of age and the global physical condition of the person. In medicine, co-occurring non-necessary signs are grouped in a *syndrome*, that is to say "a group of signs and symptoms that occur together and characterize a particular abnormality or condition." (MW, *Syndrome*). For example, the Samter's syndrome
> Samter's Triad or Aspirin Sensitive Asthma, is a chronic medical condition that consists of asthma, recurrent sinus disease with nasal polyps, and a sensitivity to aspirin and other non-steroidal anti-inflammatory drugs (NSAIDs).[1]

This grouping of signs is the basis of conclusive medical reasoning: if a patient suffers from asthma and is sensitive to aspirin, then he would very probably also has a problem with nasal polyps. He should be checked for this third condition.

When grouped, a series of separately non-conclusive signs might constitute a body of conclusive evidence. An area of the body may be *red*, because it has been rubbed; *hot*, because of incipient sunburn; *painful* or *swollen* because of an accidental blow. But if it is at once *red*, *painful*, *hot* and *swollen*, then, we can say that it is inflamed, **S. Convergence**

To guess the intentions of the enemy, the soldier observes their acts and movements and then reasons from a cluster of converging signs:
> *The writer Roland Dorgelès had "the singular privilege of baptizing a war", as "the Phony War", which refers to the strangely calm situation on the front between September 3rd 1939, the date of the declaration of war, and May 10th 1940, the date of the invasion of Belgium, the Netherlands, Luxembourg and France by Nazi Germany. His book,* The Phony War, *brings together a series of reports on the front during this period. In April 1940 he was in Alsace, at an observation post.*

---

[1] https://aerd.partners.org (08-31-2017)

Nature; "Naturalistic Fallacy" ▶ *Weight of Circumstances*

Seen from above, it looks as though the enemy lines are dominated as if from a balcony. [...] The sergeant who never loses sight of them, now knows their habits, knows where they come from and where they go.
'There, he points out, they are digging a sap. Look at the stirred earth... This gray house has certainly been consolidated... look at the embrasures... And those tiles over there? Their workers at this moment are mainly occupied there. This morning I counted sixty of them, returning from the building site, with lamps: so they must be digging underground.'
From dawn to darkness, our watchmen remain leaning over the telescope.

Roland Dorgelès, [*The Phony War*], 1957[1]

The whole art of Sherlock Holmes resides in the observation, interpretation and combination of clues, **S. Deduction**. The clue is a trace of the action from which a *modus operandi* can be inferred. If the window panes have been shattered into pieces, and these are found lying inside the drawers which have been torn from the cupboards and thrown onto the room, then we can be sure that the room was first ransacked and the windows were then fractured to give the false impression that the window had been broken to enable entry into the room. So, they entered the room through the door; so they had a key. Who has the keys? Certain individuals exploit and investigate clues is a professional capacity. Sign-based arguments are field-dependent. On the basis of a clue, the detective knows how to reconstruct the scenario of a crime; the historian knows how to judge the authenticity of a document; the archeologist knows how to reconstruct the city map, the paleontologist knows how to determine the age of the skeleton (Ginzburg 1999). The informed reasoning of these professionals should be considered to be exemplary of practice-oriented argumentation studies.

Nature; "Naturalistic Fallacy" ▶ Weight of Circumstances

Negation ▶ Denying

Non-Contradiction Principle

1. In logic
In logic, the non-contradiction principle prohibits the affirmation of contradictory propositions. In other words:
— The conjunction "**P** and **not-P**" expresses a contradiction, and, as such, is a self-destructing statement, which is necessarily *false*.
— The disjunction "**P** or **not-P**" is necessarily *true*.
One of the two propositions **P** and **not-P** must be true, both cannot be true simultaneously. The same thing cannot *be* and *not be*. This principle is consid-

---
[1] Roland Dorgelès, *La Drôle de Guerre 1939-1940*. Paris: Albin Michel, 1957, p. 9; p. 194.

ered by classical logic as a *law of thought*, and as an *axiom* by contemporary logicians. A logical system respecting the principle of non-contradiction does not contain any *antinomies*; it is said to be *consistent*.

Negation — Using the truth-table method, the negation operator is defined as follows:

P	¬P
T	F
F	T

This table expresses the principle of the excluded middle. It reads:
line 1: "when P is true, then not-P is false"
line 2: "when P is false, then not-P is true"

2. In natural language
The application of the non-contradiction principle to everyday language is complex, because it presupposes that **P** is plainly true or false, not *far from true* or *practically false*, not true or not *according to the circumstances*.
Many argumentative forms appeal to the non-contradiction principle, albeit under different names: S. *Ad Hominem*; **Dialectic; Contradiction; Coherence.**

The non-contradiction principle applies not only to logical, argumentative discourse, but also to any kind of discourse; inconsistent *narrations* or *descriptions* for example, are rejected as such.

According to the basic Aristotelian dialectical rule, any discourse resulting in a contradiction is irrational and must be abandoned. Hegelian dialectic sees in the ongoing treatment of contradictions the motor of History. The cynical politician can lay claim to Hegel to hide his opportunism:

> Stalin's speech on the five-year plan serves as an ardent apology for contradiction as a *"vital value"* and an *"instrument of struggle"*. One of Lenin's great strengths was his ability never to feel a prisoner of what he had preached as true the day before [...] Mussolini's famous word *"Let us beware of the mortal trap of coherence"* could be signed by all those who intend to pursue a work within currents they cannot foresee.
> Julien Benda, [*The Betrayal of the Intellectuals*], [1927][1]

The affirmation of a paradox as an *oxymoron* makes it possible to withstand the contradiction: "*O wound without scar!*". Such paradoxical assertion is not seen as absurd or fallacious and eliminated as such, but triggers a quest to identify the deeper, symbolic meaning of the words *wound* and *scar* used in this context.

---

[1] Julien Benda, *La Trahison des Clercs*, [1927]. Excerpt from the *Preface* to the 1946 edition. Paris: Grasset, 1975, p. 78-79.

Novelty ▶ **Progress**

Number ▶ **Consensus; Authority**

O

Object of Discourse

The concept of *object of discourse* (Fr. "objet de discours"; also translated as *discursive object* or *discourse object*) was introduced by Jean-Blaise Grize, in relation to the schematization@ process. An object of discourse is basically a thing, a situation, as characterized by its plasticity, that is to say, permanently re-designed throughout the discourse or the interaction.

1. Cluster of a word

At the language level, the *cluster of an object* ["faisceau d'objet"] is investigated on the basis of the *word* designating that object. It is defined as:

> [the] set of aspects normally attached to the object. Its elements are of three kinds: properties, relations and patterns of action. So the cluster of "rose" brings together properties like '*to be red*' [...], relationships like [...] '*to be more beautiful than*', patterns of action like '*to fade*'." (Grize 1990, p. 78-79)

The cluster attracted by an object is defined at the notional level and does not coincide either with linguistic categories such as those used in semantic analysis (*id.*, p. 79), with lexicographical elements used in dictionaries, with elements psychologically associated with the object, or with ontological features claiming to grasp the being of the object, **S. Category**. The cluster of a word results from an aggregation of discourses using this word (*id.*, p. 78), **S. Orientation; Words as Arguments; Inference; Polyphony**. This concept can be compared with the stereotypes associated with a word, or, better, to the set of its favorite linguistic *collocations*, as established in corpus linguistics.

2. Cluster of an object of discourse

At the discourse level, the elements which make up the cluster attached to a *specific* object of discourse are not known a priori, but are empirically constructed, on the basis of the examination of the actual discourse under analysis. A specific object of discourse is developed via the progressive aggregation of the contextual properties attributed to it, the beings it is associated with, the events in which it participates, etc.

The study of objects of discourse focuses on their plasticity, as they are progressively produced and transformed in discourse: their mode of introduction, the evolution of the contexts to which they are attached. It overlaps with the grammatical study of *designation paradigms* (Mortureux 1993); a designation paradigm is the set of words and expressions constituting the anaphoric chain related to an evolving object of discourse. It is part of the study of textual cohesion and coherence, and overlaps essential observations of the rhetoric on the displacements of meaning.

Objects of discourse may be opposed to "logical objects". Classical logic references stable objects; according to the principle of identity every occurrence of the sign (signifier) "**a**" is strictly equivalent to another one. As a consequence, any variation in the scope of the reference of "a" introduced in the development of discourse are considered fallacious. **S. Fallacy; Ambiguity.**

3. Objects of discourse in argumentative situations

A discourse may concern a large number of objects, and to study the development of each one might turn out to be unworkable; boundaries must be set. As far as argumentation studies are concerned, they must focus on the most relevant objects, that is on *conflicting* objects, and primarily on those mentioned in the formulation of the argumentative question. Just as *peaceful*, undisputed, affirmations are considered to be *true*, undisputed objects are considered to be *real* and stable in their reference.

Conflicting *objects* are associated with conflicting *claims*. The observation of their discursive development, and the correlative establishment of their contrastive characterization is a simple and practical method used to expose their argumentative relevance.

The following data is taken from a discussion between students, and concerns the conditions a person must fulfill to obtain French citizenship; the key question "*who? who can obtain French citizenship?*" immediately structures the debate. The two antagonistic positions taken by the participants are clearly mirrored by the two systems of designations they use to construct this "*who?*".

All the students agree that there is an unproblematic group, who should have an automatic right to French citizenship, that being, "*the persecuted*".

One group of students supports the claim that "*the process of obtaining citizenship should be facilitated*". Immigrants are constructed *as people having a right to French citizenship*; this group is further specified as:

workforce; people who came to work in periods of prosperity
people we asked to come;
people we welcomed;
people who have been there for a very long time
their relatives
their children   - born in France
                  - born in other countries

Another group of students support the claim that "the process of obtaining citizenship should be toughened". In this set of co-oriented discourses, immigrants are constructed *as people having no right to French citizenship*, and these individuals are referred to as:

undocumented immigrants
people with problems; having or creating problems
illegal immigrants;
immigrants by "practicality" (economic migrants)
"anyone", (that is indiscriminate foreign people asking for citizenship).

In reality, one can certainly observe that among the people applying for French citizenship there are certainly *both* undocumented people *and* people who came to France many years ago in order to work. Despite this, each group of students schematizes *immigrants* (as a group) as *either* one *or* the other.
For another example of diverging constructions of causality as an object of discourse, S. **Cause — Effect**.

This method shows how a specific light is cast on an object of discourse, how it is "spotlighted" (Grize), or given a discursive "presence", in Perelman & Olbrechts-Tyteca's terminology ([1958], 115-120).

Objection

In the same way as refutations, objections are reactive, non-preferred second-turn interventions, opposing the conclusions of the first turn, the target discourse.
From the point of view of their contents, objections can be seen as *politely mitigated refutations*, which nonetheless have the full strength of a refutation; the choice of presenting a refutation as an objection would be an insignificant price which logic pays in the name of civility.
Objections can also be seen as *weak, indecisive* refutations, which are easily disposed of. To refute is to shoot down, while to object is just an attempt to stop, at best to weaken, the position under scrutiny.
The status of a rebuttal as an objection or a refutation depends on the *kind of dialogue* which develops between the participants. In a logical language game, I cannot claim that all swans are white and simultaneously concede that this particular swan is black. *Conclusive* counter-arguments do count as refutations. In

ordinary language, I'll argue that all swans are indeed white, while conceding the existence of black swans as exotic exceptions.

The same kind of argument can be treated as a refutation or as a concession. In the same way as a refutation, for example, an objection might underline a negative consequence of the interlocutor's proposal:
> — But if you build the new school here, the students' commuting time will be half an hour longer.

This counter argument can be contextually constructed as a refutation:
> — This is clearly unacceptable, classes begin at 7.30, and some students who commute already have to travel for more than an hour. The new school cannot be built here!

or as an objection:
> — We'll have to create a new bus line for commuter students, but this remains the best place to build our new school!

Objection and refutation essentially have different interactional statuses; objections are *cooperative*, while refutation is *antagonistic*. The objecting party is a dialectical figure, essential in cooperative everyday argumentative dialogue.

While refutation seeks to close the debate, without even listening to the answers, objections keep the dialogue open; they are in line with the problematic of the discourse under discussion, which are accepted as a working hypothesis. Objections are framed as quests for answers, they seek explanations, precisions and modalizations; they accept, as the case may be, to be only partially answered or integrated.

The ethos and emotional states displayed via refutation and objection are quite different. The former wants to have the final say and is associated with aggression, whilst the latter evokes a spirit of measure, collaboration and openness.

In a proleptic discourse, referring to possible negative observations, the speaker mentions "objections", not "refutations", typically using a *but* structure:
> *It could be objected that* **P** [anti-oriented discourse], *but* **R** [answer to the objection, discourse reinforced]

S. Refutation; Concession; Prolepsis

Ontological Argument ▶ *A Priori, A Posteriori*

Opponent ▶ Roles

Opposite

1. Opposite words
The term "opposite" covers a series of lexical oppositions such as:
> *male / female*: terms in a bi-dimensional opposition
> *mandatory / allowed / forbidden*: terms in a multi-dimensional opposition
> *sight / blindness*: terms in a relation of possession / privation
> *mother / son*: correlative terms

S. Contradiction; Contrary and contradictory.

The relation of opposition broadly corresponds to the lexical relation of *antonymy*.

These various relations of *opposition* are exploited in different argumentative maneuvers, bearing on terms and propositions containing opposite terms.
1. Negation, **S. Denying**
2. Rhetorical figures of opposition, **S. Opposition,**
3. Opposition between correlative@ terms
4. Opposition between propositions: **S. Contraries and contradictories**
5. The argument scheme of the opposite predicates a *contrary predicate* upon a *contrary subject*, **S. Opposite: *Topic of the* —; Laws of discourse**
6. Refutation by the observation of the opposite, rejects a predication "**A** is **P**" on the basis of the observation that a predicate, **Q**, is actually true of the same subject, **A**, and that **P** and **Q** are opposite. **S. Refutation**

Opposite: *Refutation by the—*

1. Refutation by the observation of the opposite
Two opposite predicates cannot simultaneously be attached to the same subject. In other words, if an individual says something and one can *see* and *show* that the opposite is true, what the first speaker has said is rejected. This is a clear application of the principle of non-contradiction, two opposites cannot simultaneously exist in the same subject. This topic, as trivial as it is effective, is consistent with the view the facts are the best argument: "*You say this, but I can see the opposite.*" Example:
> 1. Claim: *Peter has white hair.*
> 2. Actual observed reality: *Peter has black hair*
> 3. Rule of opposites: "white" and "black" are opposite properties (here of contrary opposite, they cannot be simultaneously true but can be simultaneously false, for example, if Peter has red hair).
> 4. Conclusion *Peter has white hair* is false and must be rejected.

This argument has very broad scope when it comes to the refutation of factual assertions; it is actually the standard rebuttal scheme. If we are able to call upon a case in which an opposite property can be predicated of the same subject, the opponent's claim in refuted.

The refutation predicates *an* opposite term of the subject, so we will use the singular, *"refutation by (the predication of an) opposite"*. In the quite different case of the *topic of the opposites*, the word is used in the plural. **S. Opposites (Topic)**.

The condition of belonging to the same family of opposites is necessary: "*Marie has a cat*" (claim) is not refuted by stating, "*Marie has a rabbit*".

The same procedure works for *contrary* and *contradictory* opposites. In the traditional system of genres, the claim "*Mary is a man*" is denied by the observed contradictory fact that Mary is a woman. Similarly, if two terms are in a relationship of *possession / deprivation*, (another form of opposites). If I am accused of having ripped off someone's ear, I can refute the accusation by asking that person to come to court and show that he still has both his ears.

2. Facts against theories

The scheme applies to predictive discourse, somebody predicts that event **E** should happen, but, in due time, anybody can see that **not-E**, as is often the case in practical discourse.

In science, the scheme is involved in the Popperian concept of experimental refutation (Popper, 1963). When the predictions made by the theory are clearly false, the theory should be rejected or seriously revised. But at least in the humanities, the finding of the opposite is much less conclusive than it appears to be in the previous examples. The theory asserts, directly or indirectly **P**. Yet, common sense urges rather to notice and report **Q**, excluding **P**. How can we solve the dilemma? Several solutions are available.

— Rejecting the theory, a costly and painful solution.

— Minimizing and marginalizing the inconvenient fact, by opposing the mass of facts explained by the theory, that support, confirm the theory.

— Reform the intuition, and decide that the theory is brilliant, precisely because it makes us see things "differently", so richer and deeper, and that in fact **P** is a kind of deep structure underlying the elementary intuition expressed by **Q**. The refutation can be resisted by choosing to reform the internal hypotheses (the theory) or the external hypotheses (what counts for a fact).

3. Refutation by the impossibility of the opposite

Refutation by the impossibility of the opposite rejects a judgment about a person, arguing that it is not possible for this person to be the subject of contrary opinion: "*To be praised for his sobriety, he must have the opportunity to be intemperate*"; it is ironic to praise poor people for their sobriety.

This is the topic "*he cannot say otherwise*", so what you say makes no sense. Suppose that the Proponent says of Peter that "*he is kind*". This quality has an opposite, "*to be mean*". In order that the statement makes sense, the quality can be attributed to the individual only if, in another state of the world, the attribution

of the opposite quality to this same individual would also make sense:
> L1: — *Peter acted in a friendly manner (so you have to be grateful to him)*
> L2: — *To say that, still he would have to have the possibility of not being friendly (i.e. of being mean), I definitely owe him nothing*

For a statement to contribute real information in a given situation, it is necessary that the opposite information be meaningful, "*everyone agrees, how not to agree*".

> In *Le Figaro* today the CEO of EDF asserts that the French nuclear park is in a very good state; well, it is difficult to see how he could have said the contrary. (*France Culture Radio News*, 04-18-2011; the CEO of EDF is in charge of the French nuclear park)

Opposites and *A Contrario*

In a broad sense, the words *opposition* and *opposite* can cover a series of rather different argumentative phenomena, S. Opposite.

1. Topic of the opposite

Cicero recognizes the enthymeme based on opposites as the archetypal enthymeme, S. **Enthymeme**. The topic of the opposite is the first on Aristotle's list of rhetorical topics:

> One line of positive proof is based upon consideration of the opposite of the thing in question. Observe whether that opposite has the opposite quality. If it has not, you refute the original position. If it has, you establish it. (*Rhet.*, II, 23; RR, p. 355)

Ryan reformulates the Aristotelian topic as:

> "**1A** — If **A** is the contrary of **B**, and **C** the contrary of **D**,
> then if **C** is not predicated of **A**, then **D** is not predicated of **B**
> **1B** — If **A** is the contrary of **B**, and **C** the contrary of **D**,
> then if **C** is predicated of **A**, then **D** is predicated of **B**" (1984, p. 97)

We follow Freese and Rhys Roberts and use the label "topic from the opposite". Ryan uses the equivalent word "contrary" in his discussion of the topic.

The clause "— is not predicated" can be read as "is not true, acceptable, possible…". Applied to the logical implication, "**P** implies **Q**", the topic validates the conclusion "**not-P** implies **not-Q**"; this conclusion is not quasi-logical, but plainly false, as a case of a negation of the antecedent (*modus tollens*), S. **Deduction**.

The problem here is that logical negation applies to *a predicate* saying something about its subject, but not to *a name*. A *bottle* and a *gloomy thought* equally qualify as *non-cows*. As the argument from the opposite is formulated in ordinary language in a given situation, the application of negation to any word is open-ended and debatable. But whoever wants to discuss that point becomes vulnerable to the accusation of "trying to pick up a senseless quarrel over semantics".

2. A dialectical resource

The topic of opposites is a dialectical resource. If the proponent holds that "**A** is **B**", then the opponent can examine what is going on with the opposites of **A** and **B**. In a dialogue format:

> Question: *Is courage a virtue?*
> Topic of the opposite:
> > Opposite of *courage*: *cowardice*
> > Opposite of "*— being a virtue*": "*— being a vice*"
>
> Let's predicate the contrary upon the contrary, and consider the result: "*cowardice is a vice*"; this proposition seems quite convincing. So, let's conclude that courage is indeed a virtue.
>
> Argumentation:
> > "*Courage is (indeed) a virtue, since cowardice is certainly a vice*".

The same topic, which has confirmed the proposal under consideration, can simultaneous disprove another:

> Question: "*are pleasant things good?*"
> Topic of the opposite:
> > Opposite of *pleasant*: *unpleasant*
> > Opposite of "*— being good*": "*— being bad*"
>
> New question: "*are unpleasant things always bad?*" The answer is no, because cod liver oil is quite unpleasant to drink (in its natural state) and nonetheless good for one's health. So, the conclusion is that the original proposition is false.
>
> Argumentation:
> > "*pleasant things are not intrinsically good, since unpleasant things can also be good*"

The topic of opposites can also be used to suggest practical actions:

> *Inhaling black coal dust made the miners sick, they will recover their health if they drink white milk*

In both its confirmatory and refutatory functions, the topic of opposites can enact poetic oratorical amplification without loosing its confirmatory value.

> *Satan leads the war against the angels, and has just undergone a cruel defeat. He calls "His potentates to council", and explains to their assembly how a new weapon of his invention — powder and gun — will permit them to take their revenge.*
>
> > He ended, and his words their drooping cheer
> > Enlighten'd, and their languish'd hope reviv'd
> > Th'invention all admir'd, and each how he
> > To be th'inventor mifs'd; *so easy' it feemed
> > Once found, which yet unfound moft would have thought
> > Impossible.*
> >
> > Milton, *Paradise Lost*, [1667], Book VI, 498-501; (My italics)[1].

---

[1] Edinburgh: Donaldson.

3. How to apply the topic

In the preceding examples, the topic is quite easy to apply, because it operates on the basis of an elementary linguistic structure "Subject + Predicate", which is easily transformed into its contrary. The topic is also easy *to reconstruct*, because the final formulation of the argumentation "**A** is **B**, then **non-A** is **non-B**", leaves the topical relation quite transparent. In other cases, the topic is more deeply embedded in the discourse, and its perception and reconstruction is more complex:

> *It took billions of years and ideal conditions before humans appeared on the planet, maybe one global warming will be enough to make it disappear*

This is clearly an inferential structure, progressing from a categorical assertion about the past to a restricted assertion about the future:

> E1, *maybe* E2

Do these two statements contain opposite predications on opposite subjects? Both statements express consecutions, "Conditions, Result", "C results in R".

> "*It took billions of years and ideal conditions before humans appeared on the planet*"
> *it took* **B** *before* **A** = **B** has been necessary for **A**
> *billions of years and ideal conditions* RESULT *humans appeared on the planet*

> "*may be one global warming will be enough to make it disappear*"
> *may be* **W** *will be enough for* **D**
> *one global warming* RESULT *[makes] it disappear*

The contraries are to be looked for on the two parallel structures "C *RESULT* R". The results are clearly opposites:

> *humans appeared on the planet*
> *to make [humanity] disappear*

Are the conditions **C** in the same relation? Condition **C2** "*one global warming*" cannot be self-evidently opposed to condition **C1** "*it took billions of years and ideal conditions*". Nonetheless, their argumentative orientations are opposed.

(i) Let's consider **C1**, *it took billions of years and ideal conditions before …*
— *billions of years* is oriented towards conclusions like "*that's a long time*"
— *ideal conditions* is oriented towards conclusions like "*it's rare, difficult to obtain*"
— The construction "*it takes X to Y*" is oriented towards "*it's a lot*".

These three orientations converge to give rise to the global inference "*this is a very complex process*".

(ii) Conversely, **C2** is oriented towards a class of conclusions of the type: "*this is a very simple process*":
— the determiner "*one*" is oriented towards unicity, "*just one*", and simplicity;
— *will be enough* is oriented towards a limitation "*no more than*", maybe "*less than expected*", for such and such accomplishment.

If this reconstruction is acceptable, then the following argumentative structure is attributed to the discourse:

> It has been really complicated to produce **R**
> so, *maybe* it will be very easy for **R** to disappear.

Such examples also suggest that the classical Aristotelian formulation of the topic may be oversimplified.

4. Trivial and non-trivial conclusions delivered by the topic

The application of the topic of opposites is a semantic reflex. Reasoning from opposites is a basic way of thought, in much the same way as causal reasoning, or reasoning by analogy or by definition. Reasoning from opposites may seem to deliver commonplace conclusions, empty reformulations of the original sentence, because analytical when both terms have the same degree of self-evidence. Nonetheless, even in this case, it does help to clarify the meaning of the words, which is no less necessary in philosophy than in general disputes:

> Temperance is beneficial; for licentiousness is hurtful. (Aristotle, *Rhet.*, II, 23; RR, p. 355)

There are, however, cases where opposite reflex may, or must, be inhibited: If a prayer says *"Peace to the people who love you"*, should we apply the topic and conclude something like *"War on those who don't"*?
Let us consider the following argumentation:

> If war is the cause of our present troubles, peace is what we need to put things right again. (*Ibid.*)

According to the principle *"we failed for lack of determination, of radicalism, etc."*, this reasoning from the opposites may counter a proponent arguing that:

> If we are in trouble, it is because we just have waged a *limited* war; this limited war is the cause of our present troubles, an *all-out war* is what we need to put things right again; only an outright victory will bring us peace.

It can also help to rebut a proposal for renewed political leadership:

> Those who sank the country into the crisis are perhaps not the best suited to get us out of the mess.

The topic can be applied as an interpretative principle to the following argument:

> We cannot trust the same failed market mechanisms to successfully steer the country out of this crisis (after Linguee, 25-10-2015)

Similarly, the conclusion of the following example is not trivial:

> For even not evil-doers should / Anger us if they meant not what they did / Then can we owe no gratitude to such / As were constrained to do the good they did us. (Aristotle, *Rhet.*, II, 23; RR, p. 355)

This is an application of the principle "to do good, one must have the capacity to do evil", S. **Opposite: refutation**. And the following one is quite suggestive:

> *Since in this world liars may win belief,* / *Be sure of the opposite likewise — that this world* / *Hears many a true word and believes it not* (*id.*, p. 357).

The *a contrario* reflex is a typical example of how argumentation leads us to contemplate things from a different perspective, in different wordings; or, as Grize would say, in a different light, **S. Schematization.**

5. *A contrario*

Lat. *contrarius*, "contrary". Two constructions might be used to refer to the argument, with the same meaning:
— the Latin proposition *a*: argument *a contrario sensu*, "by opposite meaning"
— or, less commonly, the Latin preposition *ex*: "*complecti ex contrario*" "conclude on the basis of the opposite meaning" (Cicero, quoted in *Dicolat*, art. *Complector*). **S. Latin labels**

The label "argument *a contrario*" can be used with the meaning of "inversion", to refer to the various kinds of argumentations which draw on contradiction, **S. Contradiction.**

Argumentation from the opposite corresponds to argumentation *a contrario*. In law, an *a contrario* argument is defined as:

A discursive process according to which a legal proposition being given which asserts an obligation (or other normative qualification) of a subject (or a class of subjects), for want of any other express provision, we must exclude the validity of a different legal proposition, which asserts this same obligation (or other normative qualification) with respect to any other subject (or class of subjects)" (Tarello 1972, p. 104) Thus, if a provision obliges all young men, who have attained the age of 20, to perform their military service, it will be concluded, *a contrario*, that young girls are not subject to the same obligation. (Perelman 1979, p. 55)

If a rule explicitly concerns a category of things, then it does not apply to the things that are not part of this category; the rule is applicable only in the defined area, to all the specified things, and only to them. This is an application of Grice's Rule of Quantity, which stipulates that the speaker must provide just the necessary amount of information, no more and no less.

This rule assumes that the system of law is well made and stable. In a period of social change and revision of the law, the argumentation *a pari*@ will be opposed to argumentation on the opposites. Women engaged in a battle for gender equality will refuse to oppose their status to men's status, and will demand that laws be applied *a pari*, be it beneficial (right to vote) or quite possibly less attractive (military service).

There is no paradox in the fact that *a pari* / *a contrario* argumentation can apply to the same material situation. Political issues are not unanimous, and cannot be solved by an automatic application of an algorithm; their discussion brings in historical considerations, values and affects.

Orientation

The concept of *orientation* (argumentative orientation, oriented statement or expression), combined with the correlative concept of *argumentative scale@* (Ducrot 1972), is fundamental to the theory of Argumentation within Language (sometimes referred to as AwL theory) developed by Oswald Ducrot and Jean-Claude Anscombre since the 1970s (Anscombre, Ducrot 1983, Ducrot 1988, Anscombre 1995a, 1995b, etc.). In this entry, the word *discourse* will refer exclusively to *(polyphonic) monologue*, not to dialogue or interaction.

The following equivalences can be helpful to grasp the general concept of *meaning as argument*, that is *orientation* towards a following statement, having the status of conclusion:

> He said **E1**. What does that mean?
> He says **E1** in the perspective of **E2**
> The reason why **E1** is said is **E2**
> The meaning of **E1** is **E2**
> **E1**, that is to say **E2**

1. *But* and the grammar of orientation

The stimulating case of *but*, has played a pivotal role in the construction of a grammar for argumentation. The privileged construction chosen to analyze this conjunction is schematized by "**E1** but **E2**":

> The restaurant is good, but expensive.

The basic observations are as follows: **E1** and **E2** are true (the restaurant is good *and* expensive); *but* refers to an opposition; this opposition is not between the predicates "*to be good*" and "*to be expensive*": one knows that "*everything good is expensive*", and tends to think that all expensive restaurants are necessarily good. The opposition is between the *conclusions* drawn from **E1** and **E2**, functioning as *arguments*: if the restaurant is good, then, *let's have dinner there*; if it is expensive, *let's go to another place*; and the final decision is the latter. *But* here articulates two statements oriented towards contradictory conclusions, and retains the conclusion derived from the second argument.

Under such analysis, the meaning of *but* is instructional: connectives are provide guidance for the interpretation of the speeches they articulate. They give the receiver the instruction to infer, to reconstruct from the left context **E1** a proposition **C** opposed to something, **not-C**, that can be inferred from the context to the right of **E2** (following **E2**). It is up to the listener to rebuild an argumentative opposition.

In the context of dialogical argumentation, these "instructors" themselves fall into the scope of an argumentative question, conditioning the reconstruction of the conclusions derived from **E1** and **E2**. The preceding *but* came under a question like "*Why not try this restaurant?*". If the question was "*Which restaurant should we buy to make the best investment?*", the interpretation would be totally different:

"*This restaurant is good* (="delivers outstanding financial performance") *but is expensive* (to buy)" the inferred, implicit conclusion would be "*so, let's invest our money somewhere else*". The argumentative question structuring the text creates the field of relevance and provides the interpretive constraints. This question and the relevant conclusions - answers are said to be "implicit" only insofar as the data supporting the analysis of *but* are generally limited to a pair of statements, the analyst considering that intuition can supply the missing context.

2. Linguistic constraint on the (argument, conclusion) sequence

As classical approaches, this theory considers argumentation essentially as a combination of statements "argument + conclusion". The crucial difference lies in the concept of the link authorizing the "step" from argument to conclusion, that is the "topic". The coherence of discourse is attributed to a semantic principle called a topos@ which binds the predicate of the argument to the predicate of the conclusion.

Ducrot defines "the argumentative value of a word" as "the orientation that this word gives to discourse" (Ducrot 1988, p. 51). The linguistic meaning of the word *clever* must not be sought in its descriptive value of a capacity measured by the intellectual quotient of the person concerned, but in the *orientation* it gives to a statement, namely, the constraints it imposes on the subsequent discourse. For example:

> Peter is clever, he will solve this problem

is opposed to the chain perceived as incoherent:

> * Peter is clever, he will not be able to solve this problem.

Such an argumentation is convincing indeed, because its conclusion, "*solving problems*" belongs to the set of predicates semantically correlated with "*to be clever*". A set of pre-established conclusions is already given in the semantic definition of the predicate of the statement used as argument.

The two morphemes@, *little* / *a little* give *opposed* argumentative orientations to the statement they modify, **S. Orienting words:**

> he has taken a little food, he is improving
> he has taken little food, he is getting worse

Building upon such intuitions, Ducrot defines the *argumentative orientation* (or *argumentative value*) of a statement as "the set of possibilities or impossibilities of discursive continuation determined by its use" (*ibid.*).

The argumentative orientation of a statement **S1** is defined as the selection operated by this statement among the class of statements **S2** likely to follow it in a grammatically well-formed discourse. Theoretically, a first statement **S1** can be followed by any other statement **S2**, both being independent linguistic units. According to the Argumentation within Language theory, the use of the first statement **S1** introduces restrictions imposing certain characteristics upon the second statement **S2**; that is, it excludes some continuations and favors others.

The linguistic constraints imposed by the argument upon the conclusion are particularly visible on quasi-analytical sequences, such as *"this proposition is absurd, so it must be rejected"*. By the very meaning of words, to say that a proposition *is absurd*, is to say that *it must be rejected*. This apparent conclusion is a pseudo-conclusion, for it merely expresses the definiens of the word *absurd*, "which should not exist" as testified by the dictionary:

> A.- [Speaking of a manifestation of human activity: speech, judgment, belief, behavior, action] Which is manifestly and immediately felt as contrary to reason in common sense. Sometimes almost synonymous with the impossible in the sense of "which cannot or should not exist". (*TLFi, [Absurd]*).

In a formula as famous as it is objectionable, Roland Barthes wrote that "language is neither reactionary nor progressive; language is quite simply fascist; for fascism does not prevent speech, it compels speech" ([1977], p. 366). Barthes' perspective is certainly different from Ducrot's. Nonetheless, in Ducrot's perspective, the argument literally compels the conclusion — playing with words, one might say that the inference is compulsive. This is common argumentative experience, in ordinary language, hearing the argument is enough to guess the conclusion.

Ducrot's theory is constructed on the linguistic observation that, regardless of its informational content, any statement specifies its possible continuations and excludes others. Not just any statement can follow any other statement, not only for informational reasons, but also for semantic and grammatical reasons. There are semantic constraints on discourse construction.

At the sentence level, this idea is expressed in the purely syntactic language of the restriction of selection. In its non-metaphorical use, the statement *"Pluto barks"* assumes that Pluto is a dog. Literally taken, *barking* carries a restriction of selection determining the class of *entities* it admits as subject. Similarly, but at the discourse level, **S1** operates a selection upon the class of the statements **S2** that can succeed it. An argumentation is a pair of statements (**S1, S2**), such that **S2**, called the *conclusion*, respects the orientation conditions imposed by **S1**, called the *argument*.

3. Meaning as intention

The AwL theory rejects the conceptions of meaning as adequacy to reality, whether logical (theories of truth conditions) or analogical (theories of prototypes). It is built on a quasi-spatial conception of meaning as *sense, direction*: what the statement **S1** (as well as the speaker publicly) *means*, in a specific context, is the conclusion **S2** to which this statement is oriented. The art of arguing here is the art of managing discourse transition.

The relation "argument **S1** - conclusion **S2**" is reinterpreted in a language production perspective (Fr. *perspective énonciative*) where the meaning of the argument statement is contained in and revealed by the next statement. The understanding of what is meant by the statement *"nice weather today!"* is not developing a corresponding mental image or cognitive scheme, but accessing the intentions displayed by the speaker, that is, *"let's go to the beach"*. This is in perfect agreement with the Chinese proverb, *"when the wise man points to the stars, the fool looks at the finger"*.

The meaning of **S1** is **S2**. The meaning here is defined as the *final cause* of the speech act. The AwL thus updates a terminology referring to the *conclusion* of a syllogism as its *intention*. This reflects the fact that a reformulation connector such as *that is to say* can introduce a conclusion:

L1:   — *This restaurant is expensive.*
L2:   — *That is/ you mean / in other words/ you do not want us to go there?*

The theory has developed in three main directions, *argumentative expressions*, or *orienting@* words; connectives as argumentative *indicators@*; and the concept of semantic *topos@*.

4. Some consequences

4.1 Reason in discourse

Tarski maintains that it is not possible to develop a coherent concept of truth within ordinary language, **S. Probable**. In Ducrot's vision of argumentation, the question of the *validity* of an argument is re-interpreted as grammatical validity. An argumentation is valid if the conclusion *grammatically agrees* with its argument (if it respects the restrictions imposed by the argument). It follows that the *rationality and reasonableness* attached to the argumentative derivation are no more than an insubstantial reflection of a routine discursive concatenation of meanings, or, as Ducrot says, a mere "illusion", **S. Demonstration**. This is coherent with the structuralist project reducing the order of discourse to that of language (Saussurian *langue*). Ordinary discourse is seen as unsuited to expressing truth and reality. It follows that discourse is denied any rational or reasonable capacity.

4.2 A re-definition of homonymy and synonymy

As the theory is based exclusively on the concept of orientation, and not on quantitative data or measures, it follows that if the same segment **S** is followed in a first occurrence of the segment **Sa** and in a second occurrence of the segment **Sb** that contradicts **Sa**, then **S** does not have the same meaning in these two occurrences. As we can say *"it's hot* **(S)**, *let's stay at home* **(Sa)**" as well as *"it's hot* **(S)**, *let's take a walk* **(Sb)**" we have to admit that the statements "[are] not about the same heat in both cases" (Ducrot 1988, p. 55). This is a new definition of homonymy. By analogous considerations, Anscombre concludes that there are two verbs *to buy*, corresponding to the senses of *"the more expensive, the more I buy"* and *"the less expensive, the more I buy"* (Anscombre 1995, p. 45).

Conversely, we can assume an equivalence between statements selecting the same conclusion: if the same segment **S** is preceded, in a first occurrence by the segment **Sa**, and in a second occurrence by a different segment **Sb**, then **Sa** and **Sb** have the same meaning, because *they serve the same intention*: "*it's hot* (**Sa**), *I'll stay at home* (**S**)" vs. "*I have work* (**Sb**), *I'll stay at home* (**S**)". It is a new definition of *synonymy*, in relation to the same conclusion.

Finally, "if segment **S1** only makes sense from segment **S2**, then sequence **S1** + **S2** constitutes a single utterance", a single linguistic unit (Ducrot 1988: 51). One could probably go a step further, and consider that they make up a single *sign*, **S1** becoming a kind of signifier of **S2**. This conclusion reduces the proper "order of discourse" back to that of the statement, even of the *sign*.

5. Orientation and inferring license

Ducrot opposes his "semantic" point of view to what he calls the "traditional or naive" view of argumentation (Ducrot 1988, p. 72-76), without referring to specific authors. Let's consider Toulmin's layout@ of argument.

— Argumentation is basically a pair of statements (**S1**, **S2**), having respectively the status of argument and of conclusion.

— Each of these statements has an autonomous meaning, and refers to a distinct specific fact, each of these facts being independently assessable.

— There is a relation of implication, a physical or social extra-linguistic law between these two facts (Ducrot 1988, p. 75).

This concept of argumentation can be schematized as follows. Curved arrows, going from the discourse level to the reality level, enact the referring process.

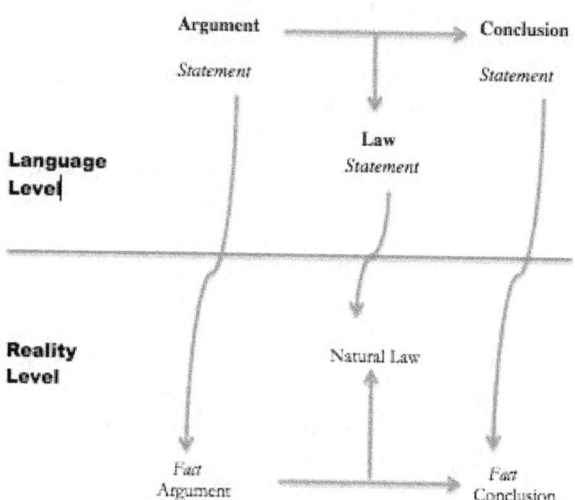

This conception may be "naive" insofar as it postulates that language is a transparent and inert medium, a pure mirror of reality. This is not the case for natural language (Récanati 1979); such conditions are only met by controlled languages like the languages of the sciences, in relation to realities that they *construct* as much as they *refer* to them.

Contrary to this view, the AwL theory emphasizes the strength of purely linguistic constraints. The orientation of a statement is precisely its capacity to project its meaning not only *on*, but also *as* the following statement, so that what appears as "the conclusion" is only a re-formulation of the "argument". For the AwL theory, discourse is an arguing machine, systematically committing the *vicious@ circle* fallacy.

To sum up, the AwL theory opposes ancient or neoclassical theories and practices of argumentation, as a semantic theory of language opposes theories and techniques of conscious discursive planning, operating according to referential data and principles. For classical theories, argumentative discourse is likely to be evaluated and declared valid or fallacious. For semantic theory, an argumentation can be evaluated only at the grammatical level, as a concatenation (**E1**, **E2**) that is acceptable or not, coherent or not. In this theory, the compelling character of an argument is entirely a matter of language. It is no different from the coherence of discourse. To reject an argument is to break the thread of the ideal discourse. This position redefines the notion of argumentation; Anscombre speaks thus of argument "in our sense" (1995b, p.16).

6. Reasoning combines the two kinds of inferences

The transition from argument to conclusion can be based on a natural or social law or on a semantic coupling of the argument with the conclusion. These two kinds of inferences are currently connected in ordinary discourse:

> You talk about the birth of the gods (1). You say, then, that at one time the gods did not exist; so you deny the existence of the gods (2), which is a blasphemy and punished by the law. So you will be punished (3) *a pari*, according to the law punishing those who speak of the death of the gods.

First, a semantic law deduces (2) from (1), **S. Inference**; second, a social law, having nothing to do with language or discourse, goes on from (2) to (3), the punishment being finally determined by an *a pari* alignment. Social law can be naturalized by somehow integrating the meaning of the words:

> You are an impious man, impiety is punished with death, so you must die.

It is difficult to tell to what extent the very meaning of the word *impious* has integrated the law "*impiety is punished with death*". Nonetheless, the link with social reality is clear: if I wish to reform the legislation, my revolt is not a semantic revolt. **S. Definition; Layout.**

Orientation Reversal

The argumentative orientation of an utterance can be reversed by the substitution of one morpheme for another, for example, by substituting *little* for *a little*, **S. Orientation; Orienting words**. The adverb *precisely*, in one of its uses, can also operate a reversal of argumentative orientation:

> S1: — *Peter does not want to go out, he's depressed.*

S2: — *Well*, precisely, *he would breathe the clean country air, it would clear his head.*

"*He is depressed*" justifies the decision not to go out; *precisely* accepts the fact that Peter is depressed, but re-orients it towards the opposite conclusion: Peter should go out (Ducrot 1982), by applying to it the different rule: "*When one is depressed, one wants to stay home*" vs. "*Going out is good for depression*".

The orientation reversal is based on the letter of what the **S1** says; **S2** replies to **S1** "*Your argument does not support your claim, it even points to the contrary; you give arguments against your position*". **S2** opposes to **S1** his or her own saying, and thus affects his conversational face. This can be considered to be a typical reply "to the letter" (*ad litteram*), a strategy of discourse destruction, **S. Matter; Destruction; Objection; Refutation.**

Classical rhetoric has identified many phenomena of reversal of the same order, such as irony:

> *Everything is possible with the SNCF (French Railway Company), that is the best slogan you ever found!*
> Said by a traveler to a train conductor when the train had been held between two stations for two hours.

The slogan is oriented towards "*the SNCF is capable of being incredibly positive and pleasant for you*"; the circumstances show that "*the SNCF is capable of being incredibly negative*", **S. Irony.**

Some of these strategies have been identified and named in rhetoric:
— Exploiting the various acceptances of a term to reverse its argumentative orientation: *antanaclasis*.
— Turning over an expression, to the same effect: *antimetabole*.
— Reversing the qualification of an act: *antiparastasis*.
— Reversing the orientation of a term by substituting another quasi-synonymous term or description: *paradiastole*.

1. Antanaclasis

Antanaclasis is a phenomenon of repetition of a polysemous or homonymous term or expression. In its second occurrence, the term has a meaning and a direction different from that which it had in the first occurrence. In other words, the signifier S^0 has the meanings **Sa** and **Sb**. In its first occurrence S^0 has the meaning **Sa** with the orientation **Oa** and, in the second occurrence the meaning **Sb** with the orientation **Ob**.

The resumption of the signifier S^0 must take place in the same discursive unit, whether a statement, a paragraph, a turn or pair of turns. It can be performed either by the same speaker in the same speech unit or by a second speaker in a second turn.

Within the same-speaker intervention, the antanaclasis introduces an ambiguity@, since the same word is used to designate different things. In a syllogism, the antanaclasis introduces in fact two terms under the cover of the same signifier S^0, and thus produces a syllogism not of three but of four terms, that is, to a *paralogism*@.

In interaction, the two meanings of the term are used in two consecutive turns of speech, the second invalidating the first. The antanaclasis is a kind of ironic echoing and aggressive retaliation. The word tolerance *refers* to a virtue; the French expression *maison de tolérance*, "house of tolerance", refers to a legal, licensed... tolerated brothel:

S1: — *A little tolerance please!* (tolerance is a virtue)
S2: — *Tolerance, there are houses for that* (tolerance allows vice).

In French *une foire* refers to "a fair", a commercial exhibition; or "a mess", a state of general noise and confusion.

S1: — *We could not book you a hotel, all are fully booked, there is a* foire ("a fair") downtown
S2: — *It seems that it is often la* foire *downtown*. Fr. *foire* = "mess"

In the second example, the second term reorients what was said as an excuse to a reproach: *"you cannot get organized"*; this word-for-word resumption undermines S1's speech. The use of derivative words authorizes maneuvers of this type, **S. Derivatives**. The one who finds his work *aliénant* (F), that is "alienating" (as in *"assembly line work alienates workers"*), is accused of being an *aliéné* (F), that is an insane person:

> The ideological policeman of collectivism can say almost the same to the opponent: "*For those who come to protest against alienation, in our society we have lunatic asylums* (F. *asiles d'aliénés*)". (Thierry Maulnier, [*The Meaning of Words*], 1976[1])

The antanaclasis reorientation differs from that operated by *precisely*. This adverb takes a statement oriented towards a given conclusion, grants the statement (accepts the information), and transforms it in order to make it back the opposite conclusion. In the preceding case, it could be *"Well, precisely, the fair was announced a long time ago, you should have taken precautions."* The antanaclasis does not take the excuse seriously, it disorients the discourse.

2. Antimetabole

Like antanaclasis, the antimetabole is a language ploy used to dismantle the opponent's speech. The discourse is resumed and restructured syntactically so as to make it lose its orientation, or even to give it an opposite orientation. Dupriez quotes the determined / determining permutation mechanisms, by which a discourse on *"the life of words"* can be ironically destroyed by the affirmation of a preference for *"the words of life"* (Dupriez 1984: 53-54):

> We do not live in a *time of change*, we live in a *change of time*.
> These announcements effects (F. *effets d'annonce*) will quickly reduce to ineffective announcements (F. *annonces sans effets*)

S. Refutation; Prolepsis; Destruction; Converse.

---

[1] Thierry Maulnier, *Le sens des mots*, Paris: Flammarion, 1976, p. 9-10.

3. Antiparastasis

This word refers to the *stasis@* theory. A charge is laid against somebody; the accused acknowledges the fact for which he or she is blamed, but does not accept the blame:

    L: — *You killed him!*
    L: — *Upon his request, I have ended his suffering.*

The first statement is an accusation, "*Shame on you, you deserve a condemnation!*"; the second introduces an argument that cancels this orientation "*what I have done is an act of courage*", or even reverses the accusation: "*what I have done deserves every respect*", **S. Motives**.

This form of counter-argumentation gives the same fact two opposite orientations. The antanaclasis is a pseudo-acceptance and an implicit reversal, whereas the antiparastasis explicitly reverses the negative orientation given to the fact by the opponent.

This choice of defense gives the speaker a militant or rebellious ethos. Such situations based on radically opposite values have a high dramatic potential, for example, the confrontation between Antigone and Creon in Sophocles' play *Antigone* enacts such a situation of antiparastasis.

4. Paradiastole

The term *paradiastole* originates from a Greek word expressing a movement of expansion and distinction. In a monologue, the paradiastole "establishes a system of nuancing and precision, generally developed upon parallel statements" (Molinié 1992, *Paradiastole*). The Latin term *distinguo* refers to a similar operation. The paradiastole refines the definition of a concept or establishes a distinction between two close concepts that, from the point of view of the speaker, should not be confused: "*sadness is not depression*". In a dialogue, paradiastole rejects a partner's word as inadequate, and substitutes for it another word, considered to be contextually more adequate, which *reorients* the discourse. *Depression* and *sadness* may be semantically close, but they can nevertheless be opposed as in:

    L1:   — *I'm depressed, I have to see a shrink.*
    L2:   — *No, you're not depressed, you're sad, and sadness is not a medical condition.*

Discourse constantly builds up such anti-oriented pairs, **S. Orienting words**

    All lovers, as we know, boast of their choice; [...] The *chatterer* [is] *good humored*; the *silent* one maintains her *virtuous modesty*

                                                      Molière, [*The Misanthrope*], 1666[1]

(What is presented as) the true strongly negative description of a person as *a chatterbox* or a *stupid person* is opposed to how she appears in her lover's eyes, as *good humored* or *maintaining her virtuous modesty*. The following example shows that

---

[1] Molière, *Le Misanthrope*, II, 4. Quoted from Moliere, *The Misanthrope*. Ed. by, Girard KS: E. Haldeman-Julius, 1922, p. 26-27. https://archive.org/details/misanthropecomed00molirich (11-04-2017).

this situation is generalized to discourse, where paradiastole no longer operates strictly between two terms, but between two discourses, opposing two points of view:

> L1: — *He's brave.*
> L2: — *I would not say that. He knows how to face danger, okay, but it seems to me that to be truly brave you also need a system of values, a clear sense of what you are fighting for... may be he is more of a hothead?*

Starting from a mere nuancing, paradiastole can evolve into a term-to-term opposition:

> L1: — *This is just ignorance*
> L2: — *No, it is simply bad faith.*

Orienting Words

The semantic concept of an argumentative morpheme, or *orienting word*, is developed by Anscombre and Ducrot as an essential part of the theory of Argumentation within Language. A morpheme (an expression) is said to be argumentative if its introduction into an utterance:

— *does not modify the factual referential value* of this statement (it has no quantifying function)
— *modifies its argumentative orientation*, that is to say, the set of conclusions compatible with this statement; the set of statements that may follow it, **S. Orientation**.

The concept has been applied to the linguistic description of "empty" words or "argumentative operators" such as *little / a little*, as well as to "full" words such as the *helpful / servile* pair.

1. Opposition of anti-oriented words

Consider the statements (1) "*Peter is helpful*" and (2) "*Peter is servile*". Do these two statements describe two different kinds of behaviors, or one and the same attitude? Both positions can be argued.

(i) Statements (1) and (2) describe two behaviors. Helping one's grandmother cut up the chicken would be helpful; accepting to carry your boss' small suitcase would be servile. As a result, a different value is attached to each behavior; a positive value is attributed to helpfulness, whilst a negative value is placed on servility. In order to determine the nature of Peter's behavior, one must scrutinize reality.

(ii) It can also be said that these two words describe a single behavior cast it in two different lights, i.e. two subjectivities, involving emotions and value judgments. I judge this behavior positively, and I say, "*Peter is helpful*"; I judge it negatively, and claim, "*Peter is servile*". Reality says nothing about helpfulness or servility. The origin of the distinction is not grounded in reality, but in the active structuring operated by the speaker's perception.

Statements (1) and (2) create opposed discursive expectations within the listener: *Helpful* is a recommendation, "*A nice guy!*", while *servile* is a rejection, "*I can't bear him*".

If the job implies contacts with a person concerned specifically about deferential behavior, then *Peter is servile* might also serve as an ironic recommendation, encompassing the disapproval of the two people: "*they will make a nice pair*".

These opposed orientations correspond to the rhetorical phenomenon known as *paradiastole*, "*the world moves backwards, words have lost their meaning: the miser is economical, the unconscious courageous*"; they are interpreted as the expression of linguistic bias by normative theories of logical inspiration. **S. Orientation Reversal.**

Antithetical designations — The opposition discourse *vs.* counter-discourse is sometimes reflected in the morphology of words, as in the previous case, **S. Antithesis; Derivatives; Ambiguity:**

> disputation *vs.* disputatiousness
> politician *vs.* politico
> philosopher *vs.* philosophizer
> scientific *vs.* scientistic

In general, parties will use different terms to refer to beings at the center of the debate: you are the *persecutor*, I am the *victim*; he is the *bad rich man*, I am the *poor but honest person*; your approach is *scientistic* while mine is *scientific*, **S. Discursive Object.**

According to what criteria can I categorize this individual as a *terrorist* or as a *resistance* fighter? Is the *resistance* fighter a successful terrorist, and the terrorist the *resistance* fighter of a lost cause? Should his acts be considered (categorized) as a coward act of terrorism or as a heroic act of resistance? Shall we say that everyone has dirty hands and that everything depends on the speaker's partisan options? Humanity can and does establish universal criteria for deciding who is who, such as "targeting civilians; using and targeting children", "torturing people", **S. Categorization.**

2. Adverbial orientation operators

The adverb *even* is argumentative in:

> Leo has a bachelor's degree and even a Master's degree.

Statements " **p**, and even **p'** " are characterized by their relative position on an argument scale:

> there is a certain [conclusion] **r** determining an argument scale where **p'** is [a stronger argument] than **p** [for the conclusion **r**]. (Ducrot 1973, p. 229)

In other words, *even* statements are inherently argumentative; they are oriented towards a conclusion **r**, that can be recovered from the context; they coordinate two arguments **p** and **p'** supporting this conclusion; and they hierarchize those two statements, presenting **p'** as stronger than **p**.

Statement (1) is argumentative; it coordinates two arguments "*Leo has a bachelor's degree*" and "*Leo has a Master's degree*"; both are oriented towards a conclusion, for example "*Leo can teach some mathematics*"; and it considers that the latter "*Leo has a Master's degree*" is a stronger argument than the former for this same conclusion. This gradation can be represented as follows on an argument scale@:

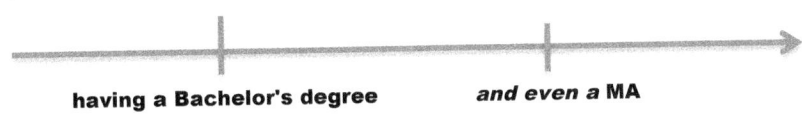

having a Bachelor's degree   **and even a MA**

The relative positions of **p** and **p'** on that scale depend on the speaker:
>We had a great meal, we even had cheese pasta.

Some gastronomes may not consider cheese pasta to be an essential component of a great meal.

Too — The theory of scales is governed by a "*plus*" principle: the higher one is on the scale, the closer one is to the conclusion. But this principle leads to a paradox:
>You reluctantly bathe in water with a temperature of twenty-two degrees, you'd be happier to bathe in water at twenty-five, at thirty, or even warmer. The hotter the water, the better for you; so you really should try bathing directly in the kettle.

Too often inverts the argumentative orientation:
>S1   — *that's cheap, buy it.*
>S2   — *(Precisely) that's too cheap.*

And sometimes reinforces this orientation:
>S1   — *It's expensive, too expensive, don't buy it*

Almost / hardly — *Almost* is a paradoxical word: "*almost* **P**" presupposes **not-P** and argues as **P**. If Leo is *almost* on time, he's *not* on time. Nonetheless, one can say:
>Excuse him, he was almost on time, he should not be sanctioned.

In other words, "*almost on time*" is co-oriented with "*on time*". The argumentative orientation of an *almost* utterance might be rejected by an inflexible superior, who rejects the positive framing being imposed upon him. The superior applies the topos of the letter of the law, **S. Strict meaning**:
>So you do confirm that he was not on time. The sanction will be applied.

This co-orientation of **P** and *almost* **P** does not apply to predicates referring to the crossing of a threshold. When transporting a seriously ill patient, the nurse might urge the ambulance driver: "*hurry, he is almost dead*" but the nurse would not say, "*hurry, he is dead*". Yet, in an alternative scenario, say that of a rather laborious assassination, the murderer can tell his accomplice, "*hurry up, he is* *almost dead, and you still haven't found anything to wrap his body in"*, and *a fortiori "hurry up, he is dead,* etc."

The permutation *almost / hardly* reverses the argumentative orientation of the statements in which they enter:
> You're *almost* healed, you can join our party!
> I'm *hardly* healed, I cannot join your party.

The appeal to the strict meaning is opposed to the raising of the thresholds produced by *almost* and *scarcely*, **S. Strict Meaning**.

Little / A Little — These two adverbs give opposing argumentative orientations to the predicates that they modify:
> (1) now, there is little trust in market mechanisms.
> (2) now, there is a little trust in market mechanisms.

(1) is oriented towards *"there is no trust at all"*, while (2) is oriented towards *"trust"*. *Little* and *a little* are not quantifiers referring to different quantities of food (*a little trust* being more than *little trust*), but give opposed orientations to a quantity that is fundamentally the same.

3. Adjectives as orientation operators

Adjectives might modify the argumentative strength or the orientation of a sentence.

De-realizing operators are defined as follows:
> A lexical word **Y** is *de-realizing* in relation to a predicate **X** if and only if the combination **XY** on the one hand is not felt as contradictory, and, on the other hand, reverses or lowers the argumentative strength of **X**. (Ducrot 1995, p. 147)

Consider the statements (Ducrot 1995, p. 148-150)
> He is a relative, and even a close relative
> He's a relative, but a distant relative

Close is a *realizing* operator (*id.*, p. 147) *"they are close relatives"* is co-oriented with *"they are relatives"*, towards conclusions such as *"they know each other well"*. They are situated as follows on the corresponding argument scale:

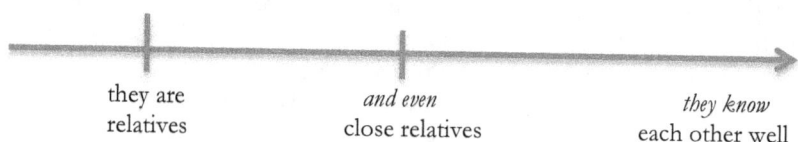

they are relatives   and even close relatives   they know each other well

Distant is a *de-realizing* operator. The sentence *"he is a distant relative of mine"*:
— can be oriented towards *"we don't know each other well"*, i.e., it has an opposite orientation to *"he is a relative of mine"*.

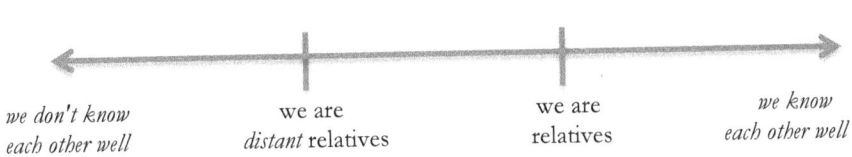

— can be oriented towards *"we know each other well"*, like *"he is a relative of mine"*, but with a lesser force:

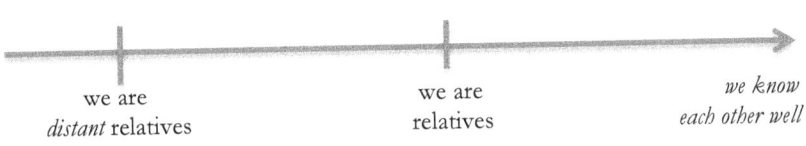

Ornamental fallacy?

The contrast between a *rhetoric of figures*@ and a *rhetoric of arguments* is a remainder and an exacerbation of the classical distinction between the two fundamental production stages of rhetorical discourse, the research of arguments and their verbal expression. The rupture between *inventio* and *elocutio* is generally attributed to Ramus (Ong, 1958). Only the *elocutio* and the *actio* would fall within the realm of rhetoric, the *inventio*, the *dispositio* and the *memoria* being independently re-assigned to thought (cognition). This opposition, which quickly became popular, between, on the one hand, an *ornate, figurative, rhetorical* discourse, and, on the other hand, an *argumentative* discourse ideally free from subjectivity or figuration, has been strongly reasserted by Locke in the modern perspective of a discourse aimed at the development of scientific thought. This antagonism has been pushed to the confrontation and mutual rejection of a *discourse of pleasure and emotion* and an *austere discourse of reason*.

1. Fallacious rhetoric?

The whole enterprise of rhetoric, as the art of constructing a persuasive discourse, has been rejected in the name of a transcendental truth, by Socrates, as staged by Plato in the *Gorgias* and the *Phaedrus*, **S. Argumentation (I); Persuasion; Probable**. In the modern age, this age-old criticism was strengthened by a new wave of criticism developed on behalf of scientific discourse. Rhetorical discourse is now routinely belittled as substituting the search for pleasure for the search for truth. Rhetoric is seen to fulfill a perverse desire for *ornament*, and, to root out this evil, ornament, and therefore figures, should be eliminated.

Persuasive rhetoric is therefore reconstructed as an ornate discourse, a discourse of passion, perverse and magical **S. Persuasion**. The figures and the tropes are defined within the framework of the *ornatus*, then, by synecdoche, the *elocutio* is assimilated to the *ornatus*, and finally rhetoric itself is reduced to the *elocutio*. It is this ornamental vision of a "makeup rhetoric" that has been opposed to the natural, healthy discourse of reasonable argument, S. **Verbiage**. The following extract from Locke serves as an authoritative reference in discourses attacking ornate language.

> 34. Seventhly, language is often abused by figurative speech. Since wit and fancy find easier entertainment in the world than dry truth and real knowledge, figurative speeches and allusion in language will hardly be admitted as an imperfection or abuse of it. I confess, in discourses where we seek rather pleasure and delight than information and improvement, such ornaments as are borrowed from them can scarce pass for faults. But yet if we would speak of things as they are, we must allow that all the art of rhetoric, besides order and clearness; all the artificial and figurative application of words eloquence hath invented, are for nothing else but to insinuate wrong ideas, move the passions, and thereby mislead the judgment; and so indeed are perfect cheats: and therefore, however laudable or allowable oratory may render them in harangues and popular addresses, they are certainly, in all discourses that pretend to inform or instruct, wholly to be avoided; and where truth and knowledge are concerned, cannot but be thought a great fault, either of the language or person that makes use of them. What and how various they are, will be superfluous here to take notice; the books of rhetoric which abound in the world, will instruct those who want to be informed: only I cannot but observe how little the preservation and improvement of truth and knowledge is the care and concern of mankind; since the arts of fallacy are endowed and preferred. It is evident how much men love to deceive and be deceived, since rhetoric, that powerful instrument of error and deceit, has its established professors, is publicly taught, and has always been had in great reputation: and I doubt not but it will be thought great boldness, if not brutality, in me to have said thus much against it. Eloquence, like the fair sex, has too prevailing beauties in it to suffer itself ever to be spoken against. And it is in vain to find fault with those arts of deceiving, wherein men find pleasure to be deceived. (Locke, *Essay*, III, X; Fraser, p. 146-147)

De Man has shown that the issue here is the status of natural language in science and philosophy, "at times, it seems as if Locke would have liked nothing better than to be allowed to forget about language altogether, difficult as this may be in an essay having to do with understanding" (1972, p. 12). But this observation does not directly invalidate Locke's thesis, for it is possible to consider that this thesis deals with *ordinary* language and its capacity to carry the new mathematical forms of scientific knowledge. In fact, in the modern age, the language in which "we preserve and develop truth and knowledge" is no longer natural language, but the languages of calculation. Nevertheless, de Man rightly emphasizes the contradictory nature of an undertaking that would engage in an analysis of reasoning in natural language by first condemning natural language.

2. Against ornate discourse

The following are the main argumentative topoi of the discourse which condemns figures as fallacious ornaments.

2.1 Fallacy of relevance and inconsistency

In an unfolding argumentative discourse, all decoration is a form of entertainment, that is to say a *distractor*. As a result, the figures show a *lack of relevance*, they are fallacious by virtue of the *ignorance of the question*@, permanently serving as a *"red herrings"* @

The figures knowingly flout three Gricean principles, the maxims of *quality*, *quality* and *relevance*. To use Klinkenberg's French term, figures are *impertinences*, that is, they are both "irrelevant" and "brazen" (Klinkenberg 2000; Klinkenberg 1990, p. 129-130). Moreover, they do not respect the *non-contradiction* principle. The metaphor@ is true *and* false, guilty of *ambiguity* and *category mistake*.

2.2 Fallacies of verbiage and emotion

The classical concept of figurative discourse is based on the possibility of choosing between two chains of signifiers to express the same idea, to refer to the same being, to the same state of the world or the same semantic content. This presupposes a *superabundance of words* compared with the strict requirements of the objective discourse. The coexistence of different signifiers to express the same thing or the same truth is at the root of the fallacy of *verbiage*@, a kind of meta-fallacy that opens the way to all others **S. Connectives**.

Furthermore, the figurative form systematically favors the intricate and the rare, the exact opposite of the *ordinary, simple* and *direct manner of speaking*. And when an apparently *plain* form appears in such elaborate discourse, it only seems plain due to a double subtlety. The unsophisticated addressee anticipates a simple expression; the sophisticated addressee knows that this expectation will be frustrated and thus anticipates the figuration. This second-level expectation is then itself frustrated by the simplicity of the expression. The ornamental figure is *offbeat*, and thus produces a *surprise*, the prodrom to emotion, opening the way for numerous *ad passiones* fallacies; *aesthetic emotions* are banned as any other passion. This link is explicit in Locke's quotation.

2.3 The language transparency fallacy

Taking scientific language as the norm, in order to guarantee a direct access to objects and their natural connections, the language of argument should be regulated, unambiguous, without defect or excess, exactly proportioned to the nature of things, in other words, *transparent, ad judicium*@. The figures, which pretend to glorify the truth, in fact *veil* it. Ornaments are worse than fallacies; they are their source and mask.

The problem is that figures are the bones and flesh of everyday expression; to get rid of them one would have to renounce natural language and argumentation in human affairs as a whole.

3. An etymological argument against the decorative view of the *ornatus*

Are the figures ornaments? The word *ornament* is a copy of the Latin *ornamentum* (adj. *ornatus*, verb *ornare*). The primary meaning of *ornamentum* is: "1. Apparatus, tackle, equipment [...] harness, collar [...] armor" (Gaffiot [1934], *Ornamentum*). The past participle adjective *ornatus* shares this fundamental meaning. The phrase: "naves omni genere armorum ornatissimae" (C. Julius Caesar [*The Gallic Wars*] 3, 14, 2) translates as "boats with ample equipment [weapons and tackles]" *(ibid.)*. Thus, an *ornatus* speech is a speech well *equipped* to fulfill its function. When dealing with a choice to be made in public affairs, a *well-equipped* rhetorical discourse is *a well-argued* discourse. The arguments are indeed part of the *ornamenta*, the *equipment* of the discourse.

Considered to be part of the *discourse equipment*, figures can be integrated into argument analysis, for example as instruments for the construction of *objects of discourse@* and *schematizations@*. In any case, they should not be seen as constituting an extraneous "level" disfiguring the pure cognitive level, but as part and parcel of all the operations constructing the argumentative discourse.

P

Paradiastole ▶ ᴓrientation Reversal

Paradoxes of Argumentation

1. Arguing for P weakens P

Arguing for P weakens P, firstly in virtue of the grounds substantiating the discourse *against the arguments*, which is often the same as the discourse *against the debate*, **S. Debate**. This discourse develops as follows:

> People don't accept living in doubt, not to being committed to some cause, not knowing, not having an opinion on everything, not challenging the other's opinions. They are ready to argue for or against all and everything. They relish dispute, and are inherently unable to dispute, as shown by the Port-Royal philosophers. Disputes are just substitutes for fights or playground games, they always produce more heat than light. Querulousness is a disease. The will to be right, to attack and defend is the transparent mask of the will to power. Our most entrenched opinions are not grounded in argument, but in our reptilian brain, we don't argue, we just reformulate our opinions. Conclusion: let us rather strive to clearly say what we have to say, etc., etc. **S. Fallacies (V)**.

Secondly, arguing for **P** weakens **P** because argument-based knowledge is inferential, i.e., *indirect* knowledge. Indirect knowledge is frequently considered inferior to *direct* knowledge, which is expressed in a simple, plain statement of fact, especially to direct, revelation-based religious knowledge, **S. Self-Evidence**. Newman expressed this idea in particularly energetic words, first in the epigraph of his *Grammar of Assent* (1870), taken from St. Ambrose, "it did not please God to save his people through dialectic" ("*Non in dialecticà complacuit Deo salvum facere populum suum*"), and further:

> Many a man will live and die upon a dogma: no man will be a martyr for a conclusion. [...] No one, I say, will die for his own calculations[1]: he dies for realities. (*Id.*, p. 73)
>
> To most men, argument makes the point in hand only more doubtful, and considerably less impressive. (*Id.*, p. 74)

Arguing along the same line, Thomas Aquinas, when discussing the question of "*whether one ought to dispute with unbelievers in public?*", envisages the following objection to a positive answer:

> *Objection 3:* Further, disputations are conducted by means of arguments. But an argument is a reason in settlement of a dubious matter: whereas things that are of faith, being most certain, ought not to be a matter of doubt. Therefore one ought not to dispute in public about matters of faith.
>
> *ST*, Part 2, 2, Quest 10, Art 7.

Arguments develop from a question; they are mirrored in counter-arguments, attested or conceivable. The same doubt is cast upon both positions. This explains the existence of the paradoxes of argumentation: to contest a position is both to accept that one's own becomes debatable and to acknowledge the attacked position as debatable. This explains why the first step in the process of legitimizing a new position involves opening *a debate about it* and, to do so, one must first find some opponents.

2. Producing a question legitimates the variety of answers

Should there be a "scientific and public debate" on the issue of whether there were gas chambers in Nazi Germany? This is exactly what the revisionist Roger Garaudy has demanded: the organization of a debate would legitimize the various positions taken in this debate.

> *Roger Garaudy has a persisting doubt about the existence of gas chambers*
> Later in the book, Roger Garaudy evokes *Shoah*, the film of Claude Lanzmann, which he considers a "stinker". '*You are talking about "Shoah business", you say that this film only brings testimonials without demonstration. This is a way of saying that the gas chambers do not exist*', suggests the President [of the Court]. '*Certainly not*' protests Roger Garaudy. '*Until a scientific and public debate is held on the issue, doubt will be allowed*'. (*Le Monde*, 11-12 January 1998, p. 7)

Here, Garaudy claims the *third party* position. He may even say that the president fallaciously argues from ignorance — saying that **P** is not proved is not claiming that **non-P**. The refutation must take into account the contextual

---

[1] May be alluding to Galileo who accepted to publicly recant heliocentrism and the movement of earth, while privately maintaining the truth: "E pur si muove" (*And yet it moves*)

knowledge: here the affirmation is *false*, because the historical and scientific work is done, has been published and libraries stay opened late into the night. We are exactly in the situation of the Aristotelian *indisputability*, **S. Conditions of Discussion.**

3. Refuting P reinforces P; but not to, even more

To be criticized is much better than to be ignored. Being at the center of a polemic may be an ideal and comfortable position. Seeking somebody who can put forward an argument that contradicts one's own is an argumentative strategy which gives initial legitimacy to a viewpoint. Refuting and opposing counter-arguments generates a question where there was none, and this question, by feedback, legitimizes all the speeches that give it an answer. The proponent is *weak* since he or she bears the burden of proof, but *strong* because he or she creates a question.

The historian P. Vidal-Naquet describes this argumentative trap as follows in the case of the negationist discourse.

> I long hesitated before writing these pages on the alleged revisionism, about a book whose editors tell us without laughing that, *"Faurisson's arguments are serious. They must be answered"*. The reasons for *not* speaking were multiple, but of unequal value. [...] Finally, was not replying accrediting the idea that there was indeed debate, and to publicize a man who is passionately greedy of it? [...]
>
> This is the last objection that is actually the most serious one. [...] It is also true that attempting to debate would be to admit the inadmissible argument of the two *"historical schools"*, the *"revisionist"* and the *"exterminationist"*. There would be, as is expressed a leaflet of October 1980, *"supporters of the existence of homicide gas chambers"* and the others, as there are supporters of a high chronology and of a low chronology for the tyrants of Corinth. [...]
>
> Since the day that R. Faurisson, a duly qualified academic, a professor at a leading university, was first allowed to write in *Le Monde*[1], even if it was immediately refuted, the question ceased to be marginal. This became central a central question, and those who had no direct knowledge of the events in question, especially young people, had the right to ask if some people wanted to hide something from them. Hence the decision made by *Les Temps modernes*[1] and by *Esprit*[1] to reply.
>
> But how to reply, since the discussion is impossible? We have to reply as we do with a sophist, that is to say, with a man who resembles the one who speaks the truth, and whose arguments must be dismantled, piece by piece, to unmask the make-believe.
>
> Pierre Vidal-Naquet, [*A Paper Eichmann*], 1987.[1]

---

[1] Pierre Vical-Naquet, *Un Eichman de papier*. In *Les Assassins de la mémoire*. Paris: La Découverte, 1987, p. 11-13. *Le Monde*, a major French newspaper; *Les Temps Modernes*, a philosophy journal, founded by Jean-Paul Sartre; *Esprit*, a philosophy journal, founded by Emmanuel Mounier.

4. A weak refutation reinforces the attacked position

According to the *law of weakness*, a weak argument for a conclusion is an argument for the opposite conclusion, **S. Scale**. Symmetrically, a weak refutation of a thesis reinforces this same thesis.

> *Gérard Chauvy appears in court for a libel against Raymond and Lucie Aubrac, two leaders of the French Resistance against the Nazis.*
> He quoted a brief from Klaus Barbie describing them as members of the resistance turned into double agents.
> Gérard Chauvy, who has said that he discovered Klaus Barbie's memoir in 1991, was the first to give these sixty pages, which had, until then, being "circulating under the cloak", a broad public dissemination, reproducing them *in extenso* in the annexes of his work. Does he share this thesis, as the civil party maintains? *Are his apparent reservations about this memoir but one more maneuver to accredit it?* In any case, this document is at the center of the debate. (*Le Monde*, 7 February 1998, p. 10; my emphasis).

5. A strong refutation reinforces the attacked position

In 2001, Elisabeth Tessier, a renowned astrologer, successfully defended at the Sorbonne University her PhD dissertation in sociology entitled "*Epistemological Situation of Astrology*". This dissertation was received with great indignation by a large number of academics. Four Nobel Prize winners and leading academics intervened to deny that it had any scientific value, dismissing it as supporting irrationality and pseudo-science.

As a result of this intervention, the debate was re-framed as follows: on the one hand, *the authorities*, renowned professors and scientists, pitted against *a woman*. Now, a quick peripheral reasoning, backed by the proportionate measure argument (**S. Proportion**), is enough to conclude that the former are deeply disturbed by this dissertation; and the trap of the strong refutation closes on its own initiators: the very prestige of the opponents reinforces the rebutted thesis, at least in the eyes of the adepts of peripheral thinking, but they are many.

Paralogism

Within the framework of classical Aristotelian logic, a paralogism is defined as an invalid syllogism. These paralogisms of deduction are "arguments of traditional syllogistic form which break one or another of a well-known set of rules" (Hamblin 1970, p. 44).

1. Rules for the syllogism

Traditional logic has established the following rules, which make it possible to eliminate invalid syllogisms. The following syllogisms respect all the rules of the syllogism; they are valid.

(i) "A syllogism contains three terms."

(ii) "From two negative premises, nothing can be concluded":
   no M is P
   no S is M
   No conclusion

(iii) "If a premise is negative, the conclusion must be negative":
   no M is P   *the major premise is negative.*
   some S are M,
   so some S are not P   *the conclusion is negative*

(iv) "In a valid syllogism, the medium term must be distributed at least once":
   no M is P   *M is distributed (universal).*
   all S are M,
   so no S is P   *the conclusion is valid.*

(v) "If a premise is particular, the conclusion is particular":
   no M is P
   some S are M   *the minor premise is particular.*
   So, some S are not P   *the conclusion is particular.*

2. Paralogisms

A paralogism is a syllogism that does not respect one or several preceding rules. Of the 256 modes of the syllogism, 19 modes are valid; therefore, a syllogism can be fallacious in 237 different ways. The question of whether it "seems" conclusive or not is irrelevant. The term *paralogism* designates nothing other than a mistaken calculation.

The following are key forms of syllogistic paralogisms. The first form corresponds to the paralogism of homonymy, the others to an inadequate distribution of qualities and quantities.

(1) Paralogism of four terms.
(2) Paralogism concluding from two negative premises.
(3) Paralogism drawing a positive conclusion from a negative premise.
(4) Paralogism of the undistributed middle term.
(5) Paralogism of universal conclusion from a particular major.
(6) Paralogism of universal conclusion from a particular minor.

Examples:
— The following paralogism consists of four terms:
   Metals are simple bodies.
   Bronze is a metal.
   * Therefore *bronze is a simple body.*

Bronze is not a simple body but an alloy. In the minor premise, bronze is said to be a metal because it looks like an authentic metal such as iron, it can be melted and molded, etc. In the major premise, *metal* is used with its strict meaning. *Metal* is homonymous, and the syllogism actually has four terms; **S. Ambiguity**.

— The following paralogism concludes from two negative premises:
    Some B are not C   *some rich are not arrogant*
    No A is B   *no poet is rich.*
    * *Therefore* No A is C *   *no poet is arrogant.*

— The following paralogism concludes universally from a particular major:
    all A are B   *all men are mortal*
    no C is A   *no dog is man*
    * *Therefore* No C is B   * *no dog is mortal.*

In the major premise, "*all men are mortal*", the major term, *mortal*, is not distributed: this premise says nothing of all mortals, but only of certain mortals, namely, that "*they are men*". Yet the conclusion "*No dog is mortal*" claims something of all mortals: "*no mortal is a dog*". So the major term is distributed in the conclusion and not in the major premise. The conclusion thus affirms more than the premise, which is impossible.

3. Evaluation using the rules of the syllogism

Syllogisms are traditionally evaluated on the basis of a system of rules (§1), in a step-by-step process:
— Check the number of terms, and propositions.
— Identify the middle term, the major term, and the minor term.
— Determine the quantity and quality of the premises and conclusion.
— Identify the distribution of terms.
— Check the organization of the distribution of terms: check that the middle term is distributed at least once. If the major term or the minor term is distributed in the conclusion, make sure that they are also distributed in the premises; etc.

This laborious method is based on the notion, at the very least unintuitive, of the quantity of the predicates. It shifts the analyst's attention from the understanding of the structure and articulation of the syllogism, from what the syllogism asserts, to the fragmented application of a system of rules. It may develop the ability to apply an algorithm, but it is far from an everyday critical thinking process.

4. Evaluation with Venn diagrams

The use of Venn diagram provides a more intuitive and clear base for syllogism assessment. Three intersecting circles represent the three sets which correspond to the three terms. The assertion made by each premise is carried to the corresponding circle. If a premise asserts that a set (made up of a circle or a portion of a circle) contains no elements, that circle or the portion of a circle is blacked out (striped). If a premise asserts that a set (*id.*) contains one or more elements, a cross is placed in the circle or portion of a circle. A portion of a circle is therefore either black, has a cross, or remains white. If it is white, it is because nothing can be said about it.

Paralogism

The data of the premises having thus been plotted on the diagram, the result can be compared with what the conclusion asserts, the diagram shows whether the syllogism is or is not valid.

Consider the syllogism:
> Some rich people are not arrogant
> No poet is rich
> * No poet is arrogant

The three intersecting circles represent the rich (R), the poets (P) and the arrogant (A), respectively.

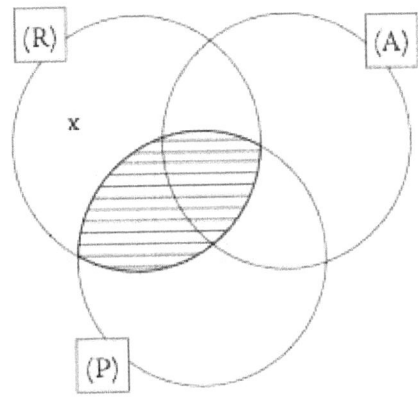

— *"Some rich are not arrogant"*: consider the circle of the rich and that of the arrogant; put a cross outside of their intersection: there are some people within this zone.

— *"No poet is rich"*: consider the circle of the poets and that of the rich, and blacken their intersection: there is nobody within this zone.

— Finally, look at the circle of poets and that of the arrogant people. The conclusion affirms that the intersection of the circle of poets with that of the arrogant is black; but we see that this is not the case; it is partly white. This syllogism is a paralogism.

Consider the syllogism:
> No **M** is **P**
> All **S** is **M**
> Therefore No **S** is **P**

The three intersecting circles represent the **M** set, the **S** set and the **P** set.

— *"No **M** is **P**"*: the intersection of the circles **M** and **P** is black (empty).

— *"Every **S** is **M**"*: the non-intersecting zone of the circles **S** and **P** is black (empty).

— Looking at the **S** circle the **P** circle, we can see that the intersection is black (empty); this is precisely what the conclusion claims, "No *S* is *P*". This syllogism is valid.

5. Paralogism of quantifier permutation

By generalization, the word *paralogism* can refer to any error made in applying the rules of formal logic. For example, the paralogism of quantification is an error committed when the existential and the universal quantifier are permuted:
> All human beings have a father; so they have the same father
> For every human **H**, there is a human **F**, such that **F** is the father of **H**
> * *Therefore* There is a human being **F** such that for every human being **H**, **F** is the father of **H**.

The following passage may contain such a paralogism, albeit complicated by a fallacious verbiage that is to say an eloquent amplification, **S. Verbiage:**
> And all the geniuses of science, including Copernicus, Kepler, Galileo, Descartes, Leibnitz, Buler, Clarke, Cauchy, speak like [Newton]. They all lived in true adoration of the harmony of the worlds and of the all-powerful hand that threw them into space and sustained them.
> And this conviction is not based on impulses, like poets. Figures, theorems of geometry give it its necessary basis. And their reasoning is so simple that children would follow it. They establish, first, that matter is essentially inert. *It follows that, if a material element is in motion, it is because another has constrained it; for every movement of matter is necessarily a communicated movement. They thus claim that since there is an immense movement in the sky, which carries away in the infinite deserts billions of suns of a weight which crushes the imagination, it is because there is an all-powerful motor.* They establish, secondly, that this movement of the heavens presupposes the solving of the problems of calculation, which have required thirty years of study, *etc.*
>   Ém. Bougaud, [*Christianity and the Present Times*], 1883.[1] (My italics).

Pathetic Argument

The expression "pathetic argument" is used to refer to phenomena linked to the pathos and the emotions in quite different ways.

Pathetic arguments are not *pathemic arguments*. *Pathemic* is a derivative from *pathos*; one can speak of a *pathemic argument* to refer to any emotion-based argument, appeal to anger, enthusiasm, pity, etc. *Pathetic* means "pitiful", it can be considered only a sub-sort of the pathemic argument. **S. Pathos.**

The label *pathetic argument* may be evaluative. A participant might refer to an argument he or she utterly rejects as a "pathetic argument" because so childish, void or desperate that, while rejected, it wins over sympathy. The label is evaluative and may apply to any kind of argument scheme; one might say "*I find this*

---

[1] Ém. Bougaud, *Le Christianisme et le temps présent*, t. I. Paris: Poussielgue Frères, 5th ed., 1883.

argument pathetic", but not "*I find this argument* a pari" (meaning, "in my view, this is an argument *a pari*"), **S. Contempt**.

1. Pathetic argument

The label "pathetic argument" refers to a variety of arguments based on negative or positive consequences. The conclusion is deemed impossible and rejected because it would frustrate the arguer; or taken for assured because agreeable to him/her.

>I fear that **P**, so **not-P**.
>I wish **P**, so **P**
>It can't rain on Sunday, our picnic would be ruined!
>This is not possible, we couldn't manage the consequences:
>— Syldavia cannot suspend its payment, that is impossible, nobody knows what might happen, actually we wouldn't know how to deal with such a situation.
>— Such pollution is unthinkable, it would make thousands of victims.
>— If this criticism were right, what would become of our discipline?

The pathetic argument applies to the field of knowledge, a style of argument quite common in the field of practical action:

>I wish that **P**, so I strive to achieve **P**, I pray for **P**, I try to bring about **P**.
>I fear **P**, so I try to avoid, prevent **P**…

But wishing **P** is different from *striving to achieve* **P**. The pathetic argument can be evaluated as pathetic "naive and desperate".

2. "Pathetic fallacy"

The label "pathetic fallacy" refers to the anthropomorphic attribution of human feelings to non-human, non-living beings; it condemns the use of the rhetorical figure of *personification*. The expression was coined by John Ruskin:

>I want to examine the nature of the other error, that which the mind admits when affected strongly by emotion. Thus, for instance, in *Alton Locke*,
>>'They rowed her in across the rolling foam
>>The cruel, crawling foam.'
>
>The foam is not cruel, neither does it crawl. The state of mind which attributes to it these characters of a living creature is one in which the reason is unhinged by grief. All violent feelings have the same effect. They produce in us a falseness in all our impressions of external things, which I would generally characterize as the 'pathetic fallacy'.
>
>John Ruskin, *Of the pathetic fallacy*, [1856][1]

---

[1] In *Modern Painters*, vol. III, part IV, London: Smith Elder, p. 160. *Alton Locke* is a novel by Charles Kingsley (1850).

Pathos

> The word *pathos* is patterned after a Greek word meaning "what we experience, *as opposed to* what we do" (Bailly, [*Pathos*]). In Latin, *pathos* is sometimes translated as *dolor*, which basically means "pain"; Cicero uses *dolor* to refer to passionate eloquence (Gaffiot [1934], *Dolor*).

In classical rhetoric, pathos is a kind of *evidence*, complementary to that drawn from logos@ and ethos@. "Evidence" here means "persuasion", in the sense of pressure and control exerted on the audience, S. **Evidence**. The word *pathos* covers a set of socio-linguistic emotions upon which the speaker might draw in order to orient his audience towards the conclusions and actions he or she advocates.

1. Ancient rhetoric: Emotions as a manipulative tool

1.1 Ethos and pathos, two levels of affects?

The Trinitarian presentation "ethos, logos, pathos" isolates each of these components, in particular ethos from pathos; but Quintilian understands *pathos* and *ethos* as two varieties of feelings (*adfectus*):

> Pathos and ēthos are sometimes of the same nature, the one to a greater and the other to a lesser degree, as love, for instance, will be pathos, and friendship ēthos, and sometimes of a different nature, as pathos in a peroration will excite the judges, and ēthos soothe them. (*IO*, VI, 2, 12)

> Of feelings, as we are taught by the old writers, there are two kinds, the first of which the Greeks included under the term πάθος (pathos), which we translate rightly and literally by the word "passion" [*adfectus*]. The other, to which they give the appellation ἦθος (ēthos), for which, as I consider, the Roman language has no equivalent term, is rendered, however, by mores, "manners"; whence that part of philosophy, which the Greeks call ἠθική (ēthikē), is called *moralis*, "moral". 9. [...] The more cautious writers, therefore, have chosen rather to express the sense than to interpret the words and have designated the one class of feelings as the more violent, the other as the more gentle and calm, under pathos they have included the stronger passions, under ēthos the gentler, saying that the former are adapted to command, the latter to persuade, the former to disturb, the latter to conciliate. 10. Some of the very learned add that the effect of pathos is but transitory. (*Id.*, 8-10)

The following table summarizes the main complementary oppositions between *ethos* and *pathos*.

ETHOS	PATHOS
has its source in *the orator's character*	has its source *in the occasion*
makes the speaker *likeable*	causes a *disturbance* in the audience
inclines the audience to *benevolence*	entails, *snatches the decision*
is *pleasing*	is *moving*
low arousal: *calm, measured, sweet*	high arousal: *vehement*
typical ethotic emotions: *affection, sympathy*	typical pathemic emotion: *love, anger, hate, fear, envy, pity*
ongoing *thymic* tonality of the discourse	*phasic* emotion episodes
convincing	*commanding*
the *introduction* focuses on ethos	the *conclusion* (end of the discourse) focuses on pathos
speech genre: *comedy*	speech genre: *tragedy*
type of causes: *ethical* (moral)	type of causes: *pathetic*
moral satisfaction	*aesthetic* satisfaction

As two complementary kinds of feelings, the ethos organizes the ongoing *thymic* basic tonality of the discourse, upon which the speaker will base the *phasic* variations of intensity which characterize episodes of emotion. The doses of ethos and pathos must be carefully balanced according to the objectives of the discourse.

1.2 The pathos: a bundle of emotions

Aristotle distinguishes in the *Rhetoric* a dozen of basic rhetorical social emotions assembled in complementary pairs (*Rhet.*, II, 1-11; RR. p. 257-310):

>anger *vs.* calm
>friendship *vs.* enmity, hatred
>fear *vs.* confidence
>shame *vs.* impudence
>kindness, helpfulness *vs.* unkindness (eliminating the feeling of kindness)
>pity *vs.* indignation
>envy *vs.* emulation.

This enumeration does not cover all the political and judicial emotions:

>Aristotle neglects, as not relevant for this purpose, a number of emotions that a more general independently conceived treatment of the emotions would presumably give prominence to. Thus *grief, pride* (of family, ownership, accomplishments), *(erotic) love, joy,* and *yearning for an absent or lost loved one* (Greek *pothos*) hardly come in for mention in the *Rhetoric* […] The same is true even for *regret*, which one would think would be of special importance for an ancient orator to know about, especially in judicial contexts. (Cooper 1996, p. 251)

1.3 Manipulating through emotions

The question of the impact of emotion on judgment is that of the equilibrium between logo-ic demonstration on the one side, and ethotic and pathemic pressures on the other side. Logical arguments transform the *representations*, and representations determine the will; but, in some situations, pathos can nonetheless outweigh the will. This makes pathos something mysterious and powerful, a little bit superhuman, a little bit demonic. Classical texts abound in such declarations opposing the pathos to the logos, that is *emotions* to *reason and judgment*, in terms of their *decision*-making capacity:

> Now nothing in oratory is more important than to win for the orator the favor of his public, and to have the latter so affected as to be swayed by something resembling a mental impulse or emotion, rather than by judgment or deliberation. For men decide far more problems by hate, or love, or lust, or rage, or sorrow, or joy, or hope, or fear, or illusion, or some other inward emotion, than by reality, or authority or any legal standard, or judicial precedent, or statute. (Cicero, *De Or.*, 178 XLII).

In a resounding passage, Quintilian opposes the pedestrian character of argument to the violent and vicious action of emotion:

> As to arguments, they generally arise out of the cause, and are more numerous on the side that has the greater justice, so that he who gains his cause by force of arguments will only have the satisfaction of knowing that his advocate did not fail him. 5. But when violence is to be offered to the minds of the judges and their thoughts are to be drawn away from the contemplation of truth, then is this peculiar duty of the orator required. This the contending parties cannot teach; this cannot be put into written instructions. (*IO* VI, 2, 4-5)

Such praises of passionate speech as capable of swaying the judge away from reality and truth is the source of the still prevailing manipulative vision of rhetoric.

2. Rhetoric and magic

One may be taken aback by such an open acknowledgements of the cynical, immoral and manipulative character of rhetorical pathemic persuasion. But one can remain skeptical as to the very possibility of such manipulation. Firstly such claims must be taken with a pinch of salt. They can be read as a form of professional advertisement intended to magnify the powers of the professional rhetorician, and push up course fees: "*follow my teaching, and you'll become a magician of spoken word!*".

More important, perhaps, as Romilly points out when referring to Gorgias, is the fact that these claims seem to transfer the virtues attributed to *magic speech* to *emotional rhetorical speech*: "what can we say about that, except that, by ways that seem irrational, the words bind and affect the listener in spite of himself?" (Romilly 1988, p. 102). This is precisely Socrates' viewpoint when he holds that the art of speech-makers:

is part of the enchanters' art and but slightly inferior to it. For the enchanter's art consists in charming vipers and scorpions and other wild things, and in curing diseases, while the other art consists in charming and persuading the members of juries and assemblies and other sorts of crowds.
Plato, *Euthydemus*, XVII, 289e, p. 130).

Magic formulas, as chanted by Tibullus, had actually the power to alter the very physical perception of reality:
> For me she [= *the witch*] made chants you can use to deceive.
> Sing them thrice, and spit thrice when you have sung.
> Then he [= *your husband*] *cannot believe anyone about us, not even*
> *if he himself has seen us on the soft bed.*
Tibullus, *Elegy* I, 2, v. 55sq (my emphasis)[1]

Pericles' persuasive speech had the same powers:
> Plutarch quotes the words of an opponent of Pericles, who was asked who, out of him and Pericles, was the strongest in the fight. His answer was: "*when I bring him down in the fight, he argues that he did not fall, and he wins by persuading all the assistants*" (Pericles, 8). (*Id.*, p. 119)

Note that the defeated Pericles addresses his persuasive speech to the public, not to his victorious opponent, who holds him firmly on the ground. The argumentative situation is in fact a three-pole situation, involving the speaker, the adversary and the judge(s), **S. Roles.**

Whatever may be, these views express an age-old classical and popular theory of the functioning of the human mind, for which *emotion*, *will* and *action* oppose, distort and victoriously compete with *reason*, *understanding* and *contemplation*.

In contrast with all these declarations, Aristotle simply warns against the overly effective use of the pathos:
> It is not right to pervert the judge by moving him to anger, envy or pity — one might as well warp a carpenter's rule before using it. (*Rhet.*, I, 1, 1354a25; RR, p. 96-97)

The judge is "the rule." The rejection of pathos is not based on moral considerations but on a cognitive imperative; to distort the rule is harmful not only to others and to the world, but first to oneself; error is more fundamental than deceit.

3. Emotion, from proof to fallacy

The standard theory of fallacies considers affects as the major pollutant of rational discursive behavior; to be valid, the argumentative discourse should be an-emotional. Pathos, the essential component of rhetorical argumentation, is therefore the typical target of this criticism. The "passions" are collected into a family of *ad passiones* fallacies, and these are to be eliminated. This is an essential

---

[1] *The Complete Poems of Tibullus: An En Face Edition*. Trans. by R. G. Dennis and M. C. J. Putnam. With an Introd. by J. H. Gaisser. Berkeley, etc: University of California Press, 2013.

point of articulation and opposition between *rhetorical* and *logical-epistemic* argumentation. With the capacity to subvert the mind and bypass rational reflection, emotions are considered to be the most powerful of rhetorical tools and, for the same reason, they are prohibited by within critical argumentation.

***Ad passiones* arguments** — The standard theory of fallacies considers that reason risks being overshadowed wherever emotion is allowed to blossom in discourse:

> I add finally, when an Argument is borrowed from any Topic which are suited to engage the Inclinations and Passions of the Hearers on the side of the Speaker, rather than to convince the Judgment, this is Argumentum *ad passiones*, an Address to the Passions: or, if it be made publicly, 'tis called an Appeal to the People. (Watts, *Logick*, 1725, quoted in Hamblin 1970, p. 164; capitalized in the text).

In an argumentative situation, emotions, like fallacies, tend to be the emotion of *the other*, the opponent: "*I try to remain cool and reasonable, why are you getting so excited?*". This is a current strategy in controversies on scientific as well as on political topics (Doury 2000). It can be considered to be a typical case of *ad fallaciam* argument, **S. Evaluation.**

These sophisms of passion are not included in the original Aristotelian list, **S. Fallacy (II)**. The label "*ad* + Latin Name" was widely used in modern times to refer to "fallacies of emotion", and traces of this use are still to be found. The *ad passiones* herbarium is well supplied: as shown by Hamblin's list of fallacious arguments in *ad*: the labels making a clear and direct reference to the affects have been underlined.

> The argumentum *ad hominem*, the argumentum *ad verecundiam*, the argumentum *ad misericordiam*, and the *argumenta ad ignorantiam*, *populum*, *baculum*, *passiones*, superstitionem, imaginationem, *invidiam* (envy), crumenam (purse), *quietem* (repose, conservatism), *metum* (fear), *fidem* (faith), socordiam (weak-mindedness), *superbiam* (pride), *odium* (hatred), *amicitiam* (friendship), *ludicrum* (dramatics), *captandum vulgus* (playing for the gallery), *fulmen* [thunderbolt], *vertiginem* (dizziness)) and *a carcere* (from prison). We feel like adding: ad nauseam but even this has been suggested before. (Hamblin, 1970, p. 41)

This list contains not only emotional arguments: for example, the appeal to ignorance (*ad ignorantiam*) is an *epistemic*, not an *emotional* argument. Others designate various forms of appeal to subjectivity, but the majority of the labels mentioned refers to personal interests and have a clear emotional content. Nonetheless, the concept of emotional language and the analytical method backing the diagnostic of these *ad passiones* fallacious appeals remain unclear.

The literature on fallacies mentions a dozen fallacies involving emotions, mainly fallacies in *ad*; as permitted by the generic *ad passiones* label, this list can be expanded to include all emotions.

fear, designated either directly (*ad metum*) or metonymically by the instrument of threat, *ad baculum, a carcere, ad fulmen, ad crumenam*
respectful fear, *ad reverentiam*
affection, love, friendship, *ad amicitiam*
joy, gaiety, laughter: *ad captandum vulgus; ad ludicrum; ad ridiculum*
pride, vanity, *ad superbiam*
calm, laziness, tranquility, *ad quietem*
envy, *ad invidiam*
popular feeling, *ad populum*
indignation, anger, hatred: *ad odium; ad personam*
modesty: *ad verecundiam*
pity: *ad misericordiam*.

It should be pointed out that basic emotions mingle with *vices* (*pride, envy, hatred, laziness*) and *virtues* (*pity, modesty, friendship*), which are *evaluated emotional states*.

One can see that the list of emotions composing the pathos in the preceding paragraph and the list of emotions stigmatized as fallacies, largely overlap. The pathemic proofs of rhetoric have become the sophisms *ad passiones* in the modern standard fallacy theory.

Four emotional fallacies: *ad hominem, ad baculum, ad populum, ad ignoratiam*

All emotions can intervene in ordinary argumentative speech, but not all of these emotions have received the same attention, the focus being placed on the emotional and subjective character of the following four fallacies.

— Arguments attacking the opponent, and other manifestations of contempt, **S. Personal Attack; Dismissal**. *Ad hominem* involves epistemic subjectivity, not emotions.

— The call to popular feelings in populist argumentation corresponds to a complex range of positive or negative emotional movements: the public is amused, enthusiastic, pleased, ashamed; the speech plays on its pride, vanity, incites hatred, etc., **S. *Ad Populum*; Laughter, Irony**.

— *Ad baculum* argument relies on various forms of threat or intimidation. *Fear*, possibly respectful, is opposed to the positive emotion of *hope*, created by the promise of a reward, **S. Threat**

— The appeal to pity, *ad misericordiam*, may serve as a fundamental example of the role of emotion in argumentation. Firstly, the speaker **S** has to back his or her appeal to pity, justifications are necessary to produce in the listener **L** a movement of pity. Then, **L** will take his or her well-constructed emotion as a good reason to help **L**, **S. Emotion**.

In conclusion, rhetoric and argumentation can be opposed on the basis of their relation to affects. If there is a concept of argument defined *within* rhetoric (*inventio*), there is also a concept of argument defined *against* rhetoric. Rhetoric is oriented towards discourse *production*, whilst argumentation is oriented towards the critical *reception* of discourse. Confronting proactive, aggressive, rhetorical precepts, critical argumentation is *defensive*.

4. Emotion rationality and action

The field of argumentation is built on the rejection of the evidence that rhetoric considers the strongest: the ethotic and pathemic proofs. This an-emotional vision of argumentation corresponds to a classical and popular vision of the functioning of the human mind, which opposes *reason, understanding,* and *contemplation* respectively to *emotion, will* and *action*. The following passage is a synthesis of this representation:

> Hitherto we have dealt with the proofs of truth, which constrains human understanding when it knows them, and for this they are effective in persuading men accustomed to follow reason. But they are incapable of compelling the will to follow them, since, like Medea, according to Ovid, "I see and approve the best; I follow the worst." This arises from the misuse of the passions of the soul, and therefore we must deal with them in so far as they produce persuasion, and this in the popular manner, and not with all this subtlety possible if one treated philosophically. (Mayans and Siscar 1786, p. 144)

Two functions are assigned to the "passions": they alter the perception of reality, put knowledge between parenthesis and, in so doing, give a decisive impulse to action. This vision or emotion as a stimulus to action seems to be grounded in an etymological argument. The word *emotion* derives from Lat. *emovere, e-* (*ex-*) "out of" and *movere*, "to move"; an emotion is something that "sets people in motion". In any case, passions are the almighty manipulative instrument of action-oriented discourse favored by rhetoric, and the main enemy of truth oriented discourse favored by logicians.

In the middle of the twentieth century, the psychologists Fraisse & Piaget considered that emotion is not an organized reaction but a disorder of conduct, resulting in a "decrease in the level of performance" (1968, p. 98):

> People get angry when they substitute violent words and gestures for efforts to find a solution to the difficulties they experience (solving a conflict, overcoming an obstacle). [...] [Anger] is also a response to the situation (hitting an object or a person who resists you), but the level of that response is lower than it should be, given the standards of a given culture. (*Ibid.*)

According to this vision, emotion would trigger low-quality behavior, and therefore poor quality reasoning. In interaction, it would be necessarily manipulative: the candidate cries in an effort to distract the examiner from his or her shortcomings, magically reframing the examination situation into a more interpersonal, private, relation.

This leads to a kind of paradox: for rhetoricians, emotions lead to action while psychologists on the other hand, consider that emotions deteriorate action. Perelman & Olbrechts-Tyteca share this vision of emotions as "obstacles" to reason, and thus consider emotions to be incompatible with sound argument.

Yet they retain the motivational quality of emotion in order to explain the relevance of argumentative discourse for action. The solution is found in a *dissociation@* opposing emotions to values:

> We should point out that the passions as obstacles must not be confused with the passions that provide a support for a positive argument. The latter will generally be designated by a less pejorative term, such as value, for instance. (Perelman & Olbrechts-Tyteca [1958], p. 475 ; my emphasis)

By this skillful operation, emotions are disposed of, and these remain pejoratively marked as obstacles to reason, while their dynamic potential is transferred to values. So the effect of argument can be extended beyond the mere production of mental persuasion to become the determiner of action, **S. Persuasion** (*id.*, p. 45).

5. Emotion, alexithymia and everyday rationality

If emotions are seen as the ideal manipulative tool, the equation "emotion = fallacy" seems more than justified, so, extending the example of scientific language to ordinary linguistic practices, a solution can be found in the pure and simple elimination of emotions. But the price to pay for the elimination of emotions from ordinary discourse is high: in everyday circumstances, the use of an-emotional discourse is actually considered to be the symptom of a mental disturbance, *alexithymia*. The word *alexithymia* is composed of three lexemes *a-lexis-thymos*, "lack - of words - for emotion"; alexithymic language is defined as a language from which all expression of feelings and emotions is banished:

> *Alexithymia*: term proposed by Sifneos to describe patients predisposed to psychosomatic disorders and characterized by: 1) the inability to verbally express the affects; 2) the poverty of imaginary life; (3) the tendency to resort to action; and (4) the tendency to focus on the material and objective aspects of events, situations and relationships. (Cosnier 1994, p. 160)

Such a discourse which is deprived of emotion is reduced to the expression of operational thinking, mirroring, "a mental mode of functioning organized on the purely factual aspects of everyday life. The discourse that makes it possible to spot it is characterized by objectivity and ignores any fantasy, emotional expression or subjective evaluation" (*id.*, p. 141).

Similarly, the repression of affect by the *neurotic* personality can lead to the same result. From a neurobiological perspective, Damasio has shown that a theory of pure logical reasoning, leaving aside the emotions, cannot account for the way people actually deal with everyday issues:

> The 'high-reason' view, which is none other than the common sense view, assumes that when we are at our decision-making best, we are the pride and joy of Plato, Descartes and Kant. Formal logic will, by itself, get us to the best available solution for any problem. An important aspect of the rationalist conception is that to obtain the best results, emotions must be kept out. Rational processing must be unencumbered by passion. (1994, p. 171)

Pure reasoning on everyday affairs can actually be observed in patients having suffered prefrontal damages:

> What the experience with patients such as Elliot suggests is that the cool strategy advocated by Kant, among others, has far more to do with the way patients with prefrontal damage go about deciding than with how normals usually operate. (*Id.*, p. 172)

The exclusion of subjectivity and emotions risks transforming argumentation into *an operative alexithymic practice*. As far as argumentation studies are interested in the treatment of everyday problems in common language, they cannot adopt as ideal discourse the discourse of neurotic, alexithymic or brain damaged individuals. The question of how emotions develop in argumentative discourse demands much more than simple a priori censorship. The adequate level of emotionality will be one of the by-product of a felicitous argumentative exchange, **S. Emotion**.

Personal Attack

> Lat. *ad personam*, Lat. *persona* referring to the actor's mask, corresponding to the interactional face or social role of the person, not precisely to his or her personal identity.

The Latin label *ad personam* is also used to refer to *personal attacks*. Personal attacks can target all aspects of the person, public or private, including his or her human dignity. Such attacks flout the rules of politeness@ and all ethical prohibitions that protect the individual, as a unique human being.

The personal attack against the adversary is, in principle, quite distinct from the *ad hominem*@ attack. The latter refers to a marked contradiction found between the positions taken by the opponent and his or hers beliefs or behavior, whereas personal attack bypasses the opponent's positions, smearing the opponent in order to devalue the argument itself. Nonetheless the label *ad hominem* is frequently used to refer to personal attacks.

Open and covered attacks — Insult is the simplest form of attack *ad personam*: "*Sir, you are only a badly educated dishonest person!*". Open personal attack can be a very efficient strategy to undermine the debate and avoid the substantial issue. The opponent will be upset, he or she will lose track of the argument and will finally resort in turn to personal attacks and insults. Third parties will then be tempted to leave the arguers to their fight, or to simply enjoy the show.

The personal attack may invoke the opponent's private life: "*you'd better take care of your children!*" said to an opponent whose children are badly behaved, is a personal attack which many would consider extremely offensive. In a debate, such a personal attack might be brought in more subtly by introducing the issue of family policy, emphasizing the need for parents to give priority to their chil-

dren's education, without openly mentioning the opponent's circumstances. The rumor and the media will explain the innuendos.
> *He cannot rule his wife and he pretends to rule Syldavia!*

Degree of relevance of the attack — Personal attacks may be more or less relevant to the issue at stake. Consider the negative descriptions of the adversary made in the context of the argumentative question, *"Should we wage war against Syldavia?"*:

    S1   — *We must take military action against Syldavia!*
    S2_1   — *Shut your mouth, stupid warmonger!*
    S2_2   — *Please, stop this bullshit!*
    S2_3   — *Poor fool, manipulated by the media!*
    S2_4   — *Poor you, last week you couldn't even locate Syldavia on the map!*

Considering the available context, **S2_1** and **S2_2** are unprovoked and irrelevant attacks against the person. That is to say that they have very little relevance to the argumentative question. But in case **S2_3**, nothing is clear; **S2** is certainly wrong in calling the opponent names, but he/she does provides an argument invalidating **S1** for his or her lack of basic geopolitical knowledge. If we apply the principle *"no argumentation without information"*, the attack is certainly not irrelevant. A distinction must be made between calling a sensible upright citizen a fool, and calling a fool a fool. But if this were the case, all slurs and attacks would be reinterpreted as well-suited literal or metaphorical descriptions of the person; hence the general prohibition of insults.

Persuading, Convincing

The opposition, or progression, from *to persuade* to *to convince*, along with the development of audiences from *particular* to *universal*, is a major focus of the *Treatise on Argumentation* (Perelman & Olbrechts-Tyteca [1958]), **S. Persuasion**

1. To *persuade* a *particular* audience, to *convince* the *universal* audience
Particular and Universal Audiences
Perelman & Olbrechts-Tyteca significantly restructure the concept of audience. First, the notion is broadened to encompass written communication, "every speech is addressed to an audience, and it is frequently forgotten that this applies to everything written as well" ([1958] p. 6-7). The focus put on this enlarged concept of audience explains the fact that the *Treatise* does not engage in the analysis of delivery (*pronunciatio*), the oral, face-to-faces, dimension of classical rhetoric, **S. Rhetoric**.
The *Treatise* goes beyond actual audiences to consider the *particular audiences* and the *universal audience*. The former is the sole object of classical rhetoric; the latter is a philosophical projection of the essential characters of the former. The concept of audience is then extended to cover self-deliberation (exploiting the resource of polyphony, **S. Interaction**):

Thus the nature of the audience to which arguments can be successfully presented will determine to a great extent the direction the arguments will take and the character, the significance that will be attributed to them. What formulation can we make of audiences, which have come to play a normative role, enabling us to judge on the convincing character of an argument? Three kinds of audiences are apparently regarded as enjoying special prerogatives as regards this function, both in current practice and in the view of philosophers. The first such audience consists of the whole of mankind, or at least, of all normal adult person; we shall refer to it as *the universal audience*. The second consists of the single *interlocutor* whom a speaker addresses in a dialogue. The third is *the subject himself*, when he deliberates or gives himself reasons for his actions. (*Id.*, p. 30)

2. A normative opposition

While the translators of classical rhetorical texts use the verbs *to persuade* or *to convince* interchangeably, Perelman & Olbrechts-Tyteca differentiate between these two verbs on the basis of the quality of the audiences:

> We are going to apply the term *persuasive* to argumentation that only claims validity for a particular audience, and the term *convincing* to argumentation that presumes to gain the adherence of every rational being. ([1958], p. 28)

This is a stipulative definition, based on a normative perspective. For the New Rhetoric, the norm of argumentation is constituted by the hierarchy of audiences who accept it. This position strongly distinguishes the new rhetoric from the standard theories of fallacies, for which the norm is given by logical laws, or by a system of rules defining rationality. S. **Norms; Rules; Evaluation.**

3. *To persuade, to convince*: the words

3.1 History

The Greek word used to refer to rhetorical evidence is *pistis*. Unlike the scientific and logical word *proof*, *pistis* belongs to a family of terms expressing the idea of "trust in others; which can be relied upon" and "proof" (Bailly, [*Pistis*]). The family of Greek terms translated as "persuasion" refer to "obeying", as well as to "persuading, seducing, deceiving" (*id.*, [*Peitho*]). The name of the goddess *Peitho*, the companion of Aphrodite, sometimes Aphrodite herself, goddess of beauty, seduction and persuasion, also belongs to this family. From this perspective, the word *pistis* is syncretic; it covers what is for us the field of influence, proof, seduction, submission and persuasion. By definition, "*rhetorical evidence is persuasive*".

The Latin verb *suadere* means "to advise"; the corresponding adjective, *suadus*, means "inviting, insinuating, persuasive" (Gaffiot [1934], *Suadeo; Suadus*). *Persuadere* is composed of *suadere* and the aspectual prefix *per-*, which indicates the completion of the process, meaning "I. Decide to do something [...] II. Persuade, convince" (*id.*, *Persuadeo*).

Convincere is composed of *con-* (*cum-*) "totally" + *vincere* "conquer": "utterly conquer" (*id.*, *Convinco*); its primary meaning is "to confound an adversary" (*ibid.*). Just like *per-* in *persuadere*, the prefix *cum-* refers to a completed action. The same meaning is expressed in *to convict*, coming from the Latin *convictus*, past participle of *convincere* meaning "to refute, convict" (MW, *Convict*, Etymology):

    1: to find or prove to be guilty. *The jury convicted them of fraud.*
    2: to convince of error or sinfulness

Both *persuadere* and *convincere* mark the completion of the action. Tradition requires that *to convince* is reserved for situations in which *beliefs* are changed without action, whilst *to persuade* is used for situations in which *action* is undertaken; the rule is based on the etymology of the words. In practice, both terms are used as synonyms. The traditional rule may be based on the principle of superfluity@, whereby there cannot be two words with the same meaning, as there cannot be two laws to the same effect. Yet two words can have the same meaning until everyday usage differentiates them.

3.2 Lexical opposition *persuasion* vs. *conviction*

The verbs *to persuade* and *to convince* belong to a lexical-semantic field including:

advising	brainwashing	bringing around	catechizing
converting	counseling	inciting	inducing
insinuating	inspiring	instilling	inviting
preaching	prevailing on	prompting	propagandizing
seducing	suggesting	talking someone into (out of) doing sth
winning somebody over to a point of view.			

This lexical basis is a rich source of semantic orientations and oppositions whose exploitation could contribute to the reflection on the diversity of the expected effects of discourse.

To persuade and *to convince* are equivalent in many contexts.

    A tries to *persuade* / *convince* **B** of something
    A addresses a *persuasive* / *convincing* argument to **B** => then **B** has new *persuasions* / *convictions*

Nonetheless, in other contexts, they are non-equivalent:

    A letter *of persuasion* — not **conviction*
    A considers that B is *persuadable (-ible)* — not **convince-able*
    The pair *persuader* / *persuadee* is not marched by a pair **convincer* / **convincee*
    Convictive and *convict* are, at least etymologically, linked to *convince*. To persuade has not produced corresponding words.

The present participle *convincing* can be used as an autonomous adjective, meaning "cogent"; a *conviction* is "a strong belief". "*Very convincing*" seems more common than "*very persuasive*"; nonetheless, both can be used to qualify not only an argumentative discourse but also many other kinds of discourse:

    very convincing accounts, reports…
    very convincing novels, tales, narratives…

>   very convincing portraits.

as well as non-verbal activities:
>   a very convincing experience
>   a very convincing scar (stage make-up).

Persuasion

1. Persuasion as the essence of rhetoric
Since Isocrates and Aristotle, argumentative rhetorical speech is commonly defined by its function, that being *persuasion*:
>   Rhetoric may be defined as the faculty of observing in any given case the available means of persuasion. (*Rhet*, I, 2, 1355b26, RR, p. 105).

According to Crassus as staged by Cicero, persuasion is the "first duty" of the orator (Cicero, *De Or.*, I, XXXI; p. 40). Perelman & Olbrechts-Tyteca, focus their definition of argumentation on how *"to induce or to increase the mind's adherence to the theses presented for its assent"* ([1958]/1969, p. 4; italics in the original) before elaborating on the notion of "adherence of minds" by means of an opposition between *persuading* and *convincing* speech, **S. Assent; Persuade and convince.**

According to these standard definitions, argumentative rhetoric is fundamentally concerned with the discourse structured by the *illocutionary* (overtly expressed in the discourse) intention of persuading, that is to communicate, explain, legitimate, and make the listeners share the speaker's point of view and the words that express it. Persuasion, as a *perlocutory* state obtained through discourse, results from the realization of these intentions.

The rhetorical tradition binds the discourse of persuasion to the production of a *plausible* representation in the audience's minds. This rhetorical representation of reality is considered to be antagonistic towards *truth* by essentialist philosophers such as Plato, **S. Probable.**

2. A rhetoric without persuasion: the *ars bene dicendi*
Chapter 15 of Book II of Quintilian's *Institutes of Oratory* is devoted to challenging the definition of rhetoric in relation to persuasion, "the most common definition therefore is that [rhetoric] is the power of persuading" (*IO*, II, 15, 3), this definition being attributed to Isocrates. Quintilian rejects all the definitions linking rhetoric to persuasion:
— As the power to persuade:
>   But money, likewise, has the power of persuasion, as do interest, and the authority and dignity of a speaker, and even his very look, unaccompanied by language, when either the remembrance of the services of any individual, or a pitiable appearance, or beauty of person, draws forth an opinion. (*Id.*, 6)

— Or as an instrument of persuasion, even with the restriction "power of persuading by speaking":
> Not only the orator, but also others, such as harlots, flatterers, and seducers, persuade or lead to that which they wish, by speaking. (*Ibid.*)

Finally, Quintilian takes up the definition of rhetoric attributed to the Stoics and Chrysippus, "rhetoricen esse bene dicendi scientiam" (*id.*, p. 84[1]), that is to say, "the art to speak well and say the Good":
> The definition that [rhetoric] is the science of speaking well […] embraces all the virtues of [rhetoric] at once and includes also the character of the true orator, as he cannot speak well unless he be a good man. (*Id.*, 34)

Its purpose is, "to think and speak rightly" (*id.*, 36).

The rhetoric of persuasive communication and the rhetoric focusing on the quality of expression have been opposed as *primary* vs. *secondary* rhetoric (Kennedy 1999), or *extrinsic* vs. *intrinsic* rhetoric (Kienpointner 2003). We can also speak of an *introverted* rhetoric, focusing on the quality of an expression based on intellectual rigor and depth of feeling. *Extroverted*, communicative rhetoric strives to be *eloquent*, while introverted rhetoric requires an alternative concept of *style*.

Note that this distinction does not correspond to the distinction forwarded in the 1960s, between a *restricted* rhetoric opposed to a *general* rhetoric, **S. Rhetoric**. Likewise, it has nothing to do with the distinction between rhetoric of *arguments* and rhetoric of *figures*, **S. Figure**.

Introverted rhetoric is a rhetoric whose communicative and interactional dimensions, hence persuasiveness, are weakened, but which nevertheless remains an argumentative rhetoric. La Bruyère expresses as follows the concept of such a rhetoric having abdicated eloquence and persuasion:
> We must only endeavor to think and speak justly ourselves, without aiming to bring others over to our Taste and Sentiments; that would be too great an enterprise. (La Bruyère, [*Of Works of Genius*], [1688])[2]

3. From persuasion to action

In an essential but often neglected complement to the basic definition of argumentation, the *Treatise on Argumentation* extends the scope of persuasion through argumentation to *action*. Argumentation would actually produce the "disposition to action":
> The goal of argumentation, as we have said before, is to create or increase the adherence of minds to the theses presented for their assent. An efficacious argument is one which succeeds in increasing this intensity of adherence among

---

[1] Quoted after Quintilien, *I. O.* = *Institution oratoire*, Trans. by J. Cousin. Paris: Les Belles Lettres,
[2] Jean de La Bruyère, *Des ouvrages de l'esprit*. In *Les Caractères ou les mœurs de ce siècle* [1688]. Quoted after *The Characters, or Manners of the Age*. London: D. Browne, etc. p. 7.
https://books.google.fr/books?id=6y9QiTEK1JAC&printsec=frontcover&dq=La+Bruyere+Ch aracters&hl=en&sa=X&redir_esc=y#v=onepage&q=think&f=false (03-19-2017)

those who hear it in such a way as to set in motion the intended action (a positive action or an abstention from action) or at least in creating in the hearers a willingness to act which will appear at the right moment. (Perelman, Olbrechts-Tyteca, [1958], p. 45)

This vision is restated a little later:

Argumentation alone [...] allows us to understand our decisions. (*Id*, p. 37)

The end point of the argumentative process, then, is not persuasion seen as a mere mental state, an "adherence of the mind". The ultimate criterion of complete persuasion is an action accomplished in the sense suggested by discourse, and emotion plays an essential role where this is acted out. Adherence beyond a certain degree would trigger action. This is a crucial point where argument, emotions@, and values@ are combined in order to give a response to the philosophical problem of action.

4. Persuasion, identification, self-persuasion?

Burke stressed that persuasion presupposes identification:

When you are with Athenians, it is easy to praise Athenians, but not when you are with the Lacedaemonians.

Here is perhaps the simplest case of persuasion. You persuade a man only insofar as you can talk his language by speech, gesture, tonality, order, image, attitude, idea, identifying your ways with his. (1950, p. 55).

According to the rhetorical doxa, the preliminary to a successful persuasive performance is based on agreements between the speaker and the audience S. **Conditions of discussion.** This negotiation of agreements could take place through a preliminary argumentative dialogue, running the risk of an infinite regression. So the orator resolves not to *explicitly agree* with his audience, but to *adapt* to it. For this reason, he or she makes a preliminary inquiry about the audience, in order to be able to correctly adapt to, or mimic, the audience. This is precisely what the theory of the ethos of audiences foresees, S. **Ethos (IV):** through ethotic suggestion, the speaker presents himself or herself as *"one of you, the people"*. Secondly, by logical proofs, the speaker gives prominence to the values and judgments accepted by his or her audience (he or she argues *ex concessis*). Thirdly, appealing to pathemic communion with the audience, empathy is shown.

Accordingly, in order that the audience identifies with the speaker, he or she must first identify with this audience. At the end of this process of adaptation, one might ask who exactly is being persuaded by whom? The extroverted rhetoric of persuasion is threatened by the solipsism of identification. It expresses only group introversion. The notion of "communion" proposed by the *Treatise*, may characterize the culmination of this process.

This rhetorical concept of identification is totally foreign to the concept of identification defined in the framework of polyphony theory, **S. Interaction, Dialogue, Polyphony.**

5. Who studies persuasion?

The characteristic difference of rhetorical argumentation cannot be defined by persuasion, for the simple reason that persuasion is an object claimed by many other disciplines, including the sciences and philosophy of cognition; neuropsychology as well as "neuro-linguistic programming".

One year before the *Treatise on Argumentation*, Vance Packard published *The Hidden Persuaders* (1957), in which he developed a criticism of rational persuasion as socially ineffective. This criticism was first elaborated in the twenties by Walter Lippman (1922) and later by Edward L. Bernays (1928). In the wake of these books, but with quite different methods, *neuromarketing* came to focus on the issue of persuasion. To take a less controversial discipline, the analysis of persuasion also belongs to social psychology. This discipline counts among its fundamental objects the theoretical and experimental study of social influences: persuasion, convictions, suggestion, grip/influence, incitement... the formation and manifestations of attitudes, representations, and correlative transformations in the ways individuals or groups behave. The whole movement of the world, the material events, including scientific discoveries and technical innovations, along with the correlative flows of language, produce and rectify the representations, thoughts, words and actions of individuals and groups. The great classical studies of social psychology of persuasion published in the last century hardly mention rhetoric or argumentation. For example, neither the word *rhetoric* nor the words *argument* or *argumentation* appear in a collection of texts on the psychology of persuasion, entitled *Persuasion* (Yzerbit and Corneille 1994). The problem of persuasion can be legitimately invoked in relation to discourse, but the study of the process of persuasion, even in term of its linguistic aspects, may in no circumstances be carried out in the sole framework of rhetorical studies (Chabrol and Radu, 2008).

6. Persuasion as a general function of language

Just as rhetorical argumentation cannot be characterized by its persuasive function, it cannot be defined as the study of the persuasive language genres, insofar as the persuasive function is not linked to a genre but is coextensive with language use, **S. Schematization.**

From the general point of view of language functions, persuasion may be considered representative of the *function of action on the recipient* (*call* function, German *Appell Funktion*, Bühler [1933], or *conative* function, Jakobson [1960]). More specifically, Benveniste contrasts *history* (narrative) with *discourse*, and considers that the intention to influence is a characteristic of the latter category, *discourse*:

> By contrast, we have in advance situated the plane of *discourse*. Discourse is to be understood in its broadest extension: *every utterance supposes a speaker and a listener, and in the first the intention of influencing the other in some way*. It is first of all the diversity of the oral discourses of every nature and of every level ... but it is also the mass of the writings that reproduce the oral discourses or borrow their turns and ends. (Benveniste [1959], p. 242, my emphasis)

Nietzsche, in his lectures on rhetoric, generalizes rhetorical force to make it "the essence of language":

> There is obviously no unrhetorical "naturalness" of language of which one could appeal; language itself is the result of purely rhetorical arts. The power to discover and to make operative that which works and impresses, with respect to each thing, a power which Aristotle calls rhetoric, is at the same time the essence of language; the latter is based just as little as rhetoric is upon that which is true, upon the essence of things. Language does not desire to instruct, but to convey to others a subjective impulse and its acceptance. (In S. L. Gilman *& al.* C. 1989, p. 21)

This trend towards the extension of rhetoric as persuasion to any kind of talk is, moreover, compatible with all the classical definitions of rhetoric as a technique capable of developing the natural capacities of individuals.

7. Persuasion and the "colonization of minds"

The concept of rhetorical persuasion is built on the key idea that persuasion is intrinsically *good*, even if men and women have an unfortunate tendency to misuse the best things. The orator is placed in the elevated position of being a "good man, speaking well" and aspiring to universalize his visions and aspirations, an aristocrat of speech, while his audience is framed in the lower, insubstantial position of the undecided, because of poor reasoning and decision-making abilities, **S. Enthymeme; Metaphor**. The audience is considered barely capable of reaching an independent decision, needing guidance and easy prey to manipulators, **S. Orator and audience.**

On the political and religious level, *persuasion* is the strictly correct term to use for *propaganda*. Propagandists and converters also introduce themselves as good persons eager to persuade, and might also count dictators and fundamentalists amongst the deeply self-convinced persuaders. **S. Dissensus.** In the early fifties, Domenach defined propaganda as the activities systematically organized "to create, transform or confirm opinions" ([1950], p. 8), while Perelman & Olbrechts-Tyteca focus on "the adherence of minds"; and to adhere is also the first step to becoming a member. A key difference between argumentation and propaganda is the means they use: argumentation uses "discursive techniques" (Perelman & Olbrechts-Tyteca [1958], p. 5), that is an *overt*, technique, while propaganda uses all the available means, both *overt* and *covert*, to achieve its goal, using not only discourse, but also images and all spectacular manifestations demanding a ritual collective action.

To *persuade* is to *convert* or, in Margaret Mead's words, to "colonize minds" (Dascal 2009), to save the audience from some evil and direct them to some good, of which they were formerly neither persuaded nor convinced.

8. Arguing in an exchange structure
The theory of rhetorical persuasion is discussed in the context of an interaction without exchange (an addressed monologue, that is to say a one-turn interaction), which gives the public a largely passive role.
Pragma-Dialectic starts not from an opinion to be conveyed to a public but from a difference of opinion between two individuals, giving each opinion an equal value and chance to prevail. This theory "takes as its object the resolution of divergences of opinion by means of argumentative discourse" (van Eemeren and Grootendorst 1992, 18). Rule 1 opens up the space for debate and controversy:
> Freedom - The parties must not obstruct the free expression of points of view or their questioning. (van Eemeren, Grootendorst, Snoeck Henkemans 2002, p. 182-183),

The debate reaches its rational goal if it can effectively eliminate either the doubt or the "inconclusively defended point of view":
> Closing - If a point of view has not been conclusively defended, the advancing party must withdraw it. If a point of view has been conclusively defended, the other party must withdraw the doubts it has expressed with respect to that point of view." (*Ibid.*)

This leads to a consensus either on the opinion, or on its "withdrawal" (from the current interaction, from the other's mind, etc.).
Interactional and cooperative approaches to argumentation consider that the viewpoint that one partner brings into the discussion and lays out for the appreciation of the other participants arguing their own perspective can be profoundly transformed by the encounter. Consensus can be achieved by merging primitive views or by co-constructing a third opinion, participants behaving like Hegelian evolutionary dialecticians progressing by synthesis of actual positions, and not as Aristotelian dialecticians, progressing by eliminating the competing position, **S. Orator; Dialectic.**

9. Externalized persuasion
Persuading, that is to say changing the audiences' *minds*, means changing the audience's *language*. The persuasion experience leaves an inflection point in the discourse of the persuadee. The new discourse produced by a persuaded audience is characterized by its argumentative co-orientation with the persuader's discourse. The persuadees ratify the persuader's interventions; they adopt the speaker's presuppositions, repeat his or her arguments, adopt his or her personal style, and, in the cases of "deep persuasion", his or her tone of voice.

That is, persuasion can be *externalized*, to be analyzed, on the basis of linguistic evidence obtained by comparing the persuading and the persuaded discourses.

Petitio Principii ▶ Vicious Circle

Plausible ▶ ⌀robable

Politeness

The verbal aspects of interpersonal relationships are regulated by a set of principles defining linguistic politeness:
> Politeness refers to all aspects of the discourse, 1. which are governed by rules, 2. which intervene in the interpersonal relationship, 3. and which have the function of preserving the harmonious character of this relationship (at worst: neutralizing potential conflicts and, at best, ensuring that each participant is as open to the other as possible). (Kerbrat-Orecchioni 1992, p. 159; 163)

Ordinary conversation is governed by the principle of *preference for agreement@*. The interactionist theory of politeness (Brown and Levinson, 1978) defines the individual by his or her *faces* and *territories*. Polite intervention respects rules of *positive politeness* and rules of *negative* politeness, both towards oneself, and towards the interlocutor. In argumentative situations, this preference for agreement is transformed into a *preference for disagreement* (Bilmes 1991). Differences are maximized, which has consequences for all the components of the system of linguistic politeness. The case of the *ad verecundiam* argument is a typical illustration of this transformation, S. Modesty.

1. Politeness oriented towards the addressee
Negative politeness recommends the *avoidance of face-threatening acts* whilst *positive politeness* recommends that *positive acts be enacted* in relation to the territories and the face of the interlocutor (Kerbrat- Orecchioni 1992, p. 184).
The argumentative situation reverses these principles. The rules of positive politeness are not applied, whilst those of negative politeness are inverted. For example, the rule "avoid encroachments on the interlocutor's private territories" (*id.*, p. 184) corresponds to a principle of non-aggression, "do not violate the territory of the other". In an argumentative situation, there is necessarily a form of aggression and territorial conflict, with encroachments and counter-encroachments being made.
Another general rule of politeness recommends that parties "[refrain] from making disparaging remarks, too sharp criticisms, too radical refutations, too violent reproaches" (*ibid.*) – to their conversational partner; whereas, in a situation of argumentation, radical refutation is sought rather than avoided and neg-

ative challenging of the opponent is a standard strategy. Praise for the interlocutor turns out to be an attack against the position he defends in the current interaction, **S. Counter-argumentation**.

The ban on personal@ attacks is a matter of politeness aimed at protecting the interlocutor, for aspects of his person that are not at stake in the debate.

2. Principles of politeness directed towards oneself

The principles of defense the speaker' territory recommend that you "protect your territory as much as you can (resist over-invasive incursions, do not let yourself be dragged through the mud, do not allow your image to be unfairly degraded, respond to criticism, attacks and insults" (*ibid.*, p. 182-183). In argumentative situations, participants vigorously apply these protecting principles.

In non-argumentative situations, the speaker territories must be protected, yet not unduly extended and praised, "our societies severely judge self-satisfaction and pro domo advocacy", except in "exceptional circumstances" (*ibid.*). These exceptional circumstances are precisely those of argumentative situations, where speakers do not hesitate to praise their persons as well as their territories, that is, their point of views and arguments. The principles of moderation and self-valorization are thus put on hold. In non-argumentative interactions, "if you have to praise yourself, at least let it be in the attenuated mode of the understatement" (*id.*, 184); you can even "slightly damage your own territory, and practice light self-criticism" (*id.*, 154). This principle requires that one be prepared to compromise and concede, all things that the arguing speaker can choose to do or not do, without being *impolite*.

The conclusion is that argumentative situations locally suspend the application of the rules of politeness in relation to the objects and persons involved in the discussion. This can even be seen as a fundamental characteristic, a defining criteria of such situations. The protagonists use a kind of "anti-system of politeness", mirroring the system of politeness. Speaking of "a system of impoliteness" however, would imply that all these interventions are felt to be impolite, which is not the case, notwithstanding the fact that, in such situations, the partners can engage in polemics about the "tone" of their interventions.

The redefinition of the system of politeness applies strictly to the aspects of the person, face and territories, which are engaged in the argumentative conflict. Outside these areas, politeness rules still apply. It is thus possible for an arguer to praise his or her personality and possessions and attack the standing or values of those of his or her opponent in an argumentative interaction where his or her behavior will, independently, be polite or impolite.

Political Arguments: Two Collections

1. Parameters of political debate

Political deliberation is a problem-solving activity. The following interrogative framework groups the most general questions that must be answered before deciding whether or not to adopt or reject a measure of general interest

> Is this measure legal? Just? Honorable? Timely? Useful? Necessary? Safe? Possible? Easy? Pleasant? What are the foreseeable consequences? (After Nadeau 1958, p. 62).

This framework functions on different modes.

— On the interrogative-deliberative mode, it guides a practical decision process:

> If you are considering such a measure, *look at whether* it is just, necessary, feasible, glorious, profitable, and whether it will have positive consequences.

In this case, the set of questions is used as a heuristic. One can take up a responsible political position on a given issue by examining each point and providing a well-argued answer to each question.

— On the prescriptive-justificatory mode, it helps to develop a global, positive or negative persuasive argumentative script about an issue:

> If you want to support (or to attack) such measure, *show that* it is (or it is not) just, necessary, etc.

— On the analytical-critical mode, it serves to test the completeness of an argumentation

> You argue that this measure is just, necessary, glorious; *but you say nothing about* its consequences and the practical modalities of its realization.

In practice, this simple, robust and effective topic applies to any practical public or private decision.

2. Arguments/fallacies of parliamentary debate: Bentham's Inventory

In *The Book of Fallacies* [1824], Bentham focuses exclusively on fallacious arguments in parliamentary debates. This collection is strongly oriented towards the refutation of conservative discourse, **S. Collection (II)**. In the same spirit, Hirschman has analyzed *The Rhetoric of Reaction* (1991).

Bentham distinguishes four main categories of fallacies: *fallacies of authority, of danger, of delay, of confusion.*

(i) Fallacies of authority, *S. Ad verecundia*; Threat; Politeness; Personal Attack.

—"The wisdom of our ancestors, or Chinese argument; *ad verecundiam.*" (p. 69)

—"Irrevocable law; *ad superstitionem*".

— "Fallacy of vows or promissory oaths; *ad superstitionem*" — "The object of this fallacy is the same as in the preceding; but to the absurdity involved in the notion of tying up the hands of generations yet to come is added, in this case, that which consists in the use sought to be made of supernatural power." (p. 104)

— "No-precedent argument; *ad verecundiam*" — "The proposition is of a novel and unprecedented complexion: the present is surely the first time that any such thing was ever heard of in this house." (p. 115)

— *"Self-assumed authority; ad ignorantiam; ad verecundiam"* (p. 116)

— *"Self-trumpeter's fallacy"* — "There are certain men in office who (...) arrogate to themselves a degree of probity, which is to exclude all imputations and all inquiry." (p. 120)

— "Laudatory personalities; *ad amicitiam*" — "The object of laudatory personalities is to effect the rejection of a measure on account of the alleged good character of those who oppose it." (p. 123)

(ii) Fallacies of danger, appealing to fear (*ad metum*) or hate (*ad odium*) to repress discussion, **S. Emotion; Threat.**

— "Vituperative personalities; *ad odium*" (p. 128). Attacking the person: "Imputation of bad design; of bad character; of bad motive; of inconsistency; of suspicious connections; imputation founded on identity of denomination." (p. 127-128)

— "Hobgoblin argument or: *No innovation!*; *ad metum*" (p. 145) — *innovation leads to anarchy.*

— *"Fallacy of distrust — What's at the bottom?"* (p. 154)

— "Official malefactor's screen (*ad metum*) — *Attack us, you attack Government.*" (p. 158)

— *"Accusation-scarer's device."* (p. 184)

(iii) Fallacies of delay, playing for time, with the intention "to postpone discussion, with a view of eluding it". Some are based on stupidity and laziness (Lat. *socordia*):

— *"The quietist,* or *'No complaint'* (*ad quietem*) — Nobody complains, therefore nobody suffers" (p.190); so, no need to change.

— "False consolation (*ad quietem*)" — *"Look at the people there, and there: think how much better off you are than they are."* (p. 194)

— "Procrastinator's argument (*ad socordiam*)" — *Wait a little, this is not the time!"* (p. 198)

— *"Snail's pace argument (ad socordiam])": "One thing at a time! Not too fast! Slow and sure!"* (p. 201)

— *"Artful diversion (ad verecundiam)"* — *"Why that?* (meaning the measure already proposed) — *Why not this? — or this?"* (p. 209)

(iv) Fallacies of confusion, "the object of which is to perplex, when discussion can no longer be avoided" (p. 213), **S. Personal attack; Ambiguity; Ad populum.**

— "Question-begging appellatives (*ad judicium*)" — The use of "eulogistic terms" and "dyslogistic or vituperative terms." (p. 214)

— "Impostor terms (*ad judicium*)" — "For instance, persecutors in matters of religion have no such word as persecution in their vocabulary; zeal is the word by which they characterize all their actions." (p. 221)

— "Vague generalities (*ad judicium*)" — a fallacy "resorted to by those who, in preference to the most particular and determinate terms and expression (...) employ others more general and indeterminate." (p. 230)

— "Allegorical idols (*ad imaginationem*)" — "substituting for men's official denomination the name of some fictitious entity, to whom (...) the attribute of excellence has been attached. Example: *Government*, for members of the governing body." (p. 258)

— "Sweeping classifications (*ad judicium*)" — "ascribing to an individual (...) any properties of another, only because the object in question is ranked in the class with that other" (p. 265) "Example 1: Kings; Crimes of Kings (...) criminals ought to be punished; kings are criminals, and Louis is a king: therefore Louis ought to be punished)" (p. 266)

— "Sham distinctions (*ad judicium*)" — "Declare your approbation of the good by its eulogistic name, and thus reserve to yourself the advantage of opposing it without reproach by its dyslogistic name (...) Example 1: Liberty and licentiousness of the press" (p. 271)

— "Popular corruption (*ad superbiam*)" — "The source of corruption is in the minds of the people; so rank and extensively seated is that corruption that no political reform can ever have any effect in removing it: This was an argument brought forward against parliamentary reform." (p. 279)

— "Anti-rational fallacies (*ad verecundiam*)" — "When reason is found or supposed to be in opposition to a man's interest, his study will naturally be to render the faculty itself and whatsoever issues from it an object of hatred and contempt" (p. 295)

— "Paradoxical assertions (*ad judicium*)" — "When of any measure, practice or principle the utility is too far above dispute to be capable of being impeached by reasoning, a rhetorician (...) in a sort of fit of desperation (...) he has assailed it with some vehement note of reprobation or strain of invective" (p. 314). "Example: Good method, a bad thing." (p. 316)

— "*Non causa pro causa* (*ad judicium*)" — "When in a system which has good points in it you have a set of abuses (...) to defend; (...) take the abuse you have to defend (...) and to them ascribe the credit of having given birth to the good effects" (p. 328)

— "Partiality-preacher's argument (*ad judicium*) — A discussion of the maxim: "*From the abuse, argue not against use.*" (p. 339)

— "The end justifies the means (*ad judicium*)" — A discussion of the maxim (p. 341).

— "Opposer-general's justification (*ad invidiam*)" — "it is not right for a man to argue against his own opinion. (...) If a member of the House of Commons, and in opposition, a measure which to him seems a proper one is brought on the carpet on the ministerial side, it is not right that he should declare it to be, in his opinion, pernicious, and use his endeavours to have it thought so, and treated as such by the House" (p. 344), and reciprocally.

— "Rejection instead of amendment (*ad judicium*)" — "this fallacy consists in urging in the character of a bar, or conclusive objection against the proposed measure, some consideration, which, if presented in the character of an amendment, might have more or less claim to notice." (p. 349)

It should be emphasized that Bentham does not express the fallacies under any "logical form", but presents them in the form of statements that are condensed argumentations, sometimes in the form of a slogan. The topoi are getting closer to the discursive clichés.

Bentham condemns these maneuvers as *prima facie* fallacies, and discusses them in more detail under the corresponding heading.

In politics, sophists@ are accused of indulging in obstructive or manipulative maneuvers, producing bad arguments in bad faith, rejecting legitimate discussion, and serving dishonest or anti-popular purposes.

Polysyllogism ▶ Sorite; Serial

Pragmatic Argument

1. The scheme

Pragmatic argument is described by argument scheme n° 13 in Aristotle's *Rhetoric*:

> Since in most human affairs the same thing is accompanied by some bad or good result, another topic consists in employing the consequence to exhort or dissuade, accuse or defend, praise or blame. (*Rhet.*, II, 23; Freese, p. 311)

Thus, since positive and negative effects can always be attributed to any action plan, public or private, under discussion or already partly implemented, the plan can always be directly *supported* and eulogized by emphasizing its positive effects (actual or alleged), or *attacked* and blamed by emphasizing its negative effects, (actual or alleged).

Pragmatic argument presupposes a chain of argumentative operations:

(0) A question: *Should we do this?*

(1) A cause-to-effect argument: the intended *action* coupled with an alleged *causal law*, will produce some mechanical *effect*.

(2) This effect is positively or negatively valued.

(3) Taking this consequence as an argument, an effect-to-cause argument transfers to the cause, that is the planned action, the positive or negative assessment of the effect,

— to *recommend* it, if the value judgment carried on it is positive: answer *Yes* to the question

— or to *reject* it, if it is negative: answer *No* to the question.

With reference to this last operation, pragmatic argumentation can be considered to be a kind of effect to cause argumentation, **S. Consequences**. In fact, it is very different from a diagnostic argumentation reconstructing a cause from a consequence. Pragmatic arguments do not reconstruct causes; they transfer to the cause value judgments already cast upon the consequences.

In scientific fields, pragmatic arguments are based upon established facts, "*You smoke*"; they rely upon a statistical-causal law "*smoking increases the risk of cancer*"; and thus lead to a conclusion "*your smoking increases your risks of getting lung cancer*". As nobody likes to have cancer, negative judgment retroacts on the cause "*I (should) quit smoking*".

In socio-political fields as in everyday reasoning the causal deduction characterizing stage (1) is reduced to a series of vaguely plausible correlated elements, that is to say, to a kind of "causal novel", and, commonly to a mere metonymic transfer "this will result in that"; S. **Metonymy**.

2. Against pragmatic arguments

The effect is the *end*, the measure corresponds to a *means* to this end, and evaluation made on the ends is immediately transferred to the means: in other words, the end *justifies* the means. As a consequence, the pragmatic argument can be countered by an objection rejecting the means on *a priori* moral grounds.

Pragmatic arguments are currently refuted by arguments about their adverse and perverse effects.

> *Nouvel Observateur*[1] — A. C., in the book you publish with C. B., "The Domestic Dragon", you support the legalization of drugs. Aren't you afraid of being seen as working for the Devil?
>
> A. C. — Rather than legalization, we prefer to speak of domestication, as this implies a progressive strategy [...]. It will not eliminate the problem of drugs. But it is a more rational solution, which will eliminate the mafias, reduce delinquency, and also reduce all the fantasies that feed drug taking itself, and are part of drug marketing.
>
> J.-P. J.— I believe that legalization would produce a pull effect, the consequences of which cannot be completely controlled. The more of a product is available on the market, the more potential consumers have access to it. This would result in a great many more people taking drugs.
>
> Le Nouvel Observateur [The New Observer], 12-18 October 1989.

A. C. argues pragmatically, emphasizing the positive effects that the legalization of the drug will have, "*eliminate the mafias, reduce delinquency, and also reduce all the fantasies*". She does not specify by which mechanism, but this is certainly not a fallacious move in a first speech turn, considering the constraints of length in interviews.

This claim could be countered by denying the postulated causal link, arguing for example, "legalization will *not* have such *reducing* effects but will just *shift* mafias and delinquents towards new occupations and fantasies towards new objects. J.-P. J. chooses to refute the claim by alleging this measure would have a perverse "pull effect", diametrically opposed to the good intentions of A. C. (note the *will / would* opposition in the argument and in its refutation).

---

[1] A French weekly political and cultural publication

This effect is said to be *perverse* because unexpected, unintended by the person proposing the measure. The opponent credits her for that: J.-P. J. does not accuse A. C. of proposing this measure *so that* "many more people will take drugs". Now, the evaluation of an effect as negative by one can be considered to be positive by the other.

    L1:   — *But this policy would blow up our research group!*
    L2:   — *This is precisely the plan.*

This case falls under Hedge's Rules 5 and 6 (1838, p. 159-162):
>5. No one has a right to accuse his adversary of indirect motive.
>6. The consequences of any doctrine are not to be charged on him who maintains it, unless he expressly avows them.

To claim that the opponent's policy would lead the country to downfall and chaos is a pragmatic refutation of the policy by its negative consequences. To claim that this policy is *intentionally implemented* by the opponents in order to lead the country to ruin and chaos, thus creating conditions conducive to their dictatorship, is to accuse them of conspiracy, and justify the use of violence against them. **S. Norm; Rule; Evaluation.**

This accusation of having a *hidden agenda* also refers to the strategy of refutation of public good reasons by hidden intentions. **S. Motives.**

Pragmatic argument is characterized by the fact that the evaluation of the measure is *indirect*. In the case of drugs legalization, a *direct* evaluation of the measure might be "*this despicable trend to solve problems by legalizing anything and everything should be stopped. So, I don't even want to consider your argument*".

A psychologist could object that drug addicts have a problem with law and moral prohibition. It follows that, legalizing the drug would in fact reinforce addiction.

Precedent

The *argument from the precedent* corresponds to the topic n° 11 of Aristotle's *Rhetoric*:
>Another line of argument is founded upon some decision already pronounced, whether on the same subject or on one like it or contrary to it. (*Rhet.*, II, 23, 11; RR, p. 365)

"Judgment" not only refers to the sentence of a court but to any assessment or decision taken in the past, in ordinary life as well as in the political sphere or in the legal domain. And if the cause has not been settled in a formal assembly, it may have been by such authorities as known *fables, parables* or *examples, proverbs* or celebrated verses (Lausberg [1960], § 426).

Judgments are made in the context of past judgments concerning cases "of the same type", that is belonging to the same legal category, **S. Categorization.** The importance granted to the precedent is a requirement of *continuity* and *consistency* between decisions made in the past and the decision to make, a particular application of the non-contradiction principle. The structural coherence of the involved discursive field is thus strengthened, and guarded against any *ad hominem* charge addressed to the institution, S. ***Ad hominem***.

In much the same way as the argument *ab exemplo@*, the argument *from precedent* motivates a decision or interpretation by relying upon data or examples drawn from tradition. It is a conservative principle, which limits innovation in all domains in which it applies. As such, it combines well with arguments appealing to "the wisdom of our ancestors" (Bentham, 1824; **S. Political argument; Authority; Progress**.

The precedent principle progresses in the following stages:

(i) A problem, **P1**, a case about which a decision has to be made.

(ii) *Research of similar problems and cases*;

(iii) A categorization: this case is similar to a prior case **P0**; it falls within the same category as **P0**, **S. Analogy (III); Categorization.**

(iv) The decision, judgment, evaluation ... **E** was made about **P0**;

(v) By application of the rule of justice@, a similar judgment has to be made about **P1**. "Similar" means here the same judgment, a judgment proportional, or opposite; or, more simply, a judgment consistent with **E**.

The invocation of the precedent can be blocked at the second stage, where it can be argued that there are essential differences between **P1** and the previous case **P0**.

The appeal to the precedent saves time and energy. The problem of judgment is automatically solved as soon as analogy is drawn between the problematic fact and an established fact.

Presence ▶ Object of discourse

Presupposition

The concept of presupposition can be approached as a logical or as a linguistic issue.

1. A logical issue

The problem of presupposition was first addressed within the field of logic. The logic of the analyzed proposition postulates that propositions such as "*all* **A**s *are* **B**s" have two truth-values, the true and the false. The problem arises when the reference of **A** and **B** is void (there are neither **A** nor **B**), as in "*unicorns can fly (are flying beings)*" or "*no unicorn is a dragon*". In this case, is the proposition "*all* **A**s *are* **B**s" true or false? Let us consider the declaration "*the king of France is bald*", as said in 1905. It is impossible to attribute a truth-value to this statement, since in 1905, and still today, the French Republic has no king (Russell 1905).

From the point of view of logical technique, it is sufficient to add the premises "there are **A**s", and "there are **B**s", or *"there is one, and just one King of France"*. An apparently mono-propositional statement such as *"the king of France is bald"* is then translated into logical language via the conjunction of three propositions, each having its own truth-value:

"there is a King of France" & *"there is only one King of France"* & *"he is bald"*.

In 1905 or 2017, the first of the three propositions is false. It follows that the conjunction of logical propositions representing the statement *"the king of France is bald"* is simply *false*. This analysis was criticized for failing to reflect the linguistic intuition of the ordinary speaker, for whom the statements *"there is a King of France"* and *"this King is bald"* do not have the same status in the original sentence. This is true, but the objection is irrelevant, since formal logic does not aim to represent linguistic intuition, but wants to solve a technical problem, and this is what it does.

2. A linguistic issue

Ordinary statements can synthetize different judgments, having different semantic and discursive statuses.

2.1 The multi-layered structure of meaning

The presupposition is defined as an element of the semantic content of the utterance that resists negation and interrogation. The statement *"Peter no longer smokes"* presupposes that *"Peter used to smoked"*, and poses that *"now, Peter does not smoke"*. The negative statement *"Peter has not stopped smoking"* and the interrogative *"has Peter stopped smoking?"* share this presupposed content *"Peter used to smoke"*. Negation and interrogation deal with the *posed* content (*"Peter smokes now"*), and do *not* concern the *presupposed* content.

This multi-layered structure of sentences is one of the major features which differentiate statements made in ordinary language from logical propositions.

2.2 Presupposition as a speech act and the "many questions@" issue

Ducrot redefines presupposition as a strategic action (an illocutionary act) made with the aim of influencing, that is to say, *restricting* the speech possibilities of the conversational partner. The act of presupposition is a discursive power grab by which the speaker performs "an act of legal value, and therefore an illocutionary act [...] [this act] transforms the speech possibilities of the interlocutor, [...] modifies the listener's right to speak" (1972, p. 91).

Consider the following question:

Interviewer — *What are you going to do to fight corruption within your own party?*

The question presupposes that *"there are corrupt people and practices within your party"*. The interviewee is given a choice:

(i) Either he or she accepts the presupposed claim and gives an answer within the range of pre-formatted, expected answers such as:
Interviewee — I'll exclude (suspend) all suspects (the members under investigation).

This answer respects the linguistic *orientation* of the question. It falls perfectly within the frame of dialogue as established by the first turn. The interviewee submits to the interviewer.

(ii) He or she might also reject the presupposed claim:
Interviewee — To my knowledge, there is no (proved) case of corruption within my party

This second answer reframes the routine consensual dialogue; the interviewee rejects the claim made by the interviewer, and the dialogue takes on a character, becoming uncompromising and polemical, opening an argumentative$_2$ situation structured by the issue *"are there (proved) cases of corruption in the party?"* The rejection of the presupposed assumption "[is] always regarded as aggressive: it personalizes the debate, which turns into a quarrel. [...] To attack the opponent's assumptions is to attack the adversary himself" (Ducrot 1972, p. 92). The presupposition seeks to impose an "ideological framework" (*id.*, p. 97) on the later dialogue, that is, to direct the partner's speech. **S. Many Questions; Conditions of Discussion; Persuasion.**

It goes without saying that presupposition phenomena are not limited to dialogue, but, as always, dialogue serves to clarify any issues. A monologue that would not respect its own presuppositions would be *inconsistent*, while, in a dialogue, the rejection of a presupposition is *contentious*. In reality, *dialogue* **(i)** develops under the same conditions as a *monologue*.

Priming and Stages ▶ Gradualism

Probable, Plausible, True

1. Truth and the predicate "— *is true*"
The predicates "— *is true*" and "— *is false*" apply to a statement or to the corresponding *judgment*, i.e., to the logical proposition expressing its content. Truth is "the adequacy between the thing and the intelligence" (Thomas Aquinas, *Summa*, Part. 1, Quest. 16, Art. 1).

According to the famous definition of truth supplied by Tarski "'*the snow is white*' is true if and only if the snow is white" (Tarski [1935]). Note that the famous proposition "*snow is white*" comes from Aristotle (*Top.*, 11, 105a), who considers it as a prototypical statement not deserving a dialectical discussion because clearly true, so impossible to problematize, **S. Dialectic; Conditions of discussion.**

For Tarski, the concept of truth can be strictly defined in formal language only; "with respect to [colloquial language] not only does the definition of truth seem to be impossible, but even the consistent use of this concept in conformity with the laws of logic" [1935], p. 153).

We shall admit that ordinary language about human affairs can use some local practical and satisfactorily defined concept of truth. "*— is true*" or "*— is false*" are said of a statement referring to an event or a state of things through a description that constitutes the meaning of the statement; this meaning is a linguistic construct, based on the common understanding that the statement must be relevant@ to the current discussion and action (Sperber & Wilson, 1995). Ordinary language is not transparent; the true statement is dependent not only on reality, but also on the linguistic system that generates it, and on the social constraints of relevance met by the speech it is part of.

Beyond the linguistic conditioning of its expression, disputability is a characteristic of the statements *"this is true, you are right"*, *"this is wrong, you are wrong, you lie"*. Truth is then a synthetic positive property attached to argumentation as such. Truth judgments oscillate between the argumentative pole of *justification*, and the pole of perceptual or intellectual *self-evidence*.

Argumentation is sometimes criticized for its alleged unsuitability for the expression and transmission of truth. A distinction must be made here between *knowledge-related* arguments and *practical* arguments. In the case of the former, the argument serves to reduce the uncertainty surrounding a claim. In the latter case, the argument seeks to develop a line of action from true or possible facts, combined with a set of values and preferences.

From the point of view of argument in dialogue, truth is a provisional property attributed to a statement that has survived critical examination, conducted, under appropriate circumstances, within interested and competent groups, on the basis of data of which the quality and completeness have been assessed. As a construction, a truth judgment can be adjusted if more and better information becomes available, or if the critical method improves, **S. Default**.

2. Probable

2.1 Enthymemes based on a *probability*, S. Enthymeme.

2.2 Abduction as reduction of uncertainty, S. Abduction

3. The Platonic dramatization: *essential truth* against *social persuasion*

In argumentative rhetoric, the question of the *likely* appears under two opposing views, either as *an arbitrary social representation accepted in lieu of an absent truth*, or as an *approach* to truth.

In Plato's *Phaedrus*, Socrates defines rhetoric as "a way of directing the soul":

> *Socrates*: Well, then, isn't the rhetorical art, taken as a whole, a way of directing the soul by means of speech, not only in the law courts and on other public

> occasions, but also in private? Isn't it one and the same art whether its subject is great or small, and no more to be held in esteem — if it is followed correctly — when its questions are serious or when they are trivial? Or what have you heard about all this? (Plato, *Phaedrus*, 261a; *CW* p. 537)

This *psychagogy* ("art of guiding the soul", probably deprived of its religious function of evoking the souls of the dead, but not of its magical connotations, immediately expresses the *control* function attributed to rhetorical persuasion, "the need for souls", which motivates religious proselytism.

Socrates dramatizes the problem of truth by radicalizing the opposition of the plausible-persuasive to the true:

> *Socrates*: [...] No one in a law court, you see, cares at all about the truth of such matters. They only care about what is convincing. This is called "the likely", and that is what a man who intends to speak according to art should concentrate on. (*Id.*, 261a; *CW* p. 549)

And the proper way of conducting souls is postponed until we know the truth about the essence of all things:

> *Socrates*: First, you must know the truth concerning everything you are speaking or writing about; you must learn how to define each thing in itself; and, having defined it, you must know how to divide it into kinds until you reach something indivisible. Second, you must understand the nature of the soul, along the same lines; you must determine which kind of speech is appropriate to each kind of soul, prepare and arrange your speech accordingly, and offer a complex and elaborate speech to a complex soul and a simple speech to a simple one. Then, and only then, will you be able to use speech artfully, to the extent that its nature allows it to be used that way, either in order to teach or in order to persuade. This is the whole point of the argument we have been making. (*Id.*, 277b-c; *CW* p. 554)

The likely is "like" the true. But to say that a representation, a story is likely, or similar to what truly is or was, we must know what truly is or was. The position of Socrates is strong, since it is based on the impossibility to saying in any sensible way "**A** *looks like* **B**", "*Peter looks like Paul*" when you do not know neither **B**, nor Paul.

When one has found the truth, one can speak truthfully and live in truth. The rhetoric adapted to this situation will no longer be a rhetoric of persuasion but a pedagogy of truth. According to Perelman & Olbrechts-Tyteca "when Plato dreams, in his *Phaedrus*, of a rhetoric which would be worthy of the philosopher, what he recommends is a technique capable of convincing the gods themselves (Plato, *Phaedrus*, 273c)". ([1958], p. 7). In *Phaedrus*, the issue is not so much about convincing the gods as it is about diverting the sensible man from other fellow ordinary men:

> And no one can acquire these abilities without great effort — a sensible man will make a laborious effort not in order to speak and act among human be-

ings, but so as to be able to speak and act in a way that pleases the god as much as possible. (Plato, *Phaedrus*, 273e; C. W. p. 550)

Socrates has thus imposed the pathos of *inaccessible truth*, with for corollary that rhetorical discourse is constructed on the basis of the likely, of verisimilitude, that is, on a pseudo-representation making it possible to forgo truth. Essentially, the function of persuasion is attached to argumentative rhetoric rather as a *stigma* marking its congenital incapacity to attain and even to approach the Truth, the Being and the Gods. The probable bears no relation to the true. To live in persuasion is to live in the world of belief and opinion, in the "cave" and not in the light of the truth. This apparently ineradicable view of rhetorical argumentation is rooted in the anti-democratic and antisocial criticism that Socrates addresses to the institutional, political and judicial discourses trying to handle the problems of the City.

4. The Aristotelian de-dramatization: the probable as oriented towards the true

The Socratic quest for truth unfolds in this atmosphere of tragic radicality. Aristotle radically de-dramatizes the whole problematic by arguing that elaborated probable opinion and truth do not conflict but are in fact complementary. This is the case for at least four reasons. On the one hand, a first range of three reasons:

> (1) The true and the approximately true are apprehended by the same faculty; it may also be noted that (2) men have a sufficient natural instinct for what is true, and (3) usually do arrive at the truth. Hence the man who makes a good guess at truth is likely to make a good guess at probabilities (Aristotle, *Rhet.*, 1355a 14-15; RR, p. 101; my numbering);

Fourth, manipulative rhetoric does not work, "things that are true and things that are just have a natural tendency to prevail over their opposites" (*id.*, 1355a20; p. 101) — a wonderfully optimistic claim; finally, to top it off, it is possible to establish an ethical control on speech: "for we must not make people believe what is bad" (*id.*, 1355a30; p. 101).

The plausible is thus defined not as any opinion bearing the mask of truth, but as a positive orientation, a first step towards truth, expressed in the form of an endoxon, that must be dialectically tested, S. **Dialectic**. It follows that "persuasion" is simply defined as a provisional state of the individual in his quest of truth, a first step toward a progressively constructed truth in progress.

Progress

1. Argument of progress

By definition, "progress moves forward"; the *argument of progress* valorizes the *most recent* as the *best*. If **F1** and **F2** belong to the same category, if **F2** comes after **F1**, then **F2** is preferable to **F1**.

The *argument from progress* rejects the authority@ of elders and their practices, which are deemed *outdated*; the contemporary practices which follow their model are dismissed as *regressive*, indeed *repulsive*.

> Cats are no longer burnt on cathedral forecourts, animal fights were banned in 1833, owls are no longer nailed on the doors of the barns, and rats are no longer crucified as targets for darting. Whatever may be said in bullfight circles, bullfighting with killing is doomed. (*Le Monde*, Sept. 21-22, 1986)

The argument is organized upon the following operations. Firstly, bullfighting is *categorized* as a case of animal abuse, whereby it is allocated to the same category as burning cats, organizing cockfights, nailing owls to doors and crucifying rats. In a second step, the practices belonging to this category are *listed in the chronological order* in which they disappeared. This factual line is then *extrapolated* to lead to the conclusion that bullfights should also be condemned in view of society's progress — and the sooner the better.

2. Argument of novelty

> Lat. *ad novitatem*; *novitas*, "novelty; condition of a man who, the first of his family, reaches an eminent position (senator)" (Gaffiot [1934], *Novitas*). *Novitas* is opposed to *nobilitas*. Its argumentative orientation can be positive (the dynamic of the *novitas* is opposed to the decadent *nobilitas*), or negative: the *homo novus*, the "New Man", coming out of nowhere, is held in suspicion.

Contemporary orientation — The contemporary interpretation links the argument of novelty to the argument of progress: *"what has just come out"* is *"super"* exciting, and *"déjà vu"* is of little value. This argument values innovation over routine, and the new over the old. It underlies the call:

> Be the first to adopt it!

According to this rule, the recently published handbook would be necessarily better than its predecessors, and, in politics, the newest candidate is already seen as the much-needed savior.

Traditional orientation — The argument of progress reverses the traditional view of the higher appreciation granted to the old, particularly in the religious sphere, "the *novitas* is [...] the index of heresy" (Le Brun 2011, §1). The argumentative orientation of the judgment *"this is a novelty!"* has been reversed.

The *argument of progress* is opposed to the *argument of decay* of civilization, which attributes all virtues to the ancients.

The *syzygy@* is a different vision of progress, as a passage from an imperfect world to a perfect and immobile world.

3. Ancients and moderns
The argument of progress structures the eternal quarrel between the Ancients and the Moderns. In its radical form, the argument affirms the absolute superiority of the latter over the former, in the domains of arts and culture as well as the sciences. Ultimately, this superiority would be that of the modern individual over his or her ancestors. In a relativized form, the argument of progress is compatible with the individual superiority of the ancients, *"we are dwarves on the shoulders of the giants"*, although not taller, we can see further ahead. This is classically refuted by the objection that the lice on the head of the giant sees no further than the giant.

Prolepsis

The speaker may choose to connect his or her own argumentative line to a counter-discourse that he or she knows or anticipates and, in any case, rejects. The prolepsis steals the argument from the mouth of the (real or fictitious) opponent, *"I know (perhaps better than you) what you are going to say"*. The counter-discourse is resumed with an indefinite degree of distortion, from a literal referenced quotation to a sketchy evocation of a possible objection, which may be framed as a self-refuting scarecrow, **S. Straw man**. At the very least, the quoted speech is extracted and re-adjusted in view of the new discursive environment, and its ethotic force is kept at bay. Through the magic of quotation, an intended *refutation* becomes a mere *objection*.

The degree to which the counter-discourse is rejected is itself variable. The counter-discourse may be radically rejected; dismissed as absurd (*"do we intend to ruin all small savers? No, quite the contrary, and for many reasons…"*), or maintained in full force, until further information becomes available. In this sense, the Modal-Rebuttal component of argumentation is a special case of prolepsis**, S. Layout**.

The proleptic structure covers not only coordinated or subordinated pairs of statements but any discourse pattern whose configuration corresponds to the staging of two anti-oriented discourses, the speaker taking responsibility for one of them; it represents the maximum development of monological argumentation, **S. Connective; Destruction; Concession; Refutation**.

Several rhetorical terms refer to this same structure:

— The *anteoccupation* refers to a refutative structure, composed of a *prolepsis*, which evokes the position of the opponent, followed by an *hypobole*, which refutes this position or expresses the position supported by the speaker (Molinié 1992, [*Anteoccupation*]). Lausberg ([1963], § 855) terms this same strategy *preoccupation* (Latin prefixes *pre-*, *ante-* "in advance").

— The *procatalepsis* and the *metathesis* refer to a discursive configuration by which the speaker "reminds listeners of past events, presents to them the facts to come, foresees objections" (*Larousse*, quoted in Dupriez 1984, p. 290; *Metathesis* has another quite distinct meaning, "swapping two sounds or letters of a word").

Proof and the Arts of Proof

The words *to prove, a proof, probation, probatory* come from the Latin *probare* and its cognate words; *probare* means "to make good; esteem, represent as good; make credible, show, demonstrate; test, inspect; judge by trial" (OED, *Prove*). All these meanings evoking a practical activity are still present in the use of proof in rhetorical theory.

1. Vocabulary of the arts of proof

The following words belong to the elementary lexicon of the arts of proof.

    to argue; an arguer, an argument, an argumentation; argumentative
    to demonstrate; a demonstration; demonstrative
    to prove; a proof, a prover; probatory
    to reason; a reason, a reasoner; reasoning; reasonable; rational
    evidence; evidential

The following remarks deal with the articulations of the ordinary lexicon of the arts of proof.

Agent names — Some names are related to their root verbs with the meaning "person who (Verb)"; so are *arguer, reasoner, prover*. *Demonstrator* a derivative from to *demonstrate_2* "show other people how something is used or done".
This can be interpreted as a mark of a subjective involvement in the mechanism of *proving, arguing, reasoning*.

Verb complementation — In "*Peter reasons about P*", **P** is the issue, the substance of the reasoning or of the argument. "*Peter has demonstrated, or proved that P*", the **P** clause is true and expresses the conclusion of the demonstration. *To argue* admits both constructions:

    Peter argues about P: **P** is the issue,
    Peter argues that P: **P** is the claim, but *to argue* does not imply that its **P** clause is true.

Aspectual distinction — The relationship of *argument* to *proof* is grammatically an aspectual distinction, that of unaccomplished / accomplished. *To argue* is no more a semantically weakened form of *to prove* that *to look for* something is a weakened form *to find* something. The proof is the "terminator" of the argument.

Semantic orientation —*Evidence, proof, argument* and *demonstration*, however, can function in co-orientation, as quasi-synonyms in many contexts. The lawyer is engaged in a brilliant *demonstration* in which he brings *conclusive* evidence and convincing *arguments*. Such discursive practices put in continuity argument/evidence and proof, the proof being the end and finality of the argument: it is "a knock-down argument" (Hamblin 1970, p 249.). Arguments are oriented towards proof.

Position markers — These terms which may be regarded as quasi-synonyms in some contexts, may clearly appear as markers of argumentative positions in the context of a debate. In the judicial field, the judge hears the statements and *arguments* of the parties; each party brings (what they consider to be) *proofs* and rejects those brought by the opponent as *quibbling*. We are no longer dealing with synonyms, but anti-oriented antonyms. The difference between *evidence/proof, argument*, and *quibble* becomes a simple matter of perspective. The probative value is now no more than the positive assessment I give to my argument and I refuse to grant to that of my opponent.

A polite although decisive *rebuttal* will be proposed as a mere *objection@* and a simple *argument*. Argument is then a "lexical softener" for *proof*, its use implies a distance, a lesser commitment of the speaker to the claim.

Dialogic Status — The distinction demonstration / proof / argument seems primarily sensitive to the presence or absence of counter-discourse. This is why the word *argument* is used to describe reasoned discourse at both ends of scientific activity, in learning activities, as well as in the sharpest controversies over open questions, where two discourses both perfectly equipped theoretically and technically, revert to the status of argument, simply because there is disagreement.

2. The proof between fact and discourse

Proofs are expressed in a language, natural or formal, and put forward in a discourse. *Formal* evidence brought by a hypothetical-deductive demonstration is often seen as the archetypical proof. Its counterpart in ordinary language would be the argument based on essentialist definition used in philosophy and theology. In other areas of activity, probationary speech requires a reference to the world, in which case, evidence is now seen as a *fact*. The proof is built by a series of experiences and calculus, as suggested by the concrete metaphors used to talk about evidence — *to produce proof, to provide evidence, to bring a proof, to make a demonstration*. This connection with reality makes the difference between proof and argumentation on one side and formal demonstration on the other.

The concept of *proof as fact* invokes non-discursive evidence of material realities, perceptible to sight and touch. The proof that I did not murder Peter is that he is alive, standing before you. Such situations seem to make language superfluous. Nonetheless, facts can become evidence through discourse alone. Evidence is relative to a problem, and discourse frames the situation in which the evidence solves the problem. Evidence may be silently brought before the relevant judges. If some facts "speak for themselves", some other times they are not so "eloquent", or even remain "silent" for many. One must speak for them, and discourse is required to make the material evidence visible. The cruel experience of Semmelweiss has certainly shown us that the de facto existence of seemingly indisputable facts does not foretell their acceptance (Plantin, 1995, chap. 7).

"The Wolf and the Lamb" — The La Fontaine fable "The Wolf and the Lamb" (*Fables*, I, X) shows how innocent people can trust material evidence, and that material evidence does *not* carry the day.

> The reason of the strongest is always the best,
> As we'll show just now.

Situation:

> A lamb was quenching its thirst
> In a pure water stream.
> A fasting wolf came by, looking for adventure;
> Attracted to this place by hunger.

The wolf starts with a violent reproach, as men do with their future victims

> —What makes you so bold as to cloud my drinking?
> Said this animal, full of rage,
> You will be punished for your audacity.

The offense is assumed (*you cloud my drinking*). The request for explanation of motives (*what makes you so bold [...]?*) appears to give the lamb the opportunity of explaining itself. Yet, the accusation is immediately followed by the sentence (*you will be punished for your temerity*). This incriminating speech is deeply mysterious, why does the wolf speak? It could simply take advantage of the food it was yearning for, and finally met, devouring the lamb like the lamb drinks the water. With exquisite courtesy, the lamb denies the presupposed fact and its denial is backed up by undisputable proof, **S. Evidence**:

> —Sir, answered the lamb, let Your Majesty
> Not get angry.
> But rather, let Him consider,
> That I am quenching my thirst
> In the stream
> More than twenty steps below Him;
> And that, consequently, in no way,
> Am I clouding his beverage.

The lamb's argument is conclusive, physical laws are such that the brook never flows upstream. But conclusive does not mean impossible to contradict:

> —You do cloud it, said the cruel beast.
> And I know you said bad things about me last year.

This second accusation is also rebutted in the same decisive way:

> — How could I have done that, when I wasn't born,

> Answered the lamb; I am still suckling my mother

Idem for the third:
> — If it wasn't you, then it was your brother.
> — I have none.

But the last accusation is irrefutable; the lamb is given no chance to refute it:
> — Then it was a relative of yours;
> For you have no sympathy for me,
> You, your shepherds and your dogs.
> I am told of that. I must avenge.

The conclusion is that good reasons do not change the course of history:
> Thereupon in the dark of the forest
> The wolf carries the lamb, and then eats it,
> Without further ado.

3. Functional heterogeneity of the discourse of the proof

Whatever the field, the discourse of proof is functionally heterogeneous. Proof fulfills a number of functions:

— *Alethic*: it establishes the truth of a fact.
— *Epistemic*: it justifies a belief; it helps to stabilize and increase knowledge.
— *Explanatory*: it accounts for facts which are not self-evident, via their integration into a coherent discourse in the correct language, be it a demonstration, or a story accounting for what took place.
— *Cognitive* and even *aesthetic*: proof must be relatively clear, and, if possible, "elegant".
— *Psychological*: it eliminates doubt and inspires confidence.
— *Rhetorical*: it is convincing.
— *Dialectical*: it eliminates the challenge, and closes the discussion.
— *Social*: it builds consensus, assuages the community affected by the problem, and strengthens its confidence in its technical capacities to produce evidence particularly, but not only, in the social and judiciary domain.
— *Conversely, evidence excludes*: those who accept proof consider that those who resist the proof must be mad, feeble minded, carried away by their passions.

4. Unity of the arts of proof

The arts of proof — reasoning, arguing, demonstrating, proving — share the following characteristics.

— *A language and discourse*: arguing, demonstrating, proving, all require a semiotic medium, a language developing in a discourse. The same can be said for reasoning, although the term focuses on the cognitive aspects of the process.
— *An intention*: Like every discourse, the flow of demonstrative, argumentative, probative, reasoned discourse is organized by an objective, i.e., an intention.
— *An interrogation*: These processes start with a problem, an uncertainty, a doubt.

— *An illation (derivation) process or inference:* The notion of inference@ is primitive. In logic an inference is defined as the logical derivation of one statement from a set of premises. The intellectual process of inference contrasts with the *intuitive* approach, for which a truth is asserted directly (without mediation) on the basis of its direct physical or intellectual perception. In the case of inference, the truth is asserted *mediately*, that is indirectly, via data or assumptions expressed by statements and supported by underlying principles, the nature of which depends on the area concerned, S. **Self-evidence.**

— *Argumentation, proof and demonstration* are referring to something external; the development of discourse is more or less governed by the external world. Anything and everything can be said, but reality creates limits. The practice of proof and argument is not pure linguistic virtuosity, it must confront objects and events.

— *Domain dependence.* As argumentation, demonstration and proof are domain dependent. The modes of production of evidence differ according to the field, the kind of technical language used and the kind of experimental method used in the considered area. The establishment of large classes of scientific proofs is the task of epistemologists. Argumentation in natural language is characterized by its capacity to combine a large variety of heterogeneous proofs, corresponding to the various argument schemes.

5. Argumentation among the arts of proof

Perelman & Olbrechts-Tyteca's *New Rhetoric* opposes "argumentation" to "calculation":

> The very nature of deliberation and argumentation is opposed to necessity and self-evidence [...] The domain of argumentation is that of the credible, the plausible, the probable, to the degree that the latter eludes the certainty of calculation. ([1958], p. 1).

This position leads us back to the Aristotelian opposition between rhetorical "means of pressure" and scientific proofs, S. **Demonstration**, without considering the possibility of bridging the gap between the two discursive regimes, or of positioning them upon the same truth oriented scale. An increasing range of contemporary discourses, however, are mixed; they seek to articulate some scientific reasoning and data, along with social values and material interests. A contemporary challenge for argumentation studies is to find a way of dealing with such mixed data. This is true of all the varieties considered to be typically argumentative in the *Treatise*: "speeches [of politicians] ... pleadings [of lawyers] ... decisions [of judges] ... treaties [of the philosophers]" (Perelman & Olbrechts-Tyteca [1958], p. 10).

The approaches of argumentation as a set of "discursive techniques" (Perelman, Olbrechts-Tyteca), as discourse orientation (Ducrot) or discursive microstructure, as dialogue or interaction, anchor the study of argumentation in ordinary linguistic practices, structured by rules and norms depending on the genre of discourse and on the frame of the situation. Argumentation studies are thus clearly distinguished from research in scientific methodology, and from the epistemological study of proof, demonstration, explanation or justification in mathematics, science, or philosophy, **S. Demonstration.**

Proper Name

The proper name argument scheme corresponds to topic n° 28 of Aristotle's *Rhetoric*, "another topic is derived from the meaning of the name. For instance, Sophocles says, '*Certainly thou art iron, like thy name*'." (*Rhet.*, II. 23, 1400b29, Freese, p. 323). The example is a pun on the proper name of the hero and the word meaning "iron".

Unlike the *nickname*, which claims to refer to a characteristic of the individual, the *proper name* is not motivated; it does not mean its bearer. When the proper name (first name or last name) of a person is homonymous with a common name, the topic of the name attributes to the person the characteristics of the homonymous thing; he or she is re-categorized as a non-human being, which may be less-than-flattering. The name functions as an *index* from which truths about the person might be inferred. Aggravated by the infinite resources of paronomasia, and rhyme, proper names can be the basis for all kinds of derivations, particularly, although not exclusively in the school playground:

> You are Peter [Lat. *Petrus*], and on this rock [Lat. *petram*] I will build my church. (Matthew, 16:18)
> June will be the end of May[1]

Being named Peter, and thus being like a rock, is being apt to be a foundation: the name is an *aptonym*, the character and destiny of the person are pre-inscribed in his or her name. Mr. *Child* is of course a pediatrician, or a teacher or perhaps he has a childish character; the aptonym reinforces the person's suitability for his task or confirms the attribution of a trait of character. Referring to John R. Searle as *Sarl*, (French acronym for "Limited Liability Company", Inc.) Derrida re-casts, fairly or not, Searle's work as a kind of business.[2] **S. Etymology, Ambiguity.**

Proponent ▶ Roles

Proportion and Proportionality

> The Greek word [*analogia*] means "proportionality", as the Latin *proportio*.

---

[1] The slogan appeared during the campaign for general elections to be hold in June 2017, Mrs Theresa May serving as Prime Minister of the United Kingdom since 2016.
[2] Jacques Derrida, *Limited Inc*. Evanston, IL: Northwestern University Press, 1988.

1. Proportion

A ***relation*** is a stable connection between two things:

    shell : fish   old age : life
    glove : hand   pilot : ship

A *proportion* is an analogy between at least two *relations* (not between individuals, as in categorical analogy); it implies at least four terms. Two pairs of beings are in a relation of proportion if, in their respective fields, they are bound by the same, or a similar relation.

    shell : fish ~ feather : bird   — *cover the body of* —
    glove : hand ~ shoe : foot   — *protect the* —
    leader : group ~ captain : ship   — *guide the* —
    old age : life ~ evening : day   — *last moment of the* —

The analogy of proportions is expressed through parallel syntactic structures:
    (Since) a ship needs a pilot, any group needs a leader!

In arithmetic, a proportion is defined as the relation between two numbers, such as '17 / 27'. The same proportional relation binds two pairs of numbers a/b and d/d if they obey the following rule:
    $3/2 = 9/6$, same proportion 1.5
    $a/b = c/d \iff ad = bc$ ($a = bc/d$, etc.)

The analogy between proportions corresponds to the linear equation with one unknown, that is to say, to the "rule of three":
    $a / b = c / x$ where $ax = bc$, $ax - bc = 0$; and $x = bc / a$

In geometry, two *similar* figures have the same shape and different dimensions. Two congruent triangles have equal angles and proportional sides.

The process of understanding is the same in the case of mathematics as it is in argumentation. The reasoning by which the value of 'x' is mathematically extracted from the arithmetical proportion is the same as the argument which extracts the necessity of a leader from the analogy of proportion between a ship's crew and a group of people more generally.

The analogy of proportion is at the basis of a specific kind of metaphor@:
    old age, evening of life.

The analogy of proportion is open to ironic self-refutation:
    A woman without a man is like a fish without a bicycle.

2. Proportionality

    Lat. *ad modum* argument, Lat. *modus*, "measure, "just measure"
    NB: Besides "moderation", the Lat. *temperantia* can mean "just measure, fair proportion"; **S. Moderation**

The *argument of proportionality* justifies a provision or an action by claiming that it is well proportioned to the facts, gradual. It is invoked *a contrario* in routine press releases such as:

Proposition

(The association, the trade union, the government...) X condemns *the disproportionate use of force used against...*

Let us consider a situation of unrest, described by the current government as a seditious uprising, led by a handful of terrorists. The government organizes a large military presence to *"show strength not to have to use it"*. This strategy of psychological war may have perverse effects. In reality, the argument of proportionality allows calculations that defeat the desired effect:

> The deployment of strength, far from minimizing the enemy, made it stronger.
> Pierre Miquel, [*The Algeria War*], 1993[1]

This conclusion is based on the topos, *"one does not use cannons to shoot flies"*. A strong refutation of a (declared) weak position entails the same kind of paradox@.

The argument of proportionality is a form of argument on the *right* measure, which can also be defined as the *intermediate* measure, **S. Moderation.**

The *proportionality* strategy can be used to avoid the risks posed by the *escalation* strategy.

Proposition

The word *proposition* may be a synonym of *proposal*, "the point to be discussed" or "demonstrated" (MW, art. *Proposition*). A proposition may be developed in a complex argumentative *discourse*, justifying the briefly expressed concrete proposition itself. **S. Argument – Conclusion.**

In classical logic, a proposition is an autonomous *statement*. *Propositional logic* considers concatenations of *unanalyzed* propositions **P, Q, R**.... *Predicate logic* considers a proposition *analyzed* in two *terms*, the *subject* and the *predicate*, "S is P".

1. Term

In logic, a distinction is made between *categorematic* and *syncategorematic* terms. *Categorematic* terms function as subject names or concept names (predicates). Used without further clarification, the word *term* refers to a *categorematic term*.

Syncategorematic terms include negation, binary logical connectives@ ("&", *and*, etc.) and quantifiers ("$\forall$", *all*, etc.). They cannot function as subject or concept names, they appear only in combination. They have no independent meaning; their meaning being defined by specific contribution they make to the meaning of the terms or proposition they combine with.

A parallel distinction in grammar contrasts with the so-called *full words* with a full semantic content (verbs, substantives, adjectives, adverbs) and the so-called *empty or grammatical words* (such as linking words, discursive particles, auxiliaries...)

---

[1] Pierre Miquel, *La Guerre d'Algérie*. Paris: Fayard, 1993, p. 190.

2. Predicate, variable, constant
A sentence may be represented by its pivotal element, the verb, accompanied by variables representing its complements. Variables are denoted 'x', 'y', or simply as empty places, "—".
— *Paul sleeps*: *To sleep* is a one place predicate, written "— *sleeps*" or "*x sleeps*". :
— *Paul eats an apple*: *To eat* is a two-place predicate, written "— *eats* —" or "*x eats* y":
— *Paul gave the apple to the lady in black*: *To give* is a three-place predicate, written "— *gives* — *to* —" or "*x gives* y *to* z".

The same object can be attached to an infinite number of predicates, for example "— *is a car*"; "— *is a means of transport*"; "— *is an object that can be bought*"; "— *is a cause of pollution*"... Discourse constantly creates new predicates, according to the interests of the speakers, as "— *was carried out on 10 June 2017*"; "— *is a car available for next Saturday's trip*".

In the case of a predicate admitting several variables, one or more empty places may be occupied by a *constant*. The predicate is then said to be partially saturated, which corresponds to a new predicate, for example, where "*Paul gives* y (something) *to* z (somebody)", "x (somebody) *gives* y (something) *to John*", "*Peter gives* y (something) *to John*".

In ordinary language, *variables* are expressed by indefinite phrases and pronouns: *any, all, some, a* (person)…".

Constants are denoted 'a', 'b'; in natural language, they are expressed by referring terms or phrases:
— *Proper names* (*Peter*), permanently attached to individuals.
— *Pronouns* (*this the other, the next one*). Their referential anchoring is based both on deictic maneuvers and on definite descriptions whose reference can be retrieved from the context. S. Object of discourse
— *Definite descriptions*, or denoting phrases (*the man with the green hat*). The noun phrase can be complexified at will: *the seated man, the man who pretends to look elsewhere*.

This simple notation renders explicit the skeleton of the sentence and is the basis of a more detailed semantic analysis of both its internal structure and external position in the broader discourse to which it belongs. Argument schemes are currently expressed in such a semi-symbolic notation.

3. Proposition
In classical logic, a proposition is a *judgment*, which can take only two values, true (**T**) or false (**F**); a proposition cannot be "more or less" true or false. A proposition is only a way of telling the truth or the false, without any consideration upon its meaning and conditions of use.

A proposition is *unanalyzed* if no information on its internal structure is available. Logical connectives and the laws of their combination are defined on the basis of such unanalyzed propositions.

A proposition is *analyzed* if its internal structure is taken into consideration. Classical logic considers that the analytic structure of a logical proposition is basically "Subject *is* Predicate", "**S** *is* **P**":

— The subject refers specifically (if a *constant*) or generally (if a *variable*) to the elements of a universe of reference.
— The predicate says something about these elements.
— The proposition categorically (without condition) affirms or denies that the predicate accepts the subject.

Capital letters **A, B, C**... **P, Q, R**... are used to denote both unanalyzed propositions and the subject and predicate in analyzed propositions.

3.1 Quality and quantity of a proposition

The *quality* of a proposition refers to its two possible dimensions, *affirmative* or *negative*.

The *quantity* of the proposition varies according to whether the subject refers to a being, certain beings or all beings of the universe of reference.

Quantifiers express the quantity. The quantifiers such as *all* (*all* **N**), or *some* (*some* **N**) express quantities. According to their quantity, propositions are:

    Universals: *all poets*
    Particular: *a poet*; *some poets*

Particular does not refer to a constant, a specific, known, individual. In its traditional form, logic does not deal with propositions predicating something from a determined individual, such as "*Peter*" or "*this poet*"; S. Syllogism

The combination of quantity and quality produces four kinds of propositions:

    A   universal affirmative *All S are P*.
    E   universal negative *No S is P*.
    I   particular affirmative *Some S are P*.
    O   particular negative *Some S are not P*

Traditionally, affirmatives are denoted by the letters **A** and **I** (two first vowels of the Latin verb **AffIrmo** "I affirm") and the negatives by the letters **E** and **O** (n**E**g**O**, "I deny").

3.2 Converse@ propositions

The converse proposition of a given proposition is obtained by swapping subject and predicate. The subject of the original proposition is the predicate of its converse proposition and the predicate of the original proposition is the subject of its converse proposition.

The *quality* (affirmative or negative) of the two propositions is the same.

The negative universal **E** and its converse are equivalent (they have the same truth conditions, cf. infra §4.2, Logical Square):

    No **P** is **Q** ↔ no **Q** is **P**.

Likewise, the particular affirmative and its converse:
Some **P** are **Q** ↔ some **Q** are **P**

3.3 Distribution of a term

A term is *distributed* if it says something of all the individuals belonging to the reference set. If not, it is *undistributed*.

The terms preceded by the quantifier *all* are *distributed*. The terms quantified by *some, many, almost all* ... are *undistributed*. For example, in a universal affirmative proposition **A**, "*All Athenians are poets*":
— The subject term, *Athenians*, is *distributed*.
— The predicate term, *poet*, is *undistributed*; the proposition only says that "some poets are Athenians".

The notion of distribution is used by the rules of evaluation of the syllogism, S. **Paralogisms**.

3.4 The presupposition of existence

Some expressions such as "*the unicorn*", "*the present king of France*", "*real-life dragons*", are misleading, insofar as they appear to be referring expressions despite the fact they do not refer to any existing being. This being the case, when such phrases are used as subjects of a proposition, this proposition cannot be said to be true or false, the present King of France is neither bald nor hairy. To avoid such perplexities, it is assumed that the universe of reference of the subject term is assumed not to be empty. S. **Presupposition**.

4. Immediate inference and logical square

4.1 Immediate inference

An immediate inference is a one-premise argument, inferring from one proposition to another:
All the **A** are **B**, *so* some **B** are **A**

The two terms of this single premise are found in the conclusion, only the quantity of the proposition changes. While syllogistic inference requires a *medium* term, "*im-mediate*" inference does not need such a transition term. It is debatable whether immediate inference is a kind of reasoning.

Immediate inference is an inference, not a *reformulation*. The reformulation relation presupposes the identity of meaning between the two utterances it links:
Some **A** are **B**, so some **B** are **A** (conversion, see §3.2).
All the **A** are **B**, so some **B** are **A** (subalternation, see infra).

In the first case, the immediate inference corresponds to an equivalence. This is not true, however, of the second.

4.2 Logical square

The logical square expresses the set of immediate inferences between analyzed propositions of the subject-predicate form according to their quality, affirmative or negative, and the quantity of their subject (**A, E, I, O**, see above).

Proposition

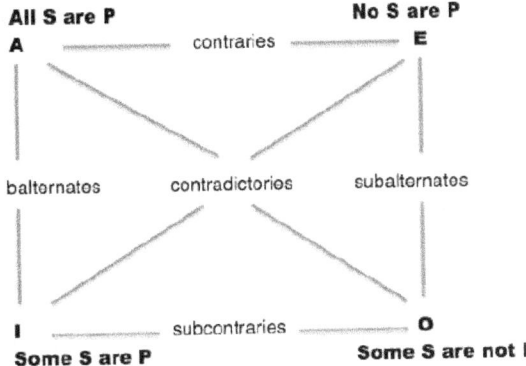

These four propositions are linked by the following relations.

— *Contrariety*, between the affirmative universal **A** and the negative universal **E**. **A** and **E** are not simultaneously true, but may be simultaneously false. In terms of immediate inference, if one is true, then the other is false.

— *Subcontrariety*, between the particular affirmative **I** and the negative particular **O**. At least one of the two propositions **I** and **O** is true. They may be simultaneously true, but cannot be simultaneously false. In terms of immediate inference, if one is false, then the other is true

— *Contradiction*, between:
- The universal negative **E** and the particular affirmative **I**.
- The universal affirmative **A** and the particular negative **O**.

E and **I** cannot be simultaneously true or simultaneously false (only one of them is true). The same will be true for **A** and **O**. In terms of immediate inference, the truth of one immediately implies the falsity of the other, and vice versa.

— *Subalternation*, between:
  A and **I**, the universal affirmative and the particular affirmative.
  E and **O**, the negative universal and the negative particular.

If the superaltern is true, its subaltern is true. Immediate inference:
  Every **S** is **P**, so some **S** are **P**.

If the subaltern is false, its superaltern is false. Immediate inference:
  It is false that some S are P, so it is false that every S is P.

The subaltern may be true and the superaltern false.

Moreover, propositions **E** and **I** are *convertible*; cf. supra, §3.2.

5. Immediate inference, quantifiers and terms

Immediate inference is an inference from *a single* premise. The *two* terms of the single premise are found in the conclusion (examples above). In the case of the syllogism@, the inference proceeds from *two* premises and *three* terms. The *middle term* functions as a "mediator", an intermediary, between the *major term* and the *minor term*. In the case of immediate inference, the conclusion is "not-mediated" by a middle term.

From a *cognitive* point of view, argumentation by definition@ assigns to an individual the properties characterizing the class to which it belongs. From a *linguistic* point of view, argumentation by definition assigns to an individual designated by a name, all the elements of the linguistic definition of this term. Argumentation by definition is therefore an immediate, substantial, *semantic inference*, on the meaning of the *terms*, **S. Definition; Inference**. Immediate inferences are formal; they are not made on the basis of *full words*, but on the basis of *quantifiers*. Both kinds of inference function as semantic reflexes in ordinary discourse, linking natural statements, according to ordinary semantic intuition combined with contextual references based on the *laws@ of discourse* and the *cooperation principles@*.

Because of their seeming obviousness, the way we handle such these inferences often goes unnoticed. This does not mean, however, that the process is always error free. Taking the correct approach to such inferences is part of the argumentative competence.

Pseudo-Simplicity ▶ Fallacies (I)

Psychological Argument (in Law) ▶ Intention of the Legislator

Q

Quasi-Logical Arguments

Perelman & Olbrechts-Tyteca introduce the class of quasi-logical arguments as the first of the three categories of "association schemes" ([1958], p. 191), that is *argument schemes*. Quasi-logical arguments can be understood "by bringing them closer to formal thought, logical or mathematical. But a quasi-logical argument differs from a formal deduction in that it always presupposes adherence to non-formal theses, which alone allows the application of the argument" (Perelman 1977, p. 65)

Six schemes are more precisely analyzed, and these bear the same name as their logical counterparts:

> Among the quasi-logical arguments, we shall first analyze those which depend on logical relations — contradiction, total or partial identity, transitivity; we shall then analyze those which depend on mathematical relations — the connection between the part and the whole, the smaller and the larger, and frequency. Many other relations could obviously be examined. (Perelman & Olbrechts-Tyteca [1958], p. 194)

Definitions@ are quasi-logical argumentations "when they are not part of a formal system, and when they nevertheless claim to identify the *definiens* and the *definiendum*, we shall consider them a form of quasi-logical argumentation" (*id.*, p. 210). They are "typical of quasi-logical argumentation" (*id.*, p. 214).

The "quasi-logical" label is symptomatic of the attitude of the authors of the *Treatise*, in that they reject logic but yet use it *a contrario* to define argumentation in general and in particular to characterize the "quasi-logical" super-category of argument schemes. The category includes all the argumentative strategies involving phenomena such as negation, scales, relations and definitional stereotypes. In fact, it is the *system of language* that is considered to be a quasi-logic.

The arguments in this category are defined by a common characteristic:
> [Quasi-logical arguments] lay claim to a certain power of conviction, in the degree that they claim to be similar to the formal reasoning of logic or mathematics. Submitting these arguments to analysis, however, immediately reveals the differences between them and formal demonstrations, for only an effort of reduction or specification of a non-formal character makes it possible for these arguments to appear demonstrative. This is why we call them quasi-logical. (*Id.*, p. 193)

According to the traditional definition, a fallacy is an argument that looks like a valid argument but is not. There is a striking similarity between this, and the definition given in the *Treatise*: quasi-logical argumentation "claim[s] to be similar" to formal reasoning, but is not.
S. Fallacies; Logic; Collections (III).

Question

1. Question as interrogation

A question may be a sentence "attempting to get the addressee to supply information" (SIL, *Question*), using the specific morphemes and syntactic transformation attached to the interrogative form.

— The fallacy of *many questions* is one of the six Aristotelian linguistic fallacies, **S. Fallacy (I)**. A *loaded question* is a question about a complex statement, containing several implicit statements. The loaded question presupposes the truth of these underlying statements, which may be disputed by the recipient of the question. **S. Many Question**. Such questions are said to be oriented, **S. Orientation**.

— Rhetoric uses a series of common@ place ontological questions to gather information, **S. Common Place**.

— A *rhetorical question*, in the traditional sense of the term, re-frames the argumentative question as a question admitting a self-evident answer, **S. Argumentative question**.

2. Question as problem

A *question* can also be the subject of a discussion, an "issue; *broadly*: a problem" (*MW*, Question). It doesn't necessarily have an interrogative form.

An *argumentative question* represents the discursive confrontation generating an argumentative situation. Such a question does not refer to a quest for information, but to a *problem*. **S. Argumentative question**.

Question: *Argumentative Question*

The concept of *argumentative question* originates in the notion of *stasis*, developed primarily by the rhetorical theory of judicial interaction, **S. Stasis**.

The concepts of an argumentative question and an argumentative situation are interdependent. An *argumentative situation* emerges when two speeches concerning the same topic begin to diverge to some extent. The contact can be made during a remote or face-to-face, oral or written, interaction. Such *potentially* argumentative situation may evolve into an *actual* argumentative situation when the divergence is topicalized and ratified by a participant. All these necessary developments delimit an argumentative space, defining what is *argumentation*, *before* the appearance of *arguments* strictly speaking (discursive segments supporting a conclusion).

The existence of a question is at the origin of the paradoxes of argumentation, **S. Paradoxes**.

1. Proposition, opposition, doubt: A question

The following example, constructed around the recurring question *"Should we legalize drugs?"* shows how the question assigns argumentative roles@, on the basis of the three fundamental argumentative speech acts, *to propose*, *to oppose* and *to doubt*.

- **The current state of law and discourse**

In Syldavia 2012,
> Drug production, importation, exportation, trade, possession, and consumption are forbidden.

This statement corresponds to the state of Syldavian legislation, generally backed by "dominant opinion", perceived as a matter of course, so needing no argument.

- **A proposition**

Another discourse is oriented towards a proposition opposed to this prohibition:
> P: — The consumption of soft drugs should be legalized, or at least tolerated.

Speaker **P** steps into the argumentative role of *proponent*, and opens the debate. All speakers aligned with this proposal serve as *allies*.

- **An opposition**

Other speakers oppose the proposal:
> O: — That's staggering!

The speaker **O** plays the argumentative role of *opponent*. Speakers willing to hold this type of rejection discourse with respect to the proposition are *allies*.

- **Doubt and question: emergence of the argumentative question**

Question: Argumentative Question

Some participants refuse to align with either position. They are in the position of third parties, synthetizing the proposition *vs.* opposition relation into an argumentative question, and transforming the discourse confrontation into a full argumentative situation:

> TP: — *All this is quite perplexing. Should the prohibition of all these drugs they call soft be maintained or not?*

The argumentative question is thus generated by the contradiction "discourse / counter-discourse", hence the schema:

> Proposition vs. Opposition → Argumentative Question (AQ)

2. The conclusion as answer to the argumentative question

When discourse develops into a confrontation, good reasons are needed and quickly provided. The proponent bears the burden of proof and, in order to meet this requirement, must put forward arguments, for example by re-categorizing soft drugs in the same category as alcohol or anxiolytics:

> P. — *Soft drugs are not more dangerous than alcohol or anxiolytics; alcohol is not subject to any general prohibition, and anxiolytics are subject to medical prescription.*

This argument supports the slogan:

> P. — *Yes! We should have at least a more tolerant approach to soft drugs!*

Produced under the general scope of the argumentative question, this *conclusion* gives an *answer* to this question.

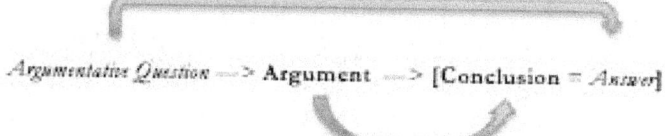

The opponent must show that the proponent's speech is untenable. First, he rejects the arguments of the proponent:

> O: — *No! Alcohol has nothing to do with drugs. We know how to drink in this country; alcohol is part of our culture, drugs are not. And if you legalize soft drugs next you'll have to tolerate hard drugs!*

> O: — *In Syldavia, they tried to legalize drugs, and the experience failed. Enough with social experimentation detrimental to young people!*

Conclusion:

> — *Let us reject this crazy proposal of legalization!*

Secondly, **O** presents a counter-argument in favor of another position. This may correspond to maintaining the status quo:

> — *Honest citizens live peacefully thanks to the prohibition; the situation is under control as it is.*

Under the standard regime, the doxa "goes without saying"; but once the argumentative situation has been opened, it requires justification.

Argumentative questions are distinct from *informative* questions. The latter permits direct, unequivocal relevant answers:
    S0: — *When did you arrive? In which hotel are you staying?*
    S1: — *Yesterday, and I stay in Grand Brand Hotel.*
    S0_2: — *Oh, that's wonderful! And what are you doing tonight?*

Whereas the answer to the former necessitates an argument:
    S0 — *Does the fight against terrorism authorize restrictions upon freedom of expression?*
    S1 — *Yes.*
    S0_2 — *Oh, that's wonderful. Now, let's turn to the next question.*

3. Argumentative situation

Representation — In a stabilized argumentative situation, proponents and opponents are also called upon to make positive arguments and to refute the antagonistic position. This situation can be roughly represented as follows:

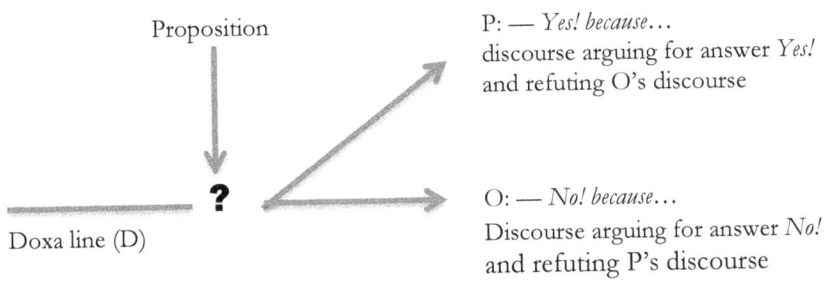

Argument is seen as a mode of constructing answers to a question for which incompatible answers have been given.

Coherence principle — Under the coherence assumption, all the semiotic acts produced in this situation are oriented towards the consolidation of the Answer-Conclusion.

The argumentative situation as an open situation — The argumentative question is essentially open; the legitimacy (interest, respectability…) of the pro and contra interventions is acknowledged, at least factually. Sometimes the participants agree on a mutually satisfactory answer (= conclusion), other times they don't. In many cases, an element of doubt remains attached to the surviving, ratified, answer, and the question may re-emerge. In other words, the answer is provisional; it cannot be completely separated from the question and the set of pro and counter-arguments that generated it. The answer is therefore an answer by *default@*; an unstable answer, which may be subject to revision.
Considering that the development of the exchange will generally alter the original positions as expressed in the opening sequence, and third parties will play

decisional roles, it follows that 1) the conclusion will not be identical with one of the positions as expressed at the opening sequence of the interaction; 2) a well-executed, successful argumentative exchange may conclude without a winner and a loser; and, 3) that the loser is not compelled to relinquish his or her doubts.

A Double constraint — Arguments are built under a double constraint; on the one hand, they are oriented by a question, and, on the other hand they are under the pressure of the counter-discourse. This situation is characterized by macro-discursive phenomena:
— *Bipolarization* of discourse. Followers are attracted by the question; they identify themselves with the speakers involved; they adjust their language to reflect the words and practices of the lead speakers; in contrast, they exclude speakers and supporters of the opposing discourse (*we* vs. *them*).
— *Crystallization* of discourse. Emergence of fixed lexical collocation, of antonymic pairs, tendency to stereotype the positions, especially stabilization of ready-made argument scripts@.
—Appearance of mechanisms of resistance to refutation. Presentation of arguments in the form of *self-argued@* claims, mimicking analyticity.

Question and relevance — The question sets the relevance principle for argumentative contributions: relevance of the arguments for the conclusion, relevance of the conclusions as answers to the question. The question may itself be challenged during the debate. It may be rejected on the basis of being flawed, poorly formulated, or irrelevant in consideration with "deeper" issues. **S. Relevance; Refutation.**

Burden@ of proof — The preceding graphic sought to represent the asymmetry between discourse and counter-discourse, established by the burden of proof resting on the proponent. This allocation may change with the participants and the kind of forum where the discussion takes place.

Changing mind and language — Not only at the end of the discussion, but also during the exchange, participants can be persuaded to change their mind, alter their opinion and language, shifting from one role to another.

4. Monologization of the "Question — AnswerS" game
The vision of argumentation as a discussion between incompatible points of views about the same object is operative in both monologue and dialogs.
Dialogs can be "homologized" in two different ways. In an argumentative intervention developing a series of co-oriented arguments towards a conclusion, the arguer voices just one position, and assumes a demonstrative "no alternative" rhetoric. The monologue is *monophonic*.
In another kind of monologue, the arguer adopts different positions, and put forward several hypotheses about the same argumentative issue, without advocating any of them in particular. The discourse stages several voices, especially

the main competing voice that of the opponent. Such a monologue is *polyphonic*, **S. Interaction, Dialogue, and Polyphony.**

Monophonic interventions ignore the speeches and positions of the opponents. This means that their practical study will necessitate the construction of a corpus bringing together the various interventions supporting the different answers. According to the "minimal structuralist postulate", the plea for **P** is best understood when referred to some contestation, or neglect of **P**.

Polyphonic interventions contain a representation of the speech of the other participants. They take over, under various polyphonic modalities, the set of situational discursive data, the question and the opponent's speech and position, which are re-framed under different discursive regimes, corresponding to different images allocated to the interlocutor and different self-allocated ethos. As a result, the assertion is introduced under an interrogative veil.

These strategies of monologization of the question have been clearly identified in ancient rhetoric, where they are considered to be figures of speech, interrogation (*interrogatio*), subjection (*subjection*) and dubitation (*dubitatio*) (Lausberg, [1960], § 766-779).

The question is framed as *having one self-evident answer* (Lat. *interrogatio*) — This is the case of the *interrogatio*, or "rhetorical question" defined classically as a question having an obvious answer.

> Now, can such a person make a better president than our candidate? Certainly not.

The speaker takes possession of the argumentative question and gives an answer presented as the only possible, self-evident answer. This operation "disambiguates" the question, by imposing one sole response. **S. Ambiguity.**
The speaker takes the position of "the one who knows" and embeds the answer in the question. Third parties are framed in the position of allies who also know and applaud; opponents are challenged by a form of reasoning through ignorance. The purpose of the *interrogatio* strategy is to suggest that *"there is no problem with this issue"*.

The question is framed as *having one justified answer* (Lat. *subjectio*, "put before, under the eyes"; here "submit to" the audience) — The question is presented as requiring clarification rather than argumentation, as *explanatory* rather than *argumentative*, **S. Explanation**. The speaker takes the place of the *investigator* or the teacher who asks the right question and resolves it objectively. The interlocutor is framed as the *pupil* or the judge, sharing the direct question and admitting the proposed answers according to the logic of pedagogical co-construction.

> Here is the situation, here is the question, and here are the data. One can think of three different answers, solutions, possibilities… Solution (a) is a variant of solution (b), as we will show. For such and such a good reason, solution (c) must be preferred to solution (b). So, the correct answer is (c).

Doctoral dissertations might approximate this strategy. During the defense, a member of the jury will possibly re-dialectize the monologue, expressing differently solution (a), and reversing the evaluation of (c) over (b).

The Question is framed as *an open question*, and the speech builds the answer in real time (*dubitatio*) — The speaker now takes the place of the third party, *the ignorant* party who has his or her doubts. In a kind of reversal of roles, the interlocutor is put in the high position of an *assistant* or *counselor*. The construction of the solution is now attributed to the interlocutor-*counselor*, not to the speaker-*investigator*.

In the three cases, the monologization of the argumentative situation plays heavily upon the preference for agreement. It does not leave the floor to other participants, and can channel their voices towards the speaker's conclusion.

R

Rationality ▶ Criticism

Reciprocity

In mathematics, the relation@ of *reciprocity* corresponds to the relation of *symmetry*: a relation is symmetric in the domain in which it is defined, if for all the pairs of elements **a** and **b** both **aRb** and **bRa** hold. The relation@ "being the friend of" is symmetrical:
> Peter is the friend of Paul = Paul is the friend of Peter = Peter and Paul are friends.

1. Returning and anticipating the good
In natural language, the reciprocal relation is defined on the basis of a set of actions which binds two persons. That is to say that if **a** does something positive to/for **b**, then **b** must reciprocate, doing the same thing to/for **a**. This is the principle of returning favors. The individuals **a** and **b** are equal in this relationship. The principle of reciprocity acts as a constraint:
> If you treat me to dinner, I must treat you to dinner.

As a form of natural morality, the imperative of reciprocity is expressed by the principle:
> Do to others as you would have them do to you. (Luke Gospel, 6:31)
> Do *not* do to others what you would *not* have them to do you.

This principle is applied in the argument:
> I'm polite to you, so be polite to me.

The speaker defines him/herself and defines his or her partner as members of the same category, who must be treated in the same way, S. **Rule of Justice**.

1. Returning and anticipating the wrong: reciprocity as retaliation
> Eye for eye, tooth for tooth
> If your disappointed lover disfigured you with vitriol, the court grants you the right to treat him likewise.

The law of retaliation is a primitive rule of justice that if **A** has wronged **B** in some way, then **B** can legitimately do the same wrong to **A**. In contemporary times, we might consider nuclear deterrence, based on the certainty of reciprocal destruction, as a concrete application of this principle. This theory corresponds to a particular case of the, *"You too!"* @ argument.

Reciprocity as a legal principle allows different states to assert their equal international dignity, and possibly to justify a retaliatory measure:
> If country **A** requires a visa for the nationals of country **B**, it is right that country **B** also requires that nationals of country **A** acquire a visa.

"Red Herring"

Herrings turn red when smoked; red herrings were used by fugitives to set dogs on a false trail. The expression is used figuratively in argumentation, where the so-called "red herring" strategy is referred to as a *diversion strategy*, where a *distractor* is used to eschew the issue under discussion, and deflect the discussion towards an irrelevant issue, S. **Relevance; Ignorance of refutation**.

Reflexivity ▶ Relations

Reformulation ▶ Vicious Circle

Refutation

All the components of written or spoken discourse in situation can be used or manipulated by the opponent in order to present this discourse as untenable, S. **Destruction**.

The word *refutation* is used to designate a *reactive* act covering the explicit forms of *discursive rejection* of positions, opinions, charges or projects. The possible use of *refutation* as a synonym of *rejection* or *denial* does not imply the absence of argument. As non-preferred second pairs, denials and rejections are also characterized by the presence of *accounts*. In fully argumentative interactions, refutation is in particular characterized by its explicitness and careful elaboration. From a scientific point of view, a proposition is refuted if it is proved to be *false*; the calculation from which it derives contains an error; it affirms something that is contradictory to the observed facts. From the point of view of ordinary interaction, an argumentative line is contextually refuted if, after being discussed, it is given up by the adversary, either explicitly or implicitly. Accordingly, the question itself disappears, and the interaction progresses to another structuring topic.

As a reactive speech act, refutation can be dealt with in only a verbal (face to face) or written (text to text) dialogue. Monological discourse knows only the concession@, there are no *refutative subordinate clauses*, and *concessive clauses* reduce the refutation to an objection.

1. Structural refutation
Each component of the propositional argumentative model may be targeted by the act of refutation, **S. Argumentation (III)); Layout.**

1.1 Turning down the argument
An argument supporting a conclusion may be rejected in different ways.
(i) The argument is declared false:
>S1 — *Peter will surely arrive on Tuesday; he has been invited to Paul's birthday.*
>S2 — *But Paul's birthday is on Monday.*

(ii) The argument is rejected as irrelevant to the conclusion, **S. Relevance:**
>S1 — *He is very intelligent, he read all of Proust's work within three days.*
>S2 — *Intelligence has nothing to do with reading speed.*

(iii) The argument can be accepted as such, recognized as somehow relevant to the conclusion but may be dismissed as too weak, or of poor quality:
>S1 — *The President has spoken, the stock market will go up.*
>S2 — *Yes, and what he says goes!* (said sarcastically).

The rejection of the argument may lead to a new argumentative question (sub-debate), about the truth, strength or relevance of the former argument.

Turning down the argument does not mean renouncing the conclusion. This is often the case in factual argumentation:
>S1 — *Peter will surely arrive on Tuesday, he wants to be there for Paul's birthday.*
>S2 — *Paul's birthday is on Monday, but sure, Peter will arrive on Tuesday, I bought him his flight tickets.*

Nonetheless, in ideological debates, only the most ascetic arguers will refute questionable or bad arguments made in favor of conclusions which they consider to be good or virtuous.

1.2 Turning down the backing
The backing invoked, implicitly or not, is declared false:

S1 — *Pedro was born in the Malvinas Islands, so he is an Argentine citizen*
S2 — *The Falkland Islands are British territory.*

The adverb *precisely (not)* substitutes one backing to another (Ducrot & *al.*, 1982), **S. Orientation**:

S1 — *Noodles for dinner!*
S2 — *Again! We had noodles for lunch!*
S1 — *Exactly, we need to finish the leftovers, we don't want to waste food.*

1.3 Turning down the conclusion

Conclusions may be dismissed even though some validity is granted to the argument:

S1 — *Cannabis should be legalized; the taxes will pay off the National Health Service deficit*
S2 — *It will certainly increase tax revenues, but it will further increase the number of drug addicts. The prohibition must be maintained.*

The counter-argumentation establishes a counter-conclusion leaving the argument it opposes intact, **S. Counter-argumentation**.

2. Weak refutation protecting the affirmation

By generalizing of the law of weakness, a weak refutation confirms the attacked position, **S. Scale**. This principle applies to various interpretative schemes, whose analysis must take into account the whole corpus produced by the argumentative question.

(i) Weak refutation of a poorly re-constructed attacked position. The wise man concludes that the refutation is not worth much, to say nothing of its author, and the problem remains intact.

(ii) Weak refutation of an outstanding exposition of the attacked position. The conclusion is that the attacked position is reinforced by this attempt at refutation. The interpretive calculation is based on the fact that the arguer is qualified.

— The poor refutation is standard, while the quality of the exposition, clearly indicates a good arguer. Since the given refutation is taken to be the best possible (according to Grice's maxims), and since it is weak or even ridiculous, the conclusion will be that, "*since even such an arguer finds nothing else to say, then, the criticized position must actually be correct*", even if this derivation is *ad ignorantiam*, **S. Counter-argumentation**.

— The poor refutation is bizarre. It contains obvious errors that warn the careful reader; there is a contrast between the quality and care of the exposition and the scanty character of the refutation. Moreover, this refutation is not put forward in the usual argumentative style of the author. For example, a fine theologian develops in a dialectical and detailed manner, a position condemned by the official authorities of his religion, and refutes it only by arguments drawn from various authorities (which the reader may be aware are considered questionable), so the careful reader is led to think that this oddity is strategic. The speech is apparently refuted, only to be better asserted in reality, the negation serving then to cover the author. This case of indirectness has been theorized by Strauss (1953). If, under special historical, social, or religious circumstances, a discourse is banned, it is nevertheless possible to give it a voice under the cover of its refutation, the negation then serving to protect the speaker from tyrannical authorities.

This strategy of confirmation, or *argumentation by weak refutation*, is dangerous to maintain. The authorities are not necessarily naive nor uninformed, and they may be well aware of the intended purpose of the pseudo refutation, which will be rightly interpreted as a denial of a belief which is actually held by the speaker: "*How can you so be such an expert about heterodox positions and such a fool when dealing with orthodoxy?*".

Such a strategy, based on the opacity of the writer's intentions, presupposes a double argumentative address, the real intentions can be captured only by a careful reader, while they remain unknown to the hasty reader, who appreciates the weak refutation because it can be easily understood, absorbed and repeated, **S. Strategy**

3. Refutation and counter-discourse

The concept of refutation is defined at the very general level of the current patterns of argumentation. The counter-discourse approach specifies the possible refutative strategies according to the specific structure of the argument. The argument *type* is flanked by a *counter-type*, an integral part of the form and substance of the argument considered; any of its defining components might be attacked. Each counter-argument outlines the specific discourse that can be opposed to an argument invoking a testimony@, an authority@, a definition@, an induction@, a causal@ claim, etc.

In the Skeptical philosophical style, these counter-discourses can also be directed *at the argumentative type itself*, as general discourse, "against authority, analogy, causality, etc.", which rejects a priori all forms of argument from authority, etc.

Related Words

Lat. *A conjugata* argument, Lat. *conjugatus*, "related, of the same family"

Three types of argument are based on the fact that two words are "related", depending on the nature of their relationship:
1. An etymological link: **S. Etymology.**
2. A morphological link: **S. Derived Words.**
3. A phonic or graphic resemblance: **S. Ambiguity.**

Relations

A relation is a two-place predicate **R** associating two objects, **a** and **b**, denoted by "aRb". Relations are characterized by three general properties, *symmetry*, *transitivity*, and *reflexivity*.

— *Symmetry, or Reciprocity*: The same relationship holds between "**a** and **b**" and "**b** and **a**".
— *Reflexivity*: The relationship connects an object to itself.
— *Transitivity*: The relationship connecting **a** to **b** and **b** to **c** also connects **a** to **c**.

1. Symmetry, or reciprocity@

A relation is *symmetric* if it relates both **a** to **b** and **b** to **a**. In other words, both "**aRb**" and "**bRa**" hold. If **a** loves **b**, **b** does not necessarily loves **a**: a love relationship is not symmetrical. "Meeting" is a symmetric relationship. The following argument is neither more nor less logical than any other, but it would make a valid point in any detective novel; it can only be rejected by accusing Peter of lying:

> If Peter confessed to having met Paul at the bar, we must assume that Paul met Peter. Paul cannot deny the obvious.

2. Reflexivity

A *reflexive* relation relates a being to itself, noted "**aRa**". "— *being contemporary of* —" is a reflexive relationship: **a** is its own strict contemporary. For the average person, the causal relationship is not reflexive; only God is *causa sui*, his own cause.

The reflexive relation can be used *ad hominem*. The principle "*charity begins at home*" for example forces the reflexivity of the relationship "**a** makes charity to **b**"; all the same, the love of others can be used to encourage self-care:

> You who love the whole of humanity, you should try to love yourself as well!

The competence of an adviser can be challenged by inciting him to make a reflexive application of his talents:

> Physician, heal thyself!

Such replies correspond to the *ad hominem* variety setting up practices against words, S. *Ad hominem*.

3. Transitivity

A relation is transitive if, when it relates **a** to **b** and **b** to **c**, it also connects **a** to **c**; in other words, "**aRb** and **bRc**" imply that "**aRc**".

If **a** loves **b**, and if **b** loves **c**, then **a** does not necessarily love **c**; a relationship of love is thus not transitive. The relation "— *be the father of* —" is not transitive, but "— *being an ancestor of* —" is transitive. If **a** is an ancestor of **b** and if **b** is an ancestor of **c**, then **a** is an ancestor of **c**.

Inferences based on the transitivity of a predicate apply whenever at least three objects are positioned on a graduated scale:
> If **a** is bigger, older, richer ... than **b**
> and **b** larger, older, richer ... than **c**,
> Then a is bigger, older, richer ... than **c**.

Inferences based on these properties are part of the unnoticed evidences exploited by everyday reasoning and argument. They are sometimes considered to be "quasi-logical", S. **"Quasi-logic"**; but being *valid* does not preclude being an *argument*.

4. Conversion@

Relevance

1. Ignorance of refutation, a fallacy of method

Lat. *ignoratio elenchi*. The Greek word [*elenkhos*] means: "1. Argument to refute ... 2. Proof in general" (Bailly, [*elenkhos*])". The Latin title of Aristotle's *Sophistical Refutations* is *De Sophisticis elenchi* (Hamblin 1970, 305).

The fallacy of "ignorance of refutation" (*ignoratio elenchi*) is defined in the context of the dialectical game, where a participant, the *Respondent* (or Proponent), is committed to a statement, and the partner, the *Questioner* (or Opponent), tries to lead the Respondent to a contradiction, to thus refute the statement he or she (the Respondent) had previously accepted. The dialectical game considers only contradictory propositions (one and only one of them is true). The opponent must conform to the rules of the method in order to truly refute (and not in appearance) the primitive affirmation, S. **Dialectic**. The fallacy of ignorance of refutation is independent of language, occurring "because the terms 'proof' or 'refutation' have not been defined, and because something is left out in their definition". (Aristotle, *R. S.*, 167a20, §5), S. **Fallacy: Aristotle**. In other word, the *misconception of refutation* is a general term covering all *methodological errors* occurring in a dialectical game.

This concept may be extended to any *argumentative language game*: "the arguer argues and does not know how to argue; thinks something is being proven or successfully refuted, when this is not the case; his or her practical concept of argument is flawed, etc." This basically occurs when the argument does not respect the principles of relevance: on the one hand, *the argument must be relevant to the conclusion* (internal relevance) and, on the other hand, *the conclusion must be relevant as a reply to the question* (external relevance), S. **Question**.

2. Relevance of the argument for the conclusion

In the context of a dialectic game, the Respondent asserts **P**. Starting from **P**, the Questioner deductively constructs a chain of propositions ending with proposition **not-P**. So the Questioner claims that this chain proves proposition **not-P**. Apparently, the Respondent has been refuted and the Questioner has won the game. But the Respondent claims that the chain of proofs backing **not-P** is not valid because the arguments put forward do not support this conclusion; so, the Respondent claims that the Questioner actually failed to demonstrate **not-P**.

This schematizes the general situation when an arguer claims to have refuted the opponent *ex datis*, that is using only beliefs and modes of inference supposedly admitted by the opponent. In the same way, in an *ex datis*@ or *ad hominem*@ procedure, the opponent can resist the refutation by breaking the inference chain leading to the conclusive step he or she is supposedly compelled to concede. In other words, he or she argues that *the arguments are not relevant to the conclusion*. This issue actually involves all the program of criticism of argumentation.

3. Relevance of the conclusion as a reply to the Question

In the general case, the proponent commits himself to **P**, the opponent constructs from **P** a chain of propositions at the end of which the proposition **Q** is reached. The proponent therefore claims that "**Q = Not-P**". The proponent argues that proposition **Q** is not contradictory to **P**, and that, accordingly, it has not been rebutted. The arguments may be relevant to the conclusion, but the conclusion does not disprove the original thesis.

To argue that an intervention is externally irrelevant is to argue that it misses the point, is off-topic etc. It may also be rejected on the grounds that it is an effort to put the adversary on a false trail, **S. Red Herring**; the accusation of paralogism is reinforced by a suspicion of sophistry.

Criticisms of internal relevance and external relevance are cumulative. They invalidate a speech by saying that it does not back the conclusion, and that, regardless of this, the conclusion has nothing to do with the issue.

4. The question is not relevant to the "real debate"

The dialectical framework is binary, the proposition to discuss is expressed in a simple and explicit proposition, and the methodology of a refutative discussion is well defined. Since the question is "**P or not-P**?", claiming that the opponent's conclusion does not logically contradict **P**, is to claim that it is not relevant to the debate.

The situation can be equally clear in an ordinary discussion. A student disputes, that is, wants to "refute" the grade he has received: *"if you don't up my grade, I'll fail the exam; please, I badly need just three extra points!"*. The argumentation by the consequences is quite valid, but the negative consequences of the bad grade are irrelevant to the determination of the grade (according to the classical scientific and educational regimes at least). The student's conclusion is irrelevant, failing to acknowledge the real issue: *"what mark does my assignment deserve in itself?"*. The student's question is different from the teacher's question, and the teacher is master of the question.

Things may be more complicated. When the proponent refutes the rebuttal by saying, "*what you disagree with has nothing to do with what I am saying*", what he actually says can be difficult to pin down, and may be constantly reformulated and reinterpreted S. **Straw man**. On the other hand, even when the original claim and its intended rebuttal have been previously set down in writing, the link between the two does not necessarily have the clarity of the binary contradiction. For example, does **S2** refute **S1,** or merely show that the issue is complex:

> S1: — *Speculators buy raw material in advance just to speculate on future price variations. Such operations on raw material should be banned by law.*
> S2: — *Nonetheless, it is essential for companies to purchase in advance the raw materials they need, to cover themselves against price fluctuations.*

Finally, in ordinary argumentation, the *issue* itself may be controversial. When none of the participants is the (natural or conventional) master of the question, each key participant will be tempted to give a definition of the question, and, will, accordingly reject the opponent's answer as irrelevant to the real issue:

> S1: — *That's not the question!*
> S2: — *This is my answer to the problems that really arise. You're not asking the right question.*

The accusation of *fallacy of conclusion irrelevant to the question* under debate can be answered by a counter accusation of *a fallacious, wrongly framed question, irrelevant* to the "real" debate.

The function of the participating third party, be it the judge, the (universal) audience or the informed participants, is to construct, manage and decide upon the question, and accordingly, to determine what is or is not relevant in the debate.

Repetition

> The *proof by repetition* is sometimes metonymically designated under the Latin name of its effect, arg. *ad nauseam*, Lat. *nausea* "nausea, disgust".

Any meaningful or pragmatically relevant segment can be repeated for a variety of purposes: something may be repeated because it has not been clearly heard or understood; the second speaker may repeat the end of the first turn to link it with his or her second turn, etc. Repetition may consist in repeating an initial phrase or speech act word for word, as is the case in formal quotations. Alternatively, repetition might slightly reformulate something which can be heard everywhere, for example a well-known argumentation borrowed from a script@. Conscious, strategic repetition of slightly modulated core contents is the key of traditional methods in education; repetition of the same action is the basis of learning-by-doing.

While most repetitions are unplanned and remain unnoticed, the *argument by repetition* or *proof by repeated assertion* is part of a strategy to impose on people a unilateral, uncritical vision of things. The focus is put upon a single key statement, presented not as a claim but as an obvious necessary truth, repetition creating a feeling of *self-evidence@*.

Although called "argument", this process is characterized by the absence of argument. It offers no reason, good or bad, to support the claim; reasons are not implied or contextually retrievable but are carefully ignored. Such strategic repetition can therefore be considered to be argumentative only if an argument is defined by its effect, *persuasion*. Repetition is instrumental to persuasion, which could itself be seen as a disposition, or a readiness to repeat under appropriate circumstances. Note that *repeating a whole complex argument* results in an argument by repetition rather than any other kind of argument: *"we will win because we are the strongest"*.

The sociologist Gustave Le Bon emphasized the power of repetition to gain people's assent:

> Affirmation pure and simple, kept free of all reasoning and all proof, is one of the surest means of making an idea enter the mind of crowds [...]
> Affirmation, however, has no real influence unless it be constantly repeated, and so far as possible in the same terms. It was Napoleon, I believe, who said that there is only one figure in rhetoric of serious importance, namely, repetition. The thing affirmed comes by repetition to fix itself in the mind in such a way that it is accepted in the end as a demonstrated truth. [...]
> To this circumstance is due the astonishing power of advertisements. When we have read a hundred, a thousand times that X's chocolate is the best, we imagine we have heard it said in many quarters, and we end by acquiring the certitude that such is the fact. (Le Bon [1895], p. 126-127)

This last remark shows that repetition produces an illusion of legitimation by the authority of great number, **S. Consensus**.

From the point of view of the evaluation of arguments, this form of repetition is regarded as a fallacy, and even as the fallacy par excellence, since it imposes the acceptance of a statement not only *without* justification but *against* all justification.

Respect

> Argument *ad reverentiam*, Lat. *reverentia* "respectful fear".

Respect is a feeling projected by authorities, whatever or whoever they may be. If organizations and individuals are legally invested with due authority in order to carry out a mission, then, in this role, they claim respect, whatever one's private opinion may be about their relevance or efficiency.

The claim to respect is in principle distinct from the claim to obedience; one can be constrained to obey by the use of lawful violence; showing respect is essentially a supplement to compliance. This means that interactions with common authorities are ruled by specific conventions of politeness, such as the concluding formula *"Yours respectfully"*, used to convey this due conventional respect to the authority addressed in a formal letter.

As an inner sentiment, respect has to be earned. Nonetheless, a behavior, intentional or not, can be felt as disrespectful, and, if a public servant or a police officer is involved, it might be qualified as an insult and punished as such. The *argument from respect* is basically used to justify a sanction for a *lack of* respect. S. **Authority; Humility.**

Any person who is in a position of authority and feeling that his prerogatives are not respected might invoke the argument from respect. The problem arises when this claim to authority is not recognized, or is considered to be oppressive, as may be the case of religious authorities. At a more abstract level, the right to respect is claimed for all beliefs in general, and for one's own beliefs in particular. Disrespect is qualified as a provocation, a scandal, a blasphemy that gravely hurts the believer's feelings, and a complaint can be filed in court to uphold the right to respect.

An argumentative situation involving an argument from respect developed around a photographic work by the American artist Andres Serrano, entitled *Immersion Piss Christ*, features a crucifix dipped in the artist's urine. It was vandalized on Sunday, April 17, 2011, at the Yvon Lambert contemporary art collection in Avignon, France.

The Archbishop of Avignon issued a statement protesting the exhibition of this work, and so justifying the destruction. *The argument of (lack of) respect* is invoked in the following passage:

> Are not the local authorities, among other things, under the obligation to ensure respect for the faith of believers of every religion? Yet such a work remains a desecration which, on the eve of Good Friday, when we remember Christ who gave his life for us while dying on the Cross, touches us deeply in our hearts.

The argument is then repeated and amplified (our emphasis):

— *The* odious profanation *of a Christ on the cross* (Title)
— Can art be of such bad taste for no other reason than to serve as *an insult*?
— I have to react to this *odious* picture which *flouts* the image of Christ on the cross, the heart of our Christian faith. Any attack on our faith hurts us, any believer is affected deep within his faith.
— Given the gravity of such *an affront*
— For me, as a Bishop, as for every Christian and every believer, this is *a provocation, a profanation* that hurts us at the very heart of our faith!
— Did the Lambert collection not perceive that these pictures *seriously wounded* all those for whom the Cross of Christ is the heart of their faith? Or did

> they want *to provoke* believers by *flouting* what for them is at the heart of their lives.
> — *A serious desecration, a scandal* affecting the faith of these believers.
> — [These pictures] *seriously harm* the faith of Christians.
> — A behavior that *hurts us* at the heart of our faith.
>
> Infocatho, [*Odious Profanation of a Christ on the Cross*], 2011[1]

In some countries, *blasphemy laws* punish what they qualify as contempt and disrespect towards the State's religion; blasphemy is punished as any other crime. Campaigns against blasphemy laws develop a counter-discourse positing that such laws are mediaeval and obscurantist; that they are incompatible with the basic democratic principle of freedom of expression; and that they make all philosophical and historical inquiry about religious belief impossible.

Some other countries have laws prohibiting hate speech or discriminatory speech, especially intended as guarantees of the equality of rights for minority communities, religious or others.

The argument of (a lack of) respect was at the heart of the case concerning the cartoons depicting the prophet Muhammad published in 2005 in a Danish satirical weekly journal. This case culminated in the 2015 terrorist attack on the French satirical journal *Charlie Hebdo*, resulting in the shooting of 11 journalists and collaborators by two Islamist terrorists.

Resumption ▶ Straw Man

Retaliation ▶ Autophagy

Rhetorical Argumentation

Ancient argumentative rhetoric is grounded in the natural speaking competence. This natural skill is developed through conceptualization and practical exercises concerning general or social issues. Such a rhetoric combines linguistic, interactional and citizenship's competencies.

1. The rhetorical address

The *rhetorical address* corresponds to *discourse* in its traditional sense, that is to say, "that which, in public, treats a subject with a certain method, and a certain length" (Littré, [*Discourse*]); a discourse is a "formal and orderly and usually extended expression of thought on a subject." (W., *Discourse*). This concept of

---

[1] "Odieuse Profanation d'un Christ en Croix", *Infocatho*. http://infocatho.cef.fr/fichiers_html/archives/deuxmil11sem/semaine15/210nx151europeb.html 09-20-2013

discourse has nothing to do with the concept of discourse as defined by Foucault (1969, 1971) or Pêcheux (Maldidier, 1990). Moreover, this meaning of *discourse* does not appear among the six meanings considered by Maingueneau in his founding presentation of "French discourse analysis" (1976, p. 11-12).

A rhetorical address is a speech delivered by a speaker or *orator* to an *audience*; it has the following main characteristics.

— The orator deals with *an issue of general interest*, typically he or she aims to *influence a decision-making process* developing under certain time constraints. Classical rhetoric considers an orator, addressing an audience. In reality, a full rhetorical situation is a situation of choice, involving as many orators or voices as there are possible choices.

— The speech is a relatively *long monologue, planned*, composed of a set of speech acts *constructing a unified representation* leading to action.

— It is produced in the context of *discursive competition* taking place between different speeches of mutual opponents, supporting incompatible proposals. The rhetorical address is given in a space of contradictory discourse, where all interventions are *positioned in view of one another*. Even if the speaker tries to erase all traces of the counter-discourses that surround him or her, the speech is nevertheless structured by the competing discourses.

— This speech is delivered to an *audience*, composed of all the people who will play a role in the *decision-making process* relevant to the matter in hand. The audience is *divided* in regard to what the right decision would be; it includes staunch *supporters* and *opponents* of each proposal, as well as *undecided* people, S. Roles. The focus traditionally put on persuasion suggests that the orator focuses upon those who doubt and question, more than upon the determined opponents. The job is to remove doubt, to create and lead the opinion, S. Logos, Ethos, Pathos.

— The rhetorical audience is both lowered and magnified. It is *lowered*, because it is defined by its lack of knowledge, its indecision and dissension. But within the New Rhetoric framework at least, the audience is also *magnified as a critical instance*, somewhere on the way to achieving a universal, deeply rooted and justified consensus.

Argumentative rhetoric has theorized, codified, evaluated and stimulated this kind of public communication, which was the only kind of public address possible before the appearance of the radio, cinema, television and the internet. Its theoretical object, the circulation of contradictory speeches within a decisional group, remains well defined. S. Argumentation (III); Persuasion.

2. The rhetorical catechism

At least until the modern age, rhetorical argumentation was the backbone of teaching and education in the Western world. In the Middle Ages, rhetorical argumentation served as one of the three *arts of speech* constituting the *trivium* (grammar, logic, rhetoric), propaedeutic to the *quadrivium* (geometry, arithmetic, astronomy, music).

For pedagogical purposes, rhetoric has constructed a standard self-representation of both the *production process* of the address, and its *product*, the address as delivered to the audience:

— *A five step production process*, invention, disposition, speech, memory, pronunciation.

— *Three genres of discourse*, deliberative, judicial, epideictic.

— *Three actors*: the rhetorical interaction is functionally *tripole* it brings together "the speaker who wants to persuade, the interlocutor who must be persuaded, and the opponent whom he must refute" (Fumaroli 1980, p. 3).

— *Three discursive means of pressure* focused on the transformation of the audience representations and desire for action. The speaker seeks:
- *To inform and teach*, by his or her *logos*@, that is the logic of the narrative and the argument.
- *To please* and *attract* by his or her style, that is the self-image, or *ethos*@, projected in the speech.
- *To move* to action, through the *pathos*@.

— According to the tradition, the acts aimed at producing these effects are concentrated in the strategic moments of the discourse:

The *introduction* is the *ethotic* moment.

The *narration* and the *argumentation* are ruled by the logos.

The *conclusion* is the *pathemic*, emotional moment, through which the speaker hopes to wrest the final decision.

3. Organizing the process

The process of constructing argumentative rhetorical discourse is traditionally described as involving five stages. The corresponding Latin words are mentioned in order to avoid possible confusion with the English words, of which they are false cognates.

(i) *Inventio*: Finding the arguments

"Invention [*inuentio*] is the devising of matter, true or plausible, that would make the case convincing" (*Ad Her.*, i, 3). The *inventio* is the cognitive step corresponding to the methodical search for arguments, guided by the technique of "topical questions", S. **Common Places**.

The Latin word *inventio* does not mean "invention" taken as a creation of something that did not exist before. The meaning is "to find, to discover" (Gaffiot [1934], *Inventio*).

Psycho-linguistic research on the production of written and oral discourse has taken over the reflection on the *inventio* techniques.

Rhetorical arguments are found on the basis of an exploration of reality, guided by a natural, substantial ontology. Religious argument has introduced a fundamental change in this vision. Good reasons are not statements expressing sense data or elaborated intellectual conceptions, but are sacred statements drawn from the foundational sacred text and, to a lesser extent, from the texts of tradition, **S. Person; Topos; Collections; Script.**

(ii) *Dispositio*: Planning the argumentation
"Arrangement [*dispositio*] is the ordering and distribution of the matter" (*ibid.*), that is, speech planning. *Inventio* and *dispositio* are the two cognitive stages of this process.

(iii) *Elocutio*: Expressing the argumentation
"Style [*elocutio*] is the adaptation of suitable words and sentences to the matter devised" (*ibid.*). The word *style* used in translation may evoke a superficial arrangement of the expression, but the *elocutio* is more than that, it corresponds to the "putting into language" of the arguments, to their semantization, corresponding to the whole linguistic expression.

The *elocutio* is characterized by four qualities, the grammatical *correctness* (*latinitas*), the *clarity* of the message (*perspicuitas*), the customization of the message to suit the audience (*aptum*) and the density and wealth of its expression (*ornatus*). A discourse may be rejected as defective on any of these levels, **S. Destruction.**

The English word *elocution* currently refers to "the skill of clear and expressive speech, especially of distinct pronunciation and articulation" (W., *Elocution*); elocution clearly belongs first to *pronuntiatio*, and only peripherally to *elocutio*, as expression and style.

(iv) *Memoria*: Memorizing the speech
The discourse must be memorized since it is intended to be delivered orally, without the use of paper documents or autocues. As the invention, the memory involves cognitive factors. The cultural import of this memorization work, which might seem anecdotal, was revealed by Yates (1966).

(v) *Pronuntiatio*: Delivering the speech
"Delivery [*pronuntiatio*] is the graceful regulation of voice, countenance, and gesture" (*ibid.*). The Latin word *pronuntiatio* refers not only to this physical process of speech production and modulation, but also expresses the idea of an *assertive speech*: a *pronuntiatio* is a "declaration, announcement, proposal" (Gaffiot [1934], *Pronuntiativus*). The judge does not *say* or *read* the verdict, he or she *pronounces* it. Rhetorical tradition sees delivery as the moment of performance, and dramatization of discourse, requiring a special education of the body, the gesture and the voice. The orator, the preacher, the actor are under the same public performance constraints although their techniques, social statuses and messages are quite different.

In short, finding arguments, ordering them, expressing them in writing: the rhetorical prescriptions are particularly suited to general academic essays. They seem clear and they are easy enough to teach — but unfortunately not so easy to put into practice.

In the *Divisions of Oratory Art*, Cicero has framed the concepts of ancient rhetoric as a succession of question-answers, "very similar to a catechism", as Bornecque notes ([1924], p. VII). Rhetoric may have suffered from such allegedly pedagogical representation where everything has to be done and said by the book.

4. Structural organization of the speech
This process leads to a finished product, the speech delivered in a specific situation. It is articulated in parts, traditionally named:
> *Introduction* (exordium) — *Narration* — *Argumentation* (a *confirmation* followed by a *refutation*) — *Conclusion*.

The argument is the central part of the speech. Contrary to a simplistic vision of discourse, there is *no opposition* between *argumentation, narration* and *description*. Argumentative narrations or descriptions, like literary narrations or descriptions, are made from a particular *viewpoint*.

5. Extensions and restrictions of rhetoric
Ancient argumentative rhetoric has been redefined on various dimensions.
— Restriction to its *expressive* dimension. Argumentative rhetoric can be oriented towards *persuasive communication* or the *quality of expression*, **S. Persuasion**.

— Generalization to its *persuasive* dimension. Nietzsche assimilates the rhetorical function to the persuasive function of language, **S. Persuasion**.

— Restriction to its *linguistic* dimension and liquidation of its *cognitive* dimensions. The apparent logic of the five components of rhetorical production was profoundly challenged in the Renaissance (Ong 1958). The three components related to *thought* (invention, disposition, memory) were separated from those related to *language* (expression and delivery). *Inventio*, the flesh and bones of argumentation, rejected out of rhetoric and language was no longer considered to be the fundamental moment of the discursive process. An orphan of the *inventio*, rhetoric redefined its object, moving away from social discourses to focus on literature and belles-lettres, and developing a passion for the autonomous study of the discourse variations and figures of style.
A language deprived of thought and a thought deprived of language: this orphaned rhetoric would become the target of violent attacks from Locke, **S. Ornamental**. In France, in the nineteenth century, Fontanier ([1827], [1831]) was the emblematic figure associated with this "restricted rhetoric" (Genette, 1970), in opposition to the so-called "general" rhetoric, which was revived by Perelman & Olbrechts-Tyteca (1958) — The question of a *revival* of an integral concept of rhetoric remaining a *topos* of rhetorical studies.

— Generalization along its *linguistic* dimension. A rhetoric *restricted* to figures of speech can itself be called *"general"*: this paradoxical expression corresponds to the Group μ approach in their *General Rhetoric* (1970). The problems of figures are taken up in a structuralist framework, and figures are reconsidered under the two basic dimensions of language, the syntagmatic and paradigmatic axes. Issues of argumentation, public speaking, interaction or communication, are not accounted for, nor indeed are the aesthetics of figures. This *General Rhetoric* was virtually the only concept of rhetoric to be considered in the French literature during the 1970s, and Perelman's New Rhetoric occupied only a marginal position. Wenzel devoted an avenging paragraph to this "alarming" view of rhetoric (1987, p. 103; see Klinkenberg, 1990, 2001). The contrast with the status of rhetoric in the United States' Speech and Communication Departments could not be greater.

— Extension to *ordinary speech*. The rhetorical approach can be extended to everyday forms of talk, insofar as they involve face management (*ethos*), data processing oriented towards a practical end (*logos*), and a correlative treatment of the affects (*pathos*) (Kallmeyer, 1996). The rhetorical trilogy can thus be regarded as the ancestor of the different theories of the functions of language (Bühler 1933, Jakobson [1960]), in a completely distinct theoretical atmosphere. This extension also retains a fundamental characteristic of rhetorical speech: to alter reality and participate in the structuration of ongoing actions. This view may resonate with Bitzer's evocation of the dialogue between fishermen at work in the Trobriand Islands, and his definition of the "rhetorical situation" as involving a degree of "urgency":

> Rhetorical situations may be defined as complexes of persons, events, objects and relations presenting an actual or potential exigence, which can be partially or completely removed if discourse, introduced into the situation can so constrain human decision or action as to bring about the significant modification of the exigence. (Bitzer [1968], p. 5)

— Extension to any *semiotic domain*. Rhetoric naturally extends to the co-verbal and paraverbal signifiers. Moreover, any strategic implementation of a semiotic system can be legitimately regarded as a rhetorical practice; the rhetoric of painting, of music, of architecture, for example.

Rich and Poor

The arguments from wealth and from poverty are two subspecies of the argument from authority. Special weight can be given to the word of the *wealthy* — because wealthy, as well as to that of the *poor* — because poor. The Rich and the Poor are then believed on their word, and their words are exploited as an argument from authority by a speaker, who validates a position by putting it in the mouth of a rich or a poor man, S. **Authority; Person**. Both arguments are extremely common and equally formidable.

Argument of wealth, or "top people" argument — The argument from wealth is the substrate of a family of discourses elaborating upon the key topic:
> She is rich, therefore what she says is true; I consider her advice to be authoritative; she made the best financial decision; she has an extraordinary artistic taste, as proved by the value of her collections — I vote for her!

This argument easily extends from the rich to the upper classes and the ruling class, the most *glamorous and lucrative* professions, etc. It could be called "the top people" argument.

Argument of poverty, the appeal to "people down there" — The argument of poverty is symmetrical to the argument of wealth. It validates a speech by the authority derived from poverty, "the poor are right":
> The poor are good, because they who have no money, and who has no money has no vice; they are not corrupt; what they say is authentic; they are the repositories of common sense; their opinions are basically sound.

As the argument of wealth, the argument of poverty extends beyond the poor to all "the people down there", that is the *exploited proletariat*, the *dominated*, the *lower 10%*... as well as to *country people*, who live *close to nature* (naturalistic argument), or to *the tramp as a wise philosopher*... Truth comes out from their mouths, as it comes from the children's mouth.

The adage *vox populi vox dei*, "the voice of the people is the voice of God", which underlies the *ad populum* argument is grounded in the argument of *poverty* and in that of *number*.

These arguments are different from the appeal to money, or the *wallet argument*, attached to the argumentation by punishment and reward, S. **Punishments and Rewards**.

Right Balance Argument ▶ Moderation and Radicalism

Roles: Proponent Opponent, Third Party

In an argumentative exchange, the participants are part of a complex system of roles and characters, according to which they speak and act. Some of these roles are general; others are specific to the argumentative situation.

1. Roles not specifically related to argumentation

1.1 Roles attached to the "participation framework" (Goffman)

The concept of a *participation framework* details and clarifies the traditional concept of a verbal exchange between a *speaker* and one or many *listeners*. The participation format is defined as a relation with two complex speech structures, the *production format* and the *reception format* (Goffman, 1981). These concepts are instrumental to the analysis of all argumentative interactions, from rhetorical addresses to everyday argumentative interactions. They are relevant to the analysis of the ethos, and the polyphonic structure of the argumentative text.

- ***The reception format*** *(id., pp. 141-142)*

The people who can actually hear the words said by a speaker occupy various statuses in relation to these words.

— The *addressed participants* are the people to whom the words are openly directed; the pronoun *you* refers to the addressed participant(s). Everyday group conversations show that successfully addressing a specific person may demand complex maneuvers.

— The *ratified participants* are the members of the group constituted around the ongoing speech event. They may be addressed or non-addressed. To get the floor in a discussion, one must be a ratified participant. *Ratified, non-addressed* participants may be addressed in the further development of the speech event.

In *a codified dialectical exchange*, the opponent is the only participant both ratified and addressed. Both participants alternately hold the floor. The referee of the debate, if there is one, is a ratified participant, who will be addressed only as a resource if a crisis looms, or during a planned slot in order to move forward, to evaluate and conclude the debate. If the debate is open to a broader audience, the members of the audience are *ratified* participants but not *addressed* participants.

In *a classical rhetorical address*, the audience is *ratified* and *addressed*. The difference with the dialectical situation is that the audience has no official right to the floor; nonetheless, it may applaud and boo its reactions (Goffman 1981).

— *Overhearers* and *eavesdroppers*. Any people passing within earshot are *non-ratified participants*. *Overhearers* accidentally hear the sounds and words of the conversation, possibly without even listening. *Eavesdroppers* intentionally spy on the conversation.

- ***The Production Format***

Speech is produced by the speaker. Goffman (1981) and Ducrot (1980) have independently shown that the speaker should not be considered to be a unified entity but as a complex articulation of different discursive personae; in Goffman's words the *Animator*, the *Author*, the *Figure* and the *Principal* (*id.*, p. 144; p. 167).

— The *Animator* (Goffman) is the *talking machine*, physically producing the discourse. In the same function, the *Speaker* is re-defined by Ducrot as "the empirical being" to which all the external determinations of speech can be attributed, "the psychological or even the physiological process which is at the origin of the utterance, the actual intentions, cognitive processes that have made the statement possible" (Ducrot 1980, p. 34).

The counterpart in the reception format of this talking machine is the *hearing machine*, that is the *listeners*, the whole range of *ratified* and *non ratified* participants, as persons who physically *hear* the speech and listen to it or not (Ducrot 1980, p. 35).

— The *Author* selects the thoughts expressed and the words to encode them. A speaker reading a book or quoting another person is the *Animator* of the words taken up without being their *Author* (Schiffrin 1990, p. 242). The pronoun *I* refers to the *Author* of the speech (except in quoted speech).

— The *Figure* corresponds to the discursive self-image of the Author, that is, the ethos, **S. Ethos**.

— The *Principal* is "(in the legalistic sense) someone whose position is established by the words that are spoken, someone whose beliefs have been told, someone who is committed to what the words say ... a person acting under a certain identity, in a certain social role" (Goffman 1987, p. 144). "The same individual can very quickly alter the social role in which he is active even though his capacity as animator and author remains constant" (*id.*, p. 145). The same Author can address a student *as a teacher, as an adult, as a citizen, as a New Yorker*, etc. Defined as "someone who believes personally in what is being said and takes the position that is implied in the remarks" (*id.*, p. 167), the Principal plays a key role in the polyphonic space, as the person taking the responsibility of what is said (Ducrot's *énonciateur*, 1980, p. 38). **S. Interaction, Dialogue, Polyphony.**

For example, in the following statement:

The weather is nice (**V1**), *but I must work* (**V2**)

the Author stages two voices:

— In **V1**, the voice of a person arguing that a nice weather is a good reason to have a walk.

— In **E2**, the voice of a person arguing that having to work is a good reason to stay home

The decisive point is that, the Principal identifies to voice **V2**; that is, the argument **E1** is dropped, and the argument **E2** validated.

There is no intrinsic superiority of argument **E2** over **E1**. The speaker authors **E1** and **E2**, and, as a principal, acts upon **E2**, not **E1**.

To sum up, "the *Animator* produces talk, the *Author* creates talk, the *Figure* is portrayed through talk, and the *Principal* is responsible for talk" (Schiffrin 1990, p. 241).

1.2 Roles attached to different types and genres of speech

To take into account the variety of discursive genres, one must introduce new roles, such as *narrator* and *narratee* for storytelling; *expert* and *profane* for explanation; *proponent*, *opponent* and *third party* for argumentation (see below).

Interactional genres bring their share of professional or occupational roles: *seller* and *customer* for shop interactions; *teacher* and *students* in classroom interactions; *physician* and *patient* for therapeutic interactions, etc.

1.3 Interactional and social roles

Linguistic roles combine with a set of social roles, in which we distinguish (after Rocheblave-Spenlé [1962]):

— Global social roles: gentleman, cool guy, cheerleader, troublemaker...
— Biosocial roles: young/old, male/female/transgender...
— Social class roles: bourgeois, aristocrat, white or blue collar...
— Professional roles: farmer, trader, truck driver...
— Community roles: religious believer, member of a trade union, a political party, a sport team...
— Family roles: husband, wife, child, father, uncle ...

2. Argument-acting roles: Proponent, Opponent, Third party

The argumentative situation is defined as a three-pole situation, that is to say a three-role situation: *Proponent*, *Opponent* and *Third Party*. Each of these poles corresponds to a specific discursive modality, a discourse of *proposition*, supported by a proponent, a discourse of *opposition*, supported by the opponent, and a discourse of *doubt or questioning*, which defines the Third Party position.

2.1 Proponent and Opponent

The terms *proponent* and *opponent* are defined in dialectical theory, which frames argumentation as a game between these two partners, **S. Dialectic**. From an interactive perspective, the argument becomes dialectical when the third party is eliminated and each actor is assigned a role (*"you will be the proponent, and I the opponent"*) that must be assumed during the whole "dialectic round" (Brunschwig 1967). The elimination of the third party goes hand in hand with the expulsion of rhetoric and the constitution of a system of objective-rational norms. In a figurative sense, one could say that the third party is then replaced by Reason or Nature, in other words by the rules of truth.

If we take a rhetorical view of argumentation, the argumentative game is defined first as an interaction between a proponent, the speaker, and a third party, who is the silent audience to be persuaded. Opponent and counter-discourse are not absent, but are consigned to the background.

By getting into a discussion, participants acknowledge the fact that none of them has enough power or authority to decide on the matter at stake, and that they are engaged in a problematic situation.

2.2 Third Party

Considering that the argumentative question as a full systemic component of argumentative interaction emphasizes the role of the third party. This figure materializes what is publicly at stake and the contact between the contradictory discourses.

In its prototypical form, the argumentative situation appears as a situation of interaction between the speech of the proponent, the counter-discourse of the opponent, both mediated by a third, mediating, interrogative discourse. The third party stabilizes or manages the question, and decides upon the external relevance of the participants' interventions. The argumentative situation, as embodied in an exemplary way in public adversarial exchanges is therefore a three-role situation. Basic argumentative situations such as political debates and cases being tried before the courts, are tripole. Argumentative speech is systematically multi-addressed, the addressee is not only, or not necessarily, the adversary-interlocutor, but in one case the public about to cast their vote, and in the other, the judge about to pronounce a verdict.

In contrast to the categorical assertions and denials of the proponent and the opponent, the third party may appear as a softer and undecided character. In reality, it is the third party who refuses to accept either of the opposing proposals or points of view, asking for more arguments, remaining doubtful and leaving the question open, in the name of making an informed decision. In this sense, and in accordance with the most classical concept of argumentation, the judge is the prototypical figure of the third party. In the third party position are all ratified participants of the argumentative situation who consider that the argumentative forces are balanced, or, more subtly, that even if one seems to prevail, the other cannot be considered to be null. In philosophy, the radicalization and reification of this position is elaborated as methodological skepticism.

Once the third party and the argumentative question have been accepted as key elements of the argumentative exchange, the proponent and opponent may be granted full responsibility for their speeches. One may answer, "*No!*" and the other "*Yes, of course!*" without either of them being systematically accused of manipulation, bad faith or other kinds of fallacious speech.

Institutions may stabilize the argumentative roles and their attribution to individuals. In an ordinary interaction, the argumentative roles correspond not to permanent roles but to footings in the sense of Goffman (1987, chapter 3), in particular, in that they are *labile*. In the same speech turn, the same person can take the role of both the third party and proponent in relation to an issue (affirming a position while expressing a certain degree of doubt about it), or speak as a Proponent on an issue and as Opponent on another.

3. Argument-actors [Fr. *actants*] and Actors [Fr. *actors*]

The individuals engaged in the argument are the physical participants, or *actors* of the argumentative situation. When clarification is needed, the expression *argument-actor* (Fr. *actant*, borrowed from semiotic theory), may be used to refer to the three basic argumentative *roles*, Proponent, Opponent, Third Party.

Any actor can occupy successively each of the three *argument-acting roles* (Fr. *rôle actantiel*), according to all the possible paths. For example, an actor may abandon his or her discourse of opposition in order to develop a discourse of doubt, switching from the argument-acting role of opponent to that of third party. An argumentative issue remains unsolved as long as the contradiction survives, even if some *actors* change their point of view. If two actors swap their *argument-acting roles*, that is to say, if they convince each other, the issue remains open.

In the case of an argumentative alliance, or co-argumentation, the same argument-acting position can be occupied simultaneously by several actors; that is to say, by several individuals producing co-oriented interventions. The study of the argument should focus as much upon co-oriented interventions as upon anti-oriented interventions.

The distinction *argument-actors* / *actors* makes it possible to revisit the famous and strangely prized slogan *"argument is war"*, accompanied with its family of connected bellicose metaphors (Lakoff, Johnson 1980). $Argument\_2$ may be a kind of war, luckily with a more limited number of casualties, but $argument\_1$, that is *argumentation*, is not. The opposition between discourses, that is between *argument-actors* [Fr. *actants*], is not necessarily confused with possible collaborations or oppositions between people, that is, between *actors* [Fr. *acteurs*]. Argumentative situations are confrontational only when the *actors* identify themselves with their argumentative roles. In the most obvious case, that of internal deliberation, the same actor may progress quite peacefully go through all the argument-acting positions. If a group deliberates upon a question involving their common interest, it fortunately happens that the associated members will together examine the various facets of the problem, that is to say the different possible answers to the question and the arguments that support them. During this process, they systematically occupy the different *argument-acting* positions, without clear identification with one of these positions, and without necessarily transforming this process into a war between the actors. The argumentative situation is not inherently polemical; but it certainly can be so when the features defining the identity of the participants are involved and are put at risk in the discussion.

Rules

Arguments can be approached on the basis of very different systems of rules.
— Rules expressing observational regularities.
— Rules expressing norms, imperatives, which are instrumental for argument evaluation.

— Rules as counsels to do things well, how to convince a person to believe or to do something.

1. General rules of interaction

Rules of interaction — Argumentative interactions in natural language follow the various systems of rules proposed for interaction in general, so for example, the rule of justification of non-preferred sequences is applied:

> A dispreferred second part is a second part of an adjacency pair that consists of a response to the first part that is generally to be avoided, and which is likely to be marked by such features as delays, prefaces and accounts.
>
> (*SIL*, Dispreferred second part)

Cooperation@ principle — The principle of cooperation expresses not only what the participants actually do (observational regularity), but also what is reasonable for them to do (rational regularity).

Principles of politeness@ — The principles of linguistic politeness regulate talk relationships on the basis of the concepts of *face* and *territory*. In ordinary conversation, these rules might inhibit the development of arguments. The overriding concern to preserve the relationship means that contradiction is difficult to express and develop, **S. Argumentative politeness**.

Language Sins — A set of commands related to the control of discourse has been developed by the theological tradition inspired by the Bible. The violation of any of these rules is characterized as a sin of language (Casagrande & Vecchio (1991), **S. Fallacies as Sins of the Tongue**.

2. Rules specifically attached to argumentative speech

Rules of the Place — Specific codes are attached to specific *argumentative places*. Parliamentary rules for example apply in Parliament; tribunal proceedings, or classroom interactions develop in line with their own specific regulatory conventions, **S. Forum**. These regulations are drawn up in accordance with a *sui generis* procedure and are applied by the competent authorities ruling in the given place. These rules frame the kind of rationality which characterizes the "genius loci", the spirit of the place.

In such places, the rules determine the topics to be dealt with, the procedures that will lead to a legitimate decision and conclusion, and the persons qualified to take the floor; they regulate the right to speak, the quantity of speech, and the succession of turns at speech. These rules might, for example prohibit overlaps and interruptions. Such *rules of the place* serve to define the rationality of the place as a *local* rationality.

"The Rules of an Honorable Controversy" — Levi Hedge, in his *Elements of Logick* (1838), presents the following seven "Rules for Honorable Controversy":

Rule 1. The terms, in which the question in debate is expressed, and the precise point at issue, should be so clearly defined, that there could be no misunderstanding respecting them.

Rule 2. The parties should mutually consider each other, as standing on a footing of equality in respect to the subject in debate. Each should regard the other as possessing equal talents, knowledge, and desire for truth, with himself; and that it is possible therefore that he may be in the wrong and his adversary in the right.

Rule 3. All expressions which are unmeaning or without effect in regard to the subject in debate should be strictly avoided.

Rule 4. Personal reflections on an adversary should in no instance be indulged.

Rule 5. No one has a right to accuse his adversary of indirect motives.

Rule 6. The consequences of any doctrine are not to be charged on him who maintains it, unless he expressly avows them.

Rule 7. As truth, and not victory, is the professed object of controversy, whatever proofs may be advanced, on either side, should be examined with fairness and candor; and any attempt to ensnare an adversary by the arts of sophistry, or to lessen the force of his reasoning, by wit, caviling, or ridicule, is a violation of the rules of honorable controversy.

(Hedge, 1838, pp. 159-162)

Some of these rules sound familiar. Rule 5 corresponds to the accusation of having a hidden motive@: "*You agree with this proposal not because you approve it but to please the director*". Rule 6 is original, and refers to the problem of the hidden agenda, or even of conspiracy, S. **Pragmatic argument**. Disputes can be said to be "honorable" in both the intellectual and social sense. This system reintroduces what is socially acceptable in a situation where the participants will not spontaneously apply the common rules of cooperation and politeness. Such considerations join the rhetorical problematic of the *prepon* and the *aptum* (Lausberg [1960], § 1055-1062).

In Hedge's system, social control is the root of the imposition of co-operation. The rules for avoiding the sins of language originate from religion, **S. Fallacies as sins of language**. In the Pragma-dialectical system, the system of rules avails itself of communicational rationality, in the spirit of Grice, **S. Cooperation**.

3. Pragma-Dialectic rules and the re-conceptualization of fallacies

These rules define "A Code of Conduct for Reasonable Discussants" (van Eemeren, Grootendorst 2004, p. 190), for partners willing to rationally resolve their difference of opinion. A *fallacy* is defined as a violation of one of these "Ten Commandments for Reasonable Discussants (*id.*, 190-196), **S. Fallacies (I)**:

Commandment 1, *Freedom rule*: Discussants may not prevent each other from advancing standpoints or calling standpoints into question

Commandment 2, *Obligation to defend rule*: Discussants who advance a standpoint may not refuse to defend this standpoint when requested to do so.

Commandment 3, *Standpoint rule:* Attacks on standpoints may not bear on a standpoint that has not actually been put forward by the other party.

Commandment 4, *Relevance rule:* Standpoints may not be defended by non-argumentation or argumentation that is not relevant to the standpoint.

Commandment 5, *Unexpressed-premise rule:* Discussants may not falsely attribute unexpressed premises to the other party, nor disown responsibility for their own unexpressed premises.

Commandment 6, *Starting-point rule*: Discussants may not falsely present something as an accepted starting point or falsely deny that something is an accepted starting point.

Commandment 7, *Validity rule*: Reasoning that in an argumentation is presented as formally conclusive may not be invalid in a logical sense.

Commandment 8, *Argument scheme rule*: Standpoints may not be regarded as conclusively defended by argumentation that is not presented as based on formally conclusive reasoning if the defense does not take place by means of appropriate argument schemes that are applied correctly.

Commandment 9, *Concluding rule*: Inconclusive defenses of standpoints may not lead to maintaining these standpoints, and conclusive defenses of standpoints may not lead to maintaining expressions of doubt concerning these standpoints.

Commandment 10, *Language use rule*: Discussants may not use any formulations that are insufficiently clear or confusingly ambiguous, and they may not deliberately misinterpret the other party's formulations.

This system is inspired by the proposals of the Erlangen school for the definition of a rational "ortholanguage", **S. Logics for dialogue**. In the spirit of Grice, these commandments introduce or impose cooperation@ where it would not be spontaneously practiced by the participants. The game is based on the notion of standpoint. It corresponds to a dialectical treatment of the difference of point of view, with a proponent affirming the point of view and responding to the attacks of an opponent who casts doubt upon it. Rule 9 recalls the aim of the game, that being to settle the difference of opinion either by eliminating the unsustainable opinion or by eliminating the doubt about a well-justified opinion.

Such a system of rules accounts for the validity judgments of the speakers (van Eemeren, Garssen, Meuffels 2009). It is also possible to identify the implicit rules to which the speakers refer for their evaluations based on observing their practices (Doury 2003, 2006).

4. On the question of the rules
S. Fallacies; Argumentation (II); Argumentation (V); Paradoxes; Dialectic; Norms.

S

Scale: Argument Scale and Laws of Discourse

The correlative concepts of *argument scale* and *laws of discourse* have been developed by Ducrot (1973) in the framework of the theory of Argumentation within Language to describe the grammar of co-oriented arguments.

Argument scale translates "échelle argumentative", word for word "argumentative$_1$ scale". Argument scales strictly deal with argument$_1$ *stricto sensu* (as premises for a conclusion), not with argument$_2$, "dispute", **S. *To argue***

1. Cooriented arguments on argument scales

An argumentative *class* is defined as follows:

> A speaker places two statements **p** and **p'** in the argumentative class determined by an utterance **r** if he considers **p** and **p'** as arguments for **r**. (Ducrot [1973], p. 17)

> S: — *Your great grandmother spent time in* The Two Maggots, *she dressed in black, she read Simone de Beauvoir, she was a true existentialist!*

S presents three convergent arguments co-oriented towards the conclusion "*She was a true existentialist*" (a mid-twentieth century popular philosophy). These arguments correspond to features borrowed from the stereotype of what existentialists are and do. **S. Categorization**.

The word *class* refers to an unordered and non-hierarchical set of elements. There is no reason to think that "*spending time in* The Two Maggots" (an existentialist Parisian café) is considered by **S** as a stronger or weaker argument than *"reading Simone de Beauvoir"*.

Two utterances **p** and **q** belong to the same *argument scale* (for a given speaker in a given situation) "if the speaker considers that **p** and **q** are both arguments for the same conclusion **r** (they belong therefore to the argumentative class of **r**), and if he considers that one of these arguments is stronger than the other" (Ducrot, [1973], p. 18).

The following scale represents a situation where **q** is stronger than **p** for the conclusion **r**.

The situation where the speaker considers that *"reading Simone de Beauvoir"* is a *stronger* argument than *"spending time in* The Two Maggots*"* for the conclusion *"to be a true existentialist"* is represented as follows:

The scales where the force of the arguments **p** and **q** is determined solely by the speaker, are called *relative*, S. Force.

Scales for which the gradation is objectively determined are called *absolute*, for example the scale of the cold:

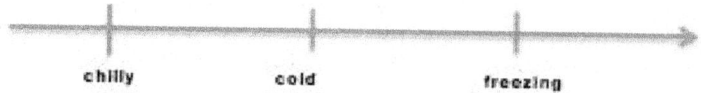

2. Laws of discourse

The functioning of argument scales is regulated by four laws: the *Lowering* Law, the law of *Negation*, the law of *Inversion*, and the law of *Weakness*.

2.1 Lowering law

According to this law "in many cases, (descriptive) negation is equivalent to *less than*" (*Id*, p.31).

Negation is asymmetric; it does not exclude just a *point* on the argument scale, but the whole *zone* including the denied argument and all arguments which are potentially stronger. Denying an argument which is positioned at a higher point on a given scale implies the affirmation of the lower argument.

Let's consider the argument scale determined by a positive answer to the argumentative question *"should we invite him to our hunting party?"*, under the presupposed context *"we are ourselves a group of decent hunters"*

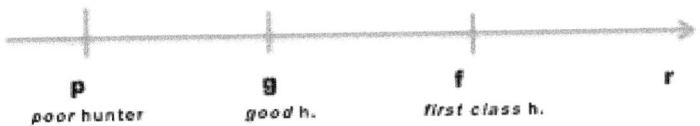

In such a context, *"he or she is not a good hunter"* means, *"he or she is a poor hunter"*, not *"he or she is a first class hunter"*.

The statement *"he or she is not a good hunter, he or she is a first class hunter"* (stress on *good* and *first class*) involves a very particular form of negation, whereby an earlier statement is refuted, **S. Denying**. The stronger argument is necessarily expressed, while the weaker argument remains implicit in the unmarked use of negation.

2.2 Law of weakness

According to this law, "if a sentence **p** is fundamentally an argument for **r**, and if, on the other hand, when certain conditions (in particular contextual conditions) are met, it appears as a weak argument (for **r**), then it becomes an argument for **not-r** (Anscombre and Ducrot 1983, p. 66):

> *He's a good hunter: he killed two partridges last year*

In particular, a weak argument must be presented in isolation, and not in conjunction with conclusive arguments. Grice's principle of exhaustiveness can also account for this fact: an isolated weak argument will be interpreted not only as weak, but also as the best possible, which results in the rejection of the attached conclusion, and consequently, in a binary situation, as a good reason to go for the opposite conclusion, **S. Cooperation**.

From an interactional point of view, putting forward a weak argument might also serve a positive purpose, serving to open a discussion and clarify the positions of the participants.

2.3 Law of negation

The law of negation posits as a regularity that, "if **p** is an argument for **r**, **not-p** is an argument for **not-r**" (Ducrot 1973, p. 27). If *"the weather is nice"* is an argument for *"let's have a walk"*, then *"the weather is not nice"* is an argument for *"let's stay at home"*. This law corresponds to the *argument by the opposite*@ (corresponding to the paralogism of negation of the antecedent).

The following example combines the law of weakness with the law of negation; a weak argument for a conclusion is reversed as a strong argument for the opposite conclusion:

> *After the Second Iraq War, which began in 2003, Saddam Hussein, former President of the Republic of Iraq, was tried and executed in 2006. Some commentators felt that the trial had not been conducted fairly, and considered that the trial was*

so rigged that even Human Rights Watch, the largest unit in the US human rights industry, had to condemn it as a total masquerade.

Tariq Ali, [*A Well-Orchestrated Lynch*], 2007[1].

According to the author, the Association *Human Rights Watch* generally approves decisions in the interests of the United States. So, the fact that they approve the sentence is a *weak* argument for the conclusion *"the sentence is fair"*. In this case, the fact that *even* the association has condemned the decision (like other persons or associations more inclined to criticize the United States) is a *strong* argument for the conclusion that the sentence is *unfair*.

Inversely, a weak refutation of **r** reinforces **r**. This strategy falls within the general framework of the *paradoxes@ of argumentation*.

2.4 Law of inversion

If **p'** is stronger than **p** with respect to **r**, then **not-p** is stronger than **not-p'** with respect to **not-r**. (Ducrot 1973, p. 239; 1980, p. 27)

— *"Leo has a Bachelor's degree"* and *"Leo has a Master's degree"* are two arguments for *"Leo is a qualified person"*.

— *"Leo has a Master's degree"* is a stronger argument that *"Leo has a Bachelor's degree"* for this conclusion: under normal circumstances, we can say:

Leo has the Bachelor degree and even a Master's degree.

While *"Peter has a Master's degree and even a Bachelor's degree"* is incomprehensible. One can say, *"he has a thesis, and even a Bachelor's degree"*, but with some irony on the value of diplomas S. *A fortiori*. If one wants to argue against Peter, to show that he is insufficiently qualified, one will say:

Peter does not have a Master's degree, let alone a Bachelor's degree.

The negation turns the *weakest* argument for *qualification* into the *strongest* argument for his *lack of qualification*.

Argument scales can express the argument *a fortiori*: *"He doesn't have a Bachelor's degree, a fortiori he doesn't have a Master's degree"*.

Schematization

The study of schematizations is the central objective of the Natural Logic developed by Jean-Blaise Grize, a student and subsequently a collaborator of Jean Piaget at the *Research Center on Genetic Epistemology* in Geneva. This logic is called "natural" as opposed to formal logic: on the one hand, it is a "logic of objects" (1996: 82) and a "logic of subjects" (Grize 1996: 96); on the other hand, it involves processes of thought that leave "traces" in natural discourse.

According to Grize, discourse is essentially argumentative, meaning that all utterances frame the world or the situation, along their subjectively relevant

---

[1] Tariq Ali, *Un Lynchage bien orchestré. Afrique Asie*, fév. 2007.

lines, to build a meaningful "schematization". "Scheme" has here a totally different meaning from "argument scheme", which would be called a "reasoned organization", in Grize's vocabulary, corresponding to the second-level phenomenon of *sentence combination*, whereas schematization is a first-level phenomenon, that of *sentence production*.

According to Grize's favorite metaphor, to argue, is to "give to see" to the audience a situation as "spotlighted" by the speaker. As every speech throws some subjective lighting on the world, argumentation is inherent to speech. In Perelman's terms, this operation consists in giving "presence" to an object (Perelman & Olbrechts-Tyteca, [1958], p. 116).

In this perspective, an argumentation is not necessarily a set of statements organized in line with the layout proposed by Toulmin. The influence of an argument and its rationality are not attached to a special kind of speech, or to the use of such and such specific "discursive techniques", as suggested by Perelman & Olbrechts-Tyteca. Any statement, any coherent succession of statements, be it viewed as descriptive, narrative or argumentative, is indeed *argumentative*, insofar as it builds a point of view mapped into a meaningful schematization. Natural Logic is defined as the study of such schematizations, the cognitive counterpart of sentence construction.

This concept is adapted to a vision of arguing as story telling, as offering a coherent and detailed presentation of the world. This might be of some comfort to all students who find themselves disheartened by the difficulty of giving a dense account of extended texts or interactions in terms of argument schemes, even where these are supplemented by an extensive repertoire of figures of speech.

If persuading is defined as shifting the partner's representations, and, accordingly, his or her behavior, then any informative statement, such as *"It is 8 a.m."* is argumentative. If the addressee has to take the 7.55 train and is savoring a last coffee, thinking it is a quarter to 8, then, the information will dramatically change his vision of the immediate future. Natural Logic is also a theory of generalized persuasion, as just "focusing on the relevant aspect of reality".

1. Schematization, a step-by-step process of constructing meaning

Argumentation is traditionally defined as a combination of utterances. Natural Logic studies argumentation as a cognitive process evidenced in natural discourse, and manifested at every stage of discourse production, from the first elaboration of an idea to the combination of utterances, which is only the final stage of the argumentative process. Schematization corresponds to a representation embodied in a complex discursive unit,

> Influencing the interlocutor is to try to modify his or her representations, by emphasizing some aspects of things, concealing others, proposing new ones, and all this by using appropriate schematization. (Grize 1990, p. 40)

Argumentation does not appear to be a chain of statements in a discourse. It emerges progressively at every stage of the production of the utterance, from the first operation of apprehension of content to the construction of a meaningful and therefore "reasoned" discourse. Any statement, any coherent succession of statements, whether or not it is considered to be argumentative or narrative in the traditional sense, is indeed argumentative to the extent that it constructs a unique point of view, that is a "schematization". This conception leads to reconsider all information as argumentation, tending to liken discursive meaning to argumentation, S. **Argumentation (I) - (IV)**.

Grize defines Natural Logic as "the study of logical-discursive operations that make it possible to construct and reconstruct a schematization" (1990, p. 65); "Its task is to account for the operations of thought allowing a speaker to construct objects and to predicate upon them at will" (1982, p. 222).

The concept of schematization defined as a "[discursive representation] oriented towards an addressee of what the author conceives or imagines of a certain reality" (1996, p. 50), "of what it is all about" (1990, p. 29). A schematization is a discourse that focuses the listener's attention upon a "micro-universe" given as "an accurate reflection of reality" (*id.*, p. 36), which constructs or "structures" (*id.*, p. 35) a synthetic, coherent, stable meaning. The purpose of schematization is "to *show* something to someone" (Grize 1996, p. 50; my emphasis); "to schematize [...] is a semiotic act: it is *to give to see*" (*id.*, p. 37; my emphasis). The object of Natural Logic is the study of the operations constructing such images.

The functioning of schematization is particularly clear in classical argumentative situations, when a discourse directly confronts a counter-discourse; the same reality is given two antagonistic descriptions:

S1 — *These replacement workers, you will pay them with* the strikers' money!
S2 — *Not the strikers' money*, the taxpayers' money.

2. Operations constructing a schematization

Natural Logic postulates the existence of "primitive notions", of a pre-linguistic nature (Grize 1996, p. 82), linked with the culture and the activities of the speakers. These pre-notions are the place of "cultural pre-constructions", i.e., received ideas and current, accepted ways of doing things. The language "semantizes" these primitive notions turning them into "objects of thought" associated with words (Grize 1996, 83).

Schematization operations are anchored in these "primitive notions" (*id.*, p. 67) and are constructed by a series of operations; "primitive notions" are actually noted by words between brackets. The following sequence is formed of the primitive image and fuzzy notions /fuzzy/ and /image/:

> It's unfortunate that the edge of the image is blurry, and it needs to be corrected. (*Ibid.*)

This construction follows these steps:
(a) The process of discourse construction begins with the selection of relevant primitive notions, to produce the objects of discourse; here *"image, edge of the image"* as well as the predicative pair *"to be blurred, not to be blurred"*. The objects thus schematized will evolve with the development of the discourse, **S. Object of discourse.**
(b) Then, the operation of characterization produces "contents of judgments" that is predications, and these are accompanied by modalizations, carried out on the objects of discourse. Here, the content of judgment is, *"that the edge of the image be quite blurry"*.
(c) A subject then asserts (positively or negatively) the content of judgment, and produces a statement, *"it is unfortunate that the edge of the image is quite blurry"*.
(d) Operations of configuration then connect several utterances and so build a discursive chain, "a reasoned organization". The preceding statement for example, is connected to another statement, *"this must be corrected"*, which is produced according to the same mechanism:

> *It's unfortunate that the edge of the image is blurry, and it needs to be corrected.*

These different linguistic-cognitive operations can be likened to the vision of language and mind developed by the philosophy of traditional logic, **S. Logic.**

>(a) Apprehension of content by the mind;
>(b) Predication, constituting unasserted propositions;
>(c) Judgment, expressed in an assertion, which can be true or false;
>(d) Concatenation of judgments, i.e. discourse construction.

The aim of this approach is to emphasize that all operations relevant to the genesis of the utterance have an argumentative import. Argumentation is as much a sentence *construction* process as a sentence *connection* process.

3. Shoring

The concept of shoring developed in Natural Logic is defined as,

> a discursive function consisting, for a given segment of speech (whose dimension can vary from a simple statement to a group of statements having a certain functional homogeneity), to accredit, to make more likely, to reinforce, etc. the content asserted in another segment of the same discourse. (Apothéloz & Miéville 1989, p. 70)

This concept corresponds to the classical problematic of argumentation as a composition of statements, a statement-argument supporting a statement-conclusion, **S. Argumentation (I).** To refer to the same phenomenon, Natural Logic also uses the expression "reasoned organizations":

> Many statements are made merely to support, to shore up the information given. This is part of the general order of argumentation and allows us to envisage more or less extensive blocks of discursive sequences as *reasoned organizations*. (Grize 1990, p. 120)

The study of reasoned organizations is an instrument for the study of representations, defined as "a network of articulated contents" (*id.* p. 119-120). It should be emphasized that, for Natural Logic, the reasoning process is not limited to the combination of utterances but includes the whole dynamic process of structuring the utterance, whether it will function as argument or conclusion in a reasoned organization.

4. Schematization and communication

Schematizations refer to a particular communication situation. They are the product of "the activity of speech [which] is used to construct objects of thought" (1990, p. 22); these objects being part of a dialogue where they are used "as shared references for interlocutors" (*ibid.*). The communication situation envisioned is intended to be "essentially dialogical in nature" (1990, p. 21), but in reality it is analogous to that of rhetorical address. It never considers the possible interactions between the respective schematizations of the participants.

> By [dialogal] I don't mean the interweaving of two discourses, but the production of a speech between two parties, a speaker [orator] ... addressing a listener. Admittedly, in most texts, the listener remains virtual. This, however, does not alter the basic problem: the speaker constructs the speech according to his or her representations of the listener, simply, if the listener is present, he or she can actually say, "*I do not agree*" or, "*I do not understand*". But if the listener is absent, the speaker must indeed anticipate his or her refusals and misunderstandings. (1982, p. 30)

Persuasion is given up, "the speaker can only propose a schematization to his or her audience, without actually 'transmitting' it" (*ibid.*).

5. "Logic of Contents" (Grize) and "Substantial Logic" (Toulmin)

Grize defines his Natural Logic in relation to formal logic:

> Alongside a logic of form, a formal logic, it is possible to envision a "logic of contents", that is, a logic taking into account the processes of thought, the development and interconnection of these contents. Formal logic based on propositions accounts for the relations between concepts, while Natural Logic proposes to highlight the construction and interconnection of the notions. (Grize 1996, p. 80)

This "logic of contents" might remind us of Toulmin's "substantial logic", S. Layout. But, unlike Toulmin, who characterizes argumentation as an arrangement of statements without discussing their internal structure, Grize considers that argumentation begins with the basic operation producing the statement itself.

Scheme

The concept of an *argument scheme* captures the specificity of the minimal concatenation of two statements (S1, S2) making up an argumentation (Arg,

Concl). An argument scheme is essentially a specific kind of sentence connection, a special case of textual coherence and cohesion; that is to say, *a general discursive inferential scheme* associating an argument statement with a conclusion statement.

An argument scheme is a discursive formula, a generic statement functioning as an *argument rule*, an inferring license. *Concrete argumentations*, or *enthymemes* are its actualization in a particular case.

In Aristotle's *Rhetoric*, all the argument lines are expressed as such generic statements, that can sometimes be formulated as proverbs or maxims. The saying, "*if you can do the hard things, you can do the easy things as well*" corresponds to the "from the biggest to the smallest" (*a maiori ad minus*) branch of the *a fortiori* scheme. Typical formulas, such as those proposed by Bentham "*let us wait a little, the moment is not favorable*" are also complete and perfectly adequate expressions of an argument scheme. **S. Collections**. This scheme can be specified in a discursive domain, S. *A fortiori*.

In the expression of the scheme, their characteristic indefinite components (subject, predicate) may also be expressed as variables. For example, the *a fortiori* scheme can be written as (according to Ryan 1984):
    If <**P is O**> is more likely (more recommendable...) than <**E is O**>,
    and <**P is O**> is false (not plausible, not recommendable)
    then <**E is O**> is false (not likely, not recommendable).

And embodied in the following argumentation:
    "If teachers do not know everything, students know even less"

In the same style, the scheme of the opposite is written as:
    If <**A is B**>, then <**not-A is not-B**>.

Derived argumentation:
    If I was of no use to you during my life, at least my death will be useful to you.

Such presentations should not be taken as a kind of "deep logical or semantic structure" of the scheme. Their unquestionable benefit is to clarify the reference of general terms.

2. Example: scheme and argumentations on waste

To detect a scheme in a text is a key moment in argument analysis. But such an identification is not easy, the key semantic components of the scheme being frequently disseminated in the text. How can we identify a scheme in a passage? Experts will say that they just recognize one when they see it; but for the non-specialist student, analyzing a passage as an occurrence of an argument scheme is not always so self-evident, and necessitates a methodical reconstruction, such as the following one:
— First, an explicit definition of the topic is needed.
— Second, the passage must be clearly delimited.

— And finally, one has to show how the scheme can be projected upon the passage; that is, one has to establish a point-to-point correspondence between the scheme and the passage under analysis. In essence, these links consist in linguistic operations of equivalence and close reformulation.

This method can be illustrated on the case of the argument from waste@, as defined and illustrated in Perelman & Olbrechts-Tyteca.

— The scheme:
> The argument of waste consists in saying that, as one has already begun a task and made sacrifices, which would be wasted if the enterprise were given up, one should continue in the same direction. ([1958], p. 279)

— First argumentation:
> this is the justification given by the banker who continues to lend to his insolvent debtor in the hope of getting him on his feet again in the long run. (*Id.*, p. 279)

— Linguistic operations associating the argument to the scheme (bijective association Scheme – Argumentation)

Argumentation *italics*: arg. wording	Linguistic Operation *italics*: arg. wording; **bold**: AS wording	Scheme (AS) **bold**: AS wording
Implicit: *a debtor* is a person to whom the banker has already *lent money*	*Lending money* is **a task; a sacrifice**	(Past:) **one has already begun a task and made sacrifices**
Insolvent debtor	*Insolvent* means that previously lent money **[will] be wasted**	(Present:) **which would be wasted if the enterprise were given up**
The banker continues to lend	*Continues to lend* = **continue in the same direction**	(Decision:) **one should continue in the same direction**

The second enthymeme is more complex:
> This is one of the reasons which, according to Saint Theresa, prompt a person to pray, even in a period of 'dryness'. One would give up, she says, if it were not "*that one remembers that it gives delight and pleasure to the Lord of the garden, that one is careful not to throw away all the service rendered, and that one remembers the benefit one hopes to derive from the great effort of dipping the pail often into the well and drawing it up empty*". (*Id.*, p. 279)

— Linguistic operations associating the argument to the scheme (same conventions):

Scheme

Argumentation	Linguistic Operation	Scheme
all the service rendered that is: *(the great effort of) dipping the pail often into the well*	*rendered* (presupposes) **already begun** *a service* (is) **a task**	one has already begun a task
the great effort of (dipping the pail often into the well)	*the great effort* (is a) **sacrifice**	and made sacrifices
in a period of "dryness"(¹) *driving it up empty*	*dryness — empty* <=> **no result**	for no result
not to throw away	*not to throw away* <=> **would be wasted**	(present) which would be wasted if the enterprise were given up
prompt a person to pray	*prompt to* <=> **urge to continue**	one should continue in the same direction

(¹) Traditional mystic metaphor for "no increase in faith" = no spiritual benefit.

3. Naming the argument schemes

Argument schemes are labeled according to their form or their content:

(i) According to their specific semantic content — Some famous arguments are named in reference to their precise content.

— *The third man argument* is an objection made by Aristotle to the Platonic theory of intelligible forms, as opposed to individuals. According to this objection, the Platonic theory implies an infinite regression. It may be seen as an argument from *vertigo*@.

— The *argument against miracles*: the likelihood that the dead person was resurrected is *lower* than the likelihood that the witness is mistaken; so we can reasonably doubt that the dead person was resurrected (Hume, 1748, §86 "Of Miracles"). This formally refers to a hierarchy of the probable, and can be represented on an argumentative scale, S. **Argumentative scale**.

— The *ontological argument* infers the existence of God from the a priori notion of a perfect being, S. **A priori; Definition**.

(ii) Depending on their form and content, S. Collections
On the use of Latin words and expressions, S. *Ab —, a/ ad — e/ ex —*

(iii) Oriented labels — Usually, the label designating an argument specifies a form or content: the argument refers to the consequences (*ad consequentiam*), to authority (*ab auctoritate*), to the consistency of human beliefs (*ad hominem*), to emotion (*ad passionem*) or to any particular emotion (*ad odium*). The speaker may admit, without inconsistency, losing face and invalidating the argument that he or she argues by the consequences, *ad hominem*, *ex datis*, upon a religious belief (*ad fidem*), or possibly from the number, *ad numerum*. These arguments can be assessed in a second, normative, stage.

Some other arguments involving the individual are designated by oriented labels. An argument cannot be dubbed an *appeal to stupidity, to superstition* or *to imagination* without invalidating it; given the current vision of emotion as antagonistic to reason, referring to a passage as containing an appeal to emotion, from *ad passiones* to *ad odium*, amounts to a rejection of the argument. Such labels contain a built-in evaluation; there is some confusion between the levels of description and evaluation.

A call to faith will or will not be judged as fallacious depending whether or not one shares the beliefs of the speaker. In such cases, the theoretical language is biased, and normative action becomes ideological.

4. Typologies of argument schemes

A general typology of argumentation schemes is an organized collection@ of argument schemes. Collections of argument schemes are locally constituted as:
— The set of arguments locally exploited by a particular speaker, in a particular discussion.
— The set of arguments attached to a question, **S. Script**.

5. Argument schemes in discourse

The concept of the argument scheme anchors the study of argumentation in the material reality of speech and discourse. The capacity to identify an argument from authority, a pragmatic argument, etc. is an essential skill for the production, interpretation and criticism of argumentative discourse, **S. Tagging**.

Some works, such as the Thomas Aquinas *Summa Theologica* or texts such as Montesquieu "On the Enslavement of Negroes" can be described as dense and dry successions of arguments. Other texts are more fluid, and seem hardly reducible to circumscribed segments that could be plausibly described as an occurrence of an argument scheme.

The schemes are relatively under-determined by the linguistic expression; there may be several plausible analyses of the same text segment, some invalidating the argument, others not. This uncertainty should not be automatically seen as an indicator of the poor quality of the argument. In this respect, contextual considerations and the kind of editing given to the analyzed passage play a crucial role.

An argumentative text or interaction can be compared to a natural meadow, the most beautiful flowers corresponding to canonical argument schemes. But it is also necessary to wonder about what the dense plant tissue around these flowers is made of. To this end, interaction analysis, discourse analysis and text linguistics serve as crucial analytic instruments, which have to be adapted to the specificities of argumentation analysis. The "topical scheme approach" comes within a larger prospect, opening with the stance taken vis-à-vis the other's discourses, the kind of argumentative situation they frame, the determination of general argumentative strategies, taking into consideration a whole range of semiotic phenomena. On a micro-level, one has to consider the operations *producing* the statements, as well as in their *coordination*: a good grammar rulebook and a good dictionary are essential if one is to construct a good argument analysis, S. **Tagging; Argumentative Question; Markers**.

Scheme, Schema, Schematization

1. Schema
The word *schema* might be used to refer to any kind of diagram used to represent and clarify the structure of an argumentative phenomenon: *Toulmin's schema*, *Convergent argument schema*, etc.
Toulmin "layout@ of argument" is also known as "Toulmin Schema" or "Toulmin Argument Pattern" (TAP).

2. Argumentation scheme
An argumentation scheme is an abstract or generic representation of a concrete argumentation, S. **Argument Scheme**

3. Schematization
Natural Logic uses the term "schematization" to designate the succession of linguistic and cognitive operations through which a reality is given a linguistic expression by a speaker, S. **Schematization**

Script

The *argument script* attached to a question includes the set of positions, arguments, counter-arguments and refutations put forward by either party when this issue is debated. They are available to any arguer entering the arena and willing to take a position on the issue.
When a new issue emerges in the media sphere, the arguments very quickly stabilize in an argument script.
The script corresponds to the state of the argumentative question. It may be implemented any number of times, on a wide variety of forums. It pre-exists and informs concrete argumentative discourses. It develops with the emergence of new issues and arguments.
Argument scripts are not the sole component of actual argument. A script essentially consists in a collection of arguments on the matter, on the merits of the case, regardless of the specific circumstances of particular encounters. A script may, however, also include generic characteristics of the speakers intervening in the debate and considerations on the conditions under which it takes place.
The argument *"the finances of the country are in a state of crisis"* is part of the script relating to refugees, as well as its standard refutation *"you lack generosity / let us be generous"*. An argument about the person, as *"you wear jewels and dare to speak about*

the financial crisis!" is not part of the script, the interlocutor not necessarily wearing jewels.

Argument Map — The argument script can be represented as an argument map.[1]

Script and invention — The existence of scripts largely modifies the classical idea that arguments are "invented", that is, that they are spontaneously created by the arguing speaker, **S. Rhetoric**. When an argument concerns concrete cases, the speaker may have to invent arguments, but when dealing with established socio-political issues, as well as in all disciplines where one can refer to a state of the question, arguments are merely selected from the relevant argument script and repeated. In such areas, arguments are not "invented", they are *available* for all participant.

The first task of the interested party is to review the script relevant to the issue he or she wishes to discuss, and then to perform his or her partition, that is, to organize a discourse which updates and amplifies the argument line he or she has selected. In other words, the arguer must define and follow a path within the parameters of the script.

This conception of argumentative activity has repercussions for argumentation education, and emphasizes, firstly, the necessity of carefully established information prior to the discussion, and, secondly, the importance of individual expression and style in argumentation.

Self-Argued Claim

1. Argumentation as a composition of statements

1.1 Argument, claim

Consider a discourse, composed of two statements, {**S1, S2**}. This sequence is argumentative if it can be paraphrased using some of the following sentences:

    S1 *backs, supports, motivates, justifies,* ... **S2**
    S1, *so, thus,* ... **S2**
    E2, *because, since, as, given that,* ... **S1**

The Argumentation within Language theory formulates the same relation in a way that has proved extremely fruitful: *the conclusion, it is what the speaker has in mind or in view, what he or she is getting at, when he or she produces the argument*:

    The speaker puts forward **D1**, in order to, with a view to... **D2**.
    The reason why he states **D1** is **D2**.
    The meaning of **D1**, that is the direction towards which it strives, the sense... of **D1** is **D2**.

---

[1] A map of a fraction of the script corresponding to the question "Can computers think?" can be found at web.stanford.edu/~rhorn/a/topic/phil/artclISSAFigure1.pdf (29-09-2013).

and, ultimately, "**D1**, *i.e., that is to say, in other words, that means,...* **D2**":
> S: — *You say you have homework*, you mean that *you will not go out with us tonight?*

A conclusion can thus be introduced not by a connector or an indicator of *consequence*, but by a connector of *reformulation*. The claim **D2** essentially "repackages" the argument, revealing the contextual meaning of the statement as argument. The interlocutor fully understands the statement-argument, only if the conclusion is grasped **S. Orientation**.

The claim is somehow integrated in the argument. This is why the conclusions may frequently remain implicit.

1.2 Argument, conclusion, inferring license

It is generally assumed that the argument-claim link is provided by a topic, an argument scheme, often left implicit; the consistency of a chain:
> The wind is rising, it will rain.

is based on the empirically observed regularity:
> Generally, when this type of wind comes up, it rains.

From an epistemic perspective, there is "more" in the argument than in the conclusion, as far as the argument is more reliable than the conclusion, which is only a hypothetical projection of the argument. From a semantic perspective however, there is "less" in the argument than in the conclusion, to the extent that the conclusion is more than an analytical development of the argument, it is the product of this argument enriched and structured by its combination with a general scheme or topic.

1.3 Argument, conclusion, inferring license, modal

This combination corresponds to Toulmin's layout@ of argument, which articulates the argumentative unit around five elements, *Data, Claim*, the two-level transition principle, *Warrant* and *Backing*, and finally, a *Modal* which refers to the argument *Rebuttal* conditions ([1958], chap. 3).

2. Self-argued conclusions

From the perspective of the theory of knowledge, in order to be valid an argumentation must be expressed in a coordinated sequence "**S1** (argument), **S2** (claim)", such that the claim is not a reformulation of the argument. It follows that it must be possible to assess each statement independently. This is the case in the following sentence, "*the wind has picked up, it will rain*", which expresses two independently observable facts, the fact that there is wind and the fact that it will rain a little later. The first fact is measured by an anemometer, the second by a rain gauge, two devices which operate according to entirely different principles.

In ordinary discourse, not only is the conclusion already present, if not contained, in the argument (cf. §1.1), but the argument statement can also be em-

bedded in the concluding statement in the form of a subordinated clause or somehow integrated in a component phrase of the statement expressing the claim:
> These people come to work in our country, welcome them!
> → let us welcome these people who come to work!

Ultimately, the argument is absorbed within the meaning of one key term of the statement:
> → let us welcome these workers!

In this case, the argument is included in the word (Empson [1940]); the statement is self-argued, it expresses a complete perspective, which presents itself as obvious, irrefutable.

Scientific language has one tier of signification, while natural language has several and relies on implicit significations. This essential fact opposes scientific languages and natural language. Arguments loaded with a preordained conclusion they "support" can be considered to be "biased", fallacious, and censored as such. But this is a rather desperate maneuver. It does not makes much sense to pretend to develop critical thinking about human affairs whilst ignoring or condemning the medium and substance which makes the very stuff of all transactions concerning human affairs, and will continue to retain this function for a long time.

Self-Evidence

Self-evidence is a sentiment of immediate certainty about a state of things; when expressed, the corresponding statement is obvious, that is, it does not require justification, and should be accepted as such, **S. Contempt.**

The term *aperception* is used to designate this form of knowledge as produced by a conscious perception, and accompanied by reflection. Knowledge by aperception is opposed to knowledge by *inference*, and therefore to knowledge acquired through argumentation, which is a kind of inference. Three kinds of aperception, that is to say three main sources of evidence, can be identified and distinguished form one another:

— Self-evidence as the fruit of the divine revelation of a transcendental reality.
— Perceptual self-evidence of sense data.
— Intellectual self-evidence given by intuition.

The simplest way to legitimate an assertion is to invoke one of these three sources, **S. Argument-Conclusion.**

The certainty manifested in a direct, simple affirmation corresponds to the certainty associated with aperception, **S. Repetition:**
> Affirmation pure and simple, kept free of all reasoning and all proof, is one of the surest means of making an idea enter the mind of crowds. The more concise an affirmation, the more destitute of every appearance of proof and

demonstration, the more weight it carries. The religious books and the legal codes of all ages have always resorted to simple affirmation. Statesmen called upon to defend a political cause, and commercial men pushing the sale of their products by means of advertising are acquainted with the value of affirmation.

Gustave Le Bon, [*The Psychology of the Crowd*]. [1895][1]

Inferential argumentative belief might be considered inferior to belief based on any kind of evidence: this observation is at the root of the paradoxes@ of argumentation.

1. Dogma: Revelation as a source of certainty

Believers consider the revelation gathered in the sacred books as a source of certainty. This revelation, which took place in the sacred time of the origins, can be renewed by a particular revelation, such as that which Blaise Pascal has described in what is now called his *Memorial*, producing an immediate and absolute "certitude":

> The year of grace 1654,
> Monday, 23 November, feast of St. Clement, pope and martyr, and others in the martyrology.
> Vigil of St. Chrysogonus, martyr, and others.
> From about half past ten at night until about half past midnight,
>   FIRE.
> GOD of Abraham, GOD of Isaac, GOD of Jacob
> not of the philosophers and of the learned.
> Certitude. Certitude. Feeling. Joy. Peace.
> GOD of Jesus Christ.
> My God and your God.
> Your GOD will be my God.
> Forgetfulness of the world and of everything, except GOD. […]
>
> Pascal, *Memorial.*[2]

2. Self-evidence of the sense data

The direct physical perception of a state of affairs immediately legitimates a claim. There is no need to argue to see and claim that the snow is white. As the adage says "facts are the best arguments"; the question *"Is snow white?"* is not debatable ("a-stasic", **S. Stasis; Evidentiality**).

From the philosophical point of view, Descartes has rejected the possibility of founding knowledge on sense data by the hypothesis of the "evil genius" (Descartes [1641], First Meditation).

3. Intellectual intuition

Descartes accepts only intellectual intuition as a source of certainty:

---

[1] Gustave Le Bon (1895). La Psychologie des Foules. Paris: Alcan. Quoted after Gustave Le Bon, *The Crowd. A Study of the Popular Mind.* New York: Macmillan, p. 126.
[2] Quoted after http://www.users.csbsju.edu/~eknuth/pascal.html (07-09-2017).

Rule 3 - Concerning objects proposed for study, we ought to investigate what we can clearly and evidently intuit or deduce with certainty, and not what other people have thought or what we ourselves conjecture. For knowledge can be attained in no other way. (Descartes [1628], Rule 3)

"Good intuition" is infallible:

By intuition I mean, not the wavering assurance of the senses, or the deceitful judgment of a misconstructed imagination, but a conception, formed by unclouded mental attention, so easy and distinct as to leave no room for doubt in regard to the thing we are understanding. (*Id*, Rule 7).

This intuition is that which makes us accept something as "beyond reasonable doubt". So for example, we can feel fairly certain that by taking a point out of a line one can draw a single second line parallel to this line; or that the square of any negative number is positive. These certainties have been called into question by the construction of imaginary numbers and non-Euclidean geometries.

4. Consequences

4.1 Conflict between sources of evidence

It may seem that the most incontestable kind of self-evidence is the direct evidence provided by sense data. Yet the following text shows that it may be judged inferior to that emanating from the authority of the sacred text. It must be noted that the author's concluding commentary ratifies this hierarchy.

The first disagreement among the Companions after the death of the Prophet concerned the reality of his death itself. After the Death of the Prophet, 'Umar ibn al Khattaab, may God be pleased with him, insisted that the Messenger of God did not die, considered any such talk a false rumor spread by the hypocrites, and threatened to punish them for it. This went on until Aboo Bakr appeared on the scene and recited the verse of the Qur'an:

'Muhammad is no more than a Messenger. Many were the Messengers who passed away before him. If he died or were slain, will you then turn back on your heels? Whoever turns back on his heels, not the least harm will he do to God; but God [on the other hand] will swiftly reward those who [serve him] with gratitude' (3: 144).

And another verse of the Qur'an:

'Truly you will die [one day], and truly they [too] will die [one day]' (39: 30).

When 'Umar heard these verses his sword fell from his hand and he himself fell to the ground. He realized that the Prophet, may God bless him and grant him peace, had passed away and that the divine revelation had come to an end. [...]

Differences over the Prophet's Burial [...]

These were two critical issues *[about "the reality of the death of the Prophet" and about "the burial of the Prophet"]*, which were swiftly resolved simply by resorting to the Qur'an and the Sunnah.

Taha Jabir al 'Alwani, 1993, *The Ethics of Disagreement in Islam*, p. 35-36.[1]

4.2 Subtracting from doubt

The argument, the basis of the argumentative derivation of a conclusion, is presented as being above doubt. It is conveniently framed as an aperceptive datum, that is to say as something which is as certain as a revelation, as sensible evidence, or intellectual intuition. It follows that the person who refuses to share this data will be considered, as disgraced, infirm or idiotic. It is therefore not necessary to refute him or her, since he or she is already defamed, **S. Destruction; Contempt.**

Extended argumentability assumes that any person can be summoned to account for his or her beliefs, and that he or she must justify them, so that it is illegitimate to postulate any kind of a priori certainty. This thesis is difficult to apply to points of view which are considered certitudes of a religious order, such as *"there is no God but God"*; mathematical, *"the square of a positive number is positive"*; or simply everyday arguments such as, *"I believe that the ground will not collapse under my feet"*, **S. Dialectic**. Self-evidence can be opposed to extended argumentability, **S. Conditions of Discussion**

Serial Argumentation

Serial argumentation (Beardsley 1975, quoted in Wreen 1999, p. 886) also called *subordinate argumentation* (van Eemeren and Grootendorst 1992), is traditionally known as *polysyllogism*, **S. Sorite; Epeicheirema**. A serial argumentation is an argumentation where an established conclusion is used as an argument for a new conclusion, up to an ultimate conclusion. Each argumentation which contributes to the global serial argumentation has its own structure', either simple or convergent. It might correspond to any kind of argument scheme.

Serial argumentation is schematized as follows:

Arg_1 => Concl_1 = Arg_2 => Concl_2 = Arg_3 => ... Arg_n => Concl_n

Difficulties arise in the reconstruction of concrete argumentations, as shown in the following an example from Bassham (2003, p.72):

Peter is stubborn, he is a Taurus, he will not know how to negotiate.

First Interpretation, as a Serial Argumentation:

(A) Peter is a Taurus so he is stubborn; (B) being stubborn, he will not know how to negotiate.

Peter is stubborn (indeed, since...) he is a Taurus; so, he will not know how to negotiate.

---

[1] Taha Jabir al 'Alwani, 1993, *The Ethics of Disagreement in Islam*. Herndon: VA: International Institute of Islamic Thought, p. 35-36. Quoted after:
archive.org/stream/157627041TheEthicsOfDisagreementByTahaJabirAlAlwani/157627041-The-Ethics-of-Disagreement-by-Taha-Jabir-Al-Alwani_djvu.txt

(A) First argumentation: (1) *Peter is a Taurus*, so (2) *he is stubborn*
    (A.i) — Technical definition of "being a Taurus": "*the Taurus sticks to his or her positions without being willing to change them*"[1]
    (A.ii) — Instantiation of the definition: "*Peter remains on his positions without being willing to change*" and conclusion of the first argumentation
    (A.iii) — Lexical definition of *stubborn*: "*who is obstinately attached to his opinions, and his decisions; which is insensitive to the reasons and arguments against it.*"
    (A.iv) — (A1) and (A.iii) are in a paraphrase relationship.
    (A.v) — Conclusion: (2) *Peter is stubborn.*

(B) Second argumentation: (2) *Peter is stubborn*; therefore (3) *he will not know how to negotiate*
    (B.i) — Technical definition of *negotiation* "*negotiation involves the confrontation of incompatible interests on various points that each interlocutor will attempt to make compatible by a set of mutual concessions.*" (Wikipedia, [*Negotiation*])
    (B.ii) — According to the above (A.iii) lexical definition, "*being stubborn*" and "*making concession*" are opposites.
    (B.iii) — Opposites cannot be predicated upon the same subject, Peter.
    (B.iv) — Conclusion: (3) *Peter will not know how to negotiate.*

This is a serial argumentation
    Arg_1 => Concl_1; *so* [Concl_1=Arg_2] => Concl_2

Second Interpretation, as a Convergent Argumentation: Two Arguments Backing the Same Conclusion:

(C) First argumentation (1) *Peter is a Taurus*, (3) *he will not negotiate*
    (C.i) — The two technical definitions (A.i) and (B.i) are contradictory.
    (C.ii) = (Biii)
    (C.iii) — Conclusion: (3) *Peter will not know how to negotiate.*

Or:

    (C.i') — Technical definition: "*the negotiator must remain flexible, calm, and exercise self-restraint.*"[2]
    (C.ii') — "*The Taurus' promptness to accumulate feelings and grudges also makes him capable of strong anger*"[3]
    (C.iii') — (C.i') and C.ii') are contradictory
    (C.iv") = (Biii)
    (C.v') — Conclusion: (3) *Peter will not know how to negotiate.*

(D) Second argument, (2) *Peter is stubborn*, (3) *he will not negotiate*
    (D.i) — (A.iii) and (B.i) are opposites, see (B.ii).
    (D.ii) = (Biii)
    (D.ii) — Conclusion: (3) *Peter will not know how to negotiate.*

---

[1] http://www.astrologie-pour-tous.com/taureau.html (09-20-2013)
[2] Jean-Paul Guedj, *50 Fiches pour négocier avec efficacité* [*50 leaflets to negotiate effectively*], Paris: Bréal, 2010, p. 123.
[3] www.astronoo.com/zodiaque/zodiaqueTaureau.html (09-20-2013).

This is a convergent argumentation:

Signs ▶ Natural Signs

Silence

Lat. *a* or *ex silentio* argument; *silentio*, "silence".

The *argument from silence* concludes that a possible event should not have happened, or does not exist, as we have no information about it; nobody talks about it; or the relevant people don't talk about it, **S. Ignorance**.
Chroniclers note the outstanding events of their time; if they do not mention an event that should have attracted their attention, the argument from silence concludes that such an alleged event never occurred. This can be seen as a particular application of the completeness principle applied to the work of these chroniclers. **S. Completeness**. Did Syldavia face terrible flooding during a given period? If such an event had occurred, the chroniclers would have mentioned it, *a fortiori@*, since they mentioned facts of less importance. Since the chroniclers do not mention a terrible flood during that period, we can assume that no such devastating event took place. The value of the argument depends on the quality of the relevant documentation available for the period concerned. It increases considerably if we know that the chroniclers regularly note atmospheric events.
The camel argument can be opposed to the argument from silence. The Qur'an never mentions camels. We would thus assume that there were no camels in seventh century Arabia! But this conclusion can be rebutted on the grounds that camels are not mentioned simply because they were so common, and not relevant to the text under consideration. This, however, is not the case of floods for the Syldavians chroniclers.

The argument of silence is used to date literary works. Marie de France wrote the *Lais* in the late twelfth century, but can the date be further specified? The editor of the *Lais* argues as follows (after Rychner, 1978)[1]:
1. "*To date more precisely the* Lais, *it should be placed in relation to the other works of the time*".
2. To do that, he invokes "*an argument* ex silentio, *to use with caution but [which] should not be overlooked*"

---

[1] Jean Rychner, *Introduction* to the *Lais* of Marie of France. Paris: Champion, 1978, p. X-XI

3. First, "*there is no evidence in the* Lais *of Marie having read Chrestien de Troie* Eneas"; *Eneas* was published in 1178.
4. "*It is difficult for me to imagine that, having read it, she would have been able to remain so completely herself and so different from him, in her style and general inspiration.*"

The conclusion follows: the *Lais* must have been written before 1178.

Slippery Slope

The slippery slope argument is another name for the argument of *direction*@. It consists in saying that such a controversial action, **A**, apparently convincingly backed by such and such arguments, should not be accepted, even if it might seem reasonable, because, if it were, the same principles and reasoning could be used to argue in favor of another action of the same kind **A+**, which is much more controversial, and then for another action **A++**, that one would find quite unacceptable. In effect, accepting **A** removes all possible limit, "when started, you cannot stop". The slippery slope counter-argument is based on the precautionary principle, aimed at preventing a risk of extension of the decision adopted.

In a debate about the legalization of drugs, a participant proposes to legalize hashish:

> A. C. — [Legalization, or rather domestication] will not eliminate the problem of drugs. But it is a more rational solution, which will eliminate the mafias, reduce delinquency, and also reduce all the fantasies that feed drug taking itself and are part of drug marketing.

The opponent to this measure opposes that

> If you legalize hashish, you will have to legalize cocaine, then heroine, then crack and cocaine, and all the worst dirty things that man can find.
>
> *Le Nouvel Observateur* [*The New Observer*][1], 12-18 Oct. 1989

For a refutation of the same position based on its perverse consequences, S. **Pragmatic Argument**; for a refutation based on the very wording and definition of the project, S. **Related Words**.

This argument is based on the following operations. The question is:
> Question: — *What should we do about the issue of drugs?*
> S1 — *We should legalize hashish, for such and such reasons.*

The opponent **S2** is reluctant to accept this proposal, even if the reasons presented by **S1** are not entirely unacceptable. **S2**, however, refuses to become involved in **S1**'s process of reasoning on the basis of the following analysis of the situation.

(i) Consideration of the context in which the proposal is made, S. **Categorization;**

---

[1] *Le Nouvel Observateur* is a French weekly political and cultural newspaper.

Taxonomies:
> The category "drug" covers hashish, heroin, crack cocaine and so on.

A grading operation within this class of drugs:
> *Heroin* is worse than *hashish*, and crack cocaine is worse than *heroin*

So, hashish is the *low* point, the *weak* point by which one enters the graduated category of drugs.

(ii) An evaluation: the decision to legalize hashish may be *debatable*, but the legalization of heroin would clearly be *unacceptable*, and the legalization of crack cocaine would be *unthinkable*, even *outrageous*. This gradation mirrors gradation (i).

(iii) A driving mechanism: the decision to legalize hashish is related to those to be made in relation to heroin and crack cocaine; the same question will inevitably arise about these harder drugs:
> Should heroin be legalized? Should we legalize crack cocaine?

Legalizing hashish would set a precedent@; the same arguments justifying the legalization of hashish ("eliminate the mafias, reduce delinquency, and also reduce all the fantasies about drug taking") may well be used to legalize heroin, and even crack cocaine. Given the success of these arguments to justify the legalization of hashish, it would be near impossible to dismiss these arguments if they were to be used to justify the legalization of heroin and crack cocaine. A precedent has been set. In short, by accepting **A**, one has taken a decisive step toward accepting **A+** and **A++**.

(iv) Conclusion: Let's reject the legalization of hashish

The structure of the slippery slope argument parallels that of the argument from waste: **S. Waste:**
> Slippery slope: *Don't get started, you won't be able to stop!*
> Argument from waste: *Since you started, you must go on!*

The gradualist strategy and the slippery slope argument consider a hierarchized class of elements, **S. Gradualism**. The question is whether the status of these elements should be changed? The Gradualist is in favor of a change of status, and engages in a step-by-step, progressive modification of the existing hierarchy. The opponent considers that the status of the top elements can in no way be altered, and will use a slippery slope argument to counter the gradualist by opposing any change, however slight, in the status of the lower element.

The driving mechanisms invoked (often implicitly) on stage 4 can be very different:

— Organic, causal: The slippery slope designation is metaphorical, and clearly illustrates the physical movement of an ever faster physical fall. One could also evoke a domino effect, where the first falling domino pushes the second, the significance of each falling domino becoming greater and greater.

— Psychological: *"he who steals an egg will steal an ox"*.

— Strategic: The key point is the attribution of bad intentions to the proponent. The opponent may consider (as in our example) that the proponent is well intentioned, and that his or her public goal is indeed the authentic goal, and that he or she does not see the potential extreme consequence in which this might result. In this case, the proponent is portrayed as a *naive or idealistic* arguer, who doesn't see the consequences of what he or she is promoting, but nevertheless maintains his or her moral integrity. This development reflects Hedge's recommendation, not to attribute to the adversary hidden and manipulative intentions (sixth rule for honorable controversy, **S. Rules).** Nonetheless, maybe with a polemical intention, the proponent might be framed as *a Machiavellian character* with the active intention of developing a gradualist strategy, with the manipulative intention of implementing step-by-step, the most extreme measure beginning with the relatively benign one; hashish would be the bait initiating a priming strategy. So, the opponent will cast upon the proponent a suspicion of private, ugly intentions, **Motives and Reasons.**

Sophism, Sophist

The words *sophism*, *sophist* refer to very distinct realities in philosophy and in ordinary language.

1. The historical sophists

In Ancient Greece, the sophists were the first to implement a philosophy of language in their interactions with their fellow citizens. By means of calculated discursive interventions called "sophisms", the sophists destabilize the peaceful current representations of the world as seen through language. They emphasize the "arbitrariness" of language (within the Saussurian meaning of "arbitrary"), and so provoke naive speakers who consider language to be transparent and unproblematic. These discourses are not intended to deceive their audience, but to highlight to them the paradoxes of their current talk.

In the *Euthydemus*, Plato stages Socrates deconstructing sophistical arguments, such as the following one. Dionysodorus is a sophist, Ctesippus his naive interlocutor:

> [Dionysodorus:] — [...] And your father turns out to be [...] a dog.
> — And so does yours, said Ctesippus.
> — You will admit all this in a moment, Ctesippus, if you answer my questions, said Dionysodorus. Tell me, have you got a dog?
> — Yes, and a brute of a one too, said Ctesippus.
> — And has he got puppies?

— Yes indeed, and they are just like him.
— And so the dog is their father?
— Yes, I saw him mounting the bitch myself, he said.
— Well then: isn't the dog yours?
— Certainly, he said.
— Then since he is a father and is yours, the dog turns out to be your father, and you are the brother of the puppies, aren't you?

Plato, *Euthydemus*, 298d-e. CW, p. 737

This argument is not intended to convince Ctesippus that he is son and brother of a dog. The sophist does not *deceive* his listeners, he leaves them *confused and infuriated*.

The sophists have devised thought-provoking social and ethical paradoxes such as the following one:

> Antiphon the Sophist claimed that the law, by obliging man to testify the truth before the courts, often compels us to do wrong to one who has done us no harm, that is, to contradict the first precept of justice.
>
> Émile Bréhier, [*History of Philosophy*], 1928.[1]

The Sophists represent, with the Skeptics, an essential intellectual movement within argumentation theory. The Sophists established the principle of debate and irreducibly contradictory discourses, as the foundational cases presented in Antiphon's second *Tetralogy*, about a prosecution for accidental homicide; the case is discussed as follows:

> First Speech for the Prosecution
> Reply to the Same Charge
> Second Speech for the Prosecution
> Second Speech for the Defense[2]

The intellectual and social contributions of the historical sophists have been stigmatized by Platonic idealism that imposed on them deformations which were suffered until Hegel in philosophy and are surviving in common language. Ancient Sophists were no more *sophists* in the contemporary sense of the word than Duns Scott was a *dunce*.

2. Contemporary usage: the sophism, an intentional paralogism

In contemporary language, a sophism is an *eristic*, that is, a *misleading* discourse. From an interactional point of view, it is an embarrassing, false, manipulative and dangerous discourse, received as evidently false but the refutation of which is difficult. As any kind of discourse can be denounced by calling it a "sophism", the concept is essential for the analysis of the polemical reception of argumentative discourse.

---

[1] Émile Bréhier, *Histoire de la Philosophie*, Vol. 1, *Antiquité et Moyen Âge* [*Antiquity and Middle Ages*] Paris: PUF, 1981, p. 74.
[2] Antiphon, *Second Tetralogy*. KJ. Maidment, ed. Quoted after www.perseus.tufts.edu/hopper/text?doc=Perseus%3Atext%3A1999.01.0020%3Aspeech%3D3

A sophism is a paralogism enveloped in malicious speech, produced to pull the rug from under the opponent's feet. The distinction between *sophism* and *paralogism* is based on a charge of shameful intent, which may or may not be properly laid out. Paralogism is on the side of error and stupidity; sophism is a paralogism serving the interests or passions of its author. Under the principle of *"who benefits from the crime?"*, such an "error" is charged with malicious intent by the recipient and the potential victim. One moves from description to the accusation embedded in the negative contemporary use of *sophist*, *sophistry*, **S. Fallacy, Paralogism.**

Sorite

The word *sorite* is formed from the Greek word [*soros*], meaning, "heap".

1. Sorite paradox
The Sorite Paradox is one of the famous paradoxes proposed by Eubulide, a Greek philosopher, and contemporary of Aristotle:

> A grain of wheat is not enough to make a heap of wheat, nor two grains, nor three grains, & c. In other words, if **n** wheat grains do not make a heap, **n + 1** not more. So no amount of grains of wheat can make up a pile of wheat.

> Similarly, and if you take a grain out of a heap of wheat, you still have a heap of wheat, and so on, down to the last grain. So a grain of wheat is itself a heap of wheat.

This paradox can be illustrated by any collective name: *cluster, crowd, flock, army, collection, bouquet, collective…*

2. Rhetorical sorite
A rhetorical sorite is a discourse reiterating the same form:

> Cursed be
> The father of the wife
> Of the blacksmith who forged the iron of the ax
> With which the woodcutter fell the oak
> In which was carved the bed
> Where was born the great-grandfather
> Of the man who drove the car
> In which your mother
> Met your father!
>
> Robert Desnos, [*The Dove of the Ark*], [1923].[1]

3. Logical sorite: a chain of syllogisms
The term *sorite* also refers to a chain of syllogisms such that the conclusion of the first serves as a premise for the following one. The sorite is also called *poly-*

---

[1] Robert Desnos, *La Colombe de l'Arche*, 1923. In *Œuvres [Works]*. Paris: Gallimard, Quarto, 1999, p. 536.

syllogism "a polysyllogism is a series of syllogisms chained together in such a way that the conclusion of one serves as a premise for the next" (Chenique 1975, p. 255). *Serial* or *subordinate* argumentation are other names for *polysyllogistic* argument, and for this kind of *sorite*, **S. Serial Argument**

The term *sorite* may also refer to an *abbreviated polysyllogism* "in which the conclusion of each syllogism is not expressed, except the last" (Chenique 1975, pp. 256-257).

The critical problem with the polysyllogism is the reliability of the inference. In a formal system, the transmission of the truth is flawless, in a default argument chain, as the reasoning progresses the cogency of the conclusions weakens. In such series, everything happens as if the weights of the rebuttals grow exponentially, up to the point where the chain will break.

Other kinds of reasoning engage in the sorite paradox, for example:

— In *analogy*: **A** is analogous to **B**, **B** to **C**, ... and **Y** to **Z**. But is **Z** always analogous to **A**? **S. Analogy**.

— In *causal chains*, when the expected, theoretically perfect, "domino effect" is counteracted.

— In *interpretive reasoning*, which is why some Arabic legal schools refuse to interpret the sacred text of the Koran. They consider that only the starting point, the letter of the Sacred Text can be considered certain, and that engaging in interpretation would trigger a slippery@ slope process, leading to some unpredictable result, potentially contradictory with the undisputable content of the Sacred Text.

Stasis

1. The word *stasis*

The word *stasis* is borrowed from the Greek; it translates in Latin as *quaestio*, and, "in modern parlance", as *issue* (Nadeau 1964, p. 366).

In medicine, the word *stasis* is defined as "a slowing or stoppage of the normal flow of a bodily fluid or semifluid" (MW, *Stasis*); a stasis results in congestion, that is, in "an excessive accumulation especially of blood or mucus in an organ" (MW, *Congest*).

As used in rhetorical argumentation, the word *stasis* is a medical metaphor; medicine is a valuable source of examples and an important analogical resource domain for argumentative theory **S. Natural Sign**. In medicine, a state of stasis occurs when the bodily humors are blocked, and medical arts have to be applied to restore the correct flow of the fluids. Similarly, in the field of human action and interaction, a situation of stasis occurs when the consensual circulation of discourse is blocked by a contradiction or a doubt, and the argumentative arts must be implemented to restore the normal, cooperative flow of dialogue. Nadeau defines the situation of stasis as "a position of balance or rest" established between two opposite discourses (*id.*, p. 369).

In a state of stasis, the equilibrium is that of an *aporia*: "the Greek verb *aporein* describes the situation of the person who, finding himself in front of an obstacle, finds no passage"; the associated psychic state is *embarrassment* (Pellegrin 1997, art *Aporia*). In philosophical usage, an *aporia* is an insoluble contradiction.

2. The classical stasis theory

The first systematic formulation of a theory of stasis is found in Hermagoras of Temnos (late 2nd century BC; Benett 2005). The technique of stasis is used by rhetoricians before Hermagoras, but he was the first to formally identify and name the concept along with four basic kinds of stasis (Nadeau 1964, p. 370). This theory is best known via the treatise *On Stasis* of Hermogene of Tarsus, a Greek rhetorician of the 2nd half of the second century (Hermogene, *AR*; Patillon 1988). Hermogene distinguishes between:

(i) On the one hand, *misconceived* questions, upon which an argumentative debate cannot be built, either because their answer is obvious, or because they are undecidable; these questions are "incapable of stasis" (*id.*, p. 385); in other words they cannot be rationally discussed.

(ii) On the other hand, we have *well-conceived* questions, which can be rationally discussed.

Hermogene organizes the different kinds of general, well-conceived questions as follows (after Patillon, p. 57 sq.).

— Stasis of *conjecture*: Is the fact established?

— Stasis on the *definition*, upon "the name of an act" (Nadeau, p. 393): Someone robs a private person in a temple; is he a *temple plunderer*?

— The next step is the *qualification* of the act; it can be *rational* (discussed on the basis of good reason) or *judicial* (discussed on the basis of an existing law).

Judicial qualification is discussed under the following lines (after Patillon, p. 59).

Judicial Qualification

The defendant *does not admit* to the mischievousness of the fact: ANTILEPSIS ("contradiction, objection", Bailly, [*Antilepsis*])

The defendant *admits* to the mischievousness of the fact: OPPOSITION

    He *assumes* responsibility: COMPENSATION

    He *rejects* responsibility:

        and blames *the victim*: COUNTER-ACCUSATION

        and blames *somebody or something else*:

            who or which *can be guilty*: REPORT OF ACCUSATION

            who or which *cannot be guilty*: EXCUSE

3. The authentic "rhetorical question"

A stasis is a question, the *node* of a conflict articulating a judicial action in order to solve it. The *Rhetoric at Herennius* defines the first stage of a judicial encounter as the determination of the issue constituting the cause (*Ad Her.*, I, 18, 17):

> The issue [*constitutio*] is determined by the joining of the primary plea of the defense with the charge of the plaintiff (*Ad Her.*, I, 18, 11)

Quintilian explains that the *first* thing he does to disentangle an argumentative situation is to find the *quæstio*, the question, or *the issue*. The question "arises" when a statement made by a party is contradicted by the other party (note that the following text presupposes that adultery was a *crime*; that it was *legal to kill* an adulterer; and, apparently, that the executor was prosecuted for killing *the man*, while he also killed the *woman*):

> 5. First, then, (what is not difficult to be ascertained, but is above all to be regarded) I settled what each party wished to establish, and then by what means, in the following way. I considered what *the prosecutor* would state first: either an admitted or contested point. If it were admitted, the question could not lie in it. 6. I passed therefore to the answer of the defendant and considered it in the same way. Sometimes, too, what was elicited there was admitted. But as soon as there began to be any disagreement, the question arose. The process was of this nature: '*You killed a man*' —'*I did kill him*'. The fact is admitted, so I pass on. 7. The defendant ought to give a reason why he killed him. '*It is lawful*', he may say, '*to kill an adulterer with an adulteress*'. It is admitted that there is such a law. We may then proceed to a third point, about which there may be a dispute. '*They were not guilty of adultery*' — '*they were*'. Hence arises the question. (*IO*, VII, 1, 5-7; my underlining).

The question, that is to say, the point to be judged, is deduced from the nature of the reply given by the accused to the accuser. When the parties agree, the facts are considered to be established or "peaceful"; they are *disputed* when disagreement arises.

At the beginning of *On Invention* Cicero criticizes Hermagoras for having too general a view of argumentative questions, including philosophical as well as scientific questions, "Can the senses be trusted? What is the shape of the world? How large is the sun?" (*On Inv.*, I, 8, VI). Cicero limits the theory of questions to those belonging to the proper domain of the orator, the epidictic, deliberative and judicial genres. Nonetheless, the concept of question has no such pre-set limit.

The concept of stasis as a *question* is the counterpart in the rhetorical domain of the Aristotelian concept of *problem* in the dialectical domain (Aristotle, *Top.*, I, 11, 104b-105a10, pp. 25-28); a question is a *rhetorical problem*. The theory of stasis is the theory of "rhetorical questions" in the proper sense:
> The *constitutio* of the *auctor ad Herennium*, then, is the functionally dual stasis of Greek rhetoric [...] the psychical counterpart of which is the articulate question, or, as Sextus Empiricus (*Against the Geometricians*, III, 4) styled it, the "rhetorical question" (Dieter 1950, p. 360).

This meaning of the expression *rhetorical question* is quite distinct from the current meaning, which designates a question to which the speaker knows the answer, whilst also knowing that his interlocutors also knows the answer, and whose value is that of a challenge to potential opponents. To avoid confusion, we'll use the expression *argumentative question*, S. **Argumentative Question**.

4. Example
Facing the accusation, "*You have stolen my moped!*", the defendant may adopt different strategies which will determine the type of debate to follow.
(1) Denying having committed the act; the fact is not ascertained ("conjectural stasis")
> I did not even touch your moped!

(2) Recognize there has been a theft, and accuse somebody else:
> It's not me, it's him!

Idem, accusing the author of the accusation:
> It's not me, it's you, who accuse me, yet who destroyed your own moped to get the insurance premium.

This strategy, like the strategy of reorientation of the fact, manifests the tendency to radical refutation, by symmetrical reversal, S. **Reciprocity; Causality**.

(3) Recognize the fact, deny it was a theft, and re-categorize the action under a more honorable label, S. **Categorization**. This can be achieved via a number of different strategies:
> But this is *my* moped; you stole it from me last year!
> But this moped belongs to me, you pretended to buy it, but have never paid me.
> I didn't steal it, I just borrowed it. I asked you for permission.

(4) Idem, but invoking various kinds of extenuating circumstances:
> The gang leader forced me.
> I was just taking my grandmother to the hospital

(5) Idem, and apologize:
> I made a mistake, Mr. President.

(4) Recusing the judges (stasis on the procedure); disqualifying the accuser:
> It is not for the victor to judge the vanquished.
> But who are you to judge me?
> Suits you *(= the accuser)* well, you the gang leader, to complain of a theft! This should be solved by a good fistfight, as usual.

(6) Recognize the fact and claim to be proud of it:
> You were drunk, I saved your life by taking your moped, and you should thank me!

Maybe because of its spectacular character, the last case is known in the rhetoric of figures as an *antiparastasis*, **S. Orientation**. All these strategies are equally interesting, and all might deserve to be known by a specific name.

Some of these strategies are mutually exclusive, **S. Kettle**.

Strategy

A strategy is a complex set of coordinated actions, planned by an actor in order to achieve a specific goal. Strategies can be *antagonistic* or *cooperative*. *Antagonistic strategies* develop in non-cooperative fields of action, such as war, game of chess, or commercial competition. Such strategies serve to secure a decisive advantage over an opponent pursuing an antagonistic goal. Antagonistic strategies are covert, and discovered by the adversary as they are implemented, **S. Manipulation**. *Co-operative strategies* are developed by partners working together to achieve a common goal, from which both will benefit. The strategic intentions are transparent to all partners. A *research strategy* for example is an action plan to solve a problem; and teachers and students will collaborate to implement a *pedagogical strategy*.

In the military field, the *strategy* is set up before combat operations and tactics during the fight; *tactics* refers to the local implementation of a global strategy.

1. Argumentative strategies

An *argumentative strategy* is a set of speech choices planned and coordinated to support a point of view. Argumentative strategies are a subspecies of language and communicative strategies, speech and text construction strategies, interactional strategies.

An argumentative strategy is *antagonistic* if it is devised in order that the speaker may take the upper hand over the opponent.

There are two cases in which argumentative strategies may be *cooperative*. Firstly, if the speakers have the same argumentative role@, share a common point of view and collaborate to support it. Secondly, they may have different roles, and without identifying themselves with these roles, they might collaborate in the construction of a shared solution.

The phrase *argumentative tactics* is not used, but could be useful to refer to *local* argumentative phenomena as part of the *global* argumentative action. The choice to use such argument schemes can be seen as a *tactical* move, an implementation of a general policy. This does not suffice, however to define an argumentative strategy which requires the application of different kinds of instruments, for example coordinating the choice of words, the choice of arguments and self presentation (as open/closed to objections for example). An argument scheme can be identified on the basis of a brief passage, while the study of a strategy requires an extended corpus which fully represents a position.

2. Some examples of argumentative strategies

The first strategic level is that of the choice of answer to be given to the question, S. Stasis.

— A *defensive strategy* merely refutes the opponent's proposals.

— A *counter-proposal strategy* ignores the opponent's proposition **P** and argues a proposition **Q** which is incompatible with **P**. In such a context, argumentation may take an explanatory turn, S. Explanation.

— An *objectivizing strategy* focuses on objects without questioning people.

— *Ruining the discussion* is a way to ruin the opponent's arguments, S. Destruction.

— Bentham has identified types of arguments whose coordinated use defines a *stalling strategy*, a conservative strategy, aiming to postpone the debate in the hope that it will never take place: "*the conditions are not yet fulfilled for you to join the European Union.*"

— *Conciliation vs. breakthrough* strategies are characterized by the acceptance *vs.* refusal of concessions, the flexibility *vs.* radicalization of the proposals presented as compatible / incompatible. *Conciliatory* strategies use information accepted by the audience, present the conclusions and its recommendations as a continuation of previous beliefs and actions. *Rupture* strategies defy the audience, reject all its representations in order to replace them with new ones. The first strategy is *reformist*, the second is *revolutionary*.

These two last strategies are used by Paul, the Apostle of Christianity. In the following passages, in order to get the ear of the Athenians that he addresses for the first time, he uses a typical rhetorical *captatio benevolentiae* strategy, and begins his discourse with a reference to their own creeds, S. Rhetoric; Beliefs of the Audience:

> 21 The one amusement the Athenians and the foreigners living there seem to have is to discuss and listen to the latest ideas. 22 So Paul stood before the

whole council of the Areopagus and made this speech: "Men of Athens, I have seen for myself how extremely scrupulous you are in all religious matters, 23 because, as I strolled round looking at your sacred monuments, I noticed among other things an altar inscribed: To An Unknown God. In fact, the unknown God you revere is the one I proclaim to you.

Acts of the Apostles, 17, 21-23[1]

Nonetheless, the message was met by skepticism on the part of the Athenians. In particular, they questioned the message about the resurrection of the dead. Later, in quite different circumstances, Paul claims a rupture between his message and "the wisdom of the wise":

17 For Christ did not send me to baptize, but to preach the gospel, not in cleverness of speech, so that the cross of Christ would not be made void. 18 For the word of the cross is foolishness to those who are perishing, but to us who are being saved it is the power of God. 19 "I will destroy the wisdom of the wise, and the cleverness of the clever I will set aside." 20 Where is the wise man? Where is the scribe? Where is the debater of this age? Has not God made foolish the wisdom of the world? 21 For since in the wisdom of God the world through its wisdom did not come to know God, God was well pleased through the foolishness of the message preached to save those who believe. 22 For indeed Jews ask for signs and Greeks search for wisdom; 23 but we preach Christ crucified, to Jews a stumbling block and to Gentiles foolishness.

First Letter of St. Paul to the Corinthians, 17-23.[2]

3. "Strategic Maneuvering"

Pragma-dialectics has introduced the concept of *strategic maneuvering* to reconcile dialectical and rhetorical demands. The rhetorical requirement is defined as a search for efficiency: each party wishes its point of view to triumph. The dialectical requirement is a quest for rationality. During an encounter, each party simultaneously pursues these two objectives. In practice, the dialectical dimension is appraised according to the pragma-dialectical rules@ for the rational resolution of a difference of opinion. The rhetorical dimension is essentially communicational and presentational. It updates the classical demand that the issue and position must be presented in the correct language or format for the target audience (van Eemeren, Houtlosser 2006).

Straw Man

The label *straw man fallacy* refers to a special case of a general phenomenon, discourse and meaning *resumption and representation*.

The argumentative space is an interactional and textual space organized by a question@, where co-oriented, anti-oriented and third parties discourses articu-

---

[1] Quoted after www.catholic.org/bible/book.php?id=51&bible_chapter=17 (05-05-2017)
[2] Quoted after www.biblescripture.net/1Corinthians.html (05-05-2017)

late and overlap. S. **Orientation; Roles**. In such a space, argumentative interventions constantly refer to and influence each other, under quite different modes, ranging from explicit quotation, to more or less distorted repetition and reformulation, to the most allusive expression.

1. Discourse resumption

1.1 Direct, explicit quotation

An explicit quotation is expressed in a passage between quotation marks, accompanied by its space-time coordinates, so as to construct an unequivocal object: what was said, by whom, when, where.

Explicit quotations can be rejected by showing that they are incomplete, unduly de-contextualized, or misunderstood, S. **Authority; Circumstances**. Accordingly, an interpretative stasis@ can emerge about what the text really means and what the author has really said. It may be argued that a direct quotation is already an interpretation, and an indirect one certainly is. S. **Interpretation**.

In the case of direct quotation, the source text does exist; in the following ones, the very existence of such a text is problematic.

1.2 Indirect quotation

Indirect quotation of a position is presented by the speaker as a reformulation which paraphrases the original saying, or rephrases it in order to clarify its meaning and intention. The indirect quotation can be dismissed as tendentious or ludicrous, that is to say, consisting in an unfair reinterpretation of the original position, implying something that was never said, S. **Orientation; Epitrope; Prolepsis**

1.3 Allusion

Allusion to another discourse is no more than a trace that makes it possible to broadly locate the source discourse, without the possibility of designating any individual author or work. Its vague character is its best guarantee against contradiction, and a veil of mystery suits some kinds of authorities well.

2. Anticipating oppositions

The discourse can also be attributed to a voice staged in a polyphonic space, as in the case of anticipated objections, S. **Interaction; Prolepsis**.

3. Straw man

The discrepancy between what the quoted party said or might have said and what the quotation makes him say is the basis for the fallacy known as the *straw man* or *scarecrow* fallacy.

Undisputable refutation must be about what the other really claimed, S. **Relevance**. This requirement has a clear meaning in the case of written and referenced statements. In ordinary discourse, no segment is totally context-free, and its meaning is always an interpretation, so it is often not clear if someone has fully *said* something or not. In an argumentative situation, what the other has *actually* said is not a prerequisite but an issue in the argument.

The *straw man fallacy* is an accusation of mischievous representation of the diverging discourse. The expression is a metaphor on the substantive *straw man*, which literally refers to:

> a mass of straw formed to resemble a man. (Thes., *Straw man*).

As a metaphor, a straw man is:

> 1. a weak or imaginary opposition (such as an argument or adversary) set up only to be easily refuted.
> 2. a person set up to serve as a cover for a usually questionable transaction (MW, *Straw man*)

Given meaning (1), the straw man or scarecrow strategy corresponds to a tendentious, repulsive, even self-refuting, reformulation, of the discourse of something previously said.

Given meaning (2), the straw man strategy corresponds to a position which masks the real position of the arguing party. This position is advanced in order to set the public, or the speaker's opponents on a false track, **S. Red Herring; Relevance.**

Strict Meaning

> Lat. *a ratione legis stricta*, or *stricta lege*. Lat. *ratio*, "reason"; *lex*, "law"; *strictus*, "tight, narrow".
> Lat. *stricto sensu*: Lat. *sensus*, "thought, idea, meaning"
> Lat. *ad litteram*, *littera* "letter"

The argument based on the *strict meaning* of a normative text, or *principle of strict application of the law*, prohibits restricting or extending the provisions of the law or regulation beyond what it clearly says. Regulatory provisions must be taken literally, *stricto sensu*, to the letter. This can be considered to be a particular case of the principle "one does not interpret what is clear".

If the legal voting age is 18 years, then one cannot forbid people to vote on the day of their birthday because they are "barely" 18 years old, nor allow them to vote the day before their birthday because they are "almost" 18 years old.

From a linguistic point of view, "*he's almost 18 years old*" is co-oriented with "*he's 18 years old*", and "*he is barely 18 years old*" is co-oriented with "*he is not 18 years old*". The principle of interpretation *stricto sensu* cancels these co-orientations. The law establishes thresholds, and admits threshold effects whereas *almost* and *barely* blur the borders. **S. Orientation; Argumentative words; Juridical Logic.**

The argument of the *generality@ of the law* posits that the law must be applied *to all the concrete cases* it covers. The principle *of strict meaning* posits that it must be applied *according to its intended meaning* to all these cases.

Ad litteram "to the letter" et *ad orationem* — In a juridical context, the *ad litteram* label refers to an argumentation that complies strictly "to the letter" of the law, as opposed to its *spirit*. Interpretations appealing "to the spirit" of the law may be based on the *intention@ of the legislator*.

In a dialectical context, an *ad litteram* reply addresses the strict *letter* of the opponent's discourse, as opposed to its meaning as intended by the opponent. *Ad litteram* is then equivalent of *ad orationem*, S. Matter.

Structures of Argumentation

The expression *argumentative structure* is used in three different ways:

1. The theoretical structure of an *argumentation* corresponds to its internal organization, that is to say to the specific form of the "argument(s) – conclusion" relation in a given text or interaction S. **Layout; Convergent, Linked, Serial**.

2. The empirical structure of an argumentative *question* materializes in an *argument map*, which features the second- or third-level sub questions derived from the main issue, as expressed by the root question, S. **Script**.

3. The structure of an argumentative *text* corresponds to what classical rhetoric calls its *disposition*, the step-by-step organization of co-oriented and anti-oriented information and argumentations, S. **Rhetoric**. The structure of an argumentative *institutionalized interaction* fundamentally consists in the institutional arrangement of successive sequences dealing with the questions and sub-questions. *Ordinary interactions* include repetitions with variations of what was previously discussed. Argumentative texts and interactions routinely include non-argumentative sub-sequences.

These different structures can be depicted by diagrams, S. **Scheme**.

Subject Matter of the Law

Lat. *pro subjecta materia* argument. Lat. *subjectus*, "adjacent, near", *materia* "topic".

The interpretative argument from *the subject matter of the law* requires that the text of a law or of a regulation be interpreted not absolutely but depending on the material concerned, that is to say its specific *object* or *objective*, S. **Juridical arguments**.

In the following case, the interpretation derived from the object of the law leads to the redefinition of the expression *territory entirely covered by snow* as meaning "places where the snow layer is sufficient for tracking game", since the objective of the law is the protection of game. The same expression would be defined in a very different way if the objective of the law was, for example, the regulation of off-piste skiing.

> What is meant by, "*territory entirely covered with snow*"? If this condition were interpreted literally, the prohibition of hunting in snow would hardly ever produce any result. [...] The purpose of the law is to prevent the destruction of game, but this destruction is not prevented if I hunt, out of the woods, on land where I can follow the track of the game, although the neighboring grounds are stripped of snow.
> It is therefore of little consequence that the snow should be melted over an area of one hundred acres of rocks or marshy land, if I go hunting on the nearby land which remains covered with snow. Is it true that, in our hypothesis, the object of the prohibition would be evaded, if we admitted a contrary interpretation? — Quite obviously yes. It is therefore necessary to agree with our opinion, since the word "*entirely*" used here is impossible to apply literally. It is therefore necessary to attribute to it only the meaning and the scope that it contains *pro subjecta materia*.
> Thus I think that there is an offense whenever one is found hunting, outside the woods, on snow-covered ground, as long as one can track the game.
> Renaissance Joseph Bonjean, [*Code of Hunting*], 1816.[1]

The subject matter corresponds to the intention of the law, itself determined with reference to *the intention@ of the legislator*, here to protect game. If the subject is indicated by a title, the argument of the subject matter of the law and the argument of the title of the law (*a rubrica*) converge to the same conclusion. S. **Intention of the legislator; Title.**

The argument of *the subject matter of the law* is a quite different thing from the argument *to the matter@*.

Subjectivity

Just as it is a structuring feature of ordinary language, subjectivity is a defining condition of argumentation. Argumentative discourse is all about people, their characters, emotions, values and interests, as well as their knowledge and beliefs.

1. The person as an issue
Essentially, when involved in an issue, an individual may be "objectified" and treated in the same way as any other discursive object@. In particular, the person may be rhetorically constructed on the basis of a priori doxical knowledge, in order that he or she serve as a basis for pro or contra arguments concerning his or her role in the issue at stake, **S. Common Place**.

2. Values and interests
Values@ and interests, even the most specific and bizarre, contribute to the definition of a person's identity; truth is one of these values. Consequently, they will intervene in all the argumentative operations involving an assessment, such as in

---

[1] Renaissance Joseph Bonjean, *Code de la chasse*, vol. 1, Liège: Félix Oudard, 1816, p. 68-69.

an argument from the absurd@ or in a pragmatic@ argument. Values and desires are at work when a consequence is defined as *absurd*, *undesirable*, or *unwanted*.

3. Group character and emotions

One's rhetorical *ethos*@ is not defined as an individual, specific, psychological identity, but as the *public character* of an individual. All the same, rhetorical *pathos*@ is composed of a set of *public emotions*, not private feelings.

Rhetorical theory considers that the *group character and emotions* play a central role in public persuasion. Critical argumentation and fallacy theories take some distance from such agglomerations of individuals, condemning the futility of their emotions, the baseless charisma and authority of their leaders, abundantly labelled and rejected as "*ad —*" fallacies@".

When it comes to these issues, a *defensive* argumentation opposes *offensive* rhetoric. By enrolling the whole person in the battle of ideas and action, rhetoric adopts an *offensive* outlook. Conversely, critical approaches to argumentation take rather a secondary, *defensive* position.

3.1 Pathemic arguments

Points of view come with *affects*; both are correlative realities. On this basis, a sustained affective activity is a defining feature of an argumentative situation. **S. Pathos; Emotion**.

3.3 Ethotic argument

Rhetoric proposes a global, multidimensional approach to the person-group social interaction. The *character of the audience* sets the intellectual and affective conditions of the interaction, as well as the strategic construction of *the orator as such*, as embodying the values and virtues formally acknowledged by the audience, which can be the seven gifts of the Catholic Holy Spirit as well as the three Aristotelian democratic virtues, or the scientific virtues claimed by a plenary session audience. **S. Ethos**.

Global ethotic advantage can be analyzed along different dimensions, from charismatic power to scientific prestige, to delegated institutional authority. Among the different form of authority we find *expert authority*@, which consists in well-defined skills, which may be the easiest to assess. Insofar as it satisfies the condition of propositionality, any kind of authority can be sourced, quoted, and valued by default as peripheral evidence. **S. Authority**.

From a normative point of view, submission to an artfully designed charismatic-authoritarian ethos is analyzed as a *fallacy of intellectual inhibition* or unjustified humility (*ad verecundiam*), **S. Modesty**.

4. Universal or local knowledge

A specific subgroup of these fallacies concerns the knowledge and systems of representation specific to the target, the persons to be convinced or refuted.

From an epistemic point of view, the person is defined as a necessary limited synthetic focus of beliefs and knowledge. Commenting on Whately on the *ad hominem*, *ad verecundiam*, *ad populum*, and *ad ignorantiam* fallacies, to which he adds the *ad baculum* and *ad misericordiam*, Walton notes that these six fallacies taken as a whole are opposed to the *ad rem* and *ad judicium* argument (argument aimed at the thing itself, S. **Matter**). This opposition is based on the fact that the fallacious arguments all contain "a 'personal' element, meaning that they are source-based in some ways directed at a source or person (a participant in an argument) rather than at just 'the thing' itself. They all have a 'subjective' quality, as opposed to the 'objective' evidence traditionally appealed to in argumentation" (Walton 1992, p. 6).

These forms of argumentation take as premises the specific representations or circumstances of a person or a group; they are deemed fallacious because of their *localism*. In contrast to this judgment, the localism of the premises is at the root of the definition of argumentation as a "logic of subjects" (Grize), S. **Schematization; Default reasoning**. Subjectivity is seen not as a potentially manipulative limitation, but as the stamp of the fact that argumentation irreducibly does not deal with *absolute* truth but with a *revisable* process of combining knowledge with human interests, in critical discussions under the supervision of a structured community,

4.1 Causal assertions and human interests, S. **Cause-Effect argument**

4.2 Arguments based on the beliefs of the target

The arguer can choose to base his arguments on the beliefs accepted and the information known by the audience, therefore limiting his discourse to reorganizing and expanding these representations, S. **Character of the audience; Beliefs of the audience; Concession;** *Ex datis*.

4.3 Arguments based on a specific body of representations

Such arguments are referred to by invalidating labels, as appeals to superstition (*ad superstitionem*), to imagination (*ad imaginationem*), to stupidity or intellectual laziness (*ad socordiam*). These forms are sometimes associated with fallacies of emotion (*ad passiones*), which is strange, unless we qualify as emotional all the beliefs, nonsensical or not, we do not approve of, S. **Faith. S. Collections of arguments.**

4.4 Arguments based on the lack of knowledge

This lack of knowledge can be attributed to a person, S. **Ignorance**, or to humanity at large, S. **Vertigo**.

5. Silencing the opponent

A set of arguments is oriented towards the invalidation or elimination of the individual as an arguer. To refute the truth of an assertion carried by a person it

is shown that it leads to contradictions from the point of view of that person, which may result in silencing the person, S. *Ad hominem*.
In order to disqualify a point of view, negative characteristics are attributed to the individuals supporting this point of view, either in the particular encounter or in general. These negative features can bear any relation to the question under discussion, S. **Personal attack**.

Subordinate Argumentation ▶ Serial Argumentation

Superfluity of the Law

Ab inutilitate (legis); Lat. *utilitas* "utility, interest", *lex* "law"; argument of uselessness (of the law).

The argument of the *superfluity of the law* is a matter of legal logic, S. **Juridical Argument**. As it is based on the principle of legislative economy, it is also referred to as *the economy argument*, or *from uselessness* argument.
This argument presupposes that the code is well drawn, so that none of its dispositions paraphrases another. The code is supposed to be *laconic*. So, an interpretation of a law that would make another law redundant, so superfluous, must be rejected: "Under interpretation **I**, passage **A** becomes equivalent to passage **B**, which then becomes redundant and useless. We must therefore favor an alternative interpretation of passage **A**". This is a form of argumentation from the absurd@ (undesirable consequences). The new interpretation will be sought, for example, in the intention of the legislator.
By extension, the argument of the superfluity of the law applies to cases where the application of a law presupposes a state of fact. If entrance to a nightclub is denied to teenagers under 16, there is no need for a law prohibiting selling or serving alcohol to them in these places. Legislation on this would be superfluous. But if it is forbidden to sell alcohol to people under 16, then might can freely attend these institutions; otherwise the law prohibiting the consumption of alcohol would be superfluous.

Let us assume that a regulation prohibits participants from *voting* on matters of direct concern to them. The question then is whether the participants can take part in *discussions* about these issues? Should it be specified in the regulation that their presence in the assembly is authorized?
— Argument by the superfluity of the regulation: Yes, they can participate. No, there is no need for a specific provision. Suffice it to observe that to vote one must be a member of the assembly; if you are forbidden to vote, it is precisely because you are part of the assembly. If you were not in the assembly, then it would be superfluous to forbid you to vote. No further clarification is needed.

— Argument *"things which go without saying are better said"*: so, let's introduce the provision, *"all those concerned do not take part in the vote but participate in the discussion sessions on the questions concerning them"*. The new regulation is safer. In the first case, the price to pay is a subtle semantic inference; in the second, a slight redundancy.

Principle of economy applied to sacred texts

The principle of economy applies to sacred texts. Consider the problem of the application of the scheme of the opposite to a prescription expressed as: *"do not do this under such and such circumstances"*. In ordinary cases, the application of the rule of the opposite leads to the conclusion that: *"outside of these circumstances, you can do it"*. Sometimes the Koranic text explicitly mentions the opposite case (Khallâf,1942; Koran, 4-23), according to the scheme:

> Do not do this under such and such conditions. Out of these conditions, do so.

Whereas in other cases, the contrary is not explicit:

> Do not do this under such and such conditions.

The question in this second case is thus whether one can appeal to the scheme of the opposite to complete the text? If one adds *"under other circumstances, do it!"*, the literal precision given in the first case is rendered useless. If it is assumed that the Sacred Text is perfect, it expresses nothing useless or superfluous. In this case, nobody has the right *to add to it*, and to conclude anything about what to do or not to do. The supreme legislator remaining silent, the judge's decision will be founded on tradition, or on some other recognized source of legislation.

Syllogism

In the Aristotelian world, the theory of the syllogism covers all reasoning, in science, dialectic or rhetoric. In science, this is in logic, the syllogism is defined as "an argument in which, certain things being laid down, something other than these necessarily comes about through them" (Aristotle, *Top.*, I, 1). The classical syllogism is a discourse composed of three propositions, the "things being laid down" are the premises of the syllogism, and "something other than these necessarily comes about through them" is the conclusion.
Syllogistic inference involves two premises, while immediate inference is based upon one premise, **S. Proposition**.
The logic of the analyzed propositions concerns the conditions of validity of the syllogism. A *valid* syllogism is a syllogism such that, if its premises are *true*, necessarily its conclusion is *true* (the conclusion of a syllogism does not have to be a *necessary truth*, it's a truth that *follows necessarily from the premises*). The premises of the syllogism cannot be true and its conclusion false; in Aristotle's words, the syllogism, is a "demonstration" "when the premises from which the rea-

soning starts are true and primary, or are such that our knowledge of them has originally come through premises which are primary and true" (*ibid.*).

1. Terms of the syllogism
The syllogism articulates three terms, the *major* term **T**, the *minor* term **t** and the *middle* term **M**:

— The *major* term **T** is the *predicate* of the conclusion. The premise containing the great term **T** is called the *major premise*.

— The *minor* term **t** is the *subject* of the conclusion. The premise containing the small term **t** is called the *minor premise*.

— The *middle* term **M** connects the major and the minor terms, and consequently disappears in the conclusion, which is of the form < **t** is **T** >.

2. Figures of the syllogism
The form of the syllogism varies according to the subject or predicate position of the middle term in the major and minor premise. There are four (maybe only three) possibilities, which constitute the four *figures* of the syllogism, S. **Figures**. For example, a syllogism where the middle term is subject in the major premise and predicate in the minor premise is a syllogism of the first figure:

Major Premise	**M - T**	man-reasonable
Minor Premise	**t - M**	horse-man
Conclusion	**t - T**	horse-reasonable

3. Modes of the syllogism
The *mode* of the syllogism depends on the *quantity* of the three propositions which constitute the syllogism. A proposition may be *universal* or *particular*, *affirmative* or *negative*, giving a total of *four* possibilities.

Each of these four possibilities for the major premise may combine with a minor premise, also admitting four possibilities, to give a conclusion that also admits four possibilities, totaling $4 \times 4 \times 4 = 64$ forms.

Moreover, each of these forms admits the four figures, making all together 256 *modes*. Some of these modes are valid, others are not.

For example, the *first figure* of the syllogism corresponds to the case where, a universal conclusion derives from two universal premises. This deduction corresponds to the valid mode:

Major Premise	All human are reasonable
Minor Premise	All Greeks are human
Conclusion	All Greeks are reasonable

This mode is known as *Barbara*, where the vowel **a** marks that the major, minor and conclusion are universal, S. **Proposition**.

4. Example: the conclusive modes of the first figure
Syllogistic deductions are clearly exposed in the language of set theory.

— Two (non-empty) sets are disjointed if their intersection is empty; they have

no elements in common.
— The two sets intersect if they have some elements in common.
— One set is included in the other if all the elements of the first set also belong to the second set.
In what follows, **M** reads as "set **M**", similarly for **P** and **S**. The first figure of the syllogism admits four conclusive modes.

A - A - A syllogism
    A   every **M** is **P**,
    A   all **S** are **M**,
    A   *hence* all **S** are **P**

A - I - I syllogism
    A   every **M** is **P**
    I   some **S** are **M**
    I   *hence* some **S** are **P**

E - A - E syllogism
    E   no **M** is **P**
    A   all **S** are **M**
    E   therefore no **S** is **P**

E - I - O syllogism
    E   no **M** is **P**
    I   some **S** is **M**
    O   therefore some **S** is not **P**

These forms of basic reasoning is put to use in categorization, **S. Categorization and Nomination**

5. Evaluation of syllogisms S. Paralogisms

6. Syllogisms with premise(s) having a concrete subject

The preceding definitions correspond to the traditional (Aristotelian) categorical syllogism, bearing on quantified variables. The word *syllogism* is also used to refer to a form of reasoning where one premise has a *concrete subject*. A concrete subject is a subject referring to a unique single individual, by means of various expressions such as *this, this being, Peter, the N who*.

Syllogisms instantiating a universal proposition are examples of such syllogisms. These assign to an individual the properties of the class to which he or she belongs:
    the **x** are **B**
    this is an **x**
    this is **B**

The following type of reasoning is based on two concrete propositions can also be called, rather metaphorically, "syllogistic". It refutes universal proposi-

tions such as "*all swans are white*":
    This is a swan   *the proposition refers to a concrete individual*
    This is black   *the proposition attach a property to the same individual*
    Applied to the same subject, "being black" and "being white" are opposite@ *predicates*
    Therefore the proposition "*all the swans are white*" is false

7. Syllogistic forms with more than two premises

By extension, one also calls a *syllogism* an argument based on several arguments either linked@, convergent@, or again having the form of an epichereme@.

A chain of propositions where the syntactic form and mode of linking more or less mimic those of a syllogism may also be called a syllogism, **S. Expression.**

The famous syllogism "*everything rare is expensive, a cheap horse is a rare thing, so a cheap horse is expensive*" is developed on the basis of two contradictory premises, so it is normal that its conclusion be absurd.

Symmetry ▶ Relations

Synecdoche ▶ Metonymy

Systemic

The *systemic argument* is an overall argument, referring to a definition of a whole as a structure in which everything fits together perfectly. Taken literally, this principle asserts that each element of the system takes its full meaning only in light of its relation to the other elements in the system, and that it must be interpreted and applied accordingly. This applies to specific laws in collections of laws, as well as to statements and passages in sacred texts and literary masterpieces. This broad principle covers a set of argumentative techniques appealing to

    Consistency@,
    Completeness@,
    Economy or (non-)Superfluity@.

The argument from title@ postulate that the text is locally coherent.

The application of the *a pari*@, *a contrario*@, *a fortiori*@, schemes depends upon the effective realization of these systemic requirements in the considered system.

In the case of law codes, the systematic aspiration is immediately confronted with the de-stabilizing, complex forces of social and historical evolutions.

Syzygy

The word *syzygy* is an adaptation of the Greek word meaning "conjunction". In *astronomy*, a syzygy occurs when three celestial bodies are aligned, like the sun, the earth and the moon during an eclipse of the Moon.

In the *traditional Catholic exegesis*, there is a syzygy correspondence between two beings, events, actions, when 1) they are not contemporaneous; 2) they bear a strong *analogy@*; 3) and the first prefigures, signifies, or announces the second. The first, the precursor, is called the *type* and the second is termed the *antitype*. The Old Testament introduces the Types, the New Testament presents the Antitypes, "the Antitype not only repeats but completes and perfects the Type. [...] Noah, Abraham, Moses ... are "Types" of Christ" (Ellrodt 1980, 38). This vocabulary is specific, it has nothing to do with the model / anti-model perspective.

The syzygy principle orders history and the world, and, as such, founds the syzygy argument, used to an ever greater extent to establish significant links between the two "Type *vs.* Antitype" spheres. It exploits the resources of analogy@, *proportion@ and proportionality* and the argument of *progress@*: what comes before is analogous to, but has less being and substance than what comes after, in a two-state world. A variant of the syzygy principle projects the *mundane* world, considered to be a Type, onto the after-life, or *eternal* world, its Antitype, where it finds its raison d'être. The syzygy argument retains its pedagogical function, which is to give the believer an idea of future conditions: the *Monarch* is the Type, of which the *Almighty Father* is the Antitype.

> For him [Man] also he [God] hath varied the figures of combinations [syzygies], placing before him small things first, and great ones afterwards, such as the world and eternity. But the world that now is, is temporary; that which shall be is eternal.
>
> *Clementine Homilies*, 3rd Century (disputed).[1]

Syzygy-like principles can still be used, perhaps jokingly, to back a historical analysis. In France, on Brumaire 18th (November 9th), 1799, Napoleon Bonaparte engaged in a coup in order to establish his dictatorship. Half a century later, his nephew, Louis-Napoleon Bonaparte also seized power by force. Karl Marx comments as follows on the relation between these two coups:

> Hegel makes this remark somewhat that all the great events and historical personages are repeated, so to speak, twice. He forgot to add: the first time as a tragedy, the second time as a farce [...]. And we find the same caricature in the circumstances in which the second edition of the 18th Brumaire appeared.
>
> Karl Marx, *The Eighteenth Brumaire of Louis Napoleon*, 1851.[2]

---

[1] *Clementine Homilies*. Edimburgh: Clark, 1870, p. 38 (Homily II, Chapter XV. Quoted after www.ccel.org/ccel/schaff/anf08.html.
[2] *Brumaire* is the second month of the French Revolutionary Calender, corresponding to October-November; the Revolutionary year began in Autumn. *"The second edition"* is the nephew's coup.

Syzygy

The principle, "history repeats itself the first time as a tragedy, the second time as a farce" is an *inverted* syzygy.

Quoted after: www.marxists.org/archive/marx/works/download/pdf/18th-Brumaire.pdf. P. 5. (09-20-2013)

T

Tagging

To analyze an argumentative text or interaction means to tag them according to three main levels.

1. Delineating the different sequences that compose the text or the interaction under analysis. Characterizing the type and degree of argumentativity of these sequences.
2. For each argumentative sequence, determining the different lines of argument and their structures; the argument(s), the conclusion(s); the role of the counter-discourses, that is, the kind of mutual criticism and evaluation implemented by each argumentative line.
3. Specifying the argument schemes.

The analysis of an argumentative sequence must be based on relatively objective criteria, that is to say explicit, stable and shareable criteria, even if not always decisive. The analysis of an argumentative passage is an argumentative activity, whose claims can be criticized and must be justified.

In formal language, one would have *markers*, that is to say, univocal and automatically identifiable material elements that would allow us to hold discourses such as: "Presence of marker(s) **S**: so, this passage is an argumentative sequence. Presence of marker(s) **A**: so, this segment is an argument. Presence of marker(s) **C**: so, this segment is a conclusion. Presence of markers **T**: the underlying argumentation scheme is of such and such a type".

Natural language arguments do not have such markers. Actual linguistic markers are systematically polysemic and polyfunctional. Their strictly argumentative function must be evaluated according to the context. It is as much the context that designates a marker as argumentative as the marker that designates the context as argumentative.

1. Delineating an argumentative sequence

1.1 Sequencing the language flow

At the most general level, if we postulate that language or speech are by nature argumentative the problem of identifying specific argumentative sequences does not arise. If we postulate that, within the macro language datum (text, interaction), only *some* sequences are *more or less* argumentative, these sequences must be cut out from larger language flow.

A sequence is *a relevant analytical unit*. The argumentative passages that are exploited in text book as well as in scientific presentations, are the product of this first sequencing operation. Sequencing and subsequencing the language flow is, most of the time, a routine and intuitive operation. The selection of a relevant argumentative passage is nonetheless the first problem the analyst must confront, and his or her choice of giving such and such frontiers to the sequence under analysis should therefore be explicit and justified. The correct implementation of this operation supposes a foray into the broader domains of case and corpora building.

In classroom interactions, for example, the sequence, *"problem resolution"* is distinct from the sequence, *"homework and instructions"*. In a meeting, the sequence, *"agenda setting"* is distinct from the sequence *"discussion and decision about agenda item n° 19"*. For a participant, identifying the sequence simply means "knowing what is presently being done".

Sequences and subsequences of any kind are defined externally by their *boundaries* and internally by their own *structure* and foreground *activity*.

Externally, the boundaries of the sequence are *transition points* characterized by topic changes, by specific closing and opening formulas, and by a re-design of the interaction format.

From the *internal* perspective, the sequence is defined by a type of linguistic activity, by a specific interaction format, and by a semantic-thematic coherence, which globally obeys a completion principle. Exactly what a complete sequence consists in depends on the kind of sequence envisaged; the internal principle of completeness of a *problem resolution* sequence is not the same as that of an *agenda setting* sequence.

1.2 Delineating argumentative sequences

Argumentative sequences are isolated according to the same internal and external criteria, argumentativity being their specific difference.

Argumentation can be *the defining activity* of the sequence. It follows that the sequence, *"discussion and decision about agenda item n° 19"* should normally be highly argumentative. The external boundaries of such an institutionalized argumentative sequence depends upon the rules of the institution.

Argumentation can be an *emerging activity* in every kind of sequence; for example in (**R**), if somebody disagrees or makes other suggestions about the agenda. The left boundary (opening) of an emerging argumentative sequence is characterized by the concretization of an opposition into a *question*@. The right boundary (closing) can be of any kind, and is sometimes easier to grasp when it is considered in contrast to the following sequence; for example, the Chair looks at the clock and says *"Well, I suggest that you further discuss this very interesting point during the coffee break. Thanks for your participation"*.

When dealing with *a local issue*, an argumentative situation may *develop and close on the spot*, possibly leaving no trace in the memory of the participants.

When dealing with *a pre-existing question*, such as a socio-scientific issue, the present discussion is just one episode in the larger development of the question as discussed in various settings and crystallized in a specific script@. In this case, the question has a story and the sequence is *only an episode*, which will not close the file.

2. Re-composing argumentative lines: coalitions; argument(s), conclusion(s); criticism and counter-discourse

The internal structure of the argumentative sequence is characterized by the type and the density of the argumentative operations it articulates. Classical analytical points include the following:

— If the text is a multi-speaker interaction or a polyphonic monologue, the analyst must attribute their own to every participant, that is the *positions* they hold and the *roles* they play in the global dispute. Positions are identified as the segment providing an answer to the argumentative question. Experience shows that this apparently elementary task may be rather challenging.

— Once the exact content of the oppositions and positions has been determined, one can observe the *systems of coalition* of the relevant positions, as well as their evolution in the dispute.

— The positions being thus localized, one looks at how the surrounding discourse ties with the proposed conclusion-answer, that is to say one *identifies the argument*(s). The markers of argumentative function, if any, help to identify the passage as an argument or a conclusion, **S. Indicators.**

— The analysis of critical strategies implemented by the participants bears upon the different modalities of counter-discourse management: direct repetition of other discourses, or evocations, reformulation, of these discourses, rebuttals of the opponents' arguments. **S. Destruction; Refutation; Objection.**

— A most interesting point is the observation of the interactions between argument's participants, not only the opponent-proponent interactions, but also their interactions with the third parties, and the public at large.

— Observations about the relation between the arguments developed on the spot by the participants and the general script attached to the question, when available, will be always very instructive.

— Further issues involve the global characterization of the argumentative line developed by key participants, as implementing such-and-such strategy@.

3. Argument schemes

The argument scheme can be *explicitly formulated* in the passage as a sort of general law. In order to identify it, one might look for generic statements, which generally function as good supports for the affirmation of values, principles or laws.

To decide upon the correct argument scheme, one must investigate whether there is an acceptable paraphrase relationship between the *generic discourse* corresponding to the argument scheme, and *the actual current* argumentative discourse under investigation (in classical terms, between the *topos* and the *enthymeme*); for a detailed example of this mapping, S. **Argument Scheme**. The operations needed to determine whether such an argument scheme is reflected in the actual argumentation depend heavily on the schema involved. The same concrete argumentative discourse may be matched by several, non-exclusive schemes.

Taxonomies and Categories

The theory of *categories* lies at the heart of *taxonomies*. In turn, taxonomies represent a series of *coordinated scientific definitions*. Correctly articulated in taxonomies, such definitions mirror valid *syllogistic reasoning*. The world organized in a taxonomy represents the deep structure of reality; reading the taxonomy is a reasoned voyage through this world. Until the development of mathematics and their application to experimental sciences in the modern period, and the emergence of formal logic at the end of the nineteenth century, the theory of categories served as an introduction to logical reasoning, that is, to scientific reasoning.

From the point of view of argumentation, this traditional system (category-taxonomy-syllogism) defines *logic* as an "art of thinking" in natural language. It is the basis for reasoning about categorization@, nomination and definition@ or analogy@ either in the explicit form of arguments bearing these names, or implicitly present in other forms of arguments.

The theory of categories was developed by Aristotle in the *Topics* was reconstructed by Porphyry (c. 234 – c. 305 AD) in the *Isagoge*, "Introduction", and transmitted in Latin to the Middle Ages, mainly by Boethius (c. 480-525).

1. Taxonomies

The category system provides the rules for the construction of correct taxonomies. A taxonomy is a reasoned hierarchized classification of beings, a nested system, represented by an arborescence. The position of an entity in a taxonomy corresponds to its definition, and its definition determines its place in the taxonomy to which it belongs.

This "classificatory thinking" has produced impressive results in the classification of natural entities. Every entity is classified at its proper level, in a global, comprehensive hierarchy, on the basis of its common and specific properties. At the very top of this great pyramid of classification, are the plant, animal and mineral *kingdoms*. Such a kingdom includes a number of *orders*; an order includes *families*; a family includes several *genuses*; and a genus includes several *species*, producing the following pattern of nested succession:

Kingdom => Order => Family => Genus => Species : {*Individuals*}

A *species* is a set of individuals. It is the basic unit of taxonomy. In the animal kingdom, the individuals which make up a species come from the same, or similar, parents, and they can interbreed.

The above series of categories creates a seven-level taxonomy. Depending on the complexity of the kingdom considered, other intermediary levels must be introduced, for example: Kingdom => Division => Class => Order, etc.

As a knowledge domain, a taxonomy requires a well-made denominative language, which is transparent for the specialist. Latin names are used to that end. The *fairy ring* mushroom, or *mousseron*, for example, is known scientifically as *marasmius oreades*. This name corresponds to the following taxonomy: Genus: *marasmius*; Family: *marasmiaceae;* Order: *Agaricales*; etc.

The simplest three-level taxonomy includes the following three levels:
   superordinate category: "— *is a mammal*"
   basic category: " — *is a dog*"
   subordinate category: "— *is a Labrador*".

Beings are identified and designated primarily by the name of their "basic" category, characterized by its *frequency* or its perceptual, cognitive or cultural salience. Non-specialists first identify an animal as *a dog*, not as *a mammal* or *a labrador*.

The concepts of *hyponym* and *hypernym* are used in semantics to refer to pairs of terms in a hierarchic relationship. The hyponym relationship corresponds to the *genus to species* relation "*rose* is an hyponym of *flower*, all roses are flowers". The hypernym relationship corresponds to the *species to genus* relation, "*flower* is hypernym of *rose*, some flowers are roses".

2. Categories

In the Aristotelian system, the goal of science is to build stable taxonomies of entities according to their common properties and specific differences. The fundamental intellectual problem is how to correctly *categorize* an individual and *hierarchize* the various categories of individuals. This task leads to more or less convincing results depending on the kind of entities considered. We already have meaningful taxonomies of mushrooms, for example, whilst we continue to lack a taxonomy of affect, emotions and moods — and we must ask whether building such a taxonomy is possible at all.

Aristotelian theory of categories provides the tools needed to build definitions for situating terms in taxonomies. It distinguishes between five categories: genus-specie-difference-property-accident. The exact logical-metaphysical status of these concepts is disputed, but the problem is clear: which logical-semantic structure can we give to statements like the following?

Suzan is a human.
Humans are animals
Humans are rational.
The horse neighs (horses neigh)
The (this) horse suffers.

The analysis in terms of categories says that:

— "*Suzan is a human*" predicates the *species*, "man", of the *individual*, Suzan.
— "*Humans are animals*" predicates a *genus*, "animal" of a species, "man".
— "*Humans are rational*" predicates a *difference*, "rational" of a species, "man". *Human* and *horse* are two species belonging to the same *genus* animal; unlike the horse and other animals, man is endowed with *reason*, which is the defining difference between man and other animals.
— "*Horses neigh*": in its generic interpretation, this statement attached to the species horse, a property, "— neighs". The *property* is a non-essential characteristic of a species; that is (all) horses neigh, and only horses neigh. The definition of man as a "*featherless biped*" is *extensionally* valid; on this basis, one can tell a human from any other being. *Essentialist* philosophy reproaches such definitions based on properties for saying nothing of what *is*, in *essence*, a human being.
— "*This horse suffers*" predicates an accident upon an individual. The accident belongs only to individuals, not to species or genus. The horse cannot be characterized, at any level, as "a suffering animal"; a particular horse can *suffer* or not, depending on the circumstances, it cannot, however, *be a mammal* or not.
— Suppose that the statement "*some clouds are grey*" and "*all sparrows are grey*" are true. Color is an accidental property of clouds, whereas it is a common characteristic shared by all sparrows, but not exclusively: elephants are also grey. This property, "*being grey*" cannot serve as a basis for clouds and sparrows to be classed within the same natural *genus*. At most, we can say that, in term of their color, indeed, some clouds are like sparrows. If one argues that clouds and sparrows belong to the same category, due to this common property, the analogy would be deemed as misleading, **S. Analogy (II); Metaphor.**
— An object is *known* when it has been successfully defined, that is, classified. It is associated with identical objects in the same category, and disassociated from objects belonging to different categories. This knowledge is not attached

to it as a *particular individual*; this is what is meant by the expression "there is no science of the contingent".

3. Syllogistic arguments and natural taxonomies

Predicates are organized in taxonomies according to their generality. The tree-structure of the system of categories allows for valid syllogistic inferences. A taxonomic space defines a syllogistic space: to reason means here to move in a controlled manner from one branch to the other in a "Porphyrian tree".

A well-constructed taxonomy relies on definitions and authorizes inferences based *on the nature of thing*s: "— *is a labrador*" implies "— *is a dog*", and both also imply "—*is a mammal*" S. **Definitions (II).** Hence the syllogism:

Labradors are dogs, dogs are mammals, SO labradors are mammals

All **L** are **D**	*Labradors are dogs*	*Labrador is a species of genus_1, dogs*
All **D** are **M**	*Dogs are mammals*	*Genus_1 is a sub-genus of genus_2, mammals*
All **L** are **M**	*So, Labradors are mammals*	*Labrador is a sub (subspecies) of genus_2 mammals*

From the definition

"*humans*definiendum *are [reasonable*difference *animals*genus*)] definiens*",

one can construct the valid syllogism:

Human are animals	all **H** are **A**
Human are reasonable	all **H** are **R**
SO, some animals are reasonable	some **A** are **R**

Conversely, if the genus **C** includes the species **E1, E2, ... En**, then we immediately infer the truth of the disjunction:

"*to be a* **C**" implies "*to be either a* **E1**, *or a* **E2** *or ... or a* **En**"
"**X** *is a mammal*" means "**X** *is either a human, or a rat, ... or a whale*".

Other implications are based on the fact that the genus is characterized by a set of properties that belong to all the species included within its scope. If "*being a mammal*" is defined as "*being a vertebrate, warm-blooded, having a constant temperature, with pulmonary respiration, nursing the cubs*" then all of these properties can be attributed to every mammal, regardless of their differences, that is, regardless of the species they belong to.

4. Arguments destabilizing socio-linguistic categories

Scientific categorization determines the exact position of a particular individual or of a class of entities in a taxonomy, where the terms have been given an essentialist definition from which it is possible to argue syllogistically.

Linguistic nomination-categorization@ assigns to an individual a current name and the category covered by that name. This operation could be considered to be *the* basic argumentative technique. The simple and stable system of scientific-Aristotelian categories is replaced by the infinitely complex system of meaning relationships in a given language. The argument can no more proceed by syllogism on essentialist definitions, but must operate by derivations out of the heterogeneous elements assembled in a linguistic definition.

Socio-linguistic categories are said to be fuzzy and poorly defined; they are actually *evolving* categories, in a process of permanent de-stabilization and re-stabilization under the pressure of historical evolution and language change. They are *debatable* and *adjustable*. S. *A pari*; Analogy (II).

"Technical" and "Non-Technical" Evidence

1. The opposition "technical" / "non-technical"

Aristotle distinguishes between two instruments of persuasion@, or rhetorical proof (*pistis*):

> Of the modes of persuasion some belong strictly to the art of rhetoric and some do not. By the latter I mean such things as are not supplied by the speaker but are there at the outset — witnesses, evidence given under torture, written contracts, and so on. By the former I mean such as we can ourselves construct by means of the principle of rhetoric. (*Rhet.*, I, 2; RR, p. 105)

Rhetoric is a *technique*, a "techne", that is "a process that strives for perfection, occurring by means of the deliberate action of a human being" (Lausberg [1960] §1). That is why rhetorical proofs are called *technical* proofs (or *artificial* proofs), as opposed to non-rhetorical, or non-technical proofs (Lausberg [1960], § 351-426).

The distinction is made in relation to the judicial situation. "Technical" proofs are discursive, oratorical means of pressure. "Non technical" proofs are proofs in the contemporary sense of the word. They correspond to the facts submitted to the court, the "given" material facts which the orator cannot manipulate, for example, the public declarations made by the witnesses, or the contracts signed between the two parties. The language of the contracts and the discourse of the witnesses is considered to be *free from rhetoric*, at least in their basic formulation, whilst the language of the orator is *rhetorical*, that is, endowed with an autonomous power of persuasion, if necessary, challenging material evidence, S. **Pathos**.

Theoretically, the technical *vs.* non-technical distinction bears on the three components of rhetorical action:

— *Technical logos* is defined by the use of topoi, signs, etc.; *non-technical* logos is defined by the parallel intervention of witnesses, contracts, etc.

— *Technical ethos* is a speech product, the self-made image of the orator embedded in discourse (Amossy 1999); non-technical ethos is the other-made image of the orator, something like his or her reputation, which can run counter to the first.

— The *technical pathos* is the strategic *emotive* communication, and the non-technical pathos corresponds to the spontaneous *emotional* communication, S. **Emotion**.

2. A now misleading terminology?
— First and most importantly, this terminology is now opaque and counter-intuitive, incompatible with the contemporary use of the word "technical".

— *"Non-technical" proofs* are just *proofs* in the ordinary contemporary sense of the word.

— All the elements called "non-technical" are the rough material submitted to the court. They may be practically sufficient to settle the issue definitively, yet, "[they] require, very frequently, to be supported or overthrown with the utmost force of eloquence" (Quintilian I. O., V, 2, 2). In other words, the speaker must interpret the testimonies, the content of the contract, the confessions, in order to transform this rough material into data, that is, to orient them towards a conclusion in favor of the party he represents. So, "non-technical proofs" are not entirely free from language and rhetoric.

— The distinction is closely linked with judicial situations. Argumentation also deals with very different situations, such as, for example argumentation developing in everyday semi-technical discourse and presenting mixed kind of proofs. For these reasons, and to highlight terminological difficulties, the terms "technical" and "non-technical", used in the sense they have in traditional rhetorical theory, are placed within quotation marks.

Such questions may seem somewhat irrelevant, but wrongly so. The opposition has a distinct structuralist nature, emotion, character, and situation being re-defined as discursive objects*. This position has proven fruitful; nonetheless, it has its limits. The issue is the definition of the object of argumentation studies: should they be *purely discursive data*, and we shall then take into account only purely linguistic phenomenon, or *contextualized interactional discourse*, taking into account the situation and the ongoing collaborative actions?

3. "Non-technical" proofs

3.1 What are "non-technical" proofs
The "non-technical" evidence is the material, the factual elements, brought before the court. They define the issue structuring the debate, and may include elements such as, "witnesses, examinations, and like matters decide on the subject that is before the judges" (Quintilian, *IO*, V, 11, 44). They might receive a secondary rhetorical treatment, but their existence is beyond the reach of the rhetorician.

The list of non-technical evidence varies somewhat. Aristotle counts as non-technical, "witnesses, evidence given under torture, written contracts, and so on." (*Rhet, op. cit.* supra). Quintilian counts as non-technical "precognitions, public reports, evidence extracted by torture, writings, oaths, and the testimony of witnesses, with which the greater part of forensic pleadings are wholly concerned". (*IO*, V, 1, 2). Traditional lists include the following elements.

Precedent, authority, rumors — An important means of argument, *appeal to authority@*, has sometimes been considered to be technical, sometimes to be non-technical. Different kinds of authority must be distinguished, and first, the authority of a judicial decision made by a judge recognized for his or her competence, and passed into judiciary custom; in other words, a precedent turned into a law, **S. Precedent**. This is clearly a "given" of the judiciary situation, that is to say, a "non-technical" element. An authoritative statement can also be understood as (a declaration of) a socially recognized personality.
"Precognitions", prejudices, rumors, public opinion is a form of authority, perhaps linked to the possibility of recourse to an appeal to the *populus* in Roman law, **S.** *Ad populum*.

Contracts, written documents — Whether or not a contract was signed is a fact, which may be decisive.

Material elements, such as the murder weapon, or the bloody tunic of the victim, etc., all make up the judiciary hardware.

Oaths, which guarantee the citizen's declaration.

Witnesses, torture —Taking into account "reports" obtained under torture is shocking, and reminds us that the old democracies and republics put up with slavery and torture. The *Rhetoric to Herennius* presents the basic arguments about the use of such reports. The first argument advocates that they be considered in court, the second that they be discounted:

> We shall speak in favor of the testimony given under torture when we show that it was in order to discover the truth that our ancestors wished investigations to make use of torture and the rack and that men are compelled by violent pain to tell all they know [...] Against the testimony given under torture, we shall speak as follows: In the first place our ancestors wished inquisitions to be introduced only in connection with unambiguous matters, when the true statement in the inquisition could be recognized and the false reply refuted; for example, if they sought to learn in which place an object was put, or if there was in question something like that which could be seen, or be verified by means of footprints, or be perceived by some like sign. We then shall say that pain should not be relied upon, because one person is less exhausted by pain, or more resourceful in fabrication than another, and also because it is often possible to know or divine what the presiding justice wishes to hear, and the witness knows that when he has said this his pain will be at an end.
>
> *Ad Her.*, II, VII, 10

The list of "non technical" proofs might be extended by elements such as the following:
— For some cultures or persuasions, *miracles* can create a form of non-technical persuasion.

— In the early Middle Ages, the *trial by ordeal*, or God's judgment, was also believed to establish the truth in a non-technical way: if the accused is able to pass through the fire more or less alive, he or she is acquitted; if he or she dies, the punishment proves the guilt.
— In contemporary times, it would be necessary to add the forensic proofs, for example the DNA tests. All these proofs would be typically called *technical*, in contrast with *rhetorical* or language-based argument.

3.2 Superiority of non-technical evidence

In current cases, the facts, documents, witnesses or material evidence, make it possible to solve the problem and settle the dispute, "when one of the parties had "non-technical" evidence, the case was clear for the judges, and needed only a few words" (Vidal 2000, p. 56). Factual evidence is clearly essential in the judiciary, language playing of course an important role for the presentation and orientation of facts. Evidence produced by rhetorical arguments is used only when factual evidence is missing, for lack of anything better.

"Non-technical" evidence is paramount in actual judiciary decision. "Technical" evidence comes to the fore in rather special cases, when any legal document, any physical evidence, testimony, etc. is lacking. Such an exceptional situation is staged in the comical anecdote where Tisias opposes Corax. Corax has agreed to teach Tisias his rhetorical skills, and to be paid according to the results obtained by his student. If Tisias wins his first trial, he will pay his master; if he loses, he will have nothing to pay. After completing his studies, Tisias sues his master, claiming that he, Tisias, owes nothing to him, Corax. Indeed, this is Tisias first trial, and either he wins or he loses. First hypothesis, he wins; *following the judge's verdict*, he owes nothing to his master Corax. Second hypothesis, he loses; *in virtue of their private agreement*, he owes Corax nothing. How can Corax meet this challenge? He repeats verbatim Tisias' pattern of argument but in reverse. First hypothesis therefore Tisias wins the case; by private agreement, Tisias must pay. Second hypothesis, Tisias loses the case; by law, Tisias must pay for the education received. In both cases, Tisias must pay. It is said that the judges kicked out the litigants.

Rhetorical proofs, operating in a language having no contact with the world, are proofs by default. This is a borderline case, not a prototypical case of the argumentative rhetorical use of language, which, in general, operates with elements of reality and conventions that constrain its unbridled use.

Testimony

A testimony is a particular kind of authoritative statements, S. **Authority**:
— Preliminary Conditions:
- The issue **I** under discussion is related to an event **E**.
  Criminal or not, **E** has an exceptional character.

- Person **W** was in a condition to see or hear something about the event **E** at issue.
- Some discussants have a limited access, or no access at all to **E**,
- So, in the discussion **I**, **W** qualifies as a witness to **E**.

— Essential Conditions: testimonies are subject under a special truth commitment:
- **W** says **T**.
- **T** is relevant to **I**; **T** is presumed to be True.

1. Criticism of testimony

Discourses against testimonies of any kind are based on two options, examination of the fact, questioning of the witness. These discourses may be schematized as follows.

— The fact is in itself not credible, not possible, not probable.
— The witness is not credible because:
- He or she could not see, hear… what he or she pretends to have seen, heard…
- He or she is partial, biased, or lies.
- He or she is not competent; he or she has been manipulated; corrupted.
- In other cases where his or her testimony could be verified, it proved wrong.
- Other witnesses say otherwise.
- He or she is the only witness, so his or her testimony is not acceptable (*testis unus, testis nullus*, "one testimony, no testimony")

Testimony and criticism of testimony play a particularly important social role in judicial matters and in matters of faith. They also play an important private role in everyday issues, insofar at they underlie the narratives of witnesses to critical life events.

2. Testimony in rhetorical argumentation

In the *Topics*, Cicero clearly posits judicial testimony as part of the data the court must rely upon, as opposed to discursive proofs, S. "Technical" and "Non-Technical". According to the modern democratic concept of testimony, witnesses are in principle granted the same status; they and their statements are subject to the same critical examination. The ancient rhetorical concept of testimony is quite different. According to Cicero, in Roman court practice, the weight of a testimony is *a priori* proportional to the social authority granted to the witness.

> For our present purpose, we define testimony as everything that is brought in from some external circumstance in order to gain conviction. Now it is not every sort of person who is worthy of consideration as a witness. To gain conviction, authority is sought; but authority is given by one's nature or by circumstances. Authority from one's nature or character depends largely on virtue; in circumstances there are many things which lend authority, such as talent, wealth, age, good luck, skill, experience, necessity, and even occasional fortuitous events. (Cicero, *Top.*, XIX, 73; Hubbell, p. 439)

In this quotation, the mentioned "circumstances" include basic elements determining the social status of the witness. "Necessity" refers to testimony collected under duress and torture. The expression "fortuitous events" refers to emotional speech, emotion being considered a guarantee of truth.

In the Roman world, testimony was actually guaranteed not only by an in-depth examination of the witness and the alleged fact but also by the precise status of a witness, if a citizen, or by the amount of pain the witness could bear, if a slave. The use of torture to gain true information is now morally condemned, and practically recognized as an ineffective means to gather information, *"Beer, cigarettes work better than waterboarding"*[1].

The concept of testimony in ancient texts covers a wider field than personal testimony about a particular fact. Testimony can also guarantee principles, and in that case, the witnesses are "the ancient authors, the oracles, the proverbs, the sayings of the illustrious contemporaries" (Vidal 2000, p. 60).

3. Testimony in matters of faith

The capacity of truth to be more compelling than any kind of pain is inherent to the Christian tradition of martyrdom. The word *martyr* comes from the Greek word meaning *witness*; the martyr is a witness of the divine Word. Martyrdom is a kind of torture; truth is certified through torture. As Pascal states: "I believe those stories only, whose witnesses let themselves be slaughtered" (*Thoughts*, p. 117).

The validation of testimony through martyrdom leads to a paradox. In reality people have been tortured and killed for a variety of beliefs and values; Giordano Bruno for example is a "martyr of atheism". The proposal must therefore be reversed, and, according to Saint Augustine's saying, "it is not the penalty but the cause that constitutes a martyr."[2] If the cause is bad (heresy), the "martyr" is an offender and has been punished as such.

Confession and testimony — Denials are weak arguments for innocence, but avowals are strong arguments for culpability. When it contradicts a denial, a testimony is believed over and above the *denial*. But consider the case of a person accusing himself or herself of a horrible murder but confronted with a testimony claiming that materially he or she cannot have committed the crime. The general question is whether avowals, *confessions* should be trusted over *testimonies* to the contrary? Is witnessing reflexive? That is, can I bear witness of myself, of my own actions? The Evangelist John reports that Christ said no, "If

---

[1] "Mattis to Trump: beer, cigarettes work better than waterboarding". http://www.military.com/daily-news/2016/11/23/mattis-trump-beer-cigarettes-work-better-waterboarding.html (07-05-2017)

[2] Augustine, *Second Discourse on Psalm 34*. In *St Augustine on the Psalms*. Trans. and An. by S. Hegbin, and F. Korrigan, Vol. II, Ps. 30-37. New York & Mahwah: The Newman Press, p. 220.

I bear witness of Myself, My witness is not true." (John 5:31)[1]; S. **Relation** (§ **Reflexivity**).

Paradox of weak testimonies — The Latin word *testis* means "witness" and "testicule". In Roman culture, as in some contemporary cultures, testimony is the reserve of men; a woman's testimony, if admitted at all, is considered weaker and less credible than that of a man. A single testimony from one man for example, is of equivalent value to that of several women. As a consequence, if a text claims only a *woman's testimony* to certify a fact, it can be argued that this is indirect proof of the authenticity of that fact, for, if the fact were forged, the text would have claimed to be supported by a man's testimony. This argument is developed from the Gospels, which, referring to the resurrection of Christ, mentions that *women* discovered the empty tomb. The cultural *weakness* of their testimony is taken to be indirect *strong* evidence for that fact.

Third Party ▶ Roles

Threat

Threat is used by the stronger to force the weaker to do things in the interest of the stronger (seeking revenge at the same un-ethical level, the weaker may only have recourse to manipulation@) At the social level, threats of punishment function in combination with promises of rewards, as a global *argument from threat and promises* of reward.

1. Argument from threat

Threat speech, or *argument from threat*, or *appeal to fear*, (Lat. *ad metum*; *metum*, "fear") has also been called:
— By metonymy, the "argument from the stick" (Lat. *ad baculum*; *baculum*, "stick"), or "from prison" (Lat. *ad carcerem*; *carcer*, "prison"), or "to the purse" (*ad crumenam*; *crumena*, "purse"; by a double metonymy).
— By metaphor, the "thunderbolt argument" (Lat. *ad fulmen*, *fulmen*, "lightning; violence").
The prospect of a more or less imminent danger scares the person off the next planned action and induces new, more or less specific kinds of behavior. The threatened person feels an emotion, ranging from apprehension to fear or panic. The feeling depends on the mode of production and treatment of the source, which may or may not be well defined (*"we feel that something will happen to us"*), and enter into a controllable causality (*"we are living in a clash of civilizations"*). If

---

[1] New King James Version (NKJV); Quoted after
https://www.biblegateway.com/passage/?search=John+5%3A31&version=NKJV (05-05-2017)

the threat is causal, generalized and uncontrollable ("*the world is falling apart*") threat discourse will result in anxiety, fear, anguish, and even mass panic.

Two kinds of threat can be identified on the basis of whether or not the source of the danger is an intentional agent:
— The source is intentional, "*threatening enemies assail our civilization*".
— The source is *non-intentional*, the danger comes from the material world, and is interpreted as causal: "*the storm threatens the crops*"; "*you are at risk of cancer*".
The non-intentional source can be *human*. The "*shapes passing by in the fog*" for example can be perceived as *threatening* despite the fact that they are actually employees returning home from the office. This is the difference between **N0** *terrorizes* **N1** and **Na** *terrifies* **Nb**. The subject of *terrorizes* is intentional, **N0** wants to frighten **N1** (*the gang is terrorizing the honest citizens*), while the subject of *terrifies* is not necessarily intentional, and not necessarily human.

Fear speech expresses, inspires and strengthens a feeling of danger and insecurity, through oriented narratives and arguments which bring together the reasons, valid or not, to be afraid. Fear strategies can take two distinct orientations. They can either leave their targets plagued by anxieties, or can propose solutions to control or suppress the danger, **S. Pathos; Emotion.**
Fear speech can be based upon a real threat (climate change), or an invented one (alien invasion). In both case, the agent can be intentional (terrorists) or non intentional purely causal (climate change). It may or may not be correlated with hate speech.
In the hands of the established power, threat and fear, like joy and reward, can be used as powerful instruments of social cohesion and social control in societies which abide by the doctrine, "*let the good rejoice and the wicked tremble*".

Threat speech is no different from fear speech, where the speaker refers to an external threat. Threat speech may also be conveyed by an individual **A** expressing his or her intention to cause damage or harm to another individual **N**, if **N** does not comply with such and such requirement as imposed by **A**. In this case, the same person occupies the roles of speaker and villain. Such threat speech has an "*either… or…*" format:
> Either you do this for me — which is, I agree, quite unpleasant for you — or I do that to you — which is really much more unpleasant for you.

Whether this second kind of threat speech should be considered an argument or not is disputable. If we are accosted on a street corner and presented with the option of keeping our money or our life, we are likely to make a *rational choice*, opting to keep our life. When asked to explain where our money has gone, the existence of such a threat will be considered to be a *good reason* and a fully satisfactory justification for the loss of the money.

At the political level, balanced threats are the basis of nuclear deterrence; and it would be quite irrational not to take into account the fears imposed on the populations affected.

2. Threat and argumentation by the consequences

Threats can be efficiently presented as an argument by the consequences, where causality is veiled under agentivity. Rather than openly taking on the role of a villain, the speaker poses as the unwitting agent of a negative event provoked by the irresponsible behavior of the future victim. The blackmailer presents himself or herself as an *advisor*, and frames the interlocutor as the one responsible of future misfortune:

> Question: *Should the company grant a salary increase to its employees?*
> Labor's representative: — *If there is no increase, we'll occupy the plant!*
> Employer's representative: — *If you persist in your unrealistic demands, we'll be forced to close down the plant and cut back jobs.*

The same change of footing is operated by the politician presenting his or her own political decision as motivated by "the order of things", **S. Weight of Circumstances.**

2. Arguments from threat and promise

The Chinese philosopher Han-Fei proposes a theory of power as an expert blending of the two *measures* (Han-Fei, *Tao*); that is the two basic material interests motivating human actions, *punishments* and *rewards*, excluding the *rationality* issue, or other kinds of value, such as *justice*.

This kind of management of human actions exploits two antagonistic psychic movements, *fear and suffering* of punishment, *desire and joy* resulting from reward. If arguing is making somebody do something, or dissuading somebody from doing something, *threat* and *promise* would thus be the two argumentative speech acts par excellence — **S. Authority; Pragmatic Argument**

The everyday expression "*the carrot and the stick*" rightly associates the appeal to financial interest, with the traditional *ad baculum* argument; which might more fittingly be called *ad baculum carotamque* argument. The latter is no more "rational" than the former, although it is certainly considered to be more acceptable by many.

Appealing to money is not the only way to get what one wants; rewards and punishments might draw on everything and anything that humans may desire. This might include, in particular, power, pleasure, and money. **S. Value.**

The *ad crumenam* argument (Lat. *crumena*, "purse"), is mentioned in *Tristram Shandy*, where it refers to the introduction of considerations about money in a debate:

> Then, added my father, making use of the argument *Ad Crumenam*, — I will lay twenty guineas to a single crown piece, [...] that this same Stevinus was some engineer or other,---- or has wrote something or other, either directly or indirectly, upon the science of fortification"

Title

Laurence Sterne, *The Life and Opinions of Tristram Shandy, gentleman*, [1760][1]

3. Appeal to superstition

Lat. *ad superstitionem, superstitio*, "superstition"

The label *appeal to superstition* was introduced by Bentham, to refer to the fallacy of "irrevocable commitment", which prohibits the revision of prevailing political dispositions ([1824], p. 402); **S. Political Arguments**.
— "*Fallacy of vows or promissory oaths*; *ad superstitionem*": "*But we swore!*"
— "*Fallacy of* irrevocable laws": "*But that wouldn't respect the constitution!*"
Superstition is invoked because of the oath supposedly taken to honor the will of a sacred Supernatural Power, or of the Founding Fathers, "who knew better", and "to whom we owe everything". Failing this duty would constitute not only a lack of respect@ for the authority@ of the Founding Fathers, but a religious or moral sin provoking some supernatural revenge. It can be assumed that such threats are the flipside to promises that submission to the Law will be duly rewarded. As a consequence, the appeal to superstition as defined here is a subspecies of appeal to threats and promises, made by transcendental powers. In this case, the argument represents a somewhat materialistic version of the argument from faith.
Non-cynical, ordinary citizens consider that politicians must honor their election commitments. It would be difficult for failed politicians to invoke the *fallacy of irrevocable commitment* to perpetually justify their alliance and agenda reversals.

Title

Lat. *a rubrica* argument; the Lat. name *rubrica* belongs to the semantic family of *rubor* "red", and means "red earth; heading". In the collections of laws "the headings of chapters were written in red color" (Gaffiot [1934], *Rubrica*).

The interpretive argument based on the title of the relevant section of the code or *argument of the title* (*a rubrica*) is a matter of legal logic, **S. Legal Arguments**. Codes and administrative regulations are divided into sections and subsections with titles and subtitles. These headings carry no legal weight, but they are relevant to the interpretation of the law, insofar as they define the scope of the following articles. The argument based on the title legitimates or suspends the application of an article depending on whether or not the case considered falls within this scope. If the College regulation contains a section entitled "*Rules of conduct in classrooms*", Article 1 providing that:

The use of mobile phones is forbidden,

---

[1] In *The Complete Work of Laurence Sterne*. Delphi Classics, 2013, p. 98.

this article cannot be invoked in order to ban mobile phones in the playground. If the prohibition is made under the heading "General provisions", it applies to the conduct during classes, etc. The highest disposition in the hierarchy prevails.

Topic, Topos, Commonplace, Argument Scheme, Argument Line

1. Topic
In general vocabulary, the word *topic* refers to (MW, *Topic*):
    1   a: one of the general forms of argument employed in probable reasoning
        b: argument, reason
    2   a: a heading in an outlined argument or exposition
        b: the subject of a discourse or of a section of a discourse

The two meanings of *topic* go from *topic$_1$* a *formal, inferential* pole (meaning (1)) to a *substantial* pole (meaning (2)). They fit with the two different meanings of *argument* used in their definition, S. *(To) Argue, Argument*:

— *Topic$_1$*: According to (1), in a "reasoning" context, *argument* means "*argument$_1$*". Correlatively, a *topic$_1$* is here an argument$_1$ scheme or an argumentation derived from such a scheme. With this meaning, it can be considered a translation of the Greek word *topos*.

— *Topic$_2$*: According to (2), in an "exposition" and "subject" context, *argument* means *argument$_3$*. Correlatively, a *topic$_2$* is an argument$_3$, that is to say, the matter, the content of a discourse.

2. Topos
In contemporary English, the word *topos* is defined as "a traditional or conventional literary or rhetorical theme or topic" (MW, *Topos*).

2.1 *Topos* as "argument scheme" and "topic$_1$"
In Greek, the word *topos* (pl. *topoi*) has the basic meaning of "place". In argumentative rhetoric, *topos* is used metaphorically to refer to "the place where arguments are found"; a *topic* is an argument scheme. So, *topos* is translated as *topic* by Freese and as *argumentative line* by Rhys Robert in their respective translations of Aristotle's *Rhetoric*.

In Latin, the corresponding word is *locus* (pl. *loci*), which also means "place", translated as *topic* by Hubble in his translation of Cicero's *Topica*, and as "presumptive proof" by Caplan in his translation of the *Rhetorica ad Herennium*. In a famous metaphor, Cicero defines the *argumentative places* (Lat. *loci*, sg. *locus*) as "the name given by Aristotle to the 'regions' from which arguments are drawn" (*Top*, I, 8, p. 387). "Region" translates Lat. *sedes*, which also means "position, ground"; the *loci* are the foundations or "pattern" of arguments (*id.*, I, 9, p. 389).

In the Argumentation within Language theory, the concept of inferential topos is re-defined as a *pair of semantically associated predicates*, **S. Topos in Semantic**

2.2 *Topos* as "topic₂"

In the substantial sense, a *topic* (*topos*, *commonplace*) is an endoxon, a formulaic element corresponding to an answer to a "topical question"; or the whole discourse developing such a formula, *"the lawyer developed the topos of the well-known peaceful character of the Syldavians"*, **S. Substantial Common Places; Doxa**. Such discourses are suspected to be fake and insincere, because traditional:

> it is not easy to distinguish fact from topos in these documents (*OD*, *Topos*)

2.3 Topos in literary analysis

The concept of *topos* (pl. *topoi*) has been introduced by Ernst-Robert Curtius in literary analysis, to refer to a substantial, traditional thought that the writer develops, comments on and magnifies in the light of the circumstances. From a cultural and psychological perspective, a topos is "an archetype, [...] a representation of the collective subconscious as defined by C. G. Jung" (Curtius [1948], vol.1, p.180). For example, the topos associating "the old man and the child" is consistently exploited in advertisements for wealth management and inheritance arrangements.

The topoi can be used to fill a compulsory discursive slot. Thus, to close a presentation, a speaker declares that *"he submits quite willingly to possible negative observations, objections or even refutations, which will be truly considered to be help to understand his own data better"*.

Curtius' proposals have given rise to an important research trend on the topoi, especially in Germany (Viehweg, 1953; Bornscheuer 1976, Breuer & Schanze 1981).

3. Common place

1. In argumentation theory, the expression *commonplace* corresponds to the Latin *locus communis*, which translates as the Greek *topos*.

— Often reduced to *place* (*locus*, pl. *loci*), an *inferential common place* is an inferential topos, or *argument scheme*.

— A substantial common place is an *endoxon, a formulary expression of common thought*. Traditional rhetoric specialized in the argumentative use of substantial common places@.

2. In the general vocabulary, the expression *commonplace* is synonymous with *cliché*, both have the same depreciative orientation. *Topos* can be used with the same value.

3. In literary analysis, a commonplace is a "substantial topos", in the sense of Curtius [1948].

4. Argument schemes

The designations, *argument type*, or *argument scheme*, or *presumptive proof* unambiguously designate a general, formal, inferential scheme.

The words *topic* and *commonplace* are ambiguous between a formal and a substantial meaning.

The expression *argument line* is somewhat ambiguous, as it can be used to refer to an argument scheme or to a whole, coherent argumentative strategy@, S. **Argumentation Scheme; Types and typologies.**

4. Argument line

The phrase *argument line* can refer:

— to an *argument scheme*;

— to a discourse developing a series of *co-oriented* or *convergent*@ arguments, developed by the same arguer to support a conclusion;

— to a corpus of discourses developing a series of co-oriented arguments, presented by allied arguers to support their common conclusion, either in the same interaction or in different verbal or written interventions.

Topos in Semantic

In the Argumentation within Language theory of Ducrot and Anscombre, the topoi are defined as general *gradual* principles, relating predicates, and "presented [by the speaker] as accepted by the group" (Ducrot 1988, p. 103; Anscombre 1995a). The word *topos* (pl. *topoi*) will be used to refer to this specific concept as distinct from the classical argument schemes@.

Topoi are pairs of predicates (noted by capital letters). The (+) or (-) factor indicates that these predicates are gradual.

+ A, + P "more... more"	"The higher one rises in the P scale, the higher one rises in the Q scale", (Ducrot 1988, p. 106): Topos: "(+) democratic regime, (+) happy citizens" Argumentation: "*Syldavia is a democratic regime, SO its citizens should be happy*"
− B, − Q "less... less..."	"the more one moves down P, the more one moves down Q": Topos: "(−) working time, (−) stress" Argumentation: "*But now you work only halftime, SO you should be less stressed*"
+ C, − R "more... less"	More we have P, less we have Q: Topos: "(+) money, (−) true friends" Argumentation: "*He is rich, SO he has many friends* (topos "+M, +F"), BUT *not so many true friends*" (topos "+M, -F").
− D, + S "less... more"	Less one makes P, more one is Q: Topos: "(−) sport, (+) heart problems" Argumentation: "*He stopped doing sport, AND (SO) now he has heart problems*"

This type of inter-predicate linkage was also observed by Perelman & Olbrechts-Tyteca in their discussion of values@ ([1958], pp. 115-128).

All predicates are gradual. For example, in a Syldavian subculture, the following topos might structure conversation about *"being a real man"* (**M**) and *"drinking BeverageB"*, (**B**); this relation is expressed by the topos "(+)**M**, (+)**B**"; advertisers claim that *"real men drink BeverageB"*; the more of BeverageB that one drinks, the more of a "real man" one will be.

The same predicate may be associated by the four topical forms, for example in the following argumentations

 (+) money (–) happiness: *"he is a rich financier, so he has many anxieties and sleeps poorly"*

 (–) money (+) happiness: *"money can't buy happiness"*; *"the poor cobbler sings all the day long"*[1].

 (–) money, (–) happiness: *"lack of money is terrible"*

 (+) money, (+) happiness: *"money can buy everything"*.

In the case of sport and health:

 (+) sport, (–) health: *"champions die young"*

 (–) sport (+) health: *"to stay healthy, refrain from sports"* (Churchill, *"no sport"*).

 (–) sport, (–) health: *"when I stop training, I feel bad"*

 (+) sport, (+) health: *"do a sport, you will feel better"*.

In such cases, the predicates are linked by four different topoi <+/- S, +/- H>; nonetheless, communities have preferences, in this case for the two last ones.

These topoi are the exact linguistic expression of the "active associative nodes for ideas" mentioned by Ong (1958, p. 122); **S. Collections (I)**. They express the possible linguistic associations between *"having money"* and *"being happy"*, between *"living a healthy life"* and *"practicing sport"*. To summarize, current talk about money and happiness prefers the (–, –) association, whilst current talk about sport and a healthy life prefers the (+, +) association.

Such associations will emerge in the discourse as *reasonable and convincing* inferences. In ordinary discourse a complex causal elaboration such as *"some/all plant protection products are the/a cause of bees disappearance"* boils down in ordinary talk to an accepted, doxical association "+PPP, – bees".

These expressions are semantic inferences, and are pseudo-reasoning insofar as they say nothing about reality; discourse is an inference machine, an argumentative machine; language can and does speak. This vision legitimates the skepticism of the theory of argumentation in the language with respect to ordinary argumentation as a form of reasoning, **S. Criticism**. Reasoning emerges from ordinary talk only under specific conditions; there might be a big step between *debating* and *learning* (Buty & Plantin 2009).

---

[1] According to La Fontaine, "The cobbler and the financier", *Fables*, Book VIII, Fable 2.

True Meaning of the Word

The appeal to the "true meaning of the word" is advanced in opposition to discourses which are said to use an incorrect, improper or superficial meaning of a given word. This appeal produces a stasis of definition, **S. Definition (III)**. The true meaning of a word can be sought in:
— Its etymological meaning;
— Its morphology;
— The meaning of the corresponding word in another language.

1. Argument from etymology

The label "argument by etymology" corresponds to different kinds of arguments, according to the meaning given to *etymology*.
1. Under the heading "argument from etymology", some modern texts discuss phenomena related to related@ words (Dupleix, 1603).
2. In contemporary use, the etymological meaning of a word is the meaning of the oldest historical root identified in the word's history. Etymological argument values the meaning of this root by considering that this ancient meaning corresponds to the *true and permanent* meaning of this word, which has been *altered by historical evolution* to produce a contemporary perverted and misleading meaning. This etymological meaning is used in argumentation by the definition, **S. Definition (III)**:

> *Atom* comes from / is a Greek word composed of the negative prefix *a-* and a noun meaning "cutting"; it means "in-divisible". So you cannot break the atom.

> *Democracy* comes from / is a Greek word composed of *demos* "people" *kratos* "rule". In Syldavia, the people don't rule, they vote and forget. Thus, Syldavia is not in a democracy.

The appeal to etymology is itself supported by an argumentation by etymology, since the word *etymology* is derived from a Greek root *ètumos* meaning "true".
Knowledge of etymology being culturally valued, the argument by etymology gives the speaker a certain ethotic posture of majesty and learned authority. It serves very well the strategy of destruction of the discourse *"you don't even know the language you claim to speak"*, **S. Destruction**.

2. Argumentation based on the structure of the word

> Lat. *notatio*, "the act of marking a sign ... to designate [...] to note", as well as "etymology" (Gaffiot [1934], *Notatio*).

Cicero in the *Topics* defines the argument "*ex notatione*" (*Topics*, VIII, 35: 78), translated as "argument by etymology". This translation takes the word *etymology* with its ancient meaning, "true". The true sense of the word under examination is now defined as the meaning reconstructed by the correct analysis of the word (and not as its original historical meaning). One of the examples of argument discussed by Cicero in this context deals with a conflict of interpretation of a compound legal term (still in use today) the *postliminium* (*Top.*, VIII, 36, p. 78). The *postliminium* is the right of a prisoner returning to his country to recover the properties and social position he held before his captivity. Cicero's discussion concerns the establishment of the correct meaning of the word, relying on its linguistic structure, without any clear allusion to its etymology in the contemporary sense of the term.

A *contradictory report* (joint report) is a report reproducing the declarations of the two parties, and not a *self-contradictory* verbal report or a report *contradicting* another.

Argumentation by the structure of the word thus connects two arguments:

— The first argument establishes the meaning of the compound word on the basis of the meaning of its terms and its morphological structure. This form of argumentation is relevant to all idioms whose meaning depends more or less on that of the terms that compose them. It is based upon linguistic knowledge and technique, S. Definition (II).

— A second argumentation exploits the "true" meaning thus established for some legal conclusion, according to the general mechanisms of argumentation by the definition, S. Definition (III).

The argument by the structure of the word functions as a way of avoiding conflict of interpretations.

3. Arguing from the meaning of the word in another language

One can look for the true meaning of the word in other languages, which for various reasons are considered closer to the "origin" or the "essence" of things. One such language is Chinese. The word *crisis*, for example, can be defined as "a time of intense difficulty or danger" (Google, *Crisis*). In search of "what crises really are", one can shift to "what the word *crisis* really, truly, means", and call on the word's Chinese equivalent. The Chinese word meaning *crisis* is a compound of two word-signs, meaning respectively "danger" and "opportunity". So crises are opportunities; and, by an argument based on the Chinese definition, we deduce that:

> The opportunistic approach of the crisis then takes on its full meaning: Not to seize the opportunity of a crisis, means miss a chance, perhaps hidden, but within reach. (Stéphane Saint Pol, [*Wei Ji, A return to the Roots*][1])

The argument presupposes that the Chinese language had preserved or elaborated a better concept of crisis, closer to the essence of the thing, and better adapted to the modern world.

Truth ▶ Probable, Plausible, True

---

[1] www.communication-sensible.com/articles/article0151.php]. (09-20-2013).

"Tu Quoque" ▶ "You too"

Two-Term Reasoning

1. Transduction
The concept of *transductive reasoning* is developed by Piaget ([1924], 185) to analyze the development of children's intelligence. Transductive reasoning is characterized as the prelogical and intuitive way of thinking of the young child, which goes directly from an individual or a particular fact to another individual or particular fact, without the intermediary of a general law. According to Grize, "the young child who says, *'It's not afternoon because there was no nap'* is based on the daily experience of napping as an ingredient of the afternoon [reasons by transduction]" (1996, p. 107).

Transductive reasoning seems to be the product of a conditioned association *"nap = afternoon"*, which gives, by application of the scheme of the opposites: *"no siesta = no afternoon"*. From this perspective, napping is a defining feature of the afternoon.

Grize observes that adults are also likely to use this kind of reasoning, "When we say that we stopped at the traffic light because it was red, [...] our thinking does not go through a general law of the kind: *"any red traffic light implies stop"* (*ibid.*). In the latter case, the statement has the form of a "semantic block" (Carel 2011), "Answer because Stimulus". Yet the adult does not apply the negation in the same way as the child; saying *"it is not a red light since I did not stop"* would be considered as a denial of reality. However, it is said that a motorist deeply imbued with respect for the Highway Traffic Act refused to believe that he collided head on with another vehicle *because he was driving down a one-way street*, implying the *material* impossibility of a fact from its *legal* prohibition.

2. Two-term reasoning
In a very different context, Gardet and Anawati speak of, "two-term reasoning" which is characteristic of "a specifically Semitic rhythm of thought which the Arab mind knew how to use with a rare happiness of expression" (Gardet and Anawati [1967], p. 89). This type of reasoning seems to be similar in nature to transductive reasoning.

> The 'dialectical' logic, connatural to the Arab genius, is organized according to modes of reasoning with two terms that proceed from the singular to the singular, by affirmation or negation, without a universal middle term. Should we say, as has sometimes been said before, that [this universal medium term], not explicitly understood, is nevertheless explicit in the reasoning mind? We don't think so. Undoubtedly, two-term reasoning can be 'translated' into a three-term syllogism [...]. Yet in the logical mechanism of thought, it is indeed the confrontation, by contrast, similarity or inclusion, of the two terms of the rea-

soning that gives the 'proof' its power of conviction. The universal middle term is not present in the mind, even in an implicit form. This is not about establishing a discursive proof, but about promoting a self-evident certainty. (Bouamrane & Gardet 1984, p. 75)

The Arab logician and theologian al-Sumnani has distinguished five rational processes, that is five argument schemes, which are characteristic of two-term reasoning. These five processes are based on "findings, and then, by a movement of the mind operating either by elimination or by analogy from the same to the contrary, or from the same to the same. It is always a question of passing from the present, actual fact, the "witness" (*shâhid*) [the argument, CP], to the absent, (*gha'ib*) [the conclusion, CP]. There is no abstract search for a universal principle" (Gardet and Anawati [1948], pp. 365-367).

NB: There are no entries for the letter U

V

Value

In the field of argumentation studies, the word *value* can refer to:
1. The truth-value of a proposition, S. **Proposition.**
2. The value of an argumentation, S. **Evaluation; Norms; Strength.**
3. The question of values and value judgments: this entry.

1. Values as a unified field
Since the beginning of the twentieth century, questions about "the good, the end, the right, obligation, virtue, moral judgment, aesthetic judgment, the beautiful, truth, and validity" (Frankena 1967, p. 229) have been taken up globally, within the framework of a general theory of values, of distant Platonic ancestry. This "wide-ranging discussion in terms of 'value', 'values', and 'valuation' [then] spread to psychology, the social sciences, the humanities and even to ordinary speech" (*ibid.*).

2. The New Rhetoric

2.1 Truth, facts and values
The concept of value is central to the new rhetoric of Perelman & Olbrechts-Tyteca (as well as to the problematic of conductive@ reasoning). Perelman presents his discovery of the theory of argumentation as a step beyond a research program on the "logic of value judgments" (Perelman 1979, §50, p. 110; 1980,

p. 457). This latter research led him to the "disappointing conclusion" that, "a logic of value judgments simply does not exist" (*ibid.*). Positivist philosophy considers that values are independent from facts, and that, therefore, they cannot be derived from facts. Positivist philosophers are thus led to the conclusion that value-based actions are *irrational*. Perelman argues that this conclusion is self-defeating, since it implies that practical reasoning and the whole field of law, both based on values, should be regarded as irrational, which is absurd@ because unacceptable

Perelman's conclusion is that, since science and logic deal with *truth* judgments, they cannot provide the rules of practical reason, which deals with *value* judgments. Such is the basic Perelmanian claim, that re-asserts the gap between the rational and the reasonable, between "the two cultures", science and humanities, **S. Demonstration; Proof.** Furthering his research program on values, Perelman sought other perspectives better suited to this specific object. He found them in Aristotle's *Rhetoric* and *Topics*, which provide techniques for an empirical study on how individuals justify their reasonable choices. Perelman was then able to redefine his theoretical objective no longer as a *logic*, but as a *(New) Rhetoric* (*ibid.*). The argumentative-rhetorical method appears to be the solution to the failure of the logical and philosophical treatments of values. Perelman consistently opposes the project of classical philosophy to develop a *calculus* of values, since it is not possible to derive a *hierarchy* of values from an *ontology* of values. More specifically, Perelman opposes Bentham on the possibility of a calculus of pleasures and pains.

The question of values is not only the source of the development of the New Rhetoric, but is its permanent foundation, as shown by the introductory chapter of Perelman's *Juridical Logic* (1979) entitled, "The New Rhetoric and Values".

2.2 Substantial values and value judgments

The *New Rhetoric* is thus structured around two issues concerning values. The first issue concerns *value judgments*, made about a concrete being or situation from the point of view of a value. The second issue concerns *substantial values* such as the true, the beautiful and the good, which are the most general of all values. Values are defined by the following distinctions and operations.

(i) A distinction is made between two kinds of substantial values, "*abstract* values such as justice or truth, and *concrete* values such as France or the Church" (Perelman & Olbrechts-Tyteca, [1958], p. 77).

(ii) Values currently conflict; their contradictions may sometimes be solved by organizing them into a hierarchy (*id.*, p. 80).

(iii) Value judgments cannot be derived from nor opposed to reality judgments. In science, if two truth-judgments about a reality are contradictory, one of them is necessarily false (principle of the excluded middle), while two contradictory value judgments, "*this is beautiful!* vs. *this is ugly!*", may both be justified by value-based arguments, developed independently from any appeal to reality.

(iv) Facts are necessary and *compel* the mind, whereas values call for an *adherence* of the mind, S. **Argumentation (I)**.

(v) Contextual considerations may be necessary to characterize a judgment as a value judgment: *"this is a car"* can be a judgment of fact or a value judgment; *"this is a real car"* is only a judgment of value (see Dominicy, n. d., p. 14-17).

(vi) In the language of the *Treatise*, it follows that substantial values and value judgments are "objects of agreement that cannot make a claim to the adherence of the *universal* audience" (*id.*, p. 76), but only to the adherence of *particular* audiences, S. **Audience**. The so-called universal values "such things as the *True*, the *Good*, the *Beautiful*, and the *Absolute*" might be regarded "as valid for the universal audience only on condition that their content not be specified" (*ibid.*). They are the "empty frame[s]" suited to all audiences, and are as such pure instruments of persuasion (*ibid.*). Natural law theorists would object to this conclusion.

(vii) Values and truth are acquired via different processes. Group values are acquired through education and language. According to Perelman, the epidictic genre specifically deals with values; it does not admit contradiction. Its specific social function is to strengthen the adherence of the group to its common founding values, "without which the discourses aimed at action could not find leverage to move and rouse their listeners" (1977, p. 33).

(viii) Values are linked to emotions@, cf. infra.

(ix) Perpetually reconstructed in epidictic encounters, where they are subject to a quasi-axiomatic treatment, values find their application in the two argumentative genres properly called, the deliberative and the judicial.

To conclude, the *Treatise*, maintains the opposition between *value judgments* and *judgments of fact* only as the result of "precarious agreements" (Perelman & Olbrechts-Tyteca [1958], p. 513) and for special debates.

2.3 The apple and the three libidos

The multiplication of values does not call into question the fact that rhetorical discourse relies on substantial values, perhaps more prosaic than "the True, the Good, the Beautiful, the Absolute", but nonetheless firmly attached to the human condition, and having highly mundane contents, namely *honos*, *uoluptas*, *pecunia*. That is, *glory* as the desire for social recognition; *pleasure* in all its forms; *money* and possessions. There is no place for knowledge in this trinity of materialistic values. By contrast, knowledge is one of the three criteria used by Eve to evaluate the fruit that put an end to the state of innocence:

> So when the woman saw that the tree was good for food, and that it was a delight to the eyes, and that the tree was to be desired to make one wise, she took of its fruit and ate; and she also gave some to her husband, and he ate.
>
> *Genesis*, 3, 6[1]

---

[1] Quoted after www.biblegateway.com/passage/?search=Genesis+3&version=RSVCE

Value

"Good for food": the *good*, pleasure of the mouth; "a delight to the eye": the *beautiful*, pleasure of the eyes; "to be desired to make one wise": the *true*, pleasure of the mind. These three values can most probably be axiomatically attached to the human condition; that is, they are widely available for immediate valorization in *pragmatic* argument, which is in fact the Devil's argument, "you will be like God, knowing good and evil." (*id.*, p. 5.).

3. Values and argument

3.1 Values as argument

According to the *Treatise*, argumentation can be based upon two classes of object, an object being defined as anything that can be agreed upon, "the first concerning the *real*, comprising facts, truths and presumptions, the other concerning the *preferable* comprising values, hierarchies lines of argument relation to the preferable" (*id.*, p. 66; emphasis in the text). In other words, with the agreement of the participants, statements about values and reality can be used as arguments. In rhetorical@ argumentation, the speaker proceeds on the basis of values shared with the audience, or presented as such, S. **Ex datis**. In an adversarial debate, the speech of the proponent and the opponent may be based on radically incompatible values. In such cases, the role@ of third parties (judges, voters, members of a jury...) becomes essential to settle the conflict of values, rather than to solve it definitely.

3.2 Topics of values

Values are treated by means of *loci*, which the *Treatise* defines as "premises of a general nature that can serve as the basis for values and hierarchies" (*id.*, p. 84). These *loci* are distinct from the "argumentative techniques", that is argument schemes and dissociation, **S. Argument scheme**. The following are considered as the "most common" *loci* (*id.*, p. 95):
— *Quantity*: "one thing is better than another for quantitative reasons" (*id.*, 85): "*the more, the better*".
— *Quality*, used to challenge *Quantity*: "the strength of numbers" (*id.*, p. 89): "*the rarer it is, the more precious it is*".
— *Order*: "the loci of order affirm the superiority of that which is earlier over that which is later" (*id.*, p. 93).
— *Existent*: "the loci relating to the existent affirm the superiority of that which exists, of the real, over the possible, the contingent, or the impossible" (*id.*, p. 94).
— *Essence*: "according a higher value to individuals to the extent that they embody [the] essence" (*id.*, p. 95), which materializes as the topos "the closer to the origin, to life, to the prototype, the better it is."
These loci can be expressed in the same way, **S. Scale**.

According to the *Treatise*, such *loci* of values correspond to the *loci* of the accident of the *Topics* of Aristotle (*id.*, p. 113). They are therefore operative over a wider field than the field of values. They can be expressed fairly well in the language of the Argumentation within Language theory, S. **Topos**.

3.3 Values and dissociation of notions

The *loci* of "the existent" provide a link between the question of values and dissociation@. Dissociation is a valuation / depreciation operation, which splits a single notion into two notions, one of which will have a positive value (orientation), and the other, a negative value (orientation).

3.4 Values and orientation

The concept of value refers to issues of subjectivity (Kerbrat-Orecchioni 1980) of affectivity and, to the inherent orientation of ordinary statements, S. **Emotion; Orientation; Bias**. Words which express values are basically antonymic pairs coupling words which have opposing argumentative orientations. The entire lexicon can be seen as an enormous reservoir of such pairs: "*pleasure* vs. *displeasure; knowledge* vs. *ignorance; beauty* vs. *ugliness; truth* vs. *lies; virtue* vs. *vice; harmony* vs. *chaos; love* vs. *hate; justice* vs. *injustice; freedom* vs. *oppression*", etc. Antonyms are also expressed by more or less fixed phrases ("free expression of self *vs.* repression of aggressive instincts", "life in the open air *vs.* life in offices").

The ratio "valorization *vs.* devaluation" can be reversed: aesthetics of *ugliness* vs. *beauty*), baroque esthetic of *inconsistency* (vs. *consistency*), etc.

Value terms are at the root of "biased" language. Unbiased talk would amount to an elimination of value statements (i. e. of subjective, emotional, oriented statements), in favor of judgments based on facts. This requirement can only be satisfied by rejecting natural language in favor of a formal, scientific or technical language, or an alexithymical expression, S. **Pathos**.

3.5 Arguments positioning an object in relation to a reference value

As mentioned by Perelman, the predication of a value upon an object follows standard argumentative procedures. For example, in France, *National sovereignty* as a political value of reference, is consecrated as such in the Article 3 of the 1789 "*Declaration of the Rights of Man and of the Citizen*":

> The principle of all Sovereignty lies essentially in the Nation. Nobody, no individual can exercise any authority that does not expressly emanate from it.[1]

An evaluation question can arise, asking for example, that an international treaty be assessed in relation to that value. For this purpose, reference can be made to the preceding axiomatic definition, as enshrined in its legal implementation and by experience drawn from analogous situations in the past. The evaluation

---

[1] https://www.legifrance.gouv.fr/Droit-francais/Constitution/Declaration-des-Droits-de-l-Homme-et-du-Citoyen-de-1789 (09-20-2017).

follows the general definition based categorization procedure, **S. Categorization; Definition**.
— National sovereignty is defined by the conditions **Ci, Cj, Ck ...** –
— Treaty **T** respects / does not respect these conditions.
— We can / cannot sign this treaty without renouncing our national sovereignty.

The evaluation has to take into account that, as in any broad axiomatic definition, national sovereignty has many corollaries and ramifications, financial sovereignty for example.

3.6 Argumentation using an evaluation
The positive pragmatic argument routinely implies a positive valorization operation, **S. Pragmatics**:
Question: Should we do **F**?
Argumentation: **F** will result in **C1**;
Positive/negative evaluation of C1: **C1** is **Vi(+)** (positive from the point of view of value **Vi**);
Therefore: Let's do **F**.

Refutations might follow several paths, for example:
(i) A counter evaluation of **C1**: **C1** is **Vi(-)**; this opens an *evaluation stasis*, an issue about the positioning of **C** in relation with value **Vi, Vi(+)** vs. **Vi(-)**.
(ii) Introduction of another consequence **C2**, considered to be negative from the point of view of the value **Vm**. In this case, the stasis is about the relative weights of **C1** being **Vi(+)** *vs.* **C2** being **Vm(-)**.
In both cases, the conclusion is the same: "don't do **F**!".

Verbiage

The Port-Royal *Logic* stigmatizes the technique of the *inventio* as stimulating the "noxious fertility of common thoughts" (Arnauld and Nicole [1662], p. 235). The same criticism applies to the techniques of *elocutio*, which stimulates and extolls the abundance of words (*copia verborum*) **S. Ornamental**, producing a verbose and redundant discourse:
> Among the causes which lead us into error, by a false luster, which prevent our recognizing it, we may justly reckon a certain grand and pompous eloquence. [...] for it is wonderful how sweetly a false reasoning flows in at the close of a period which well fits the ear, or of a figure which surprises us by its novelty, and in the contemplation of which we are delighted. (*Id.*, p. 279)

The condemnation of the techniques stimulating the abundance of ideas as well as the abundance of words amounts to a general condemnation of rhetoric, as inherently fallacious, **S. Ornamental**. Cicero defines eloquence as *copia verborum*; that is eloquence; the rejection of eloquence, re-named *verbiage*, is a turning point in the relations between rhetoric and logic as a criticism of discourse.

This fallacy of verbiage is, as it were, the mother of all fallacies. According to Whately, "a very long discussion is one of the most effective masks of the fallacies; [...] a fallacy, which, asserted without a veil [...] would not deceive a child can deceive half the world if it is diluted in a large quarto (*Elements of Logic* 1844)" (quoted by Mackie, 1967, p. 179).
S. **Superfluity; Fallacies (I); Fallacies (III).**

Vertigo

Argument *ad vertiginem*; Lat. *vertigo* "movement of rotation, vertigo".

The *argument of vertigo* is defined by Leibniz in his *New Essays Concerning Human Understanding* [1795], as a follow-up to his discussion of Locke's four kinds of argument, S. **Collections (II).**

> We might bring yet other arguments which are used, for example the one we might call *ad vertiginem*, when we reason thus: if this proof is not received we have no means of attaining certainty on the point in question, which we take as an absurdity.
> This argument is valid in certain cases, as if any one wished to deny primitive and immediate truths, for example, that nothing can be and not be at the same time, or that we ourselves exist, for if he were right there would be no means of knowing anything whatever. But when certain principles are produced and we wish to maintain them because otherwise the entire system of some received doctrine would fall, the argument is not decisive; for we must distinguish between what is necessary to maintain our knowledge and what serves as a foundation for our received doctrines or practices.
> Use was sometimes made among jurisconsults of probable reasoning in order to justify the condemnation or torture of pretended sorcerers upon the deposition of others accused of the same crime, for it was said: if this argument falls, how shall we convict them? And sometimes in a criminal case certain authors maintain that in the facts where conviction is more difficult, more slender proofs may pass as sufficient. But this is not a reason. It proves only that we must employ more care, and not that we must believe more thoughtlessly, except in the case of extremely dangerous crimes, as, for example, in the matter of high treason, where this consideration has weight, not to condemn a man, but to prevent him from doing harm; so that there may be a mean, not between guilty and not guilty, but between condemnation and banishment in judgment, where law and custom allow it.
> Leibniz, *New Essays Concerning Human Understanding* [1765]. P. 437.

In substance, the argument from vertigo urges us to accept certain kinds of proof for if we don't, we are left powerless. This is a subspecies of arguments by unacceptable consequences, S. **Absurd; Pathetic; Ignorance.**
These consequences are "absurd" and dramatic, when dealing with the first principles of knowledge, such as the principle of non-contradiction, which every person must admit on pain of being unable to say anything. Unlike the

argument *from ignorance*@, the argument *ad vertiginem* would therefore be valid insofar as the impossibility on which it is based is not a *subjective* impossibility, related with such and such a person or group, but an *objective, rational* impossibility concerning humanity as such.

However, Leibniz operates a *distinguo* between epistemic situations where our power to know is at stake, "what is necessary to maintain our knowledge", and social situations dealing with human affairs and ideology, that "[serve] as a foundation for our received doctrines or practices".

Since demonstrative reasoning cannot apply in the latter case, "probable reasoning" must be rehabilitated in this domain, for lack of a better proof. But having to make do with weaker proof in the criminal domain implies that a person can be condemned on the basis of insufficient proofs, which Leibniz finds undesirable. So, in an interesting maneuver, he proposes to re-equilibrate the *weakness* of the proofs motivating the condemnation by *softening* the condemnation itself.

Vicious Circle

1. "Vicious circle", "begging the question", "*petitio principii*"

The two expressions *vicious circle* and *begging the question* are equivalent. The expression *vicious circle* stresses the cognitive and textual, semantic aspects of the phenomenon, while *begging the question* emphasizes the dialectical interactional character of the same concept.

The Latin expression *petitio principii* is used as an equivalent of *begging the question*. In classical Latin, *petitio* means "request", and *principium* "beginning" (Gaffiot [1934], *Petitio; Principium*). A *petitio principii* is literally a "request" of the "principles". Tricot considers that the rendition as "petition of principle" is "vicious". He notes that "what we ask to grant is not a principle but the conclusion to be proved" (note 2 to Aristotle, *Top.*, VIII, 13, 162a30, p. 359.

The speaker is "begging the question", that is asking that what is in question (the disputed conclusion itself) be granted, as an argument or principle.

2. Vicious circle

In the Aristotelian system of fallacies, a vicious circle is a fallacy independent of language, S. **Fallacy (II)**. It is a process of reasoning which seeks to prove a conclusion, by giving as an argument this conclusion itself. Hence the image of the circle. Its schematic form is:

A, *since, so, because* **A**.

There are different ways in which to beg a question (Aristotle, *Top.*, VIII, 13).

2.1 Repetition

In ordinary discourse, compound statements "**A** because **A**" might be considered as begging the question from a logical point of view:

> S1 — *Mum, why do I have I to make my bed every morning?*
> S2 — *You have to do it because you have to do it. It's like that because it's not otherwise*

Nonetheless, despite its format, this is not a *vicious circle*. The answer is not an *invalid justification* but a *refusal of justification*, as testified by the associated mood, exasperation.

2.2 Reformulation

In common cases, there is a vicious circle where the conclusion is a paraphrastic reformulation of the argument:

> I like milk because it's good.
> Fortunately I like milk, because if I did not like it I would not drink it, and it would be a pity because it's so good.

When the very result to be demonstrated is postulated, "this is easily detected when put in so many words; but it is more apt to escape detection in the case of different terms, or a term and an expression, that mean the same thing" (Aristotle, *Top*, VIII, 13).

In the theory of Argumentation within Language, the concept of *orientation@* introduces a bias which is not so different from mere *petitio principii*. The statement

> Peter is clever, he will solve the problem.

The above example has a misleading deductive appearance, because the predicate *"can solve problems"* is in fact included in the definition of *"is clever"*.

The misleading inference is actually a *reformulation*. Reformulations are interesting insofar as they are never strictly synonymous with their basis. Instead, they introduce a semantic shift, which can be productive. Begging the question is deceitful only in so far as it is strictly the same term that is repeated, **S. Bias**.

Goethe claims that, in any argumentation, the argument is only a variation of the conclusion; hence it follows that argumentative rationality is simply vain rationalization:

> 550. It is always better for us to say straight out what we think without wanting to prove much; for all the proofs we put forward are really just variations on our own opinions, and people who are otherwise minded listen neither to one nor to the other.
>
> Johann Wolfgang von Goethe. *Maxims and Reflections*.[1]

2.3 *Ad hoc* general laws

The *Topics* point out the frequent case in which one assumes in the form of universal law what is in question in a particular case (*ibid.*):

> Politicians are liars / corrupt. So this politician is a liar / corrupt

---

[1] Johann Wolfgang von Goethe. *Maxims and Reflections*. Trans. by E. Stopp. London: Penguin Books, 1998. Quoted after https://issuu.com/bouvard6/docs/goethe_-_maxims_and_reflections_ No pag. Goethe gathered these maxims during all his life.

This is a most common kind of argumentation. The speaker postulates a one shot, *ad hoc* principle, in order to apply it to the case at hand. Such cases can also be analyzed as ill-constructed definitions: *"being corrupt"* is considered an essential characteristic of politicians, whereas it is only an accidental characteristic, **S. Definition, Accident.**

2.4 Mutual presupposition
Not all vicious circles are reformulations. An objection to the idea of a miracle for example, is that it establishes a vicious circle. Miracles are said to justify the doctrine, to prove that it is true and holy, but a fact is recognized as a miracle only by the doctrine it is supposed to prove. It is a form of resistance to refutation:

> **S1_1** — *This miraculous fact proves the existence of God.*
> **S2_1** — *But only those who believe in the existence of God recognize this fact as a miracle*

S2 may add that S1 does not recognize other equally surprising facts; to which the latter might reply that:

> **S1_2** — *These other facts are miracles operated by the devil to deceive people.*

2.5 Equal uncertainty
The term *diallel* is used by the Skeptics, with a meaning identical to "vicious circle":

> And the circularity mode occurs when what ought to make the case for the matter in question has need of support from that very matter; whence, being unable to assume either in order to establish the other, we suspend judgment about both. (Sextus Empiricus, *Outlines of Pyrrhonism*, I, 15, 169)

This definition introduces a new concept of the vicious circle that no longer focuses on a semantic equivalence or an epistemic relation, but on the very definition of argumentation as a technique to reduce the uncertainty of a claim by connecting it to a less doubtful statement, the argument. **S. Argumentation.** Skeptics will therefore endeavor to show that the argument is systematically no more obvious than the conclusion. In this sense, Skeptics are the first deconstructionists.

2.6 Circularity in explanation@
Circularity is welcome in definitions, but not in demonstrations or explanations. An explanation is circular, if the explanans is at least as obscure as the phenomenon it claims to account for.

W

Warrant ▶ Layout of Argument; Topic

Waste

The *argument from waste* is defined as follows by Perelman & Olbrechts-Tyteca:

> The argument of waste consists in saying that, as one has already begun a task and made sacrifices, which would be wasted if the enterprise were given up, one should continue in the same direction. This is the justification given by the banker who continues to lend to his insolvent debtor in the hope of getting him on his feet again in the long run. This is one of the reasons which, according to Saint Theresa, prompt a person to pray, even in a period of "dryness." One would give up, she says, if it were not
>
>> '... that one remembers that it gives delight and pleasure to the Lord of the garden, that one is careful not to throw away all the service rendered, and that one remembers the benefit one hopes to derive from the great effort of dipping the pail often into the well and drawing it up empty'. (1958], p. 279)

According to the tradition established by Aristotle in the *Rhetoric*, the *Treatise* introduces the scheme of waste by a definition immediately followed by two illustrations. The defining topos is given in the following passage:

> as one has already begun a task and made sacrifices, which would be wasted if the enterprise were given up, one should continue in the same direction.

The argument scheme is given as a *generic sentence*, outlining a typified situation. The agents are impersonal ("one"); "(one has) *already begun*" / "should *continue*"; "a *task*", an "*enterprise*"; "(one has made) *sacrifices*. This scheme corresponds to the following script (the elements of the affective scenario are underlined):

(i) A complex initial situation:
    (a) A *task* has been started in the <u>hope</u> of a significant benefice.
    (b) The task is long and difficult: *sacrifices* have been *made*.
    (c) Nothing has been obtained (implicit).

(ii) These hard conditions generate an interrogation:
    (d) Implicit: <u>despair</u> looms; it is possible and one is <u>*tempted*</u> to stop: "*should I continue?*" This key point is not explicitly mentioned in the scheme.
    (e) The situation is now framed as a dilemma; this is an all-or-nothing issue:
    — Either (e1) I "*give up*" and all the *efforts* will be wasted.
    — Or (e2), I go on, "*hoping*" that things will finally turn better. This key element, <u>hope</u>, is not mentioned in the scheme, it only appears in the first example. Note that (e2) can be derived from (e1) by application of the opposite scheme: to *give up* and *waste* everything / to *continue* and *win* the jackpot (implicit).

(iii) Conclusion: A decision, actually a bet: "*one should continue in the same direction*".

All these conditions are crucial, for example (e). If it was a cumulative task (like weight training), then one could justify the decision to stop by saying that, well, "it is something anyway".

The scheme is structured by a concatenation of emotions:
    hope → temptation of despair → renewed hope

The scheme of waste is linked to the slippery slope argument, "*we must not begin, because, if we start, we will not be able to stop*". This last scheme justifies an initial abstention, whereas the argument of waste is that of perseverance in action, S. **Direction; Slippery slope.**

The scheme of waste is related to the proverbial scheme, "*one does not stop in midstream*", to which one can reply "*either you stop or you drown yourself*". It is vulnerable to a counter-discourse such as, "*we have already lost enough time like that.*"

The following example introduces a formula frequently associated with this scheme when used to justify the continuation of a war "*then they would have died for nothing!*":

    "Beating a retreat is tantamount to recognizing that all our guys died for nothing!" claims [John McCain's [(1)] fan] Private Carl Bromberg, having returned home.
    [(1)] Republican Presidential nominee for the 2008 United States presidential election. *Marianne*, 1-10 March 2008, p. 59.

In this second example, the key elements of the scheme are scattered across the passage (our emphasis):

He [*the philosopher Alain*] does not believe in the war in the name of law. From the end of 1914 on, he favors a peace of compromise, and he follows very closely, through the *Tribune de Genève* (1) sent to him by the household Halévy, everything which looks like the beginning of a negotiation, however fragile. But he is under no illusion: *precisely because it is so hideous, so murderous, so blind, so total, war is very difficult to end.* It does not belong to this category of armed conflicts that cynical princes can stop if they consider that the costs exceeds the possible gains, and that the game is not worth the candle. It is led by patriots, honest men elected by their people, who are *locked up every day more and more in the aftermath of the decisions of July 1914*(2). *The sufferings have been so great, the deaths so numerous that no one dares to act as if they had not been necessary.* And how do we move forward without being labeled as a traitor? *The longer the war lasts, the longer it will last.* It kills democracy, from which it nevertheless receives what perpetuates its course.

(1) A Swiss newspaper (2) Date of the declaration of war.

François Furet, [The Past of an Illusion], 1995[1].

Weight of Circumstances

1. Weight of circumstances

Argumentation *by the weight of circumstances* invokes the nature of things or the external constraints, as imposing a deterministic solution on a social issue. The decision is presented as causally determined by the context: "*facts leave us no choice*", "*what happens in the world forces us to do so*".

> In 1960, Charles de Gaulle, the President of the French Republic, held a referendum on the question of the independence of Algeria, with which France had been at war since 1954. He urged the people to vote for Algerian independence.
>
> No one can doubt the extreme importance of the country's response. For Algeria, the right granted to its peoples to dispose of their fate will mark the beginning of a whole new era. Some may regret that prejudices, routines and fears previously prevented the assimilation of Muslims, assuming it were possible. But the fact that they constitute eight-ninths of the population, and that this proportion continues to grow in their favor; the evolution begun in people and in things by the events, and prominently by the insurrection; and, finally, what has happened and what is going on in the world — make these considerations chimerical and these regrets superfluous.
>
> Charles de Gaulle, December 20th 1960 Speech[2].

The *strong will argument* denies precisely this determinism: "*where there is a will, there is a way*". In May and June 1939, the Belgian, British, French and Netherlands armies were totally routed by the German Nazi armies. In a situation that appeared desperate to many, General Charles de Gaulle rejected the armistice

---

[1] François Furet, *Le Passé d'une illusion. Essai sur l'idée communiste au XXe siècle*. Paris: Robert Laffont & Calmann-Levy, 1995, p. 365.
[2] http://fresques.ina.fr/de-gaulle/fiche-media/Gaulle00063/speech-of-20-December-1960.html (20-09- 2013).

that Marshal Petain had just signed with the German Nazi enemy, and from London launched on the BBC his call to continue the fight:

> Of course, we were subdued by the mechanical, ground and air forces of the enemy. Infinitely more than their number, it was the tanks, the airplanes, the tactics of the Germans which made us retreat. It was the tanks, the airplanes, the tactics of the Germans that surprised our leaders to the point of bringing them to where they are today.
> But has the last word been said? Must hope disappear? Is *defeat* final? No!
> Believe me, I speak to you with full knowledge of the facts and tell you that nothing is lost for France.
> [...] Whatever happens, the flame of the French resistance must not, and will not be extinguished.
>
> Charles de Gaulle, *Text of the Appeal of June 18, 1940*[1]

Major political decisions combine the two forms of argumentation.

2. Naturalistic argument

In law, naturalistic argument refers to the hypothesis of an impotent legislator arguing that it is impossible to legislate in certain areas, or of a judge who waives the application of the law on the pretext of special circumstances, S. **Juridical Arguments**.

The naturalistic argument is also exploited in the field of religious law; Luther uses it in connection with the prohibition of the marriage of priests by the Pope of the Roman Catholic Church. According to Luther, most priests "[cannot] do without a woman", at least for their household:

> If therefore [the priest] takes a woman, and the Pope allows this, but will not let them marry, what is this but expecting a man and a woman to live together and not to fall? Just as if one were to set fire to straw, and command it should neither smoke nor burn.
> The Pope having no authority for such a command [forbidding the marriage of priests], any more than to forbid a man to eat and drink, or to digest, or to grow fat, no one is bound to obey it, and the Pope is answerable for every sin against it.
>
> Martin Luther, *Address To The Nobility of the German Nation*, [1520][2]

A priori, the *naturalistic argument* has little to do with the *naturalistic fallacy*, which systematically values the natural, S. **Fallacious (II)**. However, the accusation of *fallacious naturalism* might serve to refute the argument of the *force of circumstances*.

Whole and Parts ▶ Composition and Division

---

[1] http://lehrmaninstitute.org/history/index.html (01-20-2017).
[2] Quoted after http://sourcebooks.fordham.edu/mod/luther-nobility.asp, (01-20-2017).

Words as Arguments

1. A word as the hologram of argument

Holography is a technique that provides a two-dimensional representation of three-dimensional phenomena. In metaphorical sense, a *word* can function as the hologram of a whole *speech* (actually a *set* of co-oriented speeches) and mirror the totality of the argumentative discourse it is part of. The *line* of discourse is condensed into one of its *points*, that being the word. Such hologrammatic words are termed *oriented* (in the Argumentation within Language theory) or *biased* (in standard Fallacy theory).

Argumentations containing oriented words are considered to be *fallacious* and *sophistical* insofar as they actually presuppose the conclusions they apparently construct. The conclusion is embedded in the wording of the argument, and the reasoning is trapped in a vicious@ circle. Metaphorically, one may say that the target (the conclusion) is tailored to the measure of the arrow (the argument); the arrow cannot miss the target, and is therefore irrelevant.

This is true if an argumentation is considered to be a self-sufficient piece of reasoning, contained in an autonomous discursive episode. If argumentation is seen as an on-going process, however, the orientation of words testifies to the fact that the argumentative discourse not only *constructs* its conclusion on the spot, but also *recalls* that this conclusion has been previously established. Oriented words refer to the whole script corresponding to the arguer's discourse; they are the memory of argumentation, and the clearest example of *objects@ of discourse*. The word *biased* has a negative orientation ("prejudiced; to be avoided") while *orientation, oriented* can have a neutral-positive orientation ("taking bearings"), while admitting, if need be, a negative orientation ("biased").

The global issue is that of the argument *orientation@* and the *persuasive@ definition*. The first case involves language data and the second speech activity, in the first case the discourse is biased per se, in the second case it is made biased by the participant.

2. Designations as issues

Let us consider the pro-life *vs.* free choice debate. If a participant speaks of *babies* and the other of *fetuses*, we already know that the former is most probably *pro-life* and the latter *pro-free-choice*. The antagonistic words are "loaded" with the antagonistic conclusion towards which they are oriented. *Baby* refers to a human person, and implies that we must feel for this being all the value-loaded emotions we feel for young children, and treat him or her accordingly. *Fetus* puts these attitudes between parentheses, and technically refers to a "product of the conception of vertebrates during prenatal development, after the embryonic stage, when it begins to form and to present the distinctive characteristics of the species." (TLFi, *Fetus*). A word might be value-loaded in a discourse and not in another. In the developmental discourse of medicine, for example, *fetus* opposes to *embryo* and is a non-controversial technical designation, as is *baby* when referring to a pre-toddler child.

The idea of human selection is generally repulsive. The search for a positive designation for babies which have been *genetically selected in order to treat his or her sick brother or sister*, continues. Candidate terms include, *designer baby, medicine baby, savior baby, doctor baby*....

A similar debate is also reflected in the designation for products used as crop treatments, and suspected to be carcinogenic. The terms *agro-pharmaceutical* product or *phyto-sanitary* product sound highly chemical, and the latter has even been appropriated by a French association "Phyto-Victims". *Pesticide* has also a negative orientation, despite its etymological meaning, "pests killer" (as if the negation of a negation was interpreted as an hyper-negation). The terminological fight continues, and the industry has turned to *plant protection product* and *crop protection product*.

The orientation of ordinary words strongly differentiates natural language and logical languages. Biased language can be considered an obstacle to the objective treatment of the issue, and has thus been banned from argumentative discourse as an instrument of monological rationality. The problem is how to agree upon the purification principle, as it could significantly affect most of out common vocabulary.

Categorization operations are not too problematic for plants, animals and other natural species. Things are more complicated when it comes to beings and situations whose designations cannot be agreed upon before the debate, but is actually the very issue at stake.

In the debate about abortion for example, the discussion of the correct designation, *fetus* or *baby*, cannot be dissociated from the discussion on the merits and disadvantages of abortion itself.

In practice, the persuadee must to assent not only to a position, but also to the corresponding expression, **S. Persuasion**. It is not possible to remediate biased language by a conventionalism, consisting in agreeing on the meaning of the words *before* the debate in which they are to be used, refraining from using loaded terms, or creating neutral terms. The discussion of the *nature* of the object is not separable from the discussion of its *name*. The fact of being at the heart of a debate results in *duplication* of the designation of the object. Its objective designation and "real name" will eventually be attributed to it at the end of the debate; objectivity is not a condition but a product of the debate.

The search for "neutral" terms shows, on the one hand, the desire to put ordinary language between parentheses when it comes to serious issues, insofar as it does not correspond to a pure referential and inferential ideal, and, on the other hand, the wish to consider that the debate between rational beings consists only in clarifying semantic misunderstandings, which are the consequence of the defects of natural language. The task of argumentation would be relatively simple if we could assume that some data are accepted as such by both parties; this is true only for *peaceful* neutral facts, external to the heart of the debate. In the other case, the division of discourses is openly exposed by the use of so-called biased, loaded or oriented designations. The designation is already argumentative, **S. Schematization**. Agreeing on the designation of facts is a matter of identity, focus, emotional empathy; as there is a conversion to new beliefs, there is a conversion to new facts and words.

There are no entries for the letter **X**

Y

"You too!"

Lat. "tu quoque!"; *tu* "you", *quoque* "too"

In Latin and in English, the "you too" argument scheme is named after the statement that typically realizes the argument.
In the general case the reply:
    S1:   — *I do A because X does so.*

is a strategy of legitimation by imitation. The fact that **X** makes **A** creates a *precedent*@ legitimizing **A**, and if **S1** considers **X** as a *model*@, it gives **A** a second form of legitimacy. Such legitimations are part of the "*You too!*" argumentation; its scenario is as follows:
    S1 performs such action **A**.
    S2 blames him
    S1 replies: *"But you do it too! You do the same!"*

S2 criticizes **S1** for an action that he presents as blameworthy. **S1** can reply in a variety of ways:
(i) He can first answer to **S2** that *others* do the same: since Landru (a popular French serial killer) murdered his mistresses, why couldn't I? The degree of legitimation depends on the severity of the transgression and the number of transgressors. I run a red light in the open country, when there is no traffic and

the visibility is perfect, and I feel justified in saying *"well, this is forbidden, but everyone does it, the guy ahead went through, I just followed him"*.

(ii) In the case where the wrongdoer is not another third party but **S2**, **S1** has two possibilities:

— As in the previous case, **S1** can quietly legitimize his or her action by the (bad) example given by **S2**.

— **S1** can also reply using a counter-accusation, which seeks to put **S2** in the face of the contradiction between what he preaches and what he does, S. *Ad hominem*. **S1** acknowledges his or her misbehavior, but considers that, due to his or her own misbehavior, **S2** is in no position to teach him or her a lesson. In terms of stasis, the defendant does not recognize the legitimacy of the judge, S. **Stasis**:

> **S1**: — *It suits you well to blame me! Please, not you! I have no moral lessons to receive from you.*

The phrase *"two wrongs don't make a right"* can be understood in two different ways.

— First, as *"one does not fight evil with evil"*, that is, *"evil must be fought by legal means"*, a very important principle; many would be tempted to add the clause *"as far as possible"*. In other words, the good end — the struggle against evil — should not be pursued by evil means; such as torturing the former torturer to stop torture. This would amount to a case of autophagy@.

This principle is invoked to reject the justification of a mistreatment made to somebody by arguing, in a sort of anticipatory law of retaliation, that, had he been in our place, this is what he would have done to us, S. **Reciprocity** (after FF, *Two wrong*).

— Secondly, it can express the rule that "bad behavior does not become legitimate because widespread"; many wrongs never make a right. The common transgression (argument from number) never creates an against-the-law legitimacy, S. **Consensus**.

Note that in material life, thanks to a minor miracle, an error sometimes compensates for another. This also seems to occur in science:

> Kepler knows that Tycho Brahe [obtained] the best possible accuracy on the measurements of the positions of the planets (including the planet Mars), and this accuracy was of two minutes of degree. With the mathematical model of a circular orbit on the Mars planet that he (Kepler) used, Kepler noticed discrepancies of eight minutes of degree between the positions observed by Tycho Brahe and the calculated positions. Trusting the precision of Tycho Brahe's measurements, Kepler renounces the circular orbit of Mars. He revises the orbit of the Earth and, thanks to two compensating errors, discovers his first law: *"In the motion of a planet, the vector ray sweeps equal areas in equal times."*

"You too!"

Edgar Soulié, [*Johannes Kepler (1571-1630), the Protestant astronomer who discovered the laws of motion of the planets*], (no date).[1]

There are no entries for the letter Z

---

[1] Edgar Soulié, *Johannes Kepler (1571-1630) L'astronome protestant qui a découvert la loi du mouvement des planètes.* http://www.astrosurf.com/rtaa/rtaa2016/documents/kepler-edgar-soulie.pdf (01-09-2017).

References

Ad Her. = (1954). *Rhetorica ad Herennium*. Trans. by H. Caplan. Cambridge, MA: Harvard UP.

Adam J.-M. (1996). *L'argumentation dans le dialogue. Langue Française*, 112. 31-49.

Adorno Th. W., Frenkel-Brunswik E., Levinson D. J. & Sanford R. N. (1950). *The Authoritarian Personality*. New York: Harper & Row.
https://is.muni.cz/el/1423/jaro2017/SOC286/um/Adorno_et_al._1950_-_Authoritarian_Personality.pdf

Agazzi E. (ed.) (1980). *Modern Logic – A Survey*. Dordrecht: Kluwer.

Al-Ghazali (*Bal.*) = (1998). *La Balance Juste*. Paris: Iqra.

Al-Ghazali (*Dég.*) = (1995). *Les Dégâts des Mots*. Paris: Iqra.

Amossy R. (1991). *Les Idées Reçues. Sémiologie du Stéréotype*. Paris: Nathan.

Amossy R. (ed.) (1999a). *Images de Soi dans le Discours. La Construction de l'Éthos*. Geneva: Delachaux & Niestlé.

Amossy R. (1999b). La notion d'éthos, de la rhétorique à l'analyse de discours. *In* Amossy R. (ed.) (1999a). 9-30.

Amossy R. (2000). *L'Argumentation dans le Discours*. Paris: Nathan.

Angenot M. (2008). *Dialogue de Sourds. Traité de Rhétorique Antilogique*. Paris: Mille et Une Nuits.

Anscombre J.-C. (ed.) (1995a). *Théorie des Topoi*. Paris: Kimé.

Anscombre J.-C. (1995b). De l'argumentation dans la langue à la théorie des topoi. In Anscombre J.-C. (ed.) (1995a). 11-47.

Anscombre J.-C. & Ducrot O. (1983). *L'Argumentation dans la Langue*. Brussels: Mardaga.

Anscombre J.-C. & Ducrot O. (1986). Informativité et argumentativité. In Meyer M. (eds) (1986). 79-94.

Anselm of Canterbury (*Pros.*) = (2000). *Proslogion*. In *Complete Philosophical and Theological Treatises of Anselm of Canterbury*. Trans. by J. Hopkins and H. Richardson. Minneapolis: The Arthur J. Banning Press. http://jasper-hopkins.info/proslogion.pdf (12-

References

12-2017).

Antiphon (*Disc.*) = (1923). *Discours*. Ed. and trans. by L. Gernet. Paris: Les Belles-Lettres.

Apothéloz D. & Miéville D. (1989). Cohérence et discours argumenté. In M. Charolles (ed.) (1989). *The Resolution of Discourse*. Hambourg, Buske. 68-87.

Arendt H. (1958). What was authority? In Friedrich C. (ed.) (1958). *Authority*. Cambridge, MA: Harvard UP.

Arendt H. [1951] = (1976). *The Origins of Totalitarianism*. San Diego: Harcourt Brace

Aristotle (*PA*) = *Prior Analytics*. Trans. by A. J. Jenkinson.
http://classics.mit.edu//Aristotle/prior.html (12-12-2017).

Aristotle (*Pol.*) = (1855). *Politics*. Vol. 1. Trans. by B. Jowett. Oxford: Clarendon Press.

Aristotle (*Post. An.*) = (1960). *Posteriors Analytics*. Trans. by H. Trendennick. In Aristotle, *Posteriors Analytics – Topica*. Cambridge, MA, London, Eng.: Harvard UP.

Aristotle (*Rhet.*) =
— (2005) *Rhetoric*. In Aristotle, *Poetics* and *Rhetoric*. Introd. and notes by E. Garver. *Rhetoric* trans. by W. Rhys Roberts (1924). *Poetics*, trans. by S. H. Butcher (1911). New York: Barnes and Nobles.
— (1926). *Rhetoric*. Trans. by J. H. Freese. London: William Heinemann & New York: G.P. Putnam's Sons.

Aristotle (*Soph.*) = (1955). *On Sophistical Refutations*. Trans. by E. S. Forster. Cambridge, MA & London: Harvard UP. https://ebooks.adelaide.edu.au/a/aristotle/sophistical/

Aristotle (*Top.*) = (1960). *Topica*. In Aristotle, *Posteriors Analytics – Topica*. Trans. by E. S. Forster. Cambridge, MA & London, England: Harvard UP.

Arnauld A. & Nicole P. [1662] = (1850). *Logic, or The Art of Thinking, Being the Port-Royal Logic*. Trans. and introd. by Th. S. Baynes. Edinburgh: Sutherland and Know.

Atkinson J.-M. & Heritage J. (eds) (1984). *Structures of Social Action - Studies in Conversation Analysis*. Cambridge, MA: CUP. 79-112.

Auroux S. (1995). Argumentation et anti-rhétorique. La mathématisation de la logique classique. *Hermès* 15. 129-144.

Auroux (1990). *Encyclopédie Philosophique Universelle*. Vol. 2: *Les Notions Philosophiques*. Paris: PUF.

Auroux S. (ed.) (1992). *Histoire des idées linguistiques*. Brussels: Mardaga.

Austin J. L. (1962). *How To Do Things With Words*. Oxford: Oxford UP.

Bacon F. [1620] = (1901). *Novum Organum*. Ed. by J. Devey. New York: Collier and Son.

Bailly A. [1901]. *Abrégé du Dictionnaire Grec-Français*. Paris: Hachette.
http://home.scarlet.be/tabularium/bailly/index.html

Baker M. J. (1996). Argumentation et co-construction des connaissances. *Interaction et Cognitions* 2, 3. 157-191.

Bakhtin = Bakhtine M. (1978). *Esthétique et Théorie du Roman*. Trans. by D. Olivier. Paris: Gallimard.

Balacheff N. (1999). Apprendre la preuve. In Sallantin J. & Szczeciniarz J. J. (eds.) (1999). 197-236.

References

Barthes R. (1970). L'ancienne rhétorique. Aide-mémoire. *Communications*, 16. 195-226.

Barthes R. (1977) = (1996). *Inaugural Lecture at the Collège de France*. In Kearney R & Rainwater M. (1996). *The Continental Philosophy Reader*. New York: Routledge. 364-377.

Bassham G. (2003). Linked and independant premises. A new analysis. In Eemeren, F. H. van, Blair J. A., Willard C. A., Snoeck-Henkemans A. F. (eds) (2003). 69-73.

Beardsley M. C. [1950] = (1975). *Thinking Straight. Principles of Reasoning for Readers and Writers*. New York: Prentice-Hall.

Benett B. S. (2005). Hermagoras of Temnos. In Ballif M. & Moran M. G. (eds) (2005) *Classical Rhetorics and Rhetoricians: Critical Studies and Sources*. Westport, CT: Praeger.

Benoit W. L. (1987). On Aristotle example. *Philosohy and Rhetoric* 20. 261-267.

Benoit W. L. & Lindsey J. J. (1987). Argument fields and forms of argument in natural language. In van Eemeren, F. H., Grootendorst, R., Blair, J. A., Willard, C. A. (eds.) (1987). 215-224.

Bentham J., [1824] = (1962). *The Book of Fallacies*. In *The Works of Jeremy Bentham*. Published by J. Bowring. Vol. 2. New York, Russell & Russell.

Benveniste É. [1958] = (1971). Subjectivity in language. In *Problems in General Linguistics*. Trans. by M. E. Meek. Miami, University of Miami Press. 223-230.

Benveniste É. [1959] = (1971) The correlations of tense in the French verb. In *Problems in General Linguistics*. Trans. by M. E. Meek. Miami: University of Miami Press. 205-248.

Benveniste É. [1969] = (1973). The censor and *auctoritas*. In *Indo-European Language and Society*. Trans. by E. Palmaer and J. Lallot. Miami: University of Miami Press. http://chs.harvard.edu/CHS/article/display/3961

Berlioz J. (1980). Le récit efficace: L'*exemplum* au service de la prédication. *Rhétorique et Histoire. Mélanges de l'École Française de Rome, Moyen Âge - Temps Modernes*, 92. 113-146.

Bernays E. L. (1928). *Propaganda*. New York: Horace Liveright.

Bernier R. (1980). Le rôle de l'analogie dans l'explication en biologie. In A. Lichnerowicz & al. (eds) (1980). 167-193.

Bible = The Bible. www.biblegateway.com

Billig M. [1987] = (1989). *Arguing and Thinking. A Rhetorical Approach to Social Psychology*. Cambridge: Cambridge UP & Paris: Éditions de la MSH.

Bilmes J. (1991). Toward a theory of argument in conversation. The preference for disagreement. In van Eemeren F. H., Grootendorst R., Blair J. A. & Willard C. A. (eds.) (1991). 462-469.

Bird O. (1961). The re-discovery of the topics: Professor Toulmin's inference warrant. *Mind* 70. 76-96.

Bitzer L. (1959). Aristotle's Enthymeme Revisited. *Quarterly Journal of Speech* 45. 399-408.

Bitzer L. F. [1968] = (1974) The rhetorical situation. In Fisher W. R. (ed.) (1974). *Rhetoric: a Tradition in Transition*. East Lansing, MI: Michigan State UP. 247-260.

References

Black M. (1962). *Models and Metaphor. Studies in Language and Philosophy*. Ithaca, NY: Cornell University Press.

Black M. (1979). More about metaphor. In A. Ortony (ed.) (1979). 19-43.

Blair J. A & Johnson R. (eds) (2011). *Conductive Argument: An Overlooked Type of Defeasible Reasoning*. London: College Publications.

Blair J. A. (2012). *Groundwork in the Theory of Argumentation*. Dordrecht: Springer.

Blair J. A. & Johnson R. H. (eds) (1980). *Informal Logic: The First International Symposium*. Inverness: Edgepress.

Blanché R. (1970). *L'axiomatique*, Paris: PUF

Blanché R. (1973). *Le raisonnement*. Paris: PUF.

Boeckh P. A. [1886] = (1988). Philological hermeneutics. In Mueller-Vollmer K. (ed.) (1988). 132-147.

Boethius, *Top.* = (1978). *De Topicis Differentiis*. Trans. Notes and Essays by E. Stump. Ithaca: Cornell University Press,

Bonhomme M. (1998). *Les Figures Clés du Discours*. Paris: Le Seuil.

Booth W. C. (1974). *Modern Dogma and the Rhetoric of Assent*. Chicago: The University of Chicago Press.

Borel M.-J., Grize J.-B. & Miéville D. (1983). *Essai de Logique Naturelle*. Berne, Peter Lang.

Bori P. C. [1987] = (1991) *L'Interprétation Infinie*. Trans. by F. Vial. Paris: Le Cerf.

Bornscheuer L. (1976). *Topik. Zur Struktur der gesellschaftlichen Einbildungskraft*. Frankfurt: Suhrkamp.

Bossuet J.-B. [1677] = (1990). *Logique du Dauphin*. Paris: Éditions Universitaires.

Bouamrane C. & Gardet L. (1984). *Panorama de la Pensée Islamique*. Paris: Sindbad.

Boudon R. (1990). *L'art de se Persuader des Idées Douteuses, Fragiles ou Fausses*. Paris: Le Seuil.

Bourdieu P. (1982). *Ce que Parler Veut Dire. L'Économie des Échanges Linguistiques*, Paris: Fayard.

Bouveresse J. [1999]. *Prodiges et Vertiges De l'Analogie*. Paris: Raisons d'Agir.

Bouverot D. (ed.) (1993). *Rhétorique et Sciences du Langage*. Verbum, 1-2-3.

Bouvier A. (1999). *Philosophie des Sciences Sociales - Un point de vue argumentativiste en sciences sociales*. Paris: PUF.

Bouvier A. (ed.) (1994-1995). *Argumentation et Sciences sociales, (I) et (II). L'Année Sociologique*, 44-45.

Boyer A., Vignaux G. (ed.) (1995). *Argumentation et Rhétorique*, (I) & (II). Hermès 15-16.

Brandt P.-Y. & Apothéloz D. (1991). L'articulation raisons-conclusion dans la contre-argumentation. *La Négation. Travaux du Cercle de Recherches Sémiologiques*, 59). 88-102.

Brémond Cl. (1982). Décomposition syntagmatique: Les parties de l'*exemplum*. In Brémond Cl., *& al.* (1982). 113-143.

Brémond Cl., Le Goff J. & Schmitt J-Cl. (1982). *L'Exemplum*. Turnhout: Brepols.

Breton Ph. (1996). *L'Argumentation dans la Communication*. Paris: La Découverte.

Breton Ph. (1997). *La Parole Manipulée*. Paris: La Découverte.

Breuer D. & Schanze H. (eds) (1981). *Topik*. München: Wilhelm Fink.

Brody B. A. (1967). Logical terms, Glossary of —. In Edwards P (ed.) (1967). Vol. 5. 57-77.

Brown R. W. & Levinson S. (1978). *Politeness. Some Universal en Language Usage*. Cambridge, CUP.

Brunschwig J. (1967). Introduction. In Aristote, *Topiques*. Paris: Les Belles-Lettres.

Bühler K. [1933] = (1976). *Die Axiomatik der Sprachwissenschaften*. Einleitung und Kommentar von E. Ströker. Frankfurt am Main: Vittorio Klosterman.

Burke K. (1945). *A Grammar of Motives*. Berkeley: University of California Press.

Burke K. (1950). *A Rhetoric of Motives*. Berkeley: University of California Press.

Buty Chr. & Plantin Chr. (2009). *Argumenter en Classe de Sciences. Du débat à l'Apprentissage*. Lyon: INRP.

Carel M. (1995). *Trop:* Argumentation interne, argumentation externe et positivité. In Anscombre I. C. (ed.) *Théorie des Topoi*. Paris: Kimé. 177-206.

Carel M. (1999). Sémantique discursive et sémantique logique: Le cas de *mais*. *Modèles linguistiques*, XX, 1. 133-144.

Carel M. (2011). *L'Entrelacement Argumentatif. Lexique, Discours et Blocs Sémantiques*. Paris: Champion.

Casagrande C., Vecchio S. [1987] = (1991). *Les péchés de la langue. Discipline et éthique de la parole dans la culture médiévale*. Pref. by J. Le Goff. Trans. by Ph. Baillet. Paris: Le Cerf.

Cassin B. (2004). *Vocabulaire Européen des Philosophies - Dictionnaire des Intraduisibles*. Paris: Le Seuil.

CD = *Cambridge Dictionary*, http://dictionary.cambridge.org

Chabrol Cl. & Radu M. (2008). *Psychologie de la Communication et de la Persuasion*. Brussels: De Boeck.

Chaignet A. E., 1888). *La Rhétorique et Son Histoire*. Geneva: Slatkine Reprints.

Chakhotine S. [1939] = (1940). *The Rape of the Masses; The Psychology of Totalitarian Political Propaganda*. Trans. by E. W. Dickes. New York: Alliance Book / London: Routledge.

Charaudeau P. & Maingueneau D. (2002). *Dictionnaire d'Analyse du Discours*. Paris: Le Seuil.

Chenique F. (1975). *Éléments de Logique Classique*. T. 1. *L'Art de Penser et de Juger*. T. 2. *L'Art de Raisonner*. Paris: Dunod.

Cicero (*Acad.*) = *The Academic Questions*. Trans. by C. D. Yonge. The University of Adelaide. https://ebooks.adelaide.edu.au/c/cicero/academic-questions/ (08-17-2017)

Cicero (*De Inv.*) = (2006). *On Invention*. Trans. by H. M. Hubbell. Cambridge MA & London England: Harvard UP.

Cicero (*De Or.*) = (1878). *De Oratore*. Translated or edited by J. S. Watson. New York,

References

Harper. Quoted after Cicero, *On Oratory and Orators*. Carbondale & Edwardsville: Southern Illinois UP (1970).

Cicero (*Part.*) = *De Partitione Oratoria*. In *Cicero,* IV. Trans. by H. Rackham. First published (1942). Cambridge MA, London England: Harvard UP.

Cicero (*Top.*) = (2006) *Topics*. In Cicero, *On Invention — Best Kind of Orators – Topics*. Trans. by H. M. Hubble. Cambridge MA, London: Harvard UP. (2006).

Condillac E. B. de [1976] = (1981). *Traiter de l'Art de Raisonner*. Paris: Vrin.

Conein B., de Fornel M. & Quéré L. (eds) (1990). *Les Formes de la Conversation*. Vol. 1. Paris: CNET.

Conley T. M. (1984). The enthymeme in perspective. *Quarterly Journal of Speech* 70. 168-187.

Cooper J. M. (1996). *An Aristotelian Theory of the Emotions*. In A. O. Rorty (ed.) *Essays on Aristotle's Rhétoric*. Berkeley: University of California Press. 238-257.

Cosnier J. (1994). *Psychologie des Émotions et des Sentiments*. Paris: Retz & Nathan.

Cousin J. (1976). Note [to Quintilien *Institution Oratoire* V, 10]. Paris: Les Belles Lettres.

Cox J. R. & Willard C. A. (eds) (1982). *Advances in Argumentation Theory and Research*. Carbondale, IL: Southern Illinois University Press.

Curtius E. R. [1948] = (1953). *European Literature and the Latin Middle Ages*. Trans. by W. R. Trask. Princeton: Princeton University Press.

D.c = Dictionary.com. http://www.dictionary.com

Damasio A. R. (1994). *Descartes's Error — Emotion, Reason and the Human Brain*. New York: Avon Books.

Danblon E. (2005). *La Fonction Persuasive. Anthropologie du Discours Rhétorique: Origines et Actualité*. Paris: Armand Colin.

Dascal M. (2009). Colonizing and decolonizing minds. In I. Kuçuradi (ed.) *Papers of the 2007 World Philosophy Day*. Ankara: Philosophical Society of Turkey. 308-332.

Davidson D. (1978). What metaphors mean. In Sacks, S. (ed.) (1978). 29-45.

De Vries E., Lund K., & Baker M. J. (2002). Computer-mediated epistemic dialogue: explanation and argumentation as vehicles for understanding scientific notions. *The Journal of the Learning Sciences* 11, 1. 63-103.

Declerq G. (1993). *L'Art d'Argumenter. Structures Rhétoriques et Littéraires*. Paris: Éditions Universitaires.

Declerq G. (2002). Avatars de l'argument *ad hominem*. Éristique, sophistique, rhétorique. In M. Murat, M., G. Declercq & J. Dangel (eds) (2002). *La Parole Polémique*, Paris: Champion. 327-376.

Descartes R. [1628] = (1954). *Rules for the Direction of the Mind*. Trans. by E. Anscombe & P. T. Geach. In E. Anscombe & P. T. Geach (1954). *Descartes. Philosophical Writings. A Selection*. London: Nelson.

Descartes R. [1637] = (1987). *Discours de la Méthode*. Introd. and notes by É. Gilson. Paris: Vrin.

Descartes R. [1641] = (1911). *Meditations on First Philosophy*. In *The Philosophical Works of René Descartes*. Trans. by E. S. Haldane. Cambridge: CUP.

Descartes R. [1649] = (1989). *The Passions of the Soul*. Introd. by G. Rodis-Lewis. Trans. by S. Voss. Indianapolis, IN & Cambridge, MA: Hackett.

Dicolat = *Dictionnaire latin-français*.
http://www2c.aclille.fr/verlaine/College/Projets/Latin/dictionnaire_fr_latin/Dico-lat-C.html.

Dic = *Dictionary.com*, http://www.dictionary.com

Dieter A. O. L. (1950). 'Stasis'. *Speech Monographs*, 17, 4. 345-69.

Domenach J. M. [1950] = (1979). *La Propagande Politique*. Paris: PUF.

Dominicy M. (s.d.). Perelman et l'École de Bruxelles.
www.philodroit.be/spip.php?page=article&id_article=452&lang=fr. (09-20 -2013)

Dopp J. (1967). *Notions de Logique Formelle*. 2nd revised ed. Louvain & Paris: Béatrice Nauwelaerts.

Douay-Soublin F. (1992). La rhétorique en Europe à travers son enseignement. In S. Auroux (ed.), T. 2. 467-507.

Douay-Soublin F. (1999). La Rhétorique en France au XIXe siècle à travers ses pratiques et ses institutions: Restauration, renaissance, remise en cause. In M. Fumaroli (ed.) (1999). 1071-1214.

Doury M. (1997). *Le Débat Immobile - L'Argumentation dans le Débat Médiatique sur les Parasciences*. Paris: Kimé.

Doury M. (2000). La réfutation par accusation d'émotion. In Plantin Chr., Doury M., & Traverso V. (eds) (2010). 265-277.

Doury M. (2003). L'évaluation des arguments dans les discours ordinaires. Le cas de l'accusation d'amalgame. *Langage et société*, 105. 9-37.

Doury M. (2006). Evaluating analogy. Toward a descriptive approach to argumentative norms. In Houtlosser P. & van Rees A. (eds) (2006). *Considering Pragma-Dialectics*. Mahwah, NJ: Lawrence Erlbaum. 35-49.

Dubucs J. (1995). Les arguments défaisables. *Argumentation et Rhétorique I. Hermès* 15. 271-290.

Ducrot O. s.d. Quelques raisons de distinguer '*locuteurs*' et '*énonciateurs*'.
www.hum.au.dk/romansk/polyfoni/Polyphonie_III/Oswald_Ducrot.htm. (09-20-2013).

Ducrot O. (1972). *Dire et ne pas Dire*. Tours, Hermann.

Ducrot O. (1973). *La Preuve et le Dire*. Paris: Mame.

Ducrot O. (1975). '*Je trouve que*'. *Semantikos* 1. 62-88. Re-published in Ducrot & al. (1980). 57-92.

Ducrot, O. (1980). *Les Échelles Argumentatives*. Paris: Minuit.

Ducrot O. (1984). *Le Dire et le Dit*. Paris: Minuit.

Ducrot O. (1988). *Polifonía y Argumentación*. Cali, Universidad del Valle.

Ducrot O. (1993). Les topoi dans la Théorie de l'Argumentation dans la Langue. In C. Plantin (ed.) (1993). 233-248.

Ducrot O. (1995). Les modificateurs déréalisants. *Journal of Pragmatics*, 24. 145-165.

Ducrot O. & al. (1980). *Les Mots du Discours*. Paris: Minuit.

Ducrot O. & al. (1982). *Justement*, l'inversion argumentative. *Lexique*, 1. 151-164.

References

Ducrot O. (ed.) (1966). Logique et linguistique. *Langages* 2. 3-30.

Dufour M. (2008). *Argumenter - Cours de Logique Informelle*. Paris: Armand Colin.

Dumarsais C. Ch. [1730] = (1988). *Des Tropes ou des Différents Sens dans Lesquels on Peut Prendre un Même Mot dans une Même Langue*. Ed. by F. Douay-Soublin. Paris: Flammarion.

Dumoncel, J.-C. (1990). Évidence. In Auroux, S. (1990). 908.

Dupleix S. [1607] = (1984). *La Logique, ou Art de Discourir et Raisonner*. Paris: Fayard.

Dupont F. (2000). *L'Orateur Sans Visage. Essai sur l'Acteur Romain et son Masque*. Paris: PUF.

Dupréel E. (1939). *Esquisse d'une Philosophie des Valeurs*. Paris: Alcan.

Dupriez B. (1984). *Gradus. Les Procédés Littéraires – Dictionnaire*. Paris: UGE.

Duval R. (1992-1993). Argumenter, démontrer, expliquer. Continuité ou rupture explicative? « petit x » 31. 37-61.

Duval R. (1995). *Sémiosis et Pensée Humaine. Registres Sémiotiques et Apprentissages Intellectuels*. Berne: Peter Lang.

Edwards P. (ed.) (1967). *Encyclopedia of Philosophy*. New York: MacMillan & London: Collier.

van Eemeren F. H., Blair J. A., Willard C. A., & Snoeck-Henkemans A. F. (eds) (2003). *Proceedings of the Fifth Conference of the International Society for the Study of argumentation*. Amsterdam: SICSAT.

van Eemeren F. H., Blair, J. A., Willard, C. A. & Garssen, B. (eds.) (2007). *Proceedings of the Sixth International Conference of the International Society for the Study of Argumentation*. Amsterdam: SICSAT.

van Eemeren F. H. & Garssen B. (2009). The fallacies of composition and division revisited. *Cogency*, 1, 1. 23-42.

van Eemeren F., Garssen B., Godden D., Mitchell G. (eds) (2011). *Proceedings of the 2010 ISSA Conference*. Amsterdam: SICSAT.

van Eemeren F. H., Garssen, B. & Meuffels, B. (2009). *Fallacies and Judgements of Reasonableness – Empirical Research Concerning the Pragma-dialectical Discussion Rules*. Dordrecht, Springer.

van Eemeren F. H. & Grootendorst R. (1984). *Speech Acts in Argumentative Discussions: A Theoretical Model for the analysis of Discussions Directed Towards Solving Conflicts of Opinion*. Dordrecht: Foris.

van Eemeren F. H. & Grootendorst R. (1992). *Argumentation, Communication, Fallacies*. Mahwah, NJ: Lawrence Erlbaum.

van Eemeren F. H. & Grootendorst R. (1995). The Pragma-Dialectical Approach to Fallacies. In Hansen H. V. & Pinto R. C. (eds) (1995). *Fallacies: Classical and Contemporary Readings*. University Park, PA: Pennsylvania State UP. www.ditext.com/eemeren/ pd.html (09-20-2013).

van Eemeren F. H. & Grootendorst R. (2004). *A Systematic Theory of Argumentation: The Pragma-dialectical Approach*. Cambridge, MA: Cambridge UP.

van Eemeren F. H., Grootendorst R., Blair J. A. & Willard, C. A. (eds.) (1987). *Proceed-*

ings of the Conference on argumentation (1986). Dordrecht: Foris.

van Eemeren F. H., Grootendorst R., Blair J. A. & Willard C. A. (eds.) (1991). *Proceedings of the Second International Conference on Argumentation*. Amsterdam: SICSAT.

van Eemeren F. H., Grootendorst R., Blair J. A. & Willard C. A. (eds.) (1995). *Proceedings of the Third ISSA conference on argumentation (1994)*. Amsterdam: SICSAT.

van Eemeren F. H., Grootendorst R., Blair J. A. & Willard, C. A. (eds.) (1999). *Proceedings of the Fourth International Conference of the International Society for the Study of Argumentation*. Amsterdam: SICSAT.

van Eemeren F. H., Grootendorst R. & Snoeck Henkemans, A. F. (2002). *Argumentation: Analysis, Evaluation, Presentation*. Mahwah, NJ: Lawrence Erlbaum.

van Eemeren F. H., Grootendorst R., Snoeck Henkemans A. F., Blair J. A., Johnson R. H., Krabbe E. C. W., Plantin Chr., Walton D. N., Willard C. A., Woods J. & Zarefsky D. (1996). *Fundamentals of Argumentation Theory, A Handbook of Historical Backgrounds and Contemporary Developments*. Mahwah, NJ: Lawrence Erlbaum

van Eemeren F. H. & Houtlosser P. (2002). *Dialectic and Rhetoric*. Dordrecht: Kluwer.

van Eemeren F. H. & Houtlosser P. (2003). More About Fallacies as Derailments of Strategic Maneuvering: The Case of T*u Quoque*. "Informal Logic @ 25" Conference, OSSA, University of Windsor, Windsor, Ontario. CDRom.

van Eemeren F. H., Houtlosser P. & Snoeck Henkemans A. F. (2007). *Argumentative Indicators in Discourse. A Pragma-Dialectical Study*. Amsterdam: Springer.

van Eemeren F. H. & Houtlosser P. (eds) (2000). *The Relation Between Rhetoric and Dialectic. Argumentation*, 14-3.

van Eemeren F. H. & Houtlosser P. (eds) (2006). *Perspectives on Strategic Maneuvering. Argumentation* 20, 4.

van Eemeren F. H. & Kruiger T. (1987). Identifying Argumentation Schemes. In van Eemeren F. H., Grootendorst R., Blair J. A. & Willard, C. A. (eds.) (1987). 271-291.

Eggs E. (1994). *Grammaire du Discours Argumentatif*. Paris: Kimé.

Eggs E. (2000). Logos, ethos, pathos - L'actualité de la rhétorique des passions chez Aristote. In Plantin Chr., Doury M. Traverso V. (eds). (2000). 15-31.

Ehninger D. & Brockriede W. [1960] = (1983). Toulmin on argument - An interpretation and application. In Golden, J. L., Berquist, G. F. & Coleman, W. E. (1983). 121-130.

Ekman P. (1999). Basic emotions. In T. Dalgleish and T. Power (Eds.) *The Handbook of Cognition and Emotion*. Sussex, U.K.: John Wiley. 45–60.

Ellrodt R. (1980). Histoire et analogie de Saint Augustin à Milton. In Lichnerowicz A., Perroux F. & Gadoffre G. (eds) (1980). 39-53.

Ellul J. [1961] = (1999). *Histoire des Institutions I. L'Antiquité*. Paris: PUF.

Empson W. (1951). *The Structure of Complex Words*. London: Chatto and Windus.

EOD = English Oxford Dictionary. https://en.oxforddictionaries.com

Erasmus D., [1524] = (no date) *On the Freedom of the Will*. Trans. by E. G. Rupp. Quoted after www.sjsu.edu/people/james.lindahl/courses/Hum1B/s3/Erasmus-and-Luther-on-Free-Will-and-Salvation.pdf. (17-05-23) (no pag.)

Erduran S. & Jiménez-Aleixandre M. P. (2007). *Argumentation in Science Education. Perspectives from Classroom-Based Research*. Dordrecht: Springer.

Feigl H. & Sellars W. (1949). *Readings in philosophical analysis*. Atascadero, CAL: Ridgeview.

Feteris E. T. (1999). *Fundamentals of Legal Argumentation - A Survey of Theories on the Justification of Judicial Decisions*. Dordrecht, Kluwer.

FF = The Fallacy Files, http://www.fallacyfiles.org.

Finocchiaro M. A. (1994). The positive versus the negative evaluation of arguments. In Johnson R. H. & Blair J. A. (eds) (1994). 21-35.

Finocchiaro M. A. (1999). A critique of the dialectical approach - Part II. In Eemeren, F. H. van, Grootendorst, R., Blair, J. A. & Willard, C. A. (eds.) (1999). 195-199.

Finocchiaro M. A. (2013). *Meta-Argumentation - An Approach to Logic and Argumentation Theory*. London: College Publications.

Fisher D. H. (1970). *Historians' Fallacies: Toward a Logic of Historical Thought*. New York: Harper & Row

Fogelin R. (1985). The Logic of deep disagreement. *Informal Logic* 7, 1. 3-11.

Fogelin R. J. & Duggan T. J. (1987). Fallacies. *Argumentation* 1, 3. 255-262.

Fontanier P. [1827], [1831] = (1977). *Les Figures du Discours: Traité Général des Figures du Discours Autres Que les Tropes* (1827) — *Manuel Classique Pour l'Étude des Tropes ou Elémens de la Science des Mots* (1831). Introd. by G. Genette. Paris: Flammarion.

Foucault M. (1969). *L'archéologie du savoir*. Paris: Gallimard.

Foucault M. (1971). *L'ordre du discours*. Paris: Gallimard.

Foviaux J. (1986). *De l'empire romain à la féodalité. Droit et Institutions*. Paris: Economica.

Fraisse P. & Piaget J. (1968). Les émotions. In P. Fraisse & J. Piaget (eds) (1968). *Traité de psychologie expérimentale V: Motivation, émotion et personnalité*. Paris: PUF. 86-155.

Frank R. H. (1988). *Passions within reason. The strategic role of the emotions*. New York: Norton.

Frankena W. K. (1967). Value and valuation. In Edwards P. (ed) (1967). *The Encyclopedia of Philosophy*. New York: MacMillan.

Frege G. [1879] = (1952). *Begriffschrift. A Formula Language, Modeled upon that of Arithmetic, for Pure Thought*. In *Translations from the Philosophical Writings of Gottlob Frege*. Oxford: Blackwell. 1-20.

Freud S. [1900] = (1955). *The Interpretation of Dreams*. Trans. and ed. by J. Strachey. New York, Basic Books, 2010.

Freud S. [1923] = (1961). *The Ego and the Id*. In J. Strachey, *The Standard Edition of the Complete Psychological Works of Sigmund Freud*. Volume XIX (1923-1925): *The Ego and the Id and Other Works*. London: Hogarth Press & the Institute of Psychoanalysis. 1-66.

Freud S. [1925] = (1961). *Negation*. In J. Strachey *The Standard Edition of the Complete*

References

Psychological Works of Sigmund Freud. Volume XIX (1923-1925): *The Ego and the Id and Other Works.* London: Hogarth Press & the Institute of Psycho-analysis. 233-240.

Fumaroli M. (1980). *L'âge de l'éloquence. Rhétorique et 'res literaria' de la Renaissance au seuil de l'époque classique.* Paris: Droz.

Fumaroli (ed.) (1999). *Histoire de la Rhétorique dans l'Europe Moderne 1450-1950.* Paris: PUF.

Gabbay D. M. & Woods J. (2003). *Agenda Relevance: An Essay in Formal Pragmatics.* Amsterdam: North-Holland, 2003.

Gabbay D. M. & Woods J. (2005). *The Reach of Abduction: Insight and Trial.* Amsterdam: North-Holland.

Gadamer H.-G. [1967] = (1988). Rhetoric, hermeneutics, and the critique of ideology. In Mueller-Vollmer, K. (ed.) (1988). 256-292.

Gadoffre G. (1980). Introduction. In Lichnerowicz A., Perroux F. & Gadoffre G. (eds) (1980). 7-10.

Gadoffre G., Walker P. & Tripet A. (1980). Les hommes de la renaissance et l'analogie. In Lichnerowicz A., Perroux F. & Gadoffre G. (eds) (1980). 47-53.

Gaffiot F. (1934). *Dictionnaire Illustré Latin-Français.* Paris: Hachette.

Gardet L. & Anawati M. M. [1967] = (1986). *Mystique Musulmane. Aspects et Tendances, Expériences et Techniques.* Paris: Vrin.

Gardet L., Anawati M. M. [1948] = (1981). *Introduction à la Philosophie Musulmane. Essai de Théologie Comparée.* Paris: Vrin.

Garfinkel H. (1967). *Studies in Ethnomethodology.* Englewood Cliffs, NJ: Prentice-Hall.

Gautier M. (2004). Dialectique. In *Dictionnaire des Notions.* Paris: Encyclopædia Universalis. 268-270

Genette G. (1970). La Rhétorique restreinte. *Communications* 16. 158-171.

Gil F. (1988). *Preuves.* Paris: Aubier.

Gilman S. L., Blair C. & Parent D. J. (1989). *Friedrich Nietzsche on Rhetoric and Language.* Oxford: Oxford University Press.

Gilson É. *see* Descartes R. [1637]

Ginzburg C. (1999). *History, Rhetoric and Proof.* Hannovre & London: University Press of New England.

Goddu G. G. 2007). Against making the linked-convergent distinction. In van Eemeren F. H., Blair J. A., Willard C. A., Garssen B. (eds.) (2007). 465-469.

Goffman E. (1981). *Forms of Talk.* Philadelphia: University of Pennsylvania Press.

Goffman E. [1956] = (1987). *The Presentation of Self in Everyday Life.* London: Penguin.

Golden J. L., Berquist G. F. & Coleman W. E. (1983). *The Rhetoric of Western Thought.* 3rd ed. Dubuque: Kendall & Hunt.

Golder C. (1996). *Le Développement des Discours Argumentatifs.* Lausanne: Delachaux & Niestlé.

Google = Google Dictionary

References

Govier T. (1987). *Problems in Argument Analysis and Evaluation*. Dordrecht: Foris.

Grice H. P. (1975). Logic and conversation. In Cole P. Morgan J. L. (eds) (1975). *Syntax and Semantics - Vol.3 Speech Acts*. New York: Academic Press. 41-58

Grimshaw A. D. (ed.) (1990). *Conflict talk - Sociolinguistic Investigations on Arguments in Conversation*. Cambridge, MA: Cambridge University Press.

Grize J.-B. (1972). *Logique Moderne. 1: Logique des Propositions et des Prédicats*. Paris: Mouton & Gauthier-Villars.

Grize J.-B. (1982). *De la Logique à l'Argumentation*. Pref by G. Busino. Geneva: Droz.

Grize J.-B. (1990). *Logique et Langage*. Gap: Ophrys.

Grize J.-B. (1993). Comment fait-on pour dire 'P donc Q' ? In G. Maurand (ed.) *Le raisonnement*. Toulouse: cals. 3-12.

Grize J.-B. (1987). *Pensée Naturelle, Logique et Langage - Hommage à Jean-Blaise Grize*. Neuchâtel, Université de Neuchâtel *Cahiers Vilfredo Pareto - Revue européenne des sciences sociales* 77, XXV.

Grize J.-B. (ed.) (1971). *Logique de l'Argumentation et Discours Argumentatifs*. Travaux du CdRS 7, Centre de Recherches sémiologiques. Neuchâtel: Université de Neuchâtel.

Grize J.-B. (ed.) (1974). *Recherches sur le Discours et l'Argumentation*. Geneva: Droz.

Grize, J.-B. (1996). *Logique et Communication*. Paris: PUF.

Groupe Mu (1970) = Dubois J., Edeline F., Klinkenberg J.-M, Minguet P., Pire F., & Trinon H. (1970). *Rhétorique générale*. Paris: Larousse.

Hamblin C. L. (1970). *Fallacies*. Londres, Methuen.

Han-Fei-tse, *Tao* = (1999) *Han-Fei-tse ou Le Tao du Prince*. Introd. and trans. by J. Levi. Paris: Le Seuil.

Hedge L. (1838). *Elements of Logick, or a Summary of the General Principles and Different modes of Reasoning*. Boston: Hilliar.

Heritage J. (1987). Interactional accountability. A conversation analytic perspective. In Conein, B., de Fornel M., Quéré L. (eds) 1987. 23-49.

Hermogenes (*AR*) = (1997). *L'art Rhétorique*. Trad., introd. and notes by M. Patillon. Pref. by P. Laurens. Lausanne: L'Âge d'Homme.

Hermogenes (*OI*) = (1995). Hermogenes *On Issues*. Trans. by M. Heath. Oxford: Clarendon Press.

Hesse M. (1967). Models and analogy in science. In Edwards P. (ed.) (1967). Vol. 5. 354-359.

Hintikka J. (1979). Information-seeking dialogues: A Model. *Erkenntnis* 38. 355-368.

Hintikka J. (1987). The fallacy of fallacies. *Argumentation* 1, 3. 211-238.

Hirschman A. O. (1991). *The Rhetoric of Reaction: Perversity, Futility, Jeopardy*. Cambridge, MA & London: CUP.

Hoaglund J. (2003). Using argument types. In van Eemeren F. H., Blair J. A., Willard C. A. & Snoeck-Henkemans A. F. (eds) (2003). 491-495.

Hoaglund J. (2007). Informal Logic and Pragma-Dialectics. *In* van Eemeren F. H., Blair

References

J. A., Willard C. A. & Garssen B. (eds.) (2007). 621-624.

IEP = Fieser J. & Dowden B., s.d. *Internet Encyclopedia of Philosophy*. www.iep.utm.edu/ (09-20-2013)

Jacobs L. & Derovan D. (2007). Hermeneutics. In Skolnik & Berenbaum (eds) (2007) *Encyclopedia Judaica*. 2nd ed. vol. 9. USA: MacMillan & Jerusalem: Keter. 25-29.

Jacobs S. & Jackson S. (1982). Conversational argument. A discourse analytic approach. In J. R. Cox J. R., Willard C. A. (eds) (1982). 205-237.

Jakobson R. [1960] = (1987). Linguistics and poetics. In Pomorska K. & Rudt S. (eds) (1987). *Language in Literature – Roman Jakobson*. Cambridge, MA: Belknap & Harvard UP. 62-94.

Jakobson R. (1971). A joint conference of anthropologists and linguists. In *Selected Writings II. Words and Language*. The Hague, Paris: Mouton 1971. 554-567.

Johnson R. H. (1996). *The Rise of Informal Logic*. Newport News VA: Vale Press.

Johnson R. H. & Blair J. A. (eds) (1994). *New Essays in Informal Logic*. Windsor, Informal Logic.

Joule R. V. & Beauvois J. L. (1987). *Petit Traité de Manipulation à l'Usage des Honnêtes Gens*. Grenoble: Presses Universitaires de Grenoble.

Kahane H. (1971). *Logic and Contemporary Rhetoric: The Use of Reason in Everyday Life*. Belmont CA: Wadsworth.

Kalinowski G. (1965). *Introduction à la Logique Juridique – Éléments de Sémiotique Juridique, Logique des Normes et Logique Juridique*. Paris: Librairie Générale de Droit et de Jurisprudence.

Kallmeyer W. (ed.) (1996). *Gesprächsrhetorik – Rhetorisches Verfahren im Gesprächsprocess*. Tübingen: Gunter Narr.

Kant I. [1781] = (1998) *Critique of Pure Reason*. Trans. and ed. by P. Guyer & A. W. Wood. Cambridge, MA: Cambridge University Press.

Kelsen H., [1934] = (1967). *Pure Theory of Law*. Trans. by M. Knight. Berkeley, CA: University of California Press.

Kennedy G. A., [1980] = (1999). *Classical Rhetoric and Its Christian and Secular Tradition from Ancient to Modern Times*. 2nd revised and enlarged edition. Chapel Hill, NC: University of North Carolina Press.

Kerbrat-Orecchioni C. (1980). *L'Énonciation. De la Subjectivité dans le Langage*. Paris: Armand Colin.

Kerbrat-Orecchioni C. (1990 – 1994). *Les Interactions Verbales*. T. 1 (1990); T. 2 (1992); T. 3 (1994). Paris: Armand Colin.

Kerbrat-Orecchioni C. (2000). L'analyse des interactions verbales. La notion de 'négociation conversationnelle' - Défense et illustration. *Lalies* 20. 63-141.

Khallâf 'A. al-W. [1942] = (1997). *Les Fondements du Droit Musulman*. Trans. by Cl. Dabbak, A. Godin & M. Labidi Maiza. Pref. by A. Turki. Paris: Al Qalam.

Kienpointner M. (1987). Towards a typology of argumentative schemes. *In* van Eemeren F. H., Grootendorst R., Blair J. A. & Willard C. A. (eds.) (1987). 275-288.

Kienpointner M. (1992). *Alltagslogik. Struktur und Funktion von Argumentationsmustern*.

References

Stuttgart-Bad Cannstadt: Fromman-Holzboog.

Kienpointner M. (2003). Nouvelle Rhétorique / Neue Rhetorik. In G. Ueding (Hg.) *Historisches Wörterbuch der Rhetorik*, Bd 6. Tübingen: Niemeyer. 561-587.

Kleene S. C. [1967] = *Mathematical Logic*. New York: Wiley https://archive.org/details/KleeneMathematicalLogic (09-26-2017)

Kleiber G. 1990). *La Sémantique du Prototype - Catégorie et Sens Lexical*. Paris: PUF.

Klinkenberg J.-M. (1990). Rhétorique de l'argumentation et rhétorique des figures. In Meyer M. & Lempereur, A. (eds) (1990). 115-137.

Klinkenberg J.-M. (2000). L'Argumentation dans la figure. *Cahiers de praxématique* 35. http://journals.openedition.org/praxematique/2898 (09-26-2016)

Klinkenberg J.-M. (2001). Retórica de la argumentación y retórica de las figuras: ¿hermanas o enemigas ? *Tonos digital*, 1. um.es/tonosdigital/znum1/estudios/Klinkenberg.htm. (09-20-2017)

Kneale W. & Kneale M. [1962] = (1984). *The Development of Logic*. Oxford: Clarendon Press.

Koons R. (2005). Defeasible reasoning. *The Stanford Encyclopedia of Philosophy*. E. N. Zalta (ed.). http://plato.stanford.edu/archives/spr2005/entries/reasoning-defeasible/

Kotarbinski T. [1964] = (1971). *Leçons sur l'Histoire de la Logique*. Trans. by A. Posner. Paris: Presses Universitaires de France.

Krabbe E. C. W. (1998). Who is afraid of figures of speech ? *Argumentation* 12, 2. 281-294

Lakoff G. & Johnson, M. (1980). *Metaphors We Live By*. Chicago: Chicago University Press.

Laplanche J. & Pontalis, J.-B. (1967). *Dictionnaire de Psychanalyse*. Paris: PUF.

Lausberg H., [1960] = (1973). *Handbook of Literary Rhetoric: A Foundation for Literary Study*. Translated from German by M. T. Bliss, A. Jansen & D. E. Orton. Edited. D. E. Orton & R. D. Anderson. Foreword G. A. Kennedy. Leiden: E.J. Brill.

Lausberg H., [1963] = (1971). *Elemente der literarischen Rhetorik*. Munich: Max Hueber.

Le Bon G. [1895] = (1960). *The Crowd. A Study of Popular Mind*. New York: Macmillan.

Le Brun J. (2011). Review of Houdard S., *Les Invasions Mystiques*. In *Revue de l'Histoire des Religions*, 1. 124-128. http://rhr.revues.org/7738 (09-20-2013).

Leibniz G. W. [1765] = (1896). *New Essays Concerning Human Understanding*. Trans. by A. G. Langley. London: Macmillan.

Lempereur A. (1990). Les restrictions des deux néo-rhétoriques. In Meyer M. & Lempereur A. (eds) (1990). 139-158.

Lévinas E. [1981] = (1987). Langage quotidien et rhétorique sans éloquence. In *Hors sujet*. Fata Morgana. 201-211.

Lévy C. & Pernot L. (1997). *Dire l'Évidence. Philosophie et Rhétorique Antiques*. Paris: L'Harmattan.

Lichnerowicz A., Perroux F. & Gadoffre G. (eds) (1980). *Analogie et Connaissance*. T. 1, *Aspects historiques*. T. II, *De la Poésie à la Science*. Paris: Maloine.

Littré É. [1863] = (1972) *Dictionnaire de la Langue Française*. Paris: Hachette.

References

http://www.littre.org/ (09-20-2013).

Lloyd G. E. R. (1990). *Demystifyng Mentalities*. Cambridge, MA: Cambridge University Press.

Lo Cascio V. (2009). *Persuadere e Convincere Oggi. Nuovo Manuale dell'Argomentazione*. Acqui Terme: Academia Press.

Locke J. [1690] = (1959). *An Essay Concerning Human Understanding*. Collected and annotated by A. C. Fraser. New York: Dover.

Lorenzo-Basson M.-C. (2004). *La Vente à Domicile. Stratégies Discursives en Interaction*. PhD, Lyon 2 University, under the supervision of C. Kerbrat-Orecchioni.

Louis P. (1990). *Vie d'Aristote*. Paris: Hermann.

Mackenzie J. (1988). *Distinguo*. The response to equivocation. *Argumentation* 2-4. 465-482.

Mackie J. L. (1967). Fallacies. In Edwards P. (ed.) *The Encylopedia of Philosophy*. Vol. 3. 169-179.

Maingueneau D. (1976). *Initiation aux Méthodes de l'Analyse du Discours*. Paris: Hachette.

Maingueneau D. (1990). *L'Analyse du Discours*. Paris: Hachette.

Maingueneau D. (1999). Ethos, scénographie, incorporation. *In* Amossy R. (ed.) (1999). 75-102.

Man P. de (1978). The epistemology of metaphor. In Sacks S. (ed.) *On Metaphor*. 11-28.

Maritain J. [1923] = (1946). *An Introduction to Logic*. Trans. by I. Choquette. London: Scheed & Ward.

Mayans y Siscar G. (1786). *Rhetorica*. Valencia: Josef y Thomas de Orga (2e ed.)

McAdon B. (2003). Probabilities, necessary signs, *idia* and *topoi*. The confusing discussion of material for enthymemes in the *Rhetoric*. *Philosophy and Rhetoric*, 36, 3. 223-248.

McAdon B. (2004). Two irreconcilable conceptions of rhetorical proof in Aristotle's *Rhetoric*. *Rhetorica*, 22, 4. 307-325

McEvoy S. (1995). *L'invention Défensive — Poétique, Linguistique, Droit*. Paris: Métailié.

Meyer M. (ed.) (1986). *De la Métaphysique à la Rhétorique*. Brussels: Éditions de l'Université de Bruxelles.

Meyer M. & Lempereur A. (eds) 1990). *Figures et Conflits Rhétoriques*. Brussels: Éditions de l'Université de Bruxelles.

Milgram S. (1974). *Obedience to authority. An Experimental View*. London: Tavistock Publications. https://archive.org/stream/ObedienceToAuthority_368/milgram

Mill J. S. [1843] = (2017). *A System of Logic, Ratiocinative and Inductive*. London: John W. Parker. http://www.earlymoderntexts.com/assets/pdfs/mill1843book1.pdf

Mill J. S. [1859] = (1987). *On Liberty*. Harmondsworth: Penguin Classics.

Moeschler J. & Reboul A. (1994). *Dictionnaire Encyclopédique de Pragmatique*. Paris: Le Seuil.

Moeschler J. (1985). *Argumentation et Conversation. Éléments pour une Analyse Pragmatique du Discours*. Paris: Hatier.

References

Molinié G. (1992). *Dictionnaire de Rhétorique*. Paris: Librairie Générale Française.

Molino J. (1979) Métaphores, modèles et analogies dans les sciences. *Langages*, 54. 83-102.

Moore G. E., [1903] = (1986). *Principia Ethica*. Cambridge, MA: CUP.

Mortureux M.-F. (1993). Paradigmes désignationnels. *Semen* 8, Besançon: Presses de l'Université de Franche-Comté. 121-142. http://journals.openedition.org/semen/4132

Mueller-Vollmer K. (1988). *The Hermeneutics Reader*. New York: Continuum.

MW = *Merriam-Webster Dictionary*. www.merriam-webster.com

Nadeau R. (1958). Hermogenes on 'stock issues' in deliberative speaking. *Speech Monographs*, 25. 59-66.

Nadeau R. (1964). Hermogenes' *On Stases*. A translation with an introduction and notes. *Speech Monographs*, 31. 361-424.

Newman J. H., [1870] = (2010). *An Essay in Aid of a Grammar of Assent*. London: Burns, Oates. Quoted after http://onlinebooks.library.upenn.edu/webbin/gutbook/lookup?num=34022 (19-07-2017)

Nicolas L. (2007). *La Force de la Doxa. Rhétorique de la Décision et de la Délibération*. Paris: L'Harmattan.

Nietzsche F. *see* Gilman *& al*.

Nonnon E. (1996) Activités argumentatives et élaboration de connaissances nouvelles. *Langue Française* 112. 67-87.

O'Keefe B. J., [1977] = (1982). Two concepts of argument and arguing. In Cox, J. R., C. A., Willard (eds) (1982). 3-23.

OD = Oxford Dictionary https://en.oxforddictionaries.com (07-09-2017)

OED = Online Etymology Dictionary. http://www.etymonline.com/ (07-09-2017)

Olbrechts-Tyteca L. (1974). *Le Comique du Discours*. Brussels: Éditions de l'Université de Bruxelles.

Ong W. J. (1958). *Ramus. Method and the Decay of Dialogue*. Cambridge, MA: Harvard University Press.

Ortony A. (ed.) (1979). *Metaphor and Thought*. Cambridge, MA: Cambridge University Press.

Packard V. [1957] = (2007). *The Hidden Persuaders*. New York: Ig Publishing.

Parent X. & Livet P. (2002). Argumentation, révision et conditionnels. In P. Livet (ed.) *Révision des Croyances*. Paris: Hermès Sciences Publication. 229-258.

Pascal B. (*Geom.*) = (1859). Of the geometrical spirit. In *The Thoughts, Letters and Opuscules of Blaise Pascal*. Trans. by O. W. Wight. New York: Derby & Jackson.

Pascal B. (*Mem.*) = (1999). *Memorial*. Trans. by E. T. Knuth & O. Joseph. http://www.users.csbsju.edu/~eknuth/pascal.html (07-09-2017).

Pascal B. (*Thoughts*) = (1901). *The Thoughts of Blaise Pascal*. Trans. by C. Kegan Paul. London: Georges Bell.

Patillon M. (1988). *La Théorie du Discours d'Hermogène le Rhéteur. Essai sur la Structure de la Rhétorique Ancienne*. Paris: Les Belles-Lettres.

Patillon M. (1990). *Eléments de Rhétorique Classique*. Paris: Nathan.

Peirce C. S. [1958]. *Collected Papers of Charles Sanders Peirce*. Vol. 7, Book II: *Scientific Method*. Ed. by A. W. Burke. Cambridge, MA: Harvard University Press.

Pellegrin P. (1997). Glossaire. In Sextus Empiricus, *Esquisses Pyrrhoniennes*. Paris: Le Seuil. 527-556.

Perelman Ch. (1952). Acte et personne dans l'argumentation. In Perelman Ch. & Olbrechts-Tyteca L. (1952). 49-84.

Perelman Ch. (1961). Jugements de valeur, justification et argumentation. *Revue Internationale de Philosophie* 58, 4. 327-335.

Perelman Ch. [1963] = (1972). *Justice et Raison*. Brussels: Éditions de l'Université de Bruxelles.

Perelman Ch. (1977). *L'empire rhétorique. Rhétorique et Argumentation*. Paris: Vrin.

Perelman Ch. (1979). *Logique juridique – Nouvelle rhétorique*. Paris: Dalloz.

Perelman Ch. (1980). Logic and rhetoric. In Agazzi E. (ed.) (1980). 457-464.

Perelman Ch. & Olbrechts-Tyteca L. [1950] = (1952). Logique et rRhétorique. In Perelman Ch. & Olbrechts-Tyteca L. (1952). 1-43.

Perelman Ch. & Olbrechts-Tyteca L. (1952). *Rhétorique et Philosophie. Pour une Théorie de l'Argumentation en Philosophie*. Paris: Presses Universitaires de France.

Perelman Ch. & Olbrechts-Tyteca L. [1958] = (1969). *The New Rhetoric. A Treatise on Argumentation*. Trans. by J. Wilkinson and P. Weaver. Notre Dame IN: University of Notre Dame Press.

Piaget J. [1924] = (1967). *Le Jugement et le Raisonnement chez l'Enfant*. Neuchâtel: Delachaux & Niestlé.

Plantin Chr. (1990). *Essais sur l'argumentation*. Paris: Kimé.

Plantin Chr. (ed.) (1993). *Lieux Communs, Topoi, Stéréotypes, Clichés*. Paris: Kimé.

Plantin Chr. (1995). *L'Argumentation*. Paris: Le Seuil.

Plantin Chr., Doury M., Traverso V. (eds) (2000). *Les émotions dans les interactions*. Lyon, PUL.

Plantin Chr. (2005). *L'Argumentation: Histoire, Théories, Perspectives*. Paris: PUF.

Plantin Chr. (2009). A place for figures of speech in argumentation theory. *Argumentation* 23, 3. 325-337.

Plantin Chr. (2011). Les instruments de structuration des séquences argumentatives. *Verbum* 32, 1. 31-51.

Plantin, Chr. 2015, Emotion and Affect. In Tracy, Karen, Ilie, Cornelia & Sandel, Todd (eds.) (2015). *The International Encyclopedia of Language and Social Interaction*. Boston: John Wiley & Sons.

Plato *CW* = (1997) *Complete Works*. Ed. with Introd. and Notes by J. M. Cooper. Indianapolis & Cambridge, MA: Hackett.

Plato. *Euthydemus*. Trans. by R. Kent Sprague. In Plato, *CW*, 708-745

Plato. *Gorgias*. Trans. by D. J. Zeyl. In Plato, *CW*, 791-869

Plato. *Phaedrus*. Trans. by A. Nehamas and P. Woodruff. In Plato, *CW*, 506-556.

Pomerantz A. (1984). Agreeing and disagreeing with assessments. Some features of preferred / dispreferred turn-shapes. In Atkinson J.-M. & Heritage J. (eds) (1984). 79-112.

Popper K. (1963). *Conjectures and Refutations - The Growth of Scientific Knowledge*. London: Routledge & Kegan Paul.

Porphyry *Isa.* = (1975) Porphyry the Phœnician. *Isagoge*. Trans., introd. and notes by E. W. Warren. Toronto: The Pontifical Institute of Mediæval Studies.

PR = Rey-Debove J., Rey A., Chantreau S. & Drivaud M.-H. (1967). *Le Nouveau Petit Robert: Dictionnaire alphabétique et analogique de la langue française*. Paris: Le Robert.

Prior A. N. (1967). Traditional logic. In P. Edwards (ed.) (1967), Vol. 5. 34-45.

Quine W. V .O. (1959). *Methods of Logic*. Revised ed. New York, etc.: Holt, Rinehart & Winston.

Quine W. V. O. (1966). Logic as a source of syntactical insight. *Journal of Symbolic Logic* 31, 3. 496-497.

Quine W. V. O. (1980). *Elementary Logic*. Revised edition. Cambridge, MA. & London, England: Harvard UP.

Quine W. V. O. & Ullian, J. S. (1982). *The Web of Belief*. New York: Random House.

Quintilian, *IO* = (1903). *Quintilian's Institutes of Oratory*. Trans. by J. S. Watson. London: George Bell.

Reboul O. (1986). La figure et l'argument. In Meyer M. (ed.) (1986). 175–187.

Reboul O. (1991). *Introduction à la Rhétorique*. Paris: PUF.

Récanati F. (1979). *La Transparence et l'Énonciation*. Paris: Le Seuil.

Reiter R. (1980). A logic for default reasoning. *Artificial intelligence* 13. 81-131.

Reverso = Reverso Dictionary, German-French. http://dictionary.reverso.net/German-french/ (12-12-2017)

Richards I. A. (1936). *The Philosophy of Rhetoric*. Oxford: Oxford University Press.

Rimé B. & Scherer K. (eds) (1993). *Les Émotions*. Neuchâtel: Delachaux et Niestlé.

Rocheblave-Spenlé A.-M. [1962] = (1969) *La Notion de Rôle en Psychologie Sociale*. Paris: PUF.

Romilly J. de (1988). *Les Grands Sophistes dans l'Athènes de Périclès*. Paris: de Fallois.

Rorty A. O. (ed.) (1996). *Essays on Aristotle's Rhetoric*. Berkeley: University of California Press.

Russell B. [1905] = (1949). On denoting. In Feigl H. & Sellars W. (1949). 103-115.

Ryan E. E. (1984). *Aristotle's Theory of Rhetorical Argumentation*. Montréal: Bellarmin.

Ryle G. (1932). Systematically misleading expressions. *Proceedings of the Aristotelian Society*, 32. 139-170.

Sacks S. (ed.) (1978). *On Metaphor*. Chicago: The University of Chicago Press.

Sallantin J. & Szczeciniarz J. J. (eds) (1999). *Le Concept de Preuve à la Lumière de l'Intelligence Artificielle*. Paris: PUF.

Saussure L. de (2015). The straw Man argument as a pragmatic argumentative winner. *Conference Argumentation and Language*, Lausanne, 2015.

Schellens P. J. (1987). Types of argument and the critical reader. In van Eemeren F. H., Grootendorst, R., Blair, J. A. & Willard, C. A. (eds.) (1987). Dordrecht: Foris. 3B. 34-41.

Scherer K. R. [1984a] = (1993). Les émotions. Fonctions et composantes. In Rimé, B. Scherer K. (eds) (1993). 97-133.

Scherer K. R. (1984b). On the nature and function of emotion - A Component Process Approach. In Scherer K. R. & Ekman P. (eds) (1984). 293-317.

Scherer K. R. & Ekman P. (eds) (1984). Approaches to emotion. Hillsdale, NJ: Lawrence Erlbaum.

Schiappa E. (1993). Arguing about Definitions. *Argumentation* 7, 4. 403 – 417.

Schiappa E. (2000). Analyzing argumentative discourse from a rhetorical perspective: defining *'person'* and *'human life'* in constitutional disputes over abortion. *Argumentation* 14-3. 315-332.

Schiffrin D. (1987). *Discourse Markers*. Cambridge, MA: Cambridge UP.

Schiffrin D. (1990). The management of a cooperative self in argument. The role of opinions and stories. In A. Grimshaw (ed.) (1990). 241- 259.

Schmid M. (1980). Bewegung im TV-Studio. In *Eine Stadt in Bewegund. Materialen zu den Zürcher Unruhen*. Zürich: SD Stadt Zürich.

Schopenhauer A. [1830] = (2005). *The Art of Controversy*. Trans. by T. Bailey Saunders (1896). http://insomnia.ac/essays/the_art_of_controversy/penn_state_ebook.pdf (10-03-2017)

Sextus Empiricus *OP* = (1996). *Outlines of Pyrronism*. Trans. with introd. and com. by B. Mates. New York & Oxford: Oxford UP. http://www.sciacchitano.it/pensatori%20epistemici/scettici/outlines%20of%20pyrronism.pdf (08-17-2017)

Shelley C. (2002). Analogy counterarguments and the acceptability of analogical hypotheses. *British Journal for the Philosophy of Science* 53. 477-496.

Shelley C. (2004). Analogy counterarguments. A taxonomy for critical thinking. *Argumentation* 18, 2. 223-238.

SIL = *Summer Institute of Linguistics – Glossary of Linguistic Terms*. http://www.glossary.sil.org (12-12-17)

Sitri F. (2003). *L'Objet du Débat. La Construction des Objets de Discours dans des Situations Argumentatives Orales*. Paris: Presses de la Sorbonne Nouvelle.

Snoeck Henkemans A. F. (2003). Indicators of analogy argumentation. In van Eemeren F., Blair, J. A., Willard, C. A. Snoeck Henkemans A. F. (eds) (2003). 969-973.

Snoeck Henkemans A. F. (2000) Comments on E. Schiappa "Analyzing argumentative discourse from a rhetorical perspective: Defining *'Person'* and *'Human Life'* in constitutional disputes over abortion". *Argumentation* 14, 3. 333-338

Snoeck Henkemans A. F. (1992). *Analysing Complex Argumentation*. Amsterdam: SICSAT.

Solmsen F. (1941). The Aristotelian tradition in ancient rhetoric. *The American Journal of Philology* 62, 2. 169-190.

Sperber D. & Wilson D. (1995). *Relevance - Communication and Cognition*. Second ed., Oxford & Cambridge: Blackwell.

References

Stevenson C. L. [1938] = (1944). Persuasive definitions. In Stevenson, C. L. (1944). *Ethics and language*. New Haven & London: Yale UP.

Strauss L. (1953). *Persecution and the Art of Writing*. Social Research 8, 1-4. 488-503.

Tarello G. (1972). Sur la spécificité du raisonnement juridique. *Die Juristische Argumentation. Archiv für Rechts- und Sozialphilosophie*, 7. Actes du Congrès de Bruxelles de 1971. Wiesbaden: Franz Steiner. 103-124.

Tarski A. [1935] = (1983). The concept of truth in formalized languages. Trans. by J. H. Woodger. In Tarski A. (1983). *Logic, Semantics, Metamathematics*. 2nd edition. Ed. by J. Corcoran. Indianapolis: Hackett. 152-278.

Tesnière L. (1959). *Éléments de Syntaxe Structurale*. Paris: Klincksieck.

tfd = *thefreedictionary*. http://www.thefreedictionary.com/ (12-12-2017)

TLFi = *Trésor de la Langue Française informatisé*. http://www.cnrtl.fr. (09-20-2013)

Thomas S. N. (1986). *Practical Reasoning in Natural Language*. Englewood Cliffs, NJ: Prentice Hall.

Thomas Aquinas *(ST)* = (1947) *The Summa Theologica*. Trans. by Fathers of the English Dominican Province. New York: Benziger Bros.
http://dhspriory.org/thomas/summa/FP/FP001.html#FPQ1OUTP1

Toulmin S. E. (1958). *The Uses of Argument*. Cambridge: Cambridge University Press.

Toulmin S. E., Rieke R. & Janik A. (1984). *An Introduction to Reasoning*. New York: McMillan.

Traverso V. (2000). *La Conversation Ordinaire*. Paris: Nathan.

Tricot J. [1928] = (1973). *Traité de Logique Formelle*. Paris: Vrin.

Trottman C. (1999). *Théologie et Noétique au XVIIIe siècle*. Paris: Vrin.

Turner D. & Campolo C. (2005). Deep disagreement re-examined. *Informal Logic* 25. 1-2.

Tutescu M. (2003). *L'Argumentation. Introduction à l'Étude du Discours*. Bucarest: Editura Universitătii din Bucuresti.
http://ebooks.unibuc.ro/lls/MarianaTutescu-Argumentation/1.htm. (12-12-2017)

Ueding G. (ed.) (1992-2015). *Historisches Wörterbuch der Rhetorik*. Tübingen: Niemeyer.

Vannier G. (2001). *Argumentation et Droit. Introduction à la Nouvelle Rhétorique de Perelman* Paris: PUF.

Vax L. (1982). *Lexique Logique*. Paris: PUF.

Vega Reñon L. & Olmos Gómez P. (2011). *Compendio de Lógica - Argumentación y Retórica* Madrid: Trotta.

Vidal G. R. (2000). *La Retórica de Antifonte*. México: UNAM.

Vignaux G. (1976). *L'Argumentation. Essai d'une Logique Discursive*. Geneva: Droz.

Vignaux G. (1981). Énoncer, argumenter. Opérations du discours, logiques du discours. In Bouacha A. & Portine H. (eds) *Argumentation et Énonciation. Langue Française* 50. 91-116.

Vignaux G. (1999). *L'Argumentation. Du Discours à la Pensée*. Paris: Hatier.

Vion R. (1992). *La Communication Verbale*. Paris: Hachette.

Walton D. N. (1992). *The Place of Emotion in Argument*. University Park, PA: The Pennsylvania State University Press.

Walton D. N. (1996). *Argument Structure: A Pragmatic Theory*. Toronto: University of Toronto Press.

Walton D. N. (1997). *Appeal to Pity* - Argumentum ad Misericordiam. Albany: State University of New York Press.

Walton D. N. (1999). Francis Bacon. Human bias and the four idols. *Argumentation* 13, 4. 385–389.

Walton D. N. (2004). *Abductive Reasoning*. Tuscaloosa: University of Alabama Press.

Walton D. N. (2005). Deceptive arguments containing persuasive language and persuasive definitions. *Argumentation* 19, 2. 159-186.

Walton D., Reed C. & Macagno F. (2008). *Argumentation Schemes*. Cambridge, MA: Cambridge University Press.

WCD = *Webster's New World Dictionary of the American Language. Second College Edition*. Cleveland & New York: William Collins & World Publishing CO.

Weaver R. (1953). Abraham Lincoln and the argument from definition. In *The Ethics of Rhetoric*. South Bend, IN: Gateway. 85-114.

Weber M. [1922] = (1978) *Economy and Society – An Outline of Interpretive Sociology*. Ed. by Roth, G. & Wittich C. Berkeley and Los Angeles, CA & London England: University of California Press.
http://www.public.iastate.edu/~carlos/607/readings/weber.pdf (12-12-2017)

Weijers O. (1999). De la joute dialectique à la dispute scolastique. In *Comptes-rendus des séances de l'Académie des Inscriptions et Belles-Lettres*. 509-518.
http://www.persee.fr/web/revues/home/prescript/article/crai_0065-0536_1999_num_143_2_16013. (12-12-2017)

Wellman C. (1971). *Challenge and Response. Justification in Ethics*. Carbondale: Southern Illinois University Press.

Wenzel J. (1987). The rhetorical perspective on argument. In F. van Eemeren, R. Grootendorst, J. A. Blair & C. A. Willard (eds) (1987). Vol. 1. 101-109.

Whately R., [1828] = (1963). *Elements of Rhetoric Comprising an Analysis of the Laws of Moral Evidence and of Persuasion, with Rules for Argumentative Composition and Elocution*. Ed. by D. Ehninger. Foreword by D. Potter. Carbondale & Edwardsville: Southern Illinois UP.

Whately R. [1832] = (1854). *Elements of Logic*. Louisville, KY: Morton & Griswald.

Wikipedia, https://en.wikipedia.org/wiki/Main_Page.

Willard C. A. (1989). *A Theory of Argumentation*. Tuscaloosa: The University of Alabama Press.

Windisch U. (1987). *Le KO Verbal - La Communication Conflictuelle*. Lausanne: L'Âge d'Homme.

Wittgenstein L. (1974). *On Certainty / Über Gewissheit*. Ed. by Anscombe G. E. M. & von Wright G. H.. Trans. by Paul D. & Anscombe G. E. M. Oxford: Basil Blackwell.

Woods J. & Walton D. N. (1989). *Fallacies. Selected Papers 1972-1982*. Dordrecht: Foris.

References

Woods J. (2004). *The Death of Argument: Fallacies in Agent-Based Reasoning*. Dordrecht: Kluwer.

Woods J. (2009). Ignorance, inference and proof. Abductive logic meets the criminal law. In Tuzet G. & Canale D. (eds) *The rules of inference: Inferentialism in Law and Philosophy*. Heidelberg: Egea. 151-185.

Woods J. (2013/14). *Errors Of Reasoning: Naturalizing the Logic of Inference*. London: College Publications.

Woods J. (2014). *Aristotle's Earlier Logic*. 2nd edition, revised and expanded. London: College Publications.

Wreen M. J. (1999). A few remarks on the individuation of arguments. In van Eemeren F. H., Grootendorst R., Blair J. A. & Willard C. A. (eds.) (1999). 884-888.

Wreen M. J. (2000). Review of Douglas Walton '*Argument from Ignorance*'. *Argumentation* 14, 1. 51-56.

Yates F. A. (1966). *The Art of Memory*. London: Routledge & Kegan Paul.

Yzerbit V. & Corneille O. (eds) (1994). *La Persuasion*. Lausanne: Delachaux & Niestlé.

www.ingramcontent.com/pod-product-compliance
Lightning Source LLC
Chambersburg PA
CBHW060357230426
43663CB00008B/1306